Systematics of the Caligidae, copepods parasitic on marine fishes

CRUSTACEANA MONOGRAPHS constitutes a series of books on carcinology in its widest sense. Contributions are handled by the Series Editor and may be submitted through the office of KONINKLIJKE BRILL Academic Publishers N.V., P.O. Box 9000, NL-2300 PA Leiden, The Netherlands.

Published in this series:

CRM 001 - *Stephan G. Bullard* — Larvae of anomuran and brachyuran crabs of North Carolina
CRM 002 - *Spyros Sfenthourakis et al. (eds.)* — The biology of terrestrial isopods, V
CRM 003 - *Tomislav Karanovic* — Subterranean Copepoda from arid Western Australia
CRM 004 - *Katsushi Sakai* — Callianassoidea of the world (Decapoda, Thalassinidea)
CRM 005 - *Kim Larsen* — Deep-sea Tanaidacea from the Gulf of Mexico
CRM 006 - *Katsushi Sakai* — Upogebiidae of the world (Decapoda, Thalassinidea)
CRM 007 - *Ivana Karanovic* — Candoninae (Ostracoda) from the Pilbara region in Western Australia
CRM 008 - *Frank D. Ferrari & Hans-Uwe Dahms* — Post-embryonic development of the Copepoda
CRM 009 - *Tomislav Karanovic* — Marine interstitial Poecilostomatoida and Cyclopoida (Copepoda) of Australia
CRM 010 - *Carrie E. Schweitzer et al.* — Systematic list of fossil decapod crustacean species
CRM 011 - *Peter Castro et al. (eds.)* — Studies on Brachyura: a homage to Danièle Guinot
CRM 012 - *Patricio R. De los Ríos-Escalante* — Crustacean zooplankton communities in Chilean inland waters
CRM 013 - *Katsushi Sakai* — Axioidea of the world and a reconsideration of the Callianassoidea (Decapoda, Thalassinidea, Callianassida)
CRM 014 - *Charles H.J.M. Fransen et al. (eds.)* — Studies on Malacostraca: Lipke Bijdeley Holthuis Memorial Volume
CRM 015 - *Akira Asakura et al. (eds.)* — New Frontiers in Crustacean Biology: Proceedings of the TCS Summer Meeting, Tokyo, 20-24 September 2009
CRM 016 - *Danielle Defaye et al. (eds.)* — Studies on Freshwater Copepoda: a Volume in Honour of Bernard Dussart
CRM 017 - *Hironori Komatsu et al. (eds.)* — Studies on Eumalacostraca: a homage to Masatsune Takeda

In preparation (provisional title):

CRM01x - *Darren C. Yeo et al. (eds.)* — Advances in freshwater decapod systematics and biology

Authors' addresses:

M. Dojiri, Environmental Monitoring Division, Bureau of Sanitation, City of Los Angeles, 12000 Vista del Mar, Playa del Rey, CA 90293, U.S.A.; e-mail: mas.dojiri@lacity.org

J.-S. Ho, Department of Biology, California State University, Long Beach, CA 90840-3702, U.S.A.; e-mail: jsho@csulb.edu

Manuscript first received 18 June 2011; final version accepted 20 June 2012.

Cover: Anchicaligus nautili (Willey, 1896); see p. 97, fig. 29a, b.

Systematics of the Caligidae, copepods parasitic on marine fishes

By

M. Dojiri and J.-S. Ho

CRUSTACEANA MONOGRAPHS, 18

BRILL

LEIDEN • BOSTON

This book is printed on acid-free paper.

Library of Congress Cataloging-in-Publication Data

The Library of Congress Cataloging-in-Publication Data is available from the Publisher.

ISBN13: 978-90-04-204249

Koninklijke Brill NV incorporates the imprints BRILL, Hotei Publishing, IDC Publishers and Martinus Nijhoff Publishers.

PRINTED IN THE NETHERLANDS

CONTENTS

PREFACE

The Caligidae, commonly known as "sea lice", was originally established by Burmeister in 1835 and consists of more than 450 described species of parasitic copepods predominantly parasitic on marine fishes. Many caligid species have long been recognized to adversely impact the overall health of their hosts in the wild. With the decline in ocean fisheries during the last quarter of the 20th Century and a resultant rise in the development of brackish water and marine aquaculture around the world, it has become increasingly evident that sea lice infections of cultured fishes are a significant problem that cannot be ignored. The significant health problems including mass mortality of farmed fishes as a result of heavy sea lice infestations have resulted in huge economic losses to the fish-farming industry.

In view of the fact that these parasites are capable of killing their hosts, have a nearly worldwide distribution, and occur on economically important fishes, the systematics of the Caligidae must be considered important in fisheries biology and fish husbandry. The correct taxonomic identification of the caligid species is necessary and important in view of the possible differences in host-parasite relationships, life histories, and environmental tolerances depending on the species of parasite. Preventative measures and methods of treatments for caligids may also be dependent on the species of copepod, making the correct identification of these parasites even more important.

Unfortunately, a monographic generic revision of this family had never been published and the original descriptions of many of the type-species were not sufficient for our present-day needs. In addition, a discussion on the validity of each and every genus of the family needed to be presented. The closely related family Euryphoridae also needed to be re-evaluated. In order to fill this knowledge gap, the first author (M. Dojiri) completed a generic revision of the Caligidae for his Ph.D. dissertation; however, this work was never published. So, when the opportunity to update, revise, and publish this work arose, both authors readily agreed to team up to tackle this rather large undertaking to get this work officially into print.

For the majority of genera, we were able to obtain either the type-material or specimens of the type-species so that we could redescribe them in detail, not rely simply on the published literature of the original account or subsequent redescriptions of other authors. The taxonomic importance of specific structural details have come to light since the original descriptions were published, some dating back to the 19th Century and early to mid-20th Century. The descriptive accounts of all the species treated in this monograph were done with these structural details in the forefront of our minds.

The purposes of our revision are as follows: (1) to present a historical review of the Caligidae and the taxonomy of this family prior to our revision; (2) to provide a literature review on the life history, host-parasite relationships, functional morphology, and zoogeographic distribution of the Caligidae; (3) to give complete redescriptions of the type-species of the caligid genera; (4) to revise the generic diagnoses of the caligids; (5) to construct a workable key to the genera; and (6) to present the phylogenetic relationships within the Caligidae.

We have done our best to obtain and describe in detail specimens of each type-species of the various genera. In a few cases, examinations of type-material were not possible because the specimens could not be located or are no longer in existence [e.g., Markevitch's type material of *Markevichus* (originally described as *Caligopsis*)]. This was particularly problematic when the type-species was originally incompletely or poorly described and not reported, except for citations of the original account, again since its discovery. By necessity, for these few genera, only a few comments gleaned from the original description are provided and a statement on possible validity of each genus is presented.

We have accepted or rejected the validity of the many genera treated in this monograph as objectively as we possibly could and have presented our rationale in each case. However, we fully understand that the interpretation of the taxonomic values afforded specific characters, as exhibited by their respective character states, is a major factor in this decision-making process and the ultimate conclusion that is reached. For example, the taxonomic importance various authors have placed on the presence or absence of the sternal furca and postantennal process, the presence of dorsal aliform processes or lamelliform structures on various somites, and the progressive reduction of the fourth pair of legs has determined in the past whether a specific genus is accepted and recognized as a valid genus or whether it should be relegated to synonymy with another taxon.

Also, whether a unique combination of character states that are individually not unique to a specific genus is sufficient to distinguish one genus from other members of the family is still a matter of debate. In these instances, decisions on the generic validity crosses the line from objectivity for definitively valid genera to subjectivity dependent upon the taxonomic specialists' perception or interpretation of the taxonomic value of the unusual combination of characters. We have carefully considered unusual or unique character combinations of these few genera and have made decisions on their validity on a case-by-case basis. We are aware that other specialists in our field may come to disagree with some of the decisions that we have made here and anticipate constructive criticism and fruitful discussions regarding these few enigmatic/problematic genera in the hope that they lead to a better justified taxonomic classification for these organisms.

It should go without saying, but we feel compelled to mention it anyway, that an undertaking of this magnitude certainly cannot be attempted without the assistance of a multitude of individuals. We gratefully acknowledge them here for all their efforts and kindness, which resulted in this final monograph. Our deepest appreciation goes to our friend and mentor, Dr. Arthur G. Humes (deceased), as well as Dr. Roger F. Cressey (deceased), Dr. Patricia L. Dudley (deceased), Dr. Stewart Duncan, and Dr. Ivan Valiela for their review of the original Ph.D. dissertation of the first author (M. Dojiri), which formed the basis for this monograph. The first three scientists listed are no longer with us in body, but have left behind a wealth of information in the form of numerous publications that contributed enormously to our understanding of parasitic copepods. We thank them wholeheartedly for their suggested improvements to this body of work, but also for their overall contributions to our field of study.

We thank several people for their arrangement of specimen loans of parasitic copepods either from their own personal collections or from museums. They are (in alphabetical order): Dr. P. L. G. Benoit, Koninklijk Museum voor Midden-Afrika Musee Royal de l'Afrique Centrale, Tervuren, Belgium; Dr. G. A. Boxshall, Natural History Museum, London, England; Dr. R. F. Cressey, Dr. F. Ferrari, and Mr. C. Walter, National Museum of Natural History, Smithsonian Institution, Washington, D.C.; Dr. H.-E. Gruner, Zoologisches Museum, Berlin, Germany; Dr. K. Izawa, then at Mie University, but now at Izawa Marine Biological Laboratory, Tsu, Japan; Dr. J. Just, Zoologisk Museum, Copenhagen, Denmark; Dr. Z. Kabata, Pacific Biological Station, Nanaimo, British Columbia; Dr. G. Pretzmann, Naturhistorisches Museum, Wien, Austria; Dr. G. Ramakrishnan, Zoological Survey of India, Calcutta, India; Mr. T. Takegami, Seto Marine Biological Laboratory, Wakayama, Japan; Dr. T. Wolf, Zoologisk Museum, Copenhagen, Denmark; and Dr. W. Zeidler,

South Australian Museum, Adelaide, Australia. Without all the loans of specimens from these scientists, this monograph could not have been initiated and completed.

We thank Dr. Greg Deets, City of Los Angeles' Environmental Monitoring Division (CLA, EMD) for performing the phylogenetic analysis, preparing the phylogram, consensus tree, and character trees for publication, and providing his valuable expertise on the phylogeny of these parasites and Curtis Cash (also CLA, EMD) for his help with labeling several of the figure plates. Tom Juma and Renee Harrison of the CLA, EMD and CLA, Information Control Systems Division, respectively, are thanked for their assistance in scanning and converting the first author's Ph.D. dissertation into a writable Word file so that the entire document would not need to be re-typed. We are indebted to Robert Hollingsworth and Mike Chen of the City of Los Angeles' Bureau of Engineering for scanning the original figure plates and creating tiff files so that the figures would not need to be re-inked, saving the authors a tremendous amount of time. We thank two anonymous reviewers who provided comments and suggestions that greatly improved the final version of this book. We also gratefully acknowledge Folia Parasitologica and the Journal of Crustacean Biology for granting us copyright permission to include the illustrations of *Avitocaligus assurgericola*, *Alanlewisia fallolunulus*, and *Arrama tandani*.

The first author gratefully acknowledges Boston University and the Department of Biology for financial assistance through teaching fellowships and graduate scholarships during his Ph.D. program that was completed nearly three decades ago in 1983. Preparation of this monograph was partially aided by a grant from the Paramitas Foundation to the second author (JSH).

Finally, our special thanks are extended to our respective families, especially to Jackie Wang-Dojiri and Pao-Hsi Ho, for their support throughout the preparation of this monograph.

July 2012
MASAHIRO DOJIRI
JU-SHEY HO

INTRODUCTION

The taxonomic revision of the Caligidae is important to areas of biology other than just copepodology, particularly since species of this copepod family have been reported to have deleterious effects on their hosts (see White, 1940; Boxshall, 1977; Pike & Wadsworth, 1999; Johnson et al., 2004). In view of the fact that these parasites are capable of killing their hosts, have a nearly worldwide distribution, and occur on economically important fishes (e.g., Salmonidae, Scombridae, Sparidae, Pleuronectidae, among others), the systematics of the Caligidae must be considered important to fisheries biology and fish husbandry.

The correct identification of the caligid species is necessary and important in view of the possible differences in host-parasite relationships, life histories, and environmental tolerances depending on the species of parasite. Preventative measures and methods of treatments for caligids may also be dependent on the species of copepod, because there may exist some significant biological differences between two morphologically similar copepod species (Kabata, 1973a).

The genera and species of the Caligidae are frequently difficult to identify. This difficulty is due to our incomplete knowledge of the systematics of this family. Many previous descriptions of caligid species are not sufficiently detailed for comparative purposes. The diagnoses of several genera within the Caligidae overlap or are ill-defined. In addition, the familial limits of the Caligidae are not clear, particularly in regards to the relationship of this family to the morphologically similar Euryphoridae.

During the first decade of the 21st Century, two monumental works appeared in the field of copepodology. One of them is the publication of a two-volume work entitled: "Introduction to Copepod Diversity" by Boxshall & Halsey (2004) and the other one is the debut of a website: "The World of Copepods" (http://www.marinespecies.org/copepoda) created by T. Chad Walter and edited by Walter & Boxshall (2008). In the former work, Boxshall & Halsey provided a familial-level overview of the entire Copepoda and also enabled readers to identify specimens down to the generic level in some cases. In total, 32 genera were recognized by Boxshall & Halsey (2004) in the Caligidae, including those five genera (*Alebion*, *Euryphorus*, *Gloiopotes*,

Paralebion, and *Tuxophorus*) placed in the Euryphoridae by Kabata (1979). On the website edited by Walter & Boxshall (2008), 37 genera were listed under the family Caligidae. The difference in the number of the inclusive genera between these two works was partly due to the addition of new caligid genera proposed after 2004 and partly due to the difference in the recognition of some old genera. Although Dojiri (1983) completed a generic revision of the Caligidae, this work was never published and needs to be updated. Therefore, we feel that a review of all caligid genera is necessary along with a discussion regarding the group of genera formerly known as the family Euryphoridae.

HISTORICAL REVIEW

The development of the familial concept of the Caligidae occurred gradually (Kabata, 1979). The earliest report of a species of a caligid is attributed to Linnaeus who found specimens infesting the "salmon" and "haddock" (Baird, 1850). These copepods were the basis for the composite binomen *Monoculus piscinus* Linnaeus, a synonym of *Binoculus piscinus* (see Parker et al., 1968). Most species of "Entomostraca" (meaning "insect with a shell", and encompassing small crustaceans) were placed in the genus *Monoculus* by Müller (see Baird, 1850). Such was the fate of some specimens of copepods parasitic on "*Gadus merlangus*" from Danish waters described by O. F. Müller (1776). However, upon the discovery that their "eyes" were actually two sucking discs (lunules), Müller (1785) established the genus *Caligus* for these specimens, and included the new genus, along with *Argulus* and *Limulus*, under the Univalvia, which was included in the entomostracan group Binoculi.

Based on the possession of sessile eyes and two types of "feet" (presumably referring to mouth appendages and swimming legs), *Caligus* (under the name *Caligule*) was included in "Branchiopodes parasites" with *Cecrops* (presently recognized in its own family Cecropidae) and "*Argule*" (a common genus of Branchiura) by Lamarck (1818).

Nordmann (1832) apparently did not recognize familial divisions within the parasitic copepods. He simply listed *Caligus* under the "Entomostraceen". The morphologically similar genus *Lepeophtheirus*, distinguished from *Caligus* by the absence of lunules, was established in this work. Nordmann recognized the composite genus *Binoculus*, and transferred "*Dinemura*" (correctly spelled *Dinemoura*, a current member of the Pandaridae) and *Caligus productus* to it.

Wilson (1910) provided an overview of the classification of the Copepoda including both free-living and parasitic forms. He cited classifications by Linnaeus, Fabricius, Latreille, Lamarck, and Desmarest, all of whose classifications primarily or exclusively dealt with free-living copepods and not parasitic ones. According to Wilson, one of the first classifications of parasitic copepods was established by Burmeister (1835). This classification divided the parasitic copepods into five families: Argulina, Caligina, Ergasilina, Lernaeoda, and Penellina. Within the Caligina, Burmeister included genera such as *Cecrops*, *Pandarus*, and *Dinematura* (junior synonym of *Dinemoura*), which are currently in different families, and *Chalimus*, which is now known to be a larval stage in the life cycle of caligids, as well as other siphonostomes. Even though the familial suffix of the name Caligina has been changed to "-idae" in modern times, credit for the establishment of this family can be attributed to Burmeister (1835).

Following Burmeister's classification, Krøyer (1837, 1838) divided Caligina into two groups: those with eyes absent in the adults (i.e., *Anthosoma*, *Dichelestium*, *Nemesis*, *Laemargus*, *Cecrops*, *Dinematura*, and *Pandarus*), and those with two small eyes (i.e., *Caligus* and *Trebius*).

Milne Edwards (1840) established a group that he called "Des Crustacés Suceurs" (the suctorial Crustacea), which was further subdivided into the Siphonostomes and Lerneides. Under the order Siphonostomes, the family Peltocéphales was divided into three tribes: Arguliens, Caligiens, Pandariens. This was the first time that the pandarids were recognized as a separate taxon from the caligids. At that time, there were four genera (*Caligus*, *Chalimus*, *Trebius*, and *Nogagus*) under the Caligiens. *Trebius* is currently in the family Trebiidae, while *Nogagus* probably represents juveniles or males of other groups (pandarids?) (see Kabata, 1979). The tribe Pandariens was further subdivided into two groups, the pandarid group and the cecropid group. The genus *Euryphorus* (later to become the type-genus of its own family Euryphoridae) was included in the pandarid group.

Baird (1850) proposed a classification containing four families (i.e., Argulidae, Caligidae, Pandaridae, and Cecropidae) in the tribe Peltocephala in his third "legion" Paecilopoda and the order Siphonostoma. This marked the first time that the cecropids were recognized as distinct from the pandarids at the familial level. He recognized the lunules as sucking discs, not eyes, and correctly identified the eyes; however, he described the frontal plate as the vestigial antennule, and the real antennule as the antenna. The antenna, maxilla, and maxilliped were designated as "foot-jaws". Legs 1-4 were referred to as "natatory legs".

Dana (1852) is sometimes given credit for the establishment of the family Caligidae. Both his 1852 and 1853 papers provided the same classification scheme in which the Caligidae was placed under the Entomostraca and divided into three subfamilies: Caliginae, Pandarinae, and the Cecropinae. The genus *Euryphorus* was still thought to be a member of the Pandarinae. Dana's classification is not too different from Baird's, except that Dana demoted the three families to the subfamilial rank.

Using the oral appendages and the structure of the mouth as distinguishing features, Thorell (1859) divided Copepoda into three major taxa, Gnathostoma, Poecilostoma, and Siphonostoma. The Caligidae was included under the last named group, the Siphonostoma. Olsson (1869) apparently agreed with Thorell's Siphonostoma, which, as the name implies, included copepods possessing a mouth suited for sucking.

Heller (1865) also accepted the order Siphonostoma (Siphonostomes of Milne Edwards, 1840) and lumped the caligids, pandarids, and cecropids all under the family Caligina. Heller did recognize groups in his key of the Caligina. The first group composed of *Caligus*, *Synestius*, *Parapetalus*, *Caligodes*, *Lepeophtheirus*, *Anuretes*, *Hermilius*, and *Calistes*. The second group, which was later known as the family Euryphoridae included *Euryphorus*, *Elytrophora* (a synonym of *Euryphorus*), *Caligeria* (most likely a synonym of *Euryphorus*), and *Gloiopotes*. The third group contained the distantly related genera *Trebius*, *Dysgamus* (a composite genus predominantly containing juveniles of *Euryphorus*) (see Heegaard, 1972), and *Alebion*.

G. O. Sars (1903) disagreed with Thorell's three divisions. He recognized seven main copepod groups: Calanoida, Harpacticoida, Cyclopoida, Notodelphyoida, Monstrilloida, Caligoida, and Lernaeoida. Later authors, such as Norman & Scott (1906), Scott & Scott (1913), and Gurney (1933) accepted Sars' classification.

Wilson (1905a) in his key to the five subfamilies of the Caligidae utilized the number of thoracic somites fused with the head, presence or absence of dorsal plates, distinctness of the frontal plates, and the nature of the egg sacs as distinguishing features separating the subfamilies Trebiinae, Caliginae, Euryphorinae, Pandarinae, and the Cecropinae. This was the first time that all five groups were recognized as distinct taxa of equal taxonomic rank. Subsequently, Wilson (1932) elevated all five subfamilies to the familial rank. Wilson's work on the Caliginae (1905a), the Trebiinae and Euryphorinae (1907a), and the Pandarinae and Cecropinae (1907b) remained the most complete work on these groups until Cressey (1967a) revised the family Pandaridae sixty years later.

In his work on the parasitic copepods of fishes of the U.S.S.R., Markewitsch (often transliterated Markevich) (1956) recognized Wilson's earlier classification of 1905. Whether this Russian researcher was unaware of Wilson's (1932) publication or simply disagreed with splitting the Caligidae into five families is not known because Markewitsch did not elaborate on his classification.

Yamaguti (1963), in his book on Copepoda and Branchiura parasitic on fishes, included two new families, Dissonidae and Eirgidae, in the superfamily Caligoidea (order Caligidea). Although the monogeneric family Dissonidae is a valid family (Kabata, 1979), Eirgidae is no longer considered valid (Ho, 1970; Hameed & Pillai, 1973a).

Five subfamilies (i.e., Caliginae, Echetinae, Lepeophtheirinae, Anuretinae, and Mappatinae) of the Caligidae were established by Yamaguti (1963). However, these subfamilies can no longer be recognized. Echetinae was distinguished from the Caliginae by the "absence of lunules". However, *Echetus typicus* Krøyer, 1864 (the only member included in the entire subfamily) possesses lunules (Ho, 1966); consequently, it should have been included in the Caliginae. The subfamily Lepeophtheirinae was separated from the Caliginae also by the absence of lunules. This taxonomic feature may, however, be of more limited value than once thought, since Cressey & Cressey (1979) found a progressive reduction leading to the absence of this structure among female members of a single caligid genus *Abasia*. The subfamily Anuretinae was distinguished from the Lepeophtheirinae by the vestigial condition of the abdomen or its total absence in the members of the Anuretinae. The presence of a "prominent posterior median lobe of carapace" dorsally covering the abdomen and genital complex distinguished Mappatinae from the other two subfamilies. These three subfamilies may not be distinguishable, since they contain some genera (i.e., *Heniochophilus* in Lepeophtheirinae; *Anuretes* and *Pseudanuretes* in Anuretinae; and *Mappates* in Mappatinae) that are morphologically very similar. Therefore, the subfamilial divisions of the Caligidae proposed by Yamaguti have not been generally accepted (Kabata, 1979).

The structure of the oral area (siphonlike or nonsiphonlike mouth) and the morphology of the mandible as criteria for higher taxonomic divisions (Thorell, 1859) provided Kabata (1979) with the basis for six suborders within the podoplean line. This division, first established by Giesbrecht (1892), was resurrected along with his Gymnoplea by Kabata. The suborders recognized by Kabata (1979) are Monstrilloida, Harpacticoida, Misophrioida, Cyclopoida, Poecilostomatoida, and Siphonostomatoida. The gymnoplean line contains only one suborder (i.e., Calanoida of Sars). Although his treatment of Cyclopoida and Poecilostomatoida has met with criticism, the suborder Siphonostomatoida to which the Caligidae belongs appears to be valid (Stock, 1981). Kabata (1979) no longer recognizes the order Caligoida, which included the Caligidae, Euryphoridae, Pandaridae, Cecropidae, and Trebiidae. To avoid taxonomic confusion with older classifications, the term "caligiform" group or complex (Kabata, 1979) will be used instead of "Caligoida" and "caligoid" throughout this present revision.

A cladistic analysis of the ten copepod orders was conducted by Ho (1990) using 21 morphological characters. The ten orders analyzed by him included

TABLE I
List of genera included in the families Caligidae and Euryphoridae by Kabata (1979)

CALIGIDAE	
Abasia Wilson, 1908	*Hermilius* Heller, 1868
Alicaligus Shiino, 1955	*Lepeophtheirus* Nordmann, 1832
Anuretes Heller, 1865	*Mappates* Rangnekar, 1958
Caligodes Heller, 1865	*Parapetalus* Steenstrup & Lütken, 1861
Caligopsis Markewitsch, 1940	*Parechetus* Pillai, 1962
Caligulina Heegaard, 1972	*Pseudanuretes* Yamaguti, 1936
Caligus Müller, 1785	*Pseudocaligus* A. Scott, 1901
Caritus Cressey, 1967	*Pseudolepeophtheirus* Markewitsch, 1940
Dartevellia Brian, 1939	*Pseudopetalus* Pillai, 1962
Diphyllogaster Brian, 1899	*Pupulina* Beneden, 1892
Echetus Krøyer, 1864	*Sciaenophilus* Beneden, 1852
Heniochophilus Yamaguti & Yamasu, 1959	*Synestius* Steenstrup & Lütken, 1861
EURYPHORIDAE	
Alebion Krøyer, 1863	*Paralebion* Wilson, 1911
Euryphorus Milne Edwards, 1840	*Tuxophorus* Wilson, 1908
Gloiopotes Steenstrup & Lütken, 1861	

Platycopioida, Gelyelloida, and Mormonilloida, in addition to the seven orders of Kabata's (1979) book. Huys & Boxshall (1991) also analyzed the phylogeny of the same ten orders using 54 morphological characters. However, the concept of copepod orders was drastically changed in Boxshall & Halsey's (2004) book where families of the Poecilostomatoida are merged with those of Cyclopoida.

Kabata (1979) and previous workers recognized the Caligidae and Euryphoridae as distinct families, distinguished from each other by dorsal aliform plates on the fourth pedigerous somite in species of the Euryphoridae. Kabata recognized 24 genera in the Caligidae and 5 genera in the Euryphoridae (table I). However, Boxshall & Halsey (2004) treated the Euryphoridae as a junior synonym of the Caligidae. The taxonomic value of the dorsal plates and the validity of the Euryphoridae were not discussed by Kabata. In addition, several caligid genera still remained ill-defined (Kabata, 1979: 169). It is from this point in the taxonomic status of these two groups that this revision was undertaken.

MATERIALS AND METHODS

Most of the specimens studied were on loan from various museums and from personal collections of various individuals. The museums that have lent some type and nontype specimens for this revision are as follows: National Museum of Natural History, Smithsonian Institution, Washington, D.C., U.S.A.; Faculty of Fisheries, Mie University, Tsu, Mie Prefecture, Japan; Zoologisk Museum, Copenhagen, Denmark; Naturhistorisches Museum, Wien, Austria; Koninklijk Museum voor Midden-Afrika, Musée Royal de l'Afrique, Centrale, Tervuren, Belgium; South Australian Museum, Adelaide, South Australia; Zoological Survey of India, Calcutta, India; Natural History Museum, London, England; Zoologisches Museum, Museum für Naturkunde der Humboldt-Universität zu Berlin, Berlin, Germany.

Several researchers have lent specimens from their own personal collection. These individuals are: Dr. G. A. Boxshall, Natural History Museum, London, England; Dr. Roger F. Cressey, National Museum of Natural History, Smithsonian Institution, Washington, D.C., U.S.A.; Dr. Ju-Shey Ho (second author), California State University, Long Beach, California, U.S.A.; Dr. Z. Kabata, Pacific Biological Station, Nanaimo, British Columbia, Canada; Mr. Toshiya Takegami, Seto Marine Biological Laboratory, Shirahama, Wakayama Prefecture, Japan.

In addition, several fishes housed in the ichthyological collection at the National Museum of Natural History, Smithsonian Institution, were examined for parasitic copepods during a visit by the first author (M. Dojiri) to the museum in May of 1979. The copepods that were collected were preserved in 70% ethanol and eventually became part of the collection of the museum.

In every case, the specimens were cleared in lactic acid for at least a 24-h period before examination and were then transferred onto a wooden slide and dissected as described in Humes & Gooding (1964). Illustrations of the copepods were made with the aid of a camera lucida or a drawing tube mounted on the microscope.

The authors utilized FishBase (Froese & Pauly, 2011) to confirm and list the currently accepted scientific binomen, but also include the originally reported name of the host.

The revised diagnosis for each genus is presented. Each diagnosis is followed by a short discussion of the genus. Detailed redescriptions of species are accompanied by illustrations and remarks. Synonymies for the genera and species are provided. In a few cases, specimens of some genera were not obtainable (either because the type-specimens could not be located or were no longer in existence). Each of these unobtainable types is mentioned separately in the text under the appropriate generic heading.

The capital letter in parentheses after the explanation of each illustration (in the figure captions) refers to the scale at which it was drawn; in a few figures (those species described after the completion of Dojiri's 1983 Ph.D. dissertation), the scale bars are included next to the illustrations. The abbreviations used in this revision are as follows: a_1 = antennule, ae = aesthetasc, a_2 = antenna, he = hyaline element, pap = postantennal process, mt = mouth tube, md = mandible, mx_1 = maxillule, mx_2 = maxilla, mxpd = maxilliped, sf = sternal furca, icp = intercoxal plate, en = endopod, P_1 = leg 1, P_2 = leg 2, P_3 = leg 3, P_4 = leg 4, and P_5 = leg 5. In the formula for the armature of the legs, Roman numerals indicate spines; Arabic numerals represent setae.

EXTERNAL MORPHOLOGY

General habitus

The body of the Caligidae is typically dorsoventrally flattened, an adaptation that reduces resistance to water flowing over the body of the host (Kabata, 1979). Exceptions to this body form include females belonging to genera that have secondarily invaded the gills or the nasal cavity of the host, as in *Hermilius*, *Kabataella*, and *Synestius*.

As noted by Kabata (1979), the body of the caligid consists of four tagmata (fig. 1a): the cephalothorax, the fourth pedigerous (leg-bearing) somite, genital complex, and abdomen. The cephalothorax is formed from the fusion of the cephalon, the maxilliped-bearing somite (first thoracic somite) and the first, second, and third pedigerous somites.

Baird (1850) and later Wilson (1905a) identified four regions of the caligid carapace that are usually demarcated by suture lines. Baird named these regions the cephalic, the thoracic, and the lateral (one on each side) areas. The cephalic and thoracic areas of the carapace do not exactly correspond to the cephalic and thoracic components of the cephalothorax (Parker et al., 1968; Boxshall, 1974a; Kabata, 1979).

The cephalothorax acts as a suction cup in adhering to the host (Kabata & Hewitt, 1971; Kabata, 1979, 1981; Ho & Lin, 2004). The cup is formed by the frontal plate (frequently with a pair of suckerlike structures known as lunules) on its anterior margin, the shieldlike structure of the ventral apron on its posterior margin, and the transparent membrane throughout its circumference. Although usually dorsoventrally flat, the cephalothorax may be folded longitudinally so that the lateral areas are bent ventrally (fig. 84a). This alteration of the cephalothorax results in a median longitudinal groove in which the gill filament of the host lies (fig. 84b). This modification, seen in females of the caligid genera *Abasia*, *Arrama*, and *Hermilius*, reflects a specialization for attachment to the gill filament. *Kabataella*, which lives in the nasal cavity of pomadasyids, has a moderately folded cephalothorax.

In some genera (best exemplified by *Mappates*), the free margin of the thoracic zone of the cephalothorax dorsally covers the fourth pedigerous somite and much, or all, of the genital complex (fig. 102a). The shieldlike continuity of the cephalothorax probably ensures a more laminar flow of water across the dorsal surface of the body. Caligids without such an extension of the cephalothorax may experience a disruption of the water flow as it passes posteriorly beyond the cephalothoracic shield toward the fourth pedigerous somite and genital complex. However, this has not yet been tested.

The thoracic region of the carapace possesses deep indentations on its posterior margin. These indentations, known as the posterior sinuses, are provided with transparent membranes that apparently act as valves preventing water from entering the space between the ventral surface of the parasite and the attachment surface on the host and undoubtedly aids in attachment by increasing the force of suction and creating a stronger seal (Kabata & Hewitt, 1971; Kabata, 1979; Ho & Lin, 2004).

The major articulation of the body is between the third and fourth pedigerous somites. The fourth pedigerous somite is the only free thoracic somite exhibited by the caligid. It is usually short, but may be elongate and slender in the female, giving the appearance of a long neck (e.g., *Echetus*, *Parechetus*, and *Pseudechetus*) (e.g., fig. 63a). The anteriormost portion of the genital complex may constitute at least part of the neck. In some genera (i.e., *Alebion*, *Euryphorus*, *Gloiopotes*, and *Tuxophorus*), there is a pair of aliform plates arising from this somite (e.g., figs. 17a, 67a, 78a, and 146a, respectively) (see discussion on Euryphoridae).

Contrary to Wilson (1905a), it appears that both the fifth and sixth pedigerous somites fuse to form the genital complex (Parker et al., 1968). In all likelihood, the seta-bearing papillae (described by Parker et al., 1968; Ho & Bashirullah, 1977) (fig. 4g) located on the ventrolateral surface of the genital complex represent the fifth legs (thoracopods) in the female. The position of the process corresponds to the position of the fifth legs in the male. The hypothesis that the genital complex is a result of the fusion of the posterior two thoracic somites is also supported by the presence of the vestigial fifth and sixth pairs of legs on the genital complex of many caligid males (Ho & Lin, 2004).

The genital complex of the female is usually an inflated tagma, varying in shape from quadrangular, triangular, globose, to oblong. Boxshall (1990) and Huys and Boxshall (1991) have hypothesized that the genital complex of all siphonostomatoids consists of the fifth pedigerous somite, genital somite, and the first abdominal somite. In some genera, the genital complex of the female possesses lateral aliform expansions (e.g., *Parapetalus*, fig. 120a), digitiform posterolateral processes (e.g., *Paralebion*, *Pseudechetus*, and *Synestius*) (e.g., figs. 114a, 141a), or aliform posterolateral processes (e.g., *Caligodes* and *Parechetus*) (e.g., fig. 53a). This tagma is a sexually dimorphic character. Generally speaking, the genital complex of the female is much larger and comprises a larger portion of the body than in the male. In addition, the two sexes differ significantly in the general overall shape of this tagma.

In the past, researchers tended to use the size, the length-width ratio, and the general appearance of the genital complex of the female as an important and sometimes sole taxonomic feature upon which new species were based. This was unfortunate, since it is now known that the size and shape of the genital complex is partly dependent on the state of maturity, stage in egg-laying (Parker et al., 1968), and perhaps to the extent of contraction during the time of fixation. Moreover, the total size of the organism is an unreliable character since it was shown that size varied depending on the species of host (Cressey, 1967a; Lewis et al., 1969; Cressey & Collette, 1970).

The abdomen is frequently short, consisting of one or two free somites. However, it is not unusual for female caligids to possess a long abdomen (best exemplified by *Caligodes*, *Sinocaligus*, *Synestius*, and especially *Sciaenophilus*) (e.g., fig. 134a). Toward the other extreme are *Anuretes*, *Mappates*, and *Pseudanuretes*, which exhibit an extremely short (vestigial) abdomen (e.g., fig. 34a). This tagma may be completely fused to the genital complex in some species of *Anuretes*. Moreover, the abdomen may carry lateral expansions or processes as in females of *Abasia*, *Alebion*, *Dartevellia*, *Euryphorus*, *Midias*, *Parapetalus*, *Parechetus*, *Pseudechetus*, and *Sinocaligus* (e.g., figs. 10a, 17a, 61a, and 67a).

The body of the caligid possesses three different types of setules (types a, b, and c of Parker et al., 1968), differing in the number of branches arising from a common base. These setules have been implicated in a chemosensory function (Parker et al., 1968; Boxshall, 1974a), but some may function as rheoreceptors as suggested by Briggs (1978).

Caudal ramus

The last abdominal somite possesses the “caudal furcae” or “caudal rami” (fig. 1c). Bowman (1971) considered these structures as “uropods” since they originate from the anal somite. Although Kabata (1979, 1981) agreed with Bowman, there has not yet been a general acceptance of the use of the term “uropods” for these appendages. The term “caudal ramus”, which is widely in use, will be utilized throughout this revision.

The caudal ramus is morphologically consistent throughout the family varying in its length-width ratio, its shape, and relative length of its setae. In the majority of caligid genera, this appendage carries six setae (three small subterminal setae and three larger terminal ones). Apparently the caudal ramus along with the genital complex and abdominal somites can function as a "rudder" (Kabata & Hewitt, 1971; Ho & Lin, 2004).

Frontal plate and lunules

The antennule was considered to be 3-segmented at the "third molt" with the basal segment transforming to become the frontal plates in the adult (Wilson, 1905a). This view seems erroneous based on larval development (Heegaard, 1947a) and the nature of the articulation of the antennule and the frontal plate (Parker et al., 1968). In their illustrations, Parker et al. show sectional views displaying the relationship of the frontal plate to the antennule and the anterior margin of the cephalothorax. According to their figures, the frontal plate is continuous with the anterior margin of the cephalothorax. Kabata (1972) suggested that the frontal plates originate from the frontal sac, a flat structure everted during the expulsion of the frontal filament. Together with the bases of the antennule, the frontal plate aids in attachment. Not only does it help to maintain a tight seal at the anterior margin of the parasite, it also acts as the entrance for water that is drawn in under the cephalothorax and ejected out through the posterior sinuses during crawling and scuttling movements of the copepod on its host (Kabata & Hewitt, 1971).

A pair of sucking discs known as lunules (fig. 1b) are located on the ventral surface of the frontal plate in some genera. Parker et al. (1968) suggested that the lunules are modified from a small portion of the transparent membrane that fringes nearly the entire cephalothorax. These lunules act as accessory structures for attachment to the host.

The presence or absence of the lunules has been given a great deal of taxonomic weight in the past. For instance, the genus *Lepeophtheirus* is almost identical to the genus *Caligus*, except that the species of *Lepeophtheirus* lack lunules. Yamaguti (1963) actually established two subfamilies, Caliginae and Lepeophtheirinae, depending upon the presence or absence of these structures. Since the former euryphorid genus *Tuxophorus* is transferred to the Caligidae (see discussion on the Euryphoridae), the lunules are not possessed by any genera outside the Caligidae. Just as significant, the lunules appear to be present in only one evolutionary lineage of the caligiform complex.

As mentioned above, Yamaguti (1963) considered the lunules taxonomically very important structures, even at the subfamilial level. However, Cressey & Cressey (1979) have shown that the lunules exhibit a progressive reduction and eventual absence among the female members of *Abasia*, which are adapted for existence on the gill filaments of their host. The change from the usual caligid attachment site of the external body surface of the host to the gill filaments appears to be correlated with the reduction of the lunules in *Abasia*. Since this structure can exhibit secondary reduction and loss even within a genus and appears to be a homoplasious character (see Phylogeny of the Caligidae section), the taxonomic value of this character is somewhat reduced.

Antennule

The antennule (fig. 1d, e) is 2-segmented and apparently has a sensory function (Kabata & Hewitt, 1971). Laverack & Hull (1993), studying the sensory innervation in the antennule of juvenile male *Caligus elongatus* von Nordmann, 1832 discovered that the 13 setae on the distal end of the terminal segment are provided with various numbers of nerve fibers that suggested chemoreceptive or mechanoreceptive functions. In general, it is not a sexually dimorphic appendage. The basal segment is stout and armed with numerous setae varying in number depending on the species. Frequently, a small bifid process is located on the posterodistal corner of this segment (fig. 91e). The second segment is slender and cylindrical. Earlier researchers have described varying numbers of setae at the tip of the antennule of different caligid species. However, Parker et al. (1968) suggested that the number at the tip of this appendage may be constant. These researchers found one seta on the posterior margin of the segment at about midlength and 13 setae (one of these may be an aesthetasc) at the tip in *Caligus curtus* Müller, 1785 and a number of other species of *Caligus*.

Species of the genera *Dentigryps* (now recognized as a synonym of *Lepeophtheirus*), *Dissonus* (a member of the family Dissonidae), *Heniochophilus* (a synonym of *Anuretes*), *Mappates*, and *Pseudanuretes* exhibit this constant number (Parker et al., 1968). Moreover, Ho & Dojiri (1977) and Dojiri (1979) have also found this setal number to hold true for all the species (five in total) of *Lepeophtheirus* that they studied. The species examined in our revision also exhibit this constant number of 13 + 1 aesthetasc (or 14 setae), with the exception of *Caligodes* and *Pupulina*. In most caligids examined, the posterior group of setae situated at the apex of the second segment consists of five setae, two of which share a common base. However, in the two genera just mentioned, there is no bifid seta. Consequently, their setal formula is 12 + 1 aesthetasc (12 setae and 1 aesthetasc). In the genus *Avitocaligus*, there are 13 setae with apparently no bifid seta and no aesthetasc.

Antenna

The antenna of the female (fig. 1f) is considered to be 4-segmented by Kabata (1979), but 3-segmented by Parker et al. (1968). However, it is not too variable within a particular genus, although it may vary between species of different genera. The basal segment is short and, in many genera, possesses a posteriorly directed spinelike process. The size and shape of this process may, on occasion, be used as a specific discriminant. The second segment is robust and frequently carries a dorsal corrugated pad (fig. 1g). The terminal segment is a recurved claw, most likely comprised of the endopod fused with the terminal spiniform process; the composition of the recurved claw is reflected by the existence of a seta and, in some species, a partial suture line. There are exceptions to this basic structure. For example, *Abasia pseudorostris* Wilson, 1908, possesses a reduced antenna with a conical spiniform process (fig. 6e) instead of a claw. Furthermore, all the members of *Hermilius* and *Pseudanuretes* possess an accessory claw (fig. 85a) in addition to the terminal claw.

Although the antenna of the female does not exhibit, in most cases, taxonomically significant morphologic features, the antenna of the male displays taxonomic features particularly at the specific level. The posteriorly directed spiniform process found in the antenna of the female is not present in the male. The antenna of the male appears to be 3-segmented in most genera. The second segment of the antenna of the male possesses a number of corrugated adhesion pads (fig. 5b, c). Moreover, the terminal claw usually carries a few accessory tines (fig. 5d). This strong sexual dimorphism exhibited by the antenna is apparently due to the different functions of this structure in the two sexes (Parker et al., 1968). The antenna of the female is used as a prehensile organ that aids in the attachment to the host. On the other hand, the antenna of the male is primarily used to grasp the female during copulation.

Postantennal process

The origin and consequently the naming of the next structure (fig. 2a) have undergone a great deal of discussion. In the past, this uncertainty has, unfortunately, affected the naming of the appendages that are situated posterior to it. It has been called the "first maxilla" by A. Scott (1901a), Wilson (1905a), Heegaard (1945a, 1947a, and 1947b), and Ferris & Henry (1949), among others. Gurney (1947) and Lang (1947) called it a "cuticular spine", while Lewis (1963, 1969) used the term "postantennal process".

Lewis (1969) and, more recently, Kabata (1979, 1981) have given a review on this subject. The term "postantennal process" suggested by Lewis is now generally accepted and will be employed throughout this revision.

The postantennal process is a hooklike structure lying posterolaterally to the antenna. Three papillae (sclerites), bearing setules, are associated with this process. It may carry an accessory process as in the genus *Gloiopotes hygomianus* (fig. 79a). In place of the hook, there may be a corrugated pad as in the genus *Alebion* (fig. 18c) or a sclerotized plate as in *Caligodes* (fig. 53d). All species of *Pseudanuretes* lack the postantennal process. The presence or absence of the postantennal process, as well as its shape, may be useful as a taxonomic character important at the specific level.

Mouth tube and mandible

In the Caligidae and related families, the labrum and labium overlap to form a mouth tube (fig. 2b). The possession of the mouth tube characterizes the entire order Siphonostomatoida in which the Caligidae, and its relatives (Dissonidae, Trebiidae, Pandaridae, and Cecropidae) have been placed (Kabata, 1979). The structure of the mouth tube has been described recently by Parker et al. (1968), Kabata (1970, 1974, and 1979), and John & Nair (1974).

A small process known as the intrabuccal stylet (figs. 2c, 18e) is located on the dorsal side (inside surface) of the labrum near the junction with the frons labri. The homology and function of this structure is not known, but it is present in the caligid genera.

Kabata (1974) described the functional complex consisting of the mouth tube, mandible, and strigil, and speculated on the mode of feeding of caligids. The dentiferous strigil located at the tip of the labium plays an important role in Kabata's hypothesis. The strigil is a divided bar (fig. 2d) with many small teeth. It has been found in almost all caligid genera, except *Belizia*, *Kabataella*, and *Alebion* (i.e., *A. carchariae* Krøyer, 1863 and *A. glaber* Wilson, 1905) (figs. 18f, 25f). Kabata surmised from the morphology of the mouth tube that, during feeding, the mouth tube becomes perpendicular to the surface of the host, spreads the marginal membranes that rim the orifice to close off the feeding area, and presses the dentiferous strigil against the fish tissue. The divided bars of the strigil move back and forth in a sawing motion, tearing up the host tissue, which is then swept up by the mandibles into the buccal cavity (see Ho & Lin, 2004).

The base of the mandible is situated laterally to the base of the mouth tube. The mandible enters the buccal cavity through lateral openings in the mouth tube and the entire appendage consists of four sections marked off by sutures. The terminal portion is armed with 12 teeth on the inner margin in all caligid genera, except *Caligodes* and *Dartevellia*, which possess 13 teeth. As mentioned above, Kabata (1974) believed that the mandibles are not the main masticatory appendage, but merely function as rakes to sweep the torn pieces of host tissue into the buccal cavity.

Maxillule

The next structure, whose base is positioned immediately posterior to the base of the mouth tube, was the subject of terminological controversy just as in the postantennal process. A. Scott (1901a), Wilson (1905a), and Heegaard (1945a, 1947a), among others, called it the "second maxilla". It was referred to as the "first maxilla" by Gurney (1947) and Lang (1947). Lewis (1963), realizing the uncertainty of the homology and nature of this appendage, was noncomittal and called it the "postoral process". However, later believing that the two components of this structure (dentiform process and node) arise independently, Lewis (1967, 1968) used the terms "first maxilla" for the node (setiferous papilla) and "postoral process" for the spiniform process. Renaud & Cals (1975) agreed with Lewis.

Lewis (1969) examined and analyzed this structure from a number of species of different genera of the Caligidae and related families. He concluded that it is not a cuticular structure, but probably represents a true appendage and referred to it as the "first maxilla". He further suggested that the nature of this appendage in more "primitive" forms such as the pandarids exhibit a bipartite appearance. On the other hand, caligids and trebiids possess a dentiform process and a setule-bearing papilla.

The size and shape of the dentiform process is frequently taxonomically helpful in distinguishing taxa (compare figs. 2g, 18h, 61g, 64d, 85f, and 103a). This process may have an accessory tine as in species of *Gloiopotes* (fig. 79e) or may be bifid as in many species of *Lepeophtheirus* (figs. 92f and 99g). A corrugated pad is situated on the dentiform process in the female of *Caligodes laciniatus* (Krøyer, 1863) (fig. 54e), but usually this pad is restricted to males of certain species (e.g., *Midias lobodes* Wilson, 1911, fig. 112f). The males of many species of *Lepeophtheirus* possess a hyaline digitiform process on the medial surface of the dentiform spine (e.g., fig. 95g). The dentiform process is highly reduced in *Anchicaligus nautili* (refer to Ho, 1980) and absent in species of *Pseudanuretes* (fig. 123e).

Maxilla

Dependent upon the interpretation of the postantennal process, the terminology of both the maxillule and maxilla were affected. The maxilla was considered to be the "first maxilliped" by those who thought that the postantennal process represented the maxillule (A. Scott, 1901a; Wilson, 1905a; and Heegaard, 1945a). Gurney (1947) questioned the use of the term "first maxilliped" and believed that this appendage is the maxilla, especially since the maxillary gland opens at the base of this appendage. Lewis (1969) agreed with Gurney and called it the "second maxilla".

The maxilla (fig. 2h) is considered 2-segmented (Lewis, 1963). Kabata (1979) introduced some terms for the various components of this appendage that are adopted for this revision. The appendage is considered "brachiform" with the proximal portion known as the "lacertus" and the distal portion called the "brachium". The striated membrane near the midlength of the brachium is termed the "flabellum". The longer terminal process, "calamus", and the shorter one, "canna", carry serrated membranes.

The maxilla may help in the "settling" movements of the copepod that optimally orient the copepod in respect to its fish host, and aids in a more secure attachment (Kabata & Hewitt, 1971). Furthermore, this appendage is used "to manipulate the frontal filament during the developmental stages of most siphonostomes in which they are known (Kabata, 1979). The position of the maxilla (fig. 116) and the comblike serrations on both the calamus and canna (fig. 2i) suggest that this appendage may aid in grooming. However, this function for the maxilla has not yet been shown for the Caligidae.

Maxilliped

The maxilliped was at one time considered to be the "second maxilliped" by those who believed that the maxilla represented the "first maxilliped" (A. Scott, 1901a; Wilson, 1905a; and Heegaard, 1945a). It is now believed that copepods possess only one pair of maxillipeds (Kabata, 1979).

This appendage is the posteriormost oral appendage and represents the first thoracic somite. It is a 2-segmented, subchelate, prehensile structure with a robust basal segment (corpus maxillipedis of Kabata, 1979; or protopod, see

Ho & Lin, 2004) and a clawlike, terminal segment (endopod) (fig. 2j). This claw may be formed from the fusion of two segments as suggested by a suture (Parker et al., 1968) and separately called the shaft (proximal part) and claw (distal part) by Kabata (1979).

In the majority of cases, the maxilliped is a relatively uniform structure without much taxonomic significance. However, in a few genera the maxillipeds of the females are distinctly different from each other and from other genera (compare figs. 19b, 62a, 68g, and 86b). The maxilliped is occasionally sexually dimorphic with that of the male possessing patches of denticles or spinules on the corpus maxillipedis or a prominent myxa (e.g., figs. 5f, 60d, and 71c).

Sternal furca

The sternal furca (sternal fork) consists of a cylindrical basal portion referred to as the box and extends into two digitiform, posteroventrally directed structures, the tines. The sternal furca is located medially to and between the bases of the maxillipeds and the first pair of thoracic legs. It is present in some species of Caligidae and Trebiidae. Apparently, it is immovable in species of *Euryphorus*, but movable in species of other caligid genera (Lewis, 1966a).

The homology of the sternal furca is not yet known. Four major hypotheses have been proposed (Heegaard, 1947a; Ferris & Henry, 1949; Lang, 1951; and Lewis, 1966a). Lewis (1966a) discussed each of these hypotheses in detail, so a review will not be repeated here.

Although the function of the sternal furca is not known, Wilson (1905a), Gnanamuthu (1948), Lewis (1966a), Kabata & Hewitt (1971) and others suggested that the sternal furca prevents the parasitic copepod from slipping backwards on the host. Perhaps all "sharp-pointed" processes including the postantennal process and the dentiform process of the maxillule function as "brakes", thus effectively preventing slippage on the host (Kabata & Hewitt, 1971).

The presence or absence of the sternal furca and its general appearance can be used as specific discriminants. This structure is generally not considered to be of high taxonomic value although when used in conjunction with many other diagnostic features, it may be used to distinguish genera. There are several caligid genera that lack the sternal furca (i.e., *Abasia*, *Alebion*, *Anchicaligus*, *Belizia*, *Caritus*, *Dartevellia*, *Echetus*, *Mappates*, *Metacaligus*, *Pseudanuretes*, and *Pupulina*). Out of these 11 genera that do not possess the sternal furca, seven of them also lack the hooklike postantennal process. *Anchicaligus* and *Metacaligus* possess a small dentiform postantennal process.

The tines of the sternal furca (fig. 2k) can be bifid as in some species of *Gloiopotes*, *Tuxophorus* (fig. 148b), *Caligus* (e.g., *C. bifurcatus* Pearse, 1953), and *Lepeophtheirus* (e.g., *L. appendiculatus* Krøyer, 1863; *L. longispinosus* Wilson, 1908; *L. bifidus* Fraser, 1920; among others). A pair of small secondary tines is found at the base of the sternal furca in *Gloiopotes hygomianus* Steenstrup & Lütken, 1861 (fig. 79i).

Leg 1

Leg 1 (fig. 3a) is connected by an intercoxal plate (interpodal bar or coupler of some authors, see Parker et al., 1968) synchronizing the movements of the legs. The sympod is composed of the fused coxa and basis, and is armed with one outer and one inner seta. The exopod is 2-segmented with the first segment elongate and bearing an outer distal spine and an inner row of setules. The terminal segment carries three spines. The outermost spine (designated as spine 1) is often the longest (fig. 36a) and the middle (spine 2) and the innermost (spine 3) spines frequently carry an accessory process at their distal ends (fig. 93b). Occasionally, spine 1 is the shortest of the group (e.g., figs. 65a and 136a). These three spines are usually serrated and structurally conservative, but do show some modifications (compare figs. 3b, 20b, 55c, 80c, and 86f). A seta, designated seta 4 and in most species pinnate, is usually situated on the inner distal corner of the terminal segment, but may take a position between spines 2 and 3 in some species of *Anuretes* (fig. 36a) and *Pseudanuretes* (fig. 124f). The inner margin of the terminal exopodal segment typically is equipped with three relatively large plumose setae. These setae are reduced or absent in a few species (figs. 7g, 11i). The endopod is vestigial, consisting of a simple lobe in most caligids (fig. 3a, c). It is 2-segmented in *Alebion*, *Avitocaligus*, *Euryphorus*, *Gloiopotes*, and *Pupulina* with the terminal segment carrying three plumose setae (e.g., figs. 19e, 48a, 68i, 80b, and 128e). Legs 1 and 2 are important in producing the rapid swimming movements of caligids (Kabata & Hewitt, 1971).

Leg 2

Each side of leg 2 is also united by an intercoxal plate as in leg 1. The coxa and basis remain unfused and distinct. The coxa possesses a plumose seta near the junction with the intercoxal plate. The basis carries a seta on its distal corner near its junction with the exopod. This leg is biramous with 3-segmented rami. The dorsal surface of leg 2 (e.g., fig. 4a) bears two striated membranes. The first membrane is attached to the coxa. The second membrane is much larger than the first and originates from the first exopodal segment. The armature of the endopod is usually the same in all caligid genera. The formula for the leg is as follows:

Coxa 0-1 Basis 1-0	Exp. I-1; I-1; II, I, 5
	End. 0-1; 0-2; 6

The spines on the outer surface of the exopod differ not only in the number (e.g., terminal exopodal segment of species of *Metacaligus* with formula I, I, 5), but also in their appearance from taxon to taxon (compare figs. 3e, 20d, and 59a).

Kabata and Hewitt (1971) observed the scuttling movements of the salmon louse (*Lepeophtheirus salmonis* (Krøyer, 1837)) and discovered that legs 1 and 2 do not beat synchronously or metachronally in the same direction, but instead beat against each other to create a jet of water that travels in a posterior direction and out from under the cephalothoracic area via the posterior sinuses resulting in rapid forward motion of the copepod [see Ho & Lin (2004) for a summary].

Leg 3

Leg 3 has been extensively modified to form an "apron" or "shield", which is the posterior end of the suction cuplike cephalothorax. This apron has developed from the modification of the intercoxal plate and sympod (Parker et al., 1968). The free margin of the ventral apron may be moved forwards or backwards, thus increasing or decreasing the space between the cephalothorax of the parasite and the attachment surface on the host. The increase in volume is correlated with a lower pressure under the cephalothorax and a greater suction force. In this way, leg 3 plays an important role in the attachment to the host, but not in locomotion as in the two preceding pairs of legs (Kabata & Hewitt, 1971).

There is a long, oval platelike structure, the velum, situated along the free posterior margin of the apron. It appears separate from the basal endopodal segment in ventral view, but connected in dorsal view. The size of this structure varies from species to species, but, invariably, its free margin is always fringed with hairlike setules in all caligid species.

Leg 3 (fig. 4b) is biramous in the majority of the genera of the Caligidae. A seta (probably the seta of the basis) is situated at the outer margin of the exopod base. In addition, a long plumose seta (the coxal seta) is located medially to the endopod. The exopod is 3-segmented in most genera, although the second and third segments are fused in *Lepeophtheirus curtus* (Wilson, 1913) (fig. 98c) and *L. lewisi* (formerly known as *Dentigryps bifurcatus* Lewis, 1964). Some genera (i.e., *Anuretes*, *Mappates*, and *Pseudanuretes*) are characterized by a 2-segmented exopod (figs. 36e, 105a, and 125b).

The first segment of the exopod carries a basal swelling with a large clawlike spine (fig. 4c, d). There are four genera (*Alebion*, *Avitocaligus*, *Euryphorus*, and *Pupulina*), which possess only a small exopodal spine on the first segment (figs. 21b, 49b, 75b, and 130c, respectively). In species of *Anuretes*, the basal swelling at the base of the clawlike spine is apparently absent or has completely fused to the exopodal spine (fig. 36e). The exopodal spines of the second and third segments are modified in species of *Alebion* (fig. 21b).

The endopod is 3-segmented in *Alebion*, *Avitocaligus*, *Euryphorus*, *Gloiopotes*, and *Pupulina* (figs. 21d, 49c, 70b, 81c, and 130d). It is 2-segmented in other caligid genera, but a partial suture dividing the terminal segment is observed in some species (fig. 4c). A progressive reduction of the endopod is exhibited by species of *Abasia* from a 2-segmented endopod to an absence of this ramus (Cressey & Cressey, 1979). The endopod is also absent in *Dartevellia*, greatly modified into an unarmed lamella in *Caritus* (fig. 59b), and reduced to a single plumose seta in *Kabataella* (fig. 90a).

Leg 4

A great deal of discussion has arisen about the significance of leg 4. Matters concerning the taxonomic significance (Kabata, 1965a, 1979, and 1981), functional significance (Kabata & Hewitt, 1971; Boxshall, 1974a), and evolutionary significance (Parker et al., 1968) have been discussed.

This leg is biramous in two genera, *Euryphorus* (fig. 70d) and *Avitocaligus* (fig. 48e). In others it is uniramous or vestigial. In the majority of the caligid species, it has a sympod (coxa and basis indistinguishably fused) and a 1- to 3-segmented exopod (fig. 4e). If the first and second exopodal segments are present, they possess an outer spine at their distal ends. The terminal segment usually carries three spines at its tip (fig. 4f).

The structure of leg 4 has been used to distinguish some genera that are otherwise identical, for example the genera *Caligus* and *Pseudocaligus*, or *Lepeophtheirus* and *Pseudolepeophtheirus*. The use of the fourth leg as a generic discriminant has been questioned by Kabata (1965a, 1979, and 1981). The evolutionary tendency for the reduction and possible loss of the fourth leg is expressed by intermediate forms exhibiting 2- or 1-segmented exopods and species that possess a vestigial lobe bearing a few setae. *Caligopsis* (now known as *Markevichus*) is the only caligid genus that has been reported to lack leg 4 (Markewitsch, 1940).

Boxshall (1974a), attempting to explain the structural diversity of leg 4 in species of *Caligus* and *Lepeophtheirus*, suggested that the fourth legs may be concerned with copulation. If this were true, then the great structural diversity observed for this appendage could easily be explained as a reproductive isolating mechanism. However, this is contrary to the idea of the evolutionary trend toward the reduction of this appendage. Although superficially similar in general appearance to that of the brachiform maxilla, these two appendages function very differently. Unlike the maxilla, the fourth pair of legs are located outside of the cephalothorax and are not involved in the rapid scuttling movements of the copepod. Instead, the major function of leg 4 appears to be "brakes" that prevent backward slippage on the host (Kabata & Hewitt, 1971).

Legs 5 and 6

Legs 5 and 6, when present, are situated on the posterior portion of the genital complex. They are usually represented by setiferous lobes or processes of various sizes and shapes. Leg 5 is usually present in the females on the posterolateral corner of the genital complex (e.g., figs. 4g and 59e). In females of some genera, leg 5 may be extremely long and spikelike as in *Alebion* (fig. 21g), *Gloiopotes* (fig. 83e), and some species of *Tuxophorus* (e.g., *T. wilsoni* Kirtisinghe, 1937; *T. cervicornis* Heegaard, 1962). Some species of *Lepeophtheirus* (particularly those that were formerly assigned to the genus *Dentigryps*, e.g., *L. curtus*, fig. 98e) possess a long spikelike leg 5 in the females; however, most of the members of this genus carry small setiferous processes (fig. 94e). Leg 6, when present in the female, is situated on a ridge in proximity to the oviducal opening. In males, leg 6 is found on the tip of the genital ridge.

The length, armament, and shape of these two pairs of legs in the male can be used to help distinguish species. Legs 5 and 6 are present in the males of most caligid genera; however, they are both absent in females and males of *Paralebion elongatus* Wilson, 1911.

LARVAL DEVELOPMENT

General description

The first record of a description of a larval caligid is by Burmeister in 1823 (see Burmeister, 1835) when he described the larval stage as a new genus, and named it "*Chalimus*" (Gurney, 1934). Krøyer (1837) recognized the similarities existing between "*Chalimus*" and the adult stages of *Caligus* and suggested that the genus established by Burmeister probably represented an immature form of a species of *Caligus* or a related genus. It is now known that "*Chalimus*" consisted of larval stages of caligids. The term "chalimus" persists today and is used to designate those larval stages of caligids and other siphonostomes that are attached to the host by means of a frontal filament.

Since the 19th Century, many descriptions of larval stages (nauplius, copepodid, and chalimus) of numerous species of *Caligus* and *Lepeophtheirus* have been provided (A. Scott, 1901a; Wilson, 1905a; Russell, 1925; Gurney, 1933, 1934; White, 1942; Heegaard, 1943, 1947a; Gnanamuthu, 1948; Shiino, 1959a; Hwa, 1965; Hewitt, 1971a; Lopez, 1976; Johannessen, 1978; and recently summarized by Ho & Lin, 2004). Although several of these authors believed that they observed the complete larval development of the species under investigation, the actual number of phases and stages in the caligid development remained uncertain.

Most, if not all, species of the Caligidae have a common larval development (Hwa, 1965; Izawa, 1969; Kabata, 1972; Voth, 1972; Boxshall, 1974b). Kabata (1972) suggested that there are five phases and ten stages. The typical developmental cycle, proposed by Kabata (1972) and subsequently supported by Boxshall (1974b), consists of two naupliar stages, one copepodid stage, four chalimus stages, two preadult stages, and one adult stage. According to Ho & Lin's (2004) summary on the larval development of 11 species of *Caligus* and four species of *Lepeophtheirus*, the species of *Lepeophtheirus* show a more consistent developmental pattern than species of *Caligus*. For example, while most species of *Caligus* have nine stages in their life cycle, *C. epidemicus* Hewitt, 1971 has 11 stages, with six instead of four chalimus stages (Lin & Ho, 1993). According to a recent paper by Madinabeitia and Nagasawa

(2011), the life cycle of caligid copepods is composed of 8 stages, i.e., two nauplius, one copepodid, four chalimus, and one adult. Four semaphoronts (transitional intervals) are also recognized: an infective copepodid, chalimus copepodid, chalimus adult, and mobile adult. For a summary of the life cycle of the Caligidae and a comparison with other families of parasitic copepods, the reader is referred to Kabata (1981). As far as we are aware, studies on the life cycle of the copepods formerly included in the Euryphoridae have not been attempted. Only fragments of information are available concerning the larval stages of the former euryphorid species.

Nauplius

The first and second nauplii of the caligid copepods are generally ovoid and possess three pairs of appendages (antennule, antenna, and mandible). During the naupliar stages these three appendages function in locomotion. In addition, a pair of structures known as the balancers is present on the posterior end of the nauplius larva. The second nauplius is morphologically very similar to, but slightly longer than the first nauplius.

Lin & Ho's (1993) studies on the life history of *C. epidemicus* revealed that the process of naupliar hatching involved two steps: breaking of the chamber (egg sac wall) and rupture of the egg membrane. The process is as follows: shortly before hatching, the egg (with a full-grown embryo inside) swells by the absorption of water until it bursts the egg sac wall. Then, a few minutes later, while the egg (containing the embryo) is still lying in its own chamber in the egg sac, the nauplius within the egg membrane spreads out its appendages and, with vigorous movement, rapidly ejects itself out of the egg membrane. The newly hatched nauplii are motionless and carry the broken egg membrane on their balancers. Initially, the nauplius spreads its appendages outward, trailing from the head end, but a few seconds later it starts to swim with the typical jerky, naupliar movement. Boxshall (1974b) reported that hatching in *L. pectoralis* involved two steps similar to those described for *C. epidemicus*: transverse splitting of the egg sac membrane, followed by expansion and rupture of the egg membrane.

Copepodid

Although a number of investigators (Wilson, 1905a; Russell, 1925; and Heegaard, 1947a) recognized copepodid stages (Wilson and Russell referred to this phase as the metanauplius) in the development of the caligids they examined, Izawa (1969) disagreed with the idea of two copepodid stages. He suggested that there is probably only one in the development of these organisms. This was later supported by Kabata (1972) and Boxshall (1974b), among others.

The copepodid stage is the first stage of the caligids that attaches to its host. Apparently, there is only a definitive host in their life cycle and no intermediate host. However, absence of larval stages of *Caligus klawei* Shiino, 1959 on its host, the northern anchovy, *Engraulis mordax* Girard, led Barnett (1976) to propose that this species of parasitic copepod may have an intermediate host. The possibility of an intermediate host for at least one additional species of *Caligus* was proposed recently by Hayward et al. (2009) who found no larval stages of *Caligus* on their farmed southern blufin tuna, *Thunnus maccoyii* (Castelnau).

Lin & Ho (1993) found in their experiments that the survival time of the copepodid larva of *C. epidemicus* deprived of food (host) was about three to four days. Initially, the copepodid is much more active than the nauplius, moving swiftly in the water by means of the two pairs of 1-segmented, biramous legs. However, as time passes, it spends more and more time resting on the bottom of the culture dish. Nevertheless, when a fish (a fingerling tilapia 24-30 mm in length) is introduced, the larvae suddenly become active and attach to the fish by means of their strong, prehensile antennae. In approximately two days, the attached copepodid develops into the first chalimus.

Chalimus

The next four stages, chalimus I-IV, possess a frontal filament, which is actually formed during the copepodid stage and is extruded for attachment to the host immediately prior to the molt into the first chalimus stage.

All the chalimus stages of the species of the Caligidae, in which the larval development is known, possess the frontal filament. However, Wilson (1907a) reported that *A. glaber* does not possess a frontal filament, but attaches to the host by means of its prehensile antenna. This absence of a frontal filament in another former euryphorid has recently been supported by Ho & Lin (2010) by their observation that the last instar of *Euryphorus* does not have a frontal filament. The absence of the frontal filament would be very significant, especially in view of the reports by Wilson (1907a, 1907b) that the trebiids, pandarids, and cecropids, in addition to the dissonids (Boxshall et al., 2008), all possess a frontal filament during their chalimus stages.

The filament itself is hollow (Boxshall, 1974b) and is apparently similar to the cuticle in structure (Shields, 1967). This is apparently true for the frontal filament of *Lepeophtheirus salmonis*, but this structure is a solid fibrous filament in species of *Caligus* (Bron et al., 1991; Pike et al., 1993; see summary in Ho & Lin, 2004: 80). Although Heegaard (1947a) believed that a new filament is formed after each molt, others (e.g., Lewis, 1963; Hwa, 1965; Ben Hassine, 1983; Kim, 1993; Piasecki & MacKinnon, 1993; Lin et al., 1997; Ho & Lin, 2004) have observed that the original filament is retained and no new filament is formed.

Gurney (1934) reported that the original frontal filament was retained throughout the chalimus stage, but new material was secreted at the base of the frontal filament before each molt. This new material would form a lobelike base and would slightly increase the length of the filament. Furthermore, by counting the number of lobes at the base of the filament, the exact stage in larval development of the chalimus may be identified (Gurney, 1934; Hwa, 1965; Shields, 1967; Ben Hassine, 1983; Kim, 1993; Lin et al., 1997).

Preadult

According to Hewitt (1964a), *Lepeophtheirus polyprioni* Hewitt, 1963, possesses three postchalimus stages. These stages were distinguished by the differences in their total size, size of the fourth leg, and size of the genital complex. They were designated stages "A", "B", and "C". Kabata (1972) questioned Hewitt's observations, and also reviewed the literature concerning the number of preadult stages of caligids. Kabata (1972) and Boxshall (1974b) concluded that Izawa (1969) was correct in his interpretation that there are only two preadult stages. Lewis (1963) and Voth (1972) also reported these two postchalimus IV stages, but referred to them as chalimus V and VI.

The penultimate stage in the larval development of *Caligus* has been variously called 'preadult' by Lin and Ho (1993) and Lin et al. (1997); 'young adult' by Kim (1993) and Ho & Lin (2004); and 'chalimus V' by Piasecki & MacKinnon (1993). Essentially, this stage has the adult complement of armature on all its appendages and differs from the adult mainly in the internal structure and external appearance of the genital complex (due to the absence of fully developed reproductive organs). A recent study on the developmental stages of *Pseudocaligus fugu* Yamaguti, 1936 by Ohtsuka et al. (2009) provides additional evidence that the so-called predults in *Caligus* are not a developmental stage, but merely a freshly moulted, young adult as noted by Ho & Lin (2004).

The frontal filament is finally broken during the first preadult stage in *Lepeophtheirus*. It is also at this stage that the genital complex of the female differs in its external appearance from that of the male (Lewis, 1963; Izawa, 1969; Kabata, 1972; Voth, 1972; and Boxshall, 1974b).

The preadult I and preadult II of both sexes of *L. pectoralis* were observed attached to the host by means of a temporary frontal filament during ecdysis (Anstensrud, 1990c). This temporary filament, which varies in size and length between individuals, is formed by secretion from the frontal organ (or frontal gland) located between the two frontal plates in the anterior rim of the cephalothorax. At completion of ecdysis, in about 3 to 5 hours after the formation of the temporary frontal filament, the newly molted individual detaches itself from the frontal filament and is able to move freely on the body surface of the host (Anstensrud, 1990c).

Adult and reproduction

As noted above, the morphological difference between the last larval stage (preadult in *Caligus* and preadult II in *Lepeophtheirus*) and the adult is mainly seen in the size and shape of the genital complex. As the parasite matures, the genital complex becomes larger and begins to resemble the adult shape. No further structural changes in the appendages are discernible.

Mate guarding (precopula pairing) was observed by Anstensrud (1992) in his studies on the larval development of *L. pectoralis* and by Ho & Lin (2004) on *C. epidemicus* and *C. rotundigenitalis*. According to their report, in the case of *Caligus*, the youngest female larva being guarded by a full-grown adult male can be as early as the penultimate chalimus stage. This would be the chalimus V stage in the case of the *C. epidemicus* female and chalimus III in the *C. rotundigenitalis* female. However, in *L. pectoralis*, precopula pairing was not observed until the female larvae reached the preadult stage (Anstensrud, 1992).

Copulation between the "sixth chalimus" or adult male and first or second preadult females has been reported (Izawa, 1969; Voth, 1972; Boxshall, 1974c; Anstensrud, 1990b, c; Ho & Lin, 2004). The male caligid grasps the fourth pedigerous somite or the anterior margin of the genital complex of the female with its antennae (Lewis, 1963) or maxillipeds (Izawa, 1969). The male discharges two spermatophores that stick to the posteroventral surface of the genital complex of the female. As the eggs pass out of the opening of the oviduct, they are fertilized by sperm from the seminal receptacle (Wilson, 1905a; Lewis, 1963; Anstensrud, 1990b, c). Ovigerous females bear uniseriate egg sacs, presumably secreted by the cement glands located in the genital complex.

After all the eggs in the two sacs of *C. epidemicus* have hatched, the empty egg sacs may remain attached to the parasite for some time. However, if there are mature eggs in the oviducts, the empty sacs detach about 5 minutes after completion of hatching. Then, about 20 minutes after the old egg sacs have detached, a new sac bud appears, and almost immediately the first egg is extruded into the sac bud. Subsequently, the eggs are extruded into the new sac one at a time until the sac is filled with 20 eggs within 2-4 minutes. Hatching of those newly produced eggs does not start until about 2 days later (Lin & Ho, 1993).

After the molt into the adult from the last chalimus stage in species of *Caligus*, or the second preadult stage in species of *Lepeophtheirus*, there is no evidence that the organism undergoes any further molts (Gurney, 1934).

HOST-PARASITE RELATIONSHIPS

Many species of caligid copepods, commonly known as sea lice, have long been recognized to potentially affect the growth, fecundity, and survival of their hosts in the wild (White, 1940; Kabata, 1958; Hewitt, 1971a). With the decline in ocean fisheries during the last quarter of the 20th Century, along with a concomitant rise in the global demand for more food from the sea, the development of semi-intensive and intensive brackish water and marine aquaculture around the world was rapidly promoted. As a consequence, it became more and more evident that sea lice infections of cultured fishes were a significant problem that needed to be controlled, if not eliminated. The ranched or farmed fishes are inevitably infected by parasitic copepods that have originated from wild fishes (Pike & Wadsworth, 1999). Johnson et al. (2004) estimated that in marine and brackish water fish cultures, 61% of copepod infestations are caused by members of the family Caligidae of which 40% are caused by species of *Caligus* and 14% by species of *Lepeophtheirus*. The remaining 7% of infestations are caused by two species of *Pseudocaligus*, considered a junior synonym of *Caligus*, and one species of *Parapetalus*.

Three species of salmon lice, *Caligus elongatus* (Nordmann, 1832), *Caligus rogercresseyi* Boxhall & Bravo, 2000, and *Lepeophtheirus salmonis*, cause the most significant health problems in the marine farming of salmonids. It has been estimated that in 2006 the worldwide salmonid farming industry had a total loss of U.S. $480 million due to salmon lice infestations (Costello, 2009).

As many as 22 species of *Caligus* and five species of *Lepeophtheirus* have been reported from non-salmonid finfish cultured in brackish and marine waters (Johnson et al., 2004). Among them, *Caligus epidemicus* Hewitt, 1971, *C. orientalis* Gusev, 1951, and *C. punctatus* Shiino, 1955 are considered to have high potential to cause health problems in non-salmonid finfish culture in Asia (Ho, 2000). Although disease outbreaks caused by these copepods have been reported from Taiwan by Ho & Lin (2004), from the Philippines by Ho et al. (2004), from Australia by Hayward et al. (2009), and from Southeast Asia by Venmathi Maran et al. (2009), impacts of those parasitic copepods on fish growth, increased susceptibility of the fish host to other diseases, cost of production, and decrease in product value have not yet been quantified.

Deleterious effects

The greatest harm to the host inflicted by caligid copepods is caused during feeding of these organisms. The damage caused by the prehensile antenna and maxilliped usually is superficial (Kabata & Hewitt, 1971), resulting in erosion of the epidermal layer and eliciting only localized reactions (Boxshall, 1977). Little or no apparent damage results from the scuttling movements of the caligid as it moves along the surface of the host. Finally, the damage inflicted by the feeding action of the mouth tube has been well reviewed and documented by Pike & Wadsworth (1999), although Boxshall (1977) suggested that the damage is negligible.

There have been many reports in the literature of the gross pathological effects of caligids, in addition to the study conducted by Boxshall (1977) concerning the histopathological effects of *Lepeophtheirus pectoralis* Müller, 1776 on its host, *Platichthys flesus* (Linnaeus). Kabata (1970) cited damage caused by *Caligus macarovi* Gusev, 1951, *Caligus elongatus* and *Lepeophtheirus thompsoni* Baird, 1850 on their respective hosts. Also, Pike & Wadsworth (1999) reviewed the pathological effects on cultured salmon caused by the infection of salmon lice, *Lepeophtheirus salmonis*. The harmful effects generally range from open sores and scars to considerable damage to respiratory tissues and other areas. Reports of open sores, fin erosion, and inflammation of the tissue are not uncommon (A. Scott, 1901a; Russell, 1925; Lüling, 1953; Kubota & Takakuwa, 1963; Kabata, 1970; Mann, 1970; Johannessen et al., 1976; Boxshall, 1977; Johnson et al., 1996; Pike & Wadsworth, 1999; Ho & Lin, 2004). The occurrence of unsightly open sores not only lowers the commercial value of the fish, but may also result in secondary bacterial infection.

The histological responses involve hypertrophy of connective tissue, apparently in order to put a barrier between the parasite and the tissues of the host. Moreover, epidermal hyperplasia, fibroblast proliferation, fiber production, and macrophage and lymphocyte infiltration have been reported by Boxshall (1977) as localized reactions directed at *L. pectoralis* by its host, *Platichthys flesus*.

Physiological effects of salmon lice on their hosts have also been reported (Grimnes & Jakobsen, 1996). Dawson (1998) discovered that returning sea trout (*Salmo trutta* Linnaeus) infected with salmon lice had significantly lower total protein and serum albumin (important for water balance), or lower alkaline phosphatase levels and lower cholesterol levels, which decrease in response to acute and chronic stress in fish.

More important, however, are the observations of mass mortality apparently caused by caligid copepods. For example, White (1940) reported salmon louse (*Lepeophtheirus salmonis*) causing skin erosion, ulcerations, and atrophy of the connective tissues of the Atlantic salmon, *Salmo salar* Linnaeus, eventually resulting in the death of many of the salmon. Furthermore, salmon louse has been reported to cause extensive losses of salmonids in sea farms in Norway, Scotland, Ireland, North America, and Chile (Johnson et al., 2004).

Although the adult caligids are capable of causing appreciable damage, particularly when occurring in large numbers, the chalimus instars of these organisms can also cause significant injuries to the host. The extensive damage is primarily due to the limited feeding area of the chalimus stage. Attached to the host by its frontal filament, the radius of the feeding area is defined by the length of the frontal filament (White, 1942; Kabata, 1970; Boxshall, 1977). Although the attachment itself is damaging to the tissues of the host (Kabata, 1970), like the adult, the main harm is caused by the feeding habits of this larval stage. Extensive lesions and damage to the fins of the host have been well documented (White, 1940; Parker & Margolis, 1964; Boxshall, 1977, among others).

Food and feeding

The type of food taken may be dependent upon the species of caligid. White (1942) reported *Lepeophtheirus salmonis* feeding primarily on the tissues, and occasionally on the mucus and blood of *Salmo salar*. Brandal et al. (1976) spectrophotometrically demonstrated that *L. salmonis* feeds on blood.

An examination of the mouth tube and the mode of feeding of the mandibles has been provided by Parker et al. (1968), who believed the main macerating structure to be the mandibles. John and Nair (1974) described the functional morphology of the mouth tube of *Hermilius longicornis* Bassett-Smith, 1898. Kabata (1974) studying the mouth tube of *Caligus clemensi*, *C. curtus*, and *Lepeophtheirus salmonis* by light and scanning electron microscopy, suggested that the dominant macerating structure to be the "strigil", a divided bar bearing many teeth at the tip of the labium. It was first figured by Ho (1966; p. 756, fig. 7) for *Echetus typicus* Krøyer, 1864.

The mouth tube first appears in the free-swimming but infective copepodid stage and there are no major changes to this structure during the developmental process (Lewis, 1963; Ho & Lin, 2004). The strigil has been found in most caligids examined in this revision, except *Belizia*, *Kabataella*, and *Alebion*. Although the first appearance of the strigil in the caligid development has not been investigated in any of the larval stages, it seems reasonable to assume that the structure is formed and functional at the copepodid stage, as Lin (1991) found in his experiment on artificial infection in the laboratory in which a young Mozambique tilapia (*Oreochromis mozambicus*) 24-30 mm in total length was killed by the attachment of 25 copepodids of *Caligus epidemicus*. A similar result was observed by Hallett & Roubal (1995) in their experiment with the same species of sea louse on the small estuary perchlet, *Ambassis marianus* (Günther).

Host specificity

Caligids are generally host specific (Sproston & Hartley, 1941). However, some species have a wide range of hosts involving divergent families and genera, including both elasmobranchs and teleosts. For example, *C. elongatus* has been recorded from more than 80 species of fish, both teleosts and elasmobranchs, representing 17 orders and 43 families (Kabata, 1979). On the other hand, there are some species, e.g., *C. clavatus* Kirtisinghe, 1964, that are confined to one host genus, such as the barracuda genus *Sphyraena*; this species of copepod has been reported to be parasitic on *S. obtusata* Cuvier from Sri Lanka (Kirtisinghe, 1964), on *S. acutipinnis* Day from India (Pillai, 1967), and on *S. japonica* Cuvier from Taiwan (Ho & Lin, 2004). *Parapetalus occidentalis* (Wilson, 1908) is an example of a species-specific caligid, which has been reported from North Carolina (Wilson, 1908), Gulf of Mexico (Causey, 1955), India (Pillai, 1962), Sri Lanka (Kirtisinghe, 1964), Australia (Kabata, 1966b), Taiwan (Ho & Lin, 2001), and Thailand (Purivirojkul & Areechon, 2008). From all of these regions, the host was a single fish species, namely the cobia, *Rachycentron canadum* (Linnaeus).

Boxshall (1974d) studied the host specificity of 39 species of parasitic copepods of North Sea fishes. Among the caligids, he found four species of *Caligus* species specific. He found one species of *Lepeophtheirus* infesting more than one species of host belonging to the same genus. One species of *Caligus* was found on different genera of the same family. Finally, five caligid species were found infesting fish belonging to more than one family.

Even those caligid species that are only marginally host specific exhibit host preferences. These predilections are reflected in the incidence and intensity of infestation (Lewis et al., 1969) along with an increase in total size of the parasite (Cressey, 1967a; Lewis et al., 1969; Cressey & Collette, 1970).

SYSTEMATIC ACCOUNT

Discussion of the Euryphoridae Wilson, 1905

The subfamily Euryphorinae established by Wilson (1905a) included caligids with dorsal aliform plates on the fourth pedigerous somite. In this subfamily Wilson (1907a) included the following seven genera: *Dysgamus* Steenstrup & Lütken, 1861; *Alebion* Krøyer, 1863; *Dissonus* Wilson, 1906; *Euryphorus* Milne Edwards, 1840; *Elytrophora* Gerstaecker, 1853; *Caligeria* Dana, 1852; and *Gloiopotes* Steenstrup & Lütken, 1861. Since their establishment, *Elytrophora* has been synonymized with *Euryphorus* and *Caligeria* with *Alebion* (see Kabata, 1979). *Dysgamus* cannot be considered valid, since it was comprised of juveniles of other genera, mostly *Euryphorus* (Heegaard, 1972). *Dissonus* is now considered a member of its own family Dissonidae Kurtz, 1924, which was recently reviewed by Boxshall et al. (2008). Although Yamaguti (1963) considered himself as the establisher of this taxon, Wilson (1932) is the one who elevated the Euryphorinae to the familial rank.

In his monumental book on the parasitic copepods of British fishes, Kabata (1979) distinguished Euryphoridae from the Caligidae by the same criterion originally used by Wilson (1905a) to establish this group (i.e., the presence of the dorsal aliform plates on the fourth pedigerous somite). Kabata included only five genera in this family (table I). Three of those five genera were among the original seven genera proposed by Wilson (1905a), but inclusion of *Tuxophorus* Wilson, 1908 was initiated by Heegaard (1955) and that of *Paralebion* Wilson, 1911 was made by Yamaguti (1963).

Wilson's (1932) treatment to separate Euryphoridae from the Caligidae was followed by Shiino (1954b), Heegaard (1955), Yamaguti (1963), and Kabata (1979), but questioned by Dojiri (1983), who argued that there is an "absence of any consistent character to distinguish the two families". Although Dojiri's (1983) argument was not adopted by Pillai (1985), Benz (1984), Ho (1998), Kazachenko (2001), Ho & Lin (2004), and Kabata (2004), it was upheld by Boxshall & Halsey (2004) and Walter & Boxshall (2008). It should be pointed out here that the five "euryphorid" genera brought up for discussion by Dojiri (1983) has one taxon different from that of Kabata's (1979). *Paralebion* in

TABLE II

Comparison of five morphologic features among five "euryphorid" genera and three selected caligid genera that were cladistically analyzed by Boxshall & Justine (2005)

Genus	Dorsal aliform plates on fourth pediger	Endopod of leg 1	Armature on distal segment of leg 3 endopod	Endopod of leg 3	Leg 4
Avitocaligus	absent	2-segmented	4 setae	3-segmented	biramous
Euryphorus	present	2-segmented	4 setae	3-segmented	biramous
Pupulina	absent	2-segmented	4 setae	3-segmented	uniramous
Gloiopotes	present	2-segmented	4 setae	3-segmented	uniramous
Alebion	present	2-segmented	4 setae	3-segmented	uniramous
Paralebion	absent	vestigial	6 setae	2-segmented	uniramous
Tuxophorus	present	vestigial	6 setae	2-segmented	uniramous
Caligus	absent	vestigial	6 setae	2-segmented	uniramous

Kabata's five euryphorid genera was removed and, in place of it we see the appearance of *Pupulina* Beneden, 1892. This is the first time in the history of euryphorid taxonomy that *Pupulina* was involved in the debate on the acceptance or rejection of the family Euryphoridae.

Recently, upon their discovery of a new genus of caligid copepods, Boxshall & Justine (2005) took the opportunity to perform a cladistic analysis of the basal genera in the Caligidae. Altogether eight genera were analyzed (see table II), including their new genus *Avitocaligus* Boxshall & Justine, 2005. Notice that there are six genera (*Euryphorus*, *Pupulina*, *Gloiopotes*, *Alebion*, *Paralebion*, and *Tuxophorus*) in their ingroup that historically have been placed in the family Euryphoridae at one time or another. The cladogram obtained by Boxshall & Justine (2005) shows that *Caligus*, *Tuxophorus*, and *Paralebion* are monophyletic with character 5 (= endopod of leg 1 vestigial) as their synapomorph. This phylogenetic relationship between the "euryphorid" genera and "caligid" genera encouraged us to search for any consistent characters shared among the five genera basal to this monophyletic clade. As shown in table II, after performing a thorough search through the literature, we found three character states shared among those five basal taxa (i.e., *Avitocaligus*, *Euryphorus*, *Pupulina*, *Gloiopotes*, and *Alebion*): a 2-segmented endopod on leg 1, a 3-segmented endopod on leg 3, and 4 (instead of 6) plumose setae on the terminal segment of the endopod of leg 3. These, along with several additional characters, were utilized to conduct a phylogenetic analysis on the genera of the Caligidae, including the former euryphorid genera. Even though no other caligid genera possesses the above mentioned three character states shared among the former euryphorid genera, the cladograms resulting from the phylogenetic analysis have revealed that the family "Euryphoridae" is paraphyletic; therefore, the family must be considered synonymous with the Caligidae and the former euryphorid genera must be transferred to the Caligidae (see Caligidae Phylogeny section).

Family CALIGIDAE Burmeister, 1835

Diagnosis. — As in discussion on external morphology.
Type-genus. — *Caligus* Müller, 1785.

KEY TO THE GENERA OF THE CALIGIDAE

Several caligid genera are no longer considered valid or are designated as genus inquirenda; therefore, they have not been included in the following key, but have been discussed (see Miscellaneous Genera). The following key is based only on females.

1. Endopod of leg 1 vestigial and leg 3 endopod not 3-segmented 2
   Endopod of leg 1 distinctly 2-segmented and leg 3 endopod 3-segmented........... 3
2. Leg 3 expanded into ventral apron 7
   Leg 3 not expanded into ventral apron, but reduced to setiferous lobe......... *Arrama*
3. Dorsal aliform plates present on fourth pedigerous somite.......................... 4
   Dorsal aliform plates absent on fourth pedigerous somite 6
4. Leg 4 uniramous or vestigial 5
   Leg 4 biramous .... *Euryphorus*
5. Exopodal spines of legs 1-3 modified to large paddlelike structures .... *Alebion*
   Exopodal spines of legs 1-3 not so modified, but spines 2 and 3 of terminal exopodal segment of leg 1 bifid with a medial setiform process .... *Gloiopotes*
6. Genital complex with 2 pairs of large lamelliform plates and first abdominal somite with pair of ventral lamelliform plates .... *Avitocaligus*
   Genital complex and abdomen without lamelliform plates .... *Pupulina*
7. Lateral margins of cephalothorax greatly folded ventrally 8
   Lateral margins of cephalothorax not folded ventrally, or slightly folded with winglike posterolateral lob 9
8. Antenna with large accessory claw, extremely long exopodal spines present on leg 1 .. .. *Hermilius*
   Antenna without large accessory claw; exopodal spines on leg 1 not long...... *Abasia*
9. Lunules present 10
   Lunules absent 24
10. Genital complex with distinct posterolateral processes............................ 11
    Genital complex without distinct posterolateral processes........................ 14
11. Genital complex with two posterolateral processes 12
    Genital complex with four posterolateral processes.............................. 13
12. Lateral aliform processes present on abdomen .... *Parechetus*
    Lateral aliform processes absent on abdomen........................... *Caligodes*

13. Lateral digitiform process present on abdomen .... *Pseudechetus*
    Lateral digitiform process absent on abdomen .... *Synestius*
14. Genital complex with broad, aliform lateral expansion .... *Parapetalus*
    Genital complex without aliform lateral expansion 15
15. Two spines on basal swelling of first exopodal segment of leg 3 .... *Midias*
    One spine on basal swelling of first exopodal segment of leg 3 16
16. Abdomen with broad aliform lateral expansion .... *Sinocaligus*
    Abdomen without aliform lateral expansion 17
17. Large lenses of eyes present .... *Anchicaligus*
    Large lenses absent 18
18. Endopod of leg 3 lamelliform and unarmed .... *Caritus*
    Endopod of leg 3 unmodified and setose 19
19. Dorsal aliform plates present on fourth pedigerous somite .... *Tuxophorus*
    Dorsal aliform plates absent on fourth pedigerous somite 20
20. Elongate, slender "neck" present .... *Echetus*
    Elongate, slender "neck" absent 21
21. Abdomen as long as or longer than rest of body .... *Sciaenophilus*
    Abdomen shorter than rest of body (*Caligus bennetti* has an abdomen that is much longer than the rest of the body) 22
22. Posterior margin of thoracic zone covering fourth pedigerous somite, with only 4 setae on caudal ramus, and sternal furca reduced to 2 sclerotized knobs .... *Belizia*
    Posterior margin of thoracic zone not covering fourth pedigerous somite, with 6 setae on caudal ramus, and sternal furca comprising base with 2 tines or absent 23
23. Two terminal (middle and innermost) exopodal spines of leg 1 without accessory processes, and only 1 outer spine (instead of 2) on terminal exopodal segment of leg 2 .... *Metacaligus*
    Two terminal (middle and innermost) exopodal spines of leg 1 with or without accessory processes, and 2 outer spines on terminal exopodal segment of leg 2 .... *Caligus*
24. Abdomen with lateral alae .... *Dartevellia*
    Abdomen without alae 25
25. Pair of lateral rounded expansions (lunule-like structure) present on frontal plate adjacent to base of antennule .... *Alanlewisia*
    Pair of lateral rounded expansions (lunule-like structure) absent on frontal plate adjacent to base of antennule 26
26. Cephalothorax with large wing-like posterolateral lobes and leg 3 endopod represented by 1 seta .... *Kabataella*
    Cephalothorax without large wing-like posterolateral lobes and leg 3 endopod 1- or 2-segmented 27
27. A combination of small abdomen (frequently vestigial or absent), 2-segmented exopod of leg 3, and first endopodal segment of leg 3 without inner seta 28
    A combination of distinct abdomen (occasionally small, not vestigial or absent), 3-segmented exopod of leg 3 (except in *Lepeophtheirus curtus* and *L. lewisi* 2-segmented, but armature suggests 3 segments), and first endopodal segment of leg 3 with inner seta 30

28. Process of maxillule a rounded lobe . . . . *Mappates*
 Process of maxillule dentiform or absent 29
29. Accessory claw of antenna present . . . . *Pseudanuretes*
 Accessory claw of antenna absent . . . . *Anuretes*
30. Long, digitiform, posterolateral processes on genital complex present; leg 5 absent . . . . *Paralebion*
 Long, digitiform, posterolateral processes on genital complex absent; leg 5 present . . . . *Lepeophtheirus*

Genus **Caligus** Müller, 1785

Caligus Müller, 1785: 128; Nordmann, 1832: 16; Krøyer, 1837: 201; Milne Edwards, 1840: 446; Baird, 1850: 269; Dana, 1852: 55; Carus, 1885: 357; Bassett-Smith, 1896a: 156; Bassett-Smith, 1899: 446; Wilson, 1905a: 555; Scott & Scott, 1913: 44; Wilson, 1932: 397; Barnard, 1955a: 244; Markewitsch, 1956: 111; Yamaguti, 1963: 7; Hewitt, 1963: 64; Lewis, 1964a: 141; Lewis, 1967: 101; Margolis et al., 1975: 1; Kabata, 1979: 170; Dojiri, 1983: 57; Pillai, 1985: 233; Boxshall & Montú, 1997: 53; Kim, 1998: 649; Ho, 1998: 251; Kazachenko, 2001: 22; Kabata, 2003: 43; Boxshall & Halsey, 2004: 725; Ho & Lin, 2004: 106.
Caligulus(?) Heegaard, 1962: 171.

Female. — Cephalothorax suborbicular, subovate, or slightly oval, usually longer than wide and dorsoventrally flattened. Dorsal ribs (sutures) conspicuous and H-shaped. Posterior sinus distinct and located on posterolateral corner of thoracic zone of dorsal shield. Lunules situated ventrally on frontal plate. Fourth pedigerous somite free, small, and without dorsal plates. Genital complex varying in shape from subquadrate, subspherical, to subtriangular in outline, and sometimes with small posterolateral processes. Abdomen usually cylindrical, and comprising 1 to 5 free somites. Caudal ramus with 6 setae.

Antennule 2-segmented. First segment with varying number of plumose setae, and frequently with bifid process on posterodistal corner. Second segment cylindrical, with 1 posterior seta and 13 apical setae (1 may be an aesthetasc, and 2 may share common base). Antenna prehensile, 4-segmented. Second segment usually with posteriorly directed spinelike process. Terminal segment a recurved claw. Postantennal process a recurved clawlike structure, occasionally reduced to sclerotized plate or absent. Mouth tube with intrabuccal stylet and strigil. Mandible consisting of 4 sections, with inner margin of terminal section dentiferous. Maxillule comprising dentiform projection and adjacent setiferous papilla. Maxilla brachiform. Maxilliped subchelate and prehensile. Sternal furca of varying shapes, absent in some species.

Leg 1 with 2-segmented exopod bearing 3 terminal spines, and endopod reduced to small process. Leg 2 biramous, with 3-segmented rami. Leg 3 forming ventral shield (apron), biramous, and possessing large spine on first segment of exopod. Velum present. Fourth leg uniramous, brachiform, and 1-, 2-, or 3-segmented exopod or reduced to a small setiferous papilla. Leg 5 vestigial, represented by few setae, and located on posterolateral corner of genital complex.

Male. — Cephalothorax similar to that of female. Genital complex smaller than in female, and usually oval. Antenna apparently 3-segmented. Second segment with corrugated adhesion pads. Terminal segment a claw with varying number of accessory tines or folds. Postantennal process often times more slender and strongly curved than in female. Dentiform projection of maxillule occasionally with additional tine and/or adhesion pad. Corpus maxillipedis frequently stouter and myxa more ornamented than in female. Legs 5 and 6 vestigial.

Type-species. — *Caligus curtus* Müller, 1785.

Remarks. — Müller (1785) first established the genus *Caligus* to accommodate the type-species *C. curtus*. There are approximately 250 species included in this genus (see catalogue and synopsis of this genus by Margolis et al., 1975). The members of this genus are predominantly parasitic on marine teleosts and are morphologically similar to *Lepeophtheirus*. The main distinguishing feature is the presence of lunules in *Caligus* and their absence in *Lepeophtheirus*. Species of *Caligus* are found primarily in the warmer waters of the lower latitudes. On the other hand, *Lepeophtheirus* is found in temperate waters of the middle latitudes (Kabata, 1965b, 1970, and 1979); however, there is some degree of overlap.

There are a few species of *Caligus* that lack the posteriorly directed spiniform process on the second segment of the antenna, the postantennal process, and the sternal furca (table III). The taxonomic and functional importance of these dentiform structures have been discussed elsewhere in this revision (see discussion on external morphology).

Several generic diagnoses of *Caligus* have been presented since its establishment by Müller (Kabata, 1979; Boxshall & Halsey, 2004; Ho & Lin, 2004). Many of the 250 nominal species were not described in detail by their discoverers; consequently, several species may, in the future, be relegated to synonymy with others. This cumbersome genus may eventually be divided into two or more genera, but a badly needed revision of *Caligus* must be conducted first.

Caligulus may be synonymous with *Caligus*. The former is based on *Caligulus longispinosus* described from a scombrid, *Euthynnus alletteratus* (Rafinesque), near Queensland, Australia, by Heegaard (1962). The only

TABLE III

List of the species of *Caligus* displaying the presence (+) or absence (−) of (1) the posteriorly directed spiniform process on the second segment of the antenna (A_2), (2) the postantennal process (PAP), and (3) the sternal furca (SF). ? = undetermined from the literature

Species	A_2	PAP	SF
C. absens Ho, Lin & Chen, 2000	+	+	−
C. afurcatus Wilson, 1913	+	+	−
C. alaihi Lewis, 1968	−	+	+
C. arii Bassett-Smith, 1898	+	−	+
C. brevicaudus Pillai, 1963	?	+	+
C. brevisoris Shen, 1957	−	+	+
C. coryphaenae Steenstrup & Lütken, 1861	?	−	+
C. cunicephalus Gnanamuthu, 1950	?	+	?
C. cybii Bassett-Smith, 1898	−	+	+
C. dactylus Ho, Lin & Chen, 2007	−	+	+
C. dasyaticus Rangnekar, 1957	?	+	+
C. enormis Wilson, 1913	?	+	−
C. fistulariae Yamaguti, 1936	−	+	+
C. grandiabdominalis Yamaguti, 1954	?	+	−
C. inanis Ho & Lin, 2007	−	−	+
C. kanagurta Pillai, 1961	−	+	+
C. laticaudus Shiino, 1960	?	+	+
C. longiabdominus Shiino, 1965	?	+	+
C. pampi Ho & Lin, 2002	−	+	+
C. pelamydis Krøyer, 1863	−	+	+
C. reniformis Prabha & Pillai, 1983	+	−	−
C. rufus Wilson, 1908	+	+	−

diagnostic feature appears to be the posterior extension of the thoracic zone of the cephalothorax, which dorsally covers the fourth pedigerous somite. This feature is of dubious taxonomic merit (see discussion on *Caligulus*).

Caligus curtus Müller, 1785
(figs. 1-5)

For a detailed synonymy of *Caligus curtus*, the reader is referred to the synopsis of the genus by Margolis et al. (1975). Addendum to synopsis as follows: Bossanyi & Bull, 1971: 46; Boxshall, 1974d: 358; Kabata, 1979: 171; Dojiri, 1983: 61; Kabata, 1988: 12; Hogan & Trudeau, 1989: 10; Kabata, 2003: 46.

Material examined. — Three females from cod, *Gadus morhua* Linnaeus, from Barents Sea on loan from Dr. Z. Kabata, Pacific Biological Station, Nanaimo, British Columbia. Twenty-three females, 41 males, and 18 preadults (USNM 11618, no. 1) from "cod" (presumably *Gadus morhua*) from Vineyard Sound, Massachusetts collected by V. N. Edwards, 13 January 1886 and identified by R. Rathbun. Forty-nine females, 19 males, 19 preadults (USNM 11618, no. 2) from "cod" from Vineyard Sound, Massachusetts collected by V. N. Edwards, 13 January 1886, and identified by R. Rathbun. Thirteen females, 14 males, and 19 preadults (USNM 12650) from "cod" collected in Woods Hole, Massachusetts by V. N. Edwards on 2-14 December 1885 and identified by R. Rathbun.

Female. — Body as in fig. 1a. Total length (excluding setae on caudal ramus) 9.55 mm (8.96-9.83 mm) (n = 10). Cephalothorax subcircular, narrower anteriorly, with indentation at midlength on lateral margin, 5.11 mm (4.81-5.44 mm) × 4.39 mm (4.08-4.88 mm), and bearing shallow posterior sinuses. Frontal plate distinct. Lunules (fig. 1b) present. Tip of antennule within lateral limit of cephalothorax. Free margin of thoracic zone extending well beyond posterior limit of lateral zone.

Fourth pedigerous somite much wider than long, 0.59 mm (0.43-0.76 mm) × 1.38 mm (1.26-1.53 mm). Genital complex (fig. 1a) somewhat rectangular, longer than wide, 2.63 mm (2.22-2.95 mm) × 2.15 mm (1.93-2.36 mm). Abdomen (fig. 1a) consisting of 1 free somite, slightly longer than wide, 1.00 mm (0.90-1.10 mm) × 0.92 mm (0.83-1.03 mm).

Caudal ramus (fig. 1c) longer than wide, 412 μm (365-465 μm) × 289 μm (266-332 μm), bearing inner row of setules, 3 large and 3 small plumose setae.

Body with small setules (sensilla) as in fig. 1a.

Egg sacs (fig. 1a) long, uniseriate, 10.11 mm (9.53-10.69 mm) × 0.51 mm (0.50-0.53 mm), containing 112 eggs (104-120 eggs) (n = 2).

Antennule (fig. 1d) 2-segmented. First segment robust, 403 μm long, bearing 27 setae. Second segment (fig. 1e) much shorter than first segment, cylindrical, 265 μm long, bearing 14 setae (1 possibly an aesthetasc) (2 share common base). Antenna (fig. 1f, g) apparently 4-segmented. First segment unornamented. Second segment with posteriorly directed spinelike process. Third segment robust with dorsal adhesion pad (fig. 1g). Terminal segment a recurved claw possessing 1 seta near base and 1 near midlength. Postantennal process (fig. 2a) with recurved tine and 3 sclerotized papillae; associated papillae with multibranched setule.

Mouth tube (fig. 2b) longer than wide, 575 × 437 μm. Intrabuccal stylet (fig. 2c) on inside surface of labrum near frons labri. Dentiferous strigil (fig. 2d) on inside surface of labium. Mandible (fig. 2e, f) with 4 sections; first and third sections approximately same length. Terminal section (fig. 2f) bent at junction with third section, bearing hyaline flange on outer margin, and 12 teeth on inner margin. Maxillule (fig. 2g) comprised of papilla tipped with 3 setae (1 larger than other 2), and 1 process. Process with wide base and relatively straight, tapering tine. Maxilla (fig. 2h) brachiform. Flabellum serrated. Calamus (fig. 2i) with 3 serrated membranes. Canna

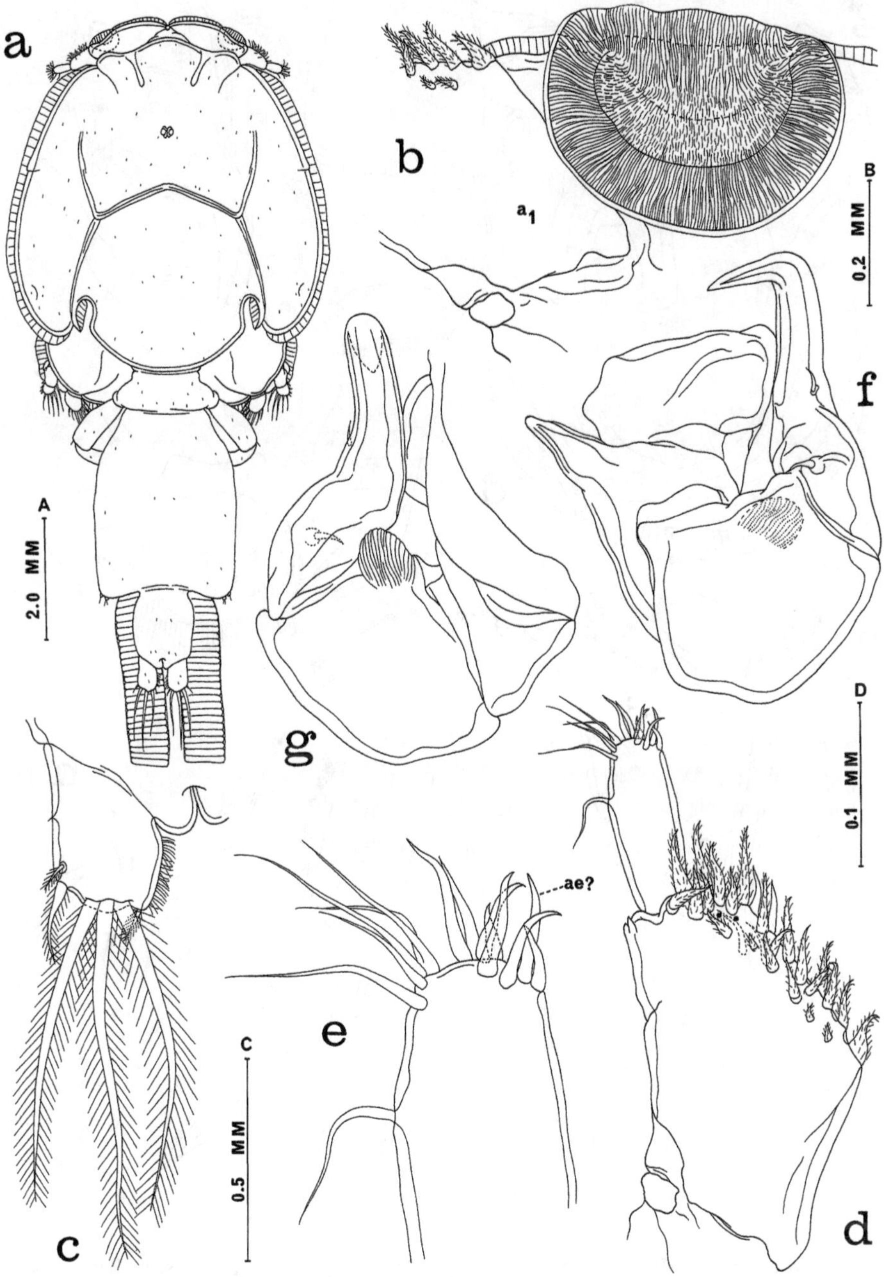

Fig. 1. *Caligus curtus* Müller, 1785, female. a, body, dorsal (scale A); b, lunule and proximal portion of antennule, ventral (B); c, caudal ramus, ventral (C); d, antennule, ventral (B); e, distal portion of antennule, ventral (D); f, antenna, ventral (B); g, same, dorsal (B).

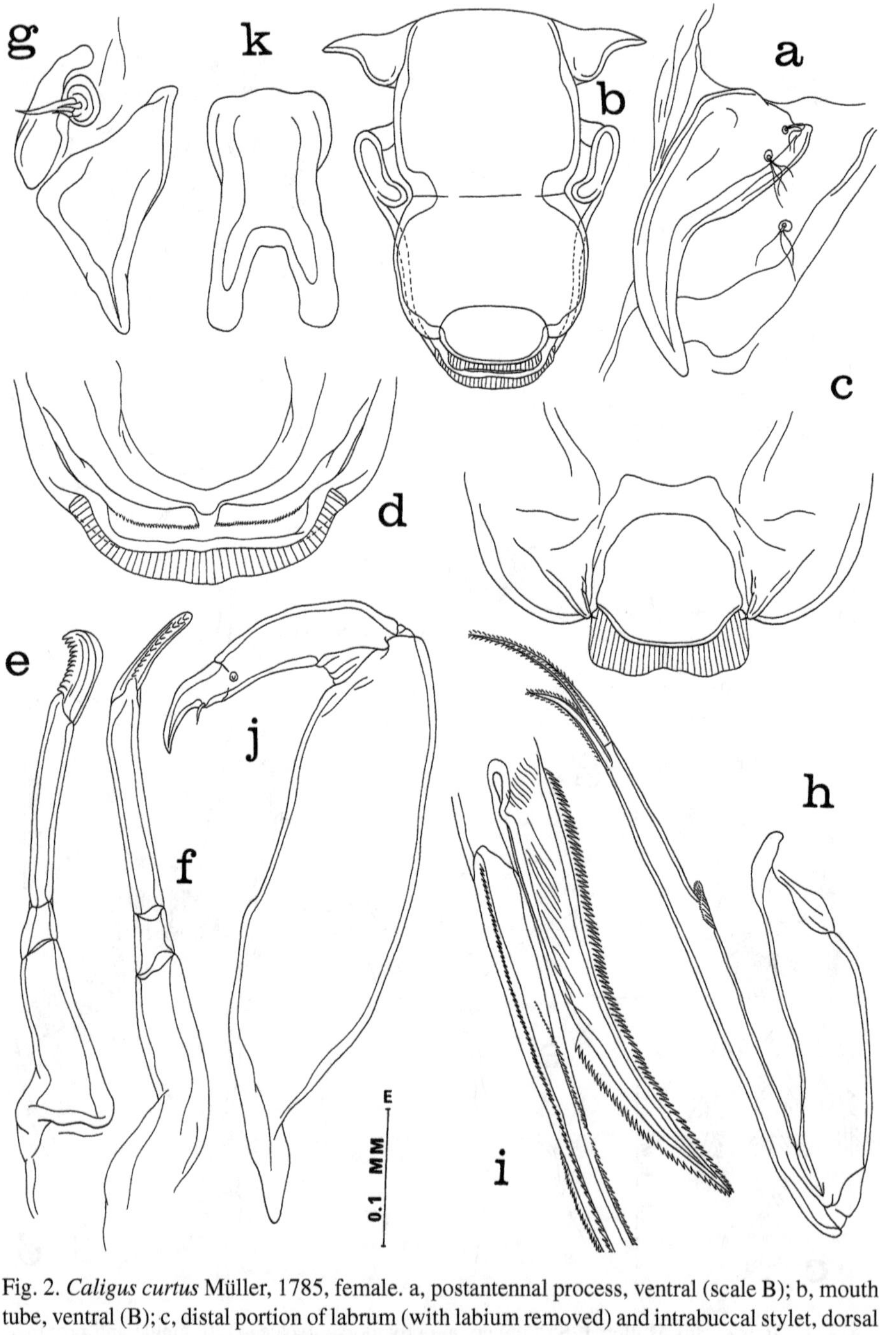

Fig. 2. *Caligus curtus* Müller, 1785, female. a, postantennal process, ventral (scale B); b, mouth tube, ventral (B); c, distal portion of labrum (with labium removed) and intrabuccal stylet, dorsal (E); d, distal portion of labium (with labrum removed) and strigil, ventral (E); e, mandible, ventral (E); f, same, inner (E); g, maxillule, ventral (B); h, maxilla, dorsal (C); i, distal portion of maxilla, dorsal (D); j, maxilliped, ventral (B); k, sternal furca, ventral (C).

(fig. 2i) bilaterally serrated; inner serrated membrane only on distal half; and proximal portion of canna with striations on dorsal surface. Maxilliped (fig. 2j) with rather slender corpus. Corpus maxillipedis with slight protrusion on inner margin. Shaft and claw 437 μm long combined; minute process arising from cuticular depression located at junction of shaft and claw. Claw with small inner seta. Sternal furca (fig. 2k) with rectangular base and blunt, slightly divergent tines.

Leg 1 (fig. 3a) biramous. Sympod with 1 outer and 1 inner plumose setae. Exopod 2-segmented. First segment with outer distal spine (with spinules at base) and inner row of setules. Terminal segment (fig. 3b) with 3 bilaterally spinulated acuminate spines; striations (similar to those found on canna of maxilla) on distal half of exopodal spines. Seta 4 smooth, but with small process at tip. Inner margin of second exopodal segment with 3 large plumose setae. Endopod (fig. 3c) very small, with 1 or 2 small papillae at tip. Area between intercoxal plates of legs 1 and 2 (fig. 3d) sclerotized, and equipped with 1 pair of protuberances (each with 1 small setule). Leg 2 (fig. 3e) biramous, with usual armature. Exopod (fig. 3f) 3-segmented. First segment with large outer spine bearing finely striated membranes on both sides and 1 row of spinules at base. Second segment with moderately small bilaterally serrated spine. Third exopodal spine smaller than second and bilaterally serrated. Last spine robust and serrated along inner margin. Dorsal surface of leg 2 (fig. 4a) with 2 striated, posteriorly directed membranes. Endopod of usual caligid structure. Sympod of leg 3 (fig. 4b) with dorsal corrugated pad on anterolateral corner. Exopod (fig. 4c) 3-segmented. Large exopodal spine (fig. 4d) terminally situated on basal swelling; both swelling and spine with striated flange. Second segment with outer row of setules, 1 outer spine, and 1 inner plumose seta. Terminal exopodal segment with outer row of setules, 3 outer spines, and 4 plumose setae. Endopod (fig. 4c) 2-segmented. First segment expanded laterally into velum on outer margin and bearing 1 plumose seta on inner margin. Velum oval and flat, setulated along free margin, and appearing separate from first endopodal segment in ventral view, but connected dorsally. Second segment with 6 plumose setae, and composed of 2 incompletely fused segments; small suture demarcating original second and third segments present. Last endopodal segment with outer row of setules. Leg 4 (fig. 4e) brachiform. Sympod slightly inflated, with 1 plumose seta on outer distal corner. Exopod (fig. 4f) 2-segmented. First segment with 1 unilaterally serrated spine. Second segment with 3 spines. Innermost spine with pectinate membrane at base, much longer than other 2, curved, and with relatively large

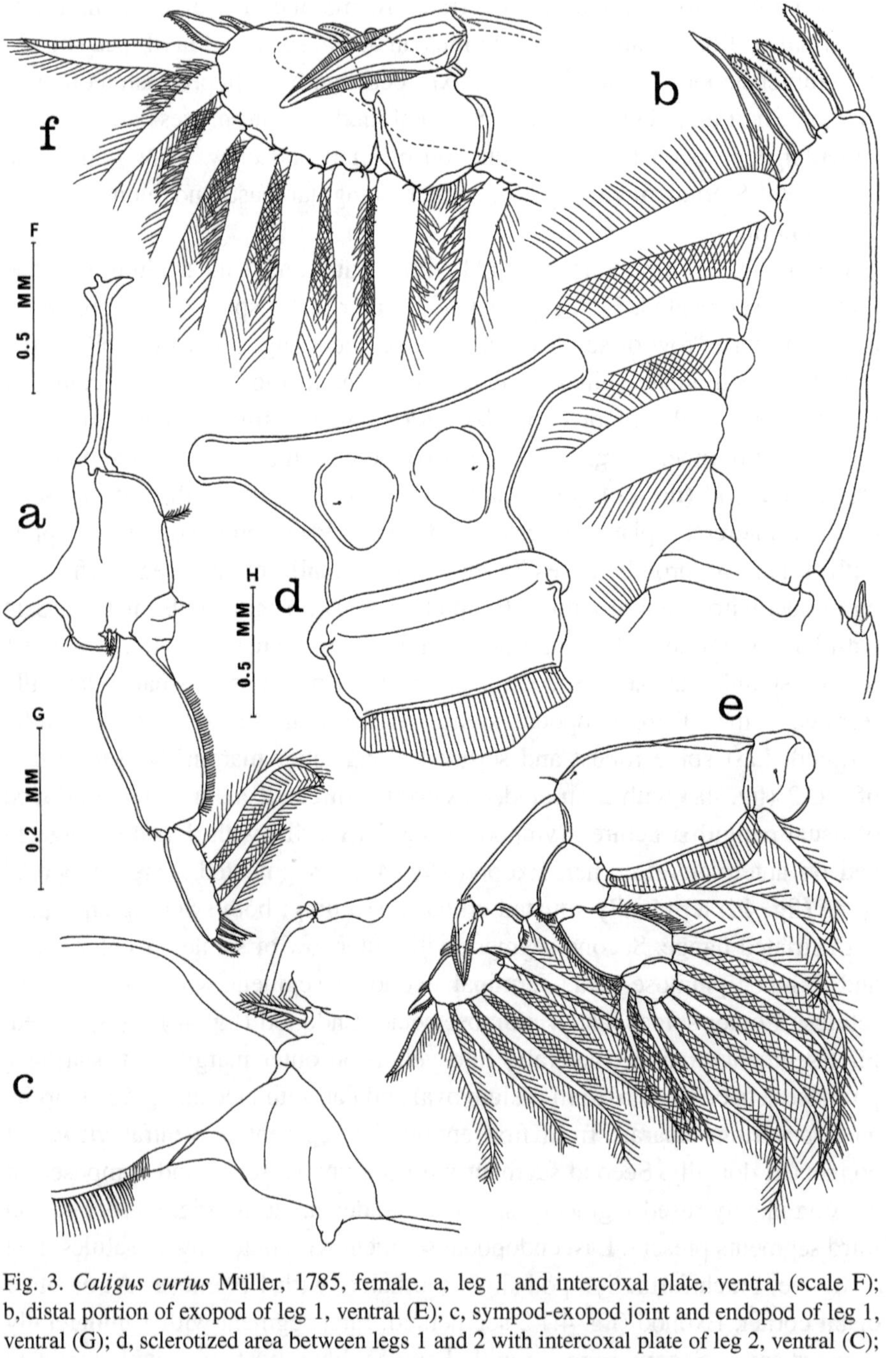

Fig. 3. *Caligus curtus* Müller, 1785, female. a, leg 1 and intercoxal plate, ventral (scale F); b, distal portion of exopod of leg 1, ventral (E); c, sympod-exopod joint and endopod of leg 1, ventral (G); d, sclerotized area between legs 1 and 2 with intercoxal plate of leg 2, ventral (C); e, leg 2, ventral (H); f, exopod of leg 2, ventral (B).

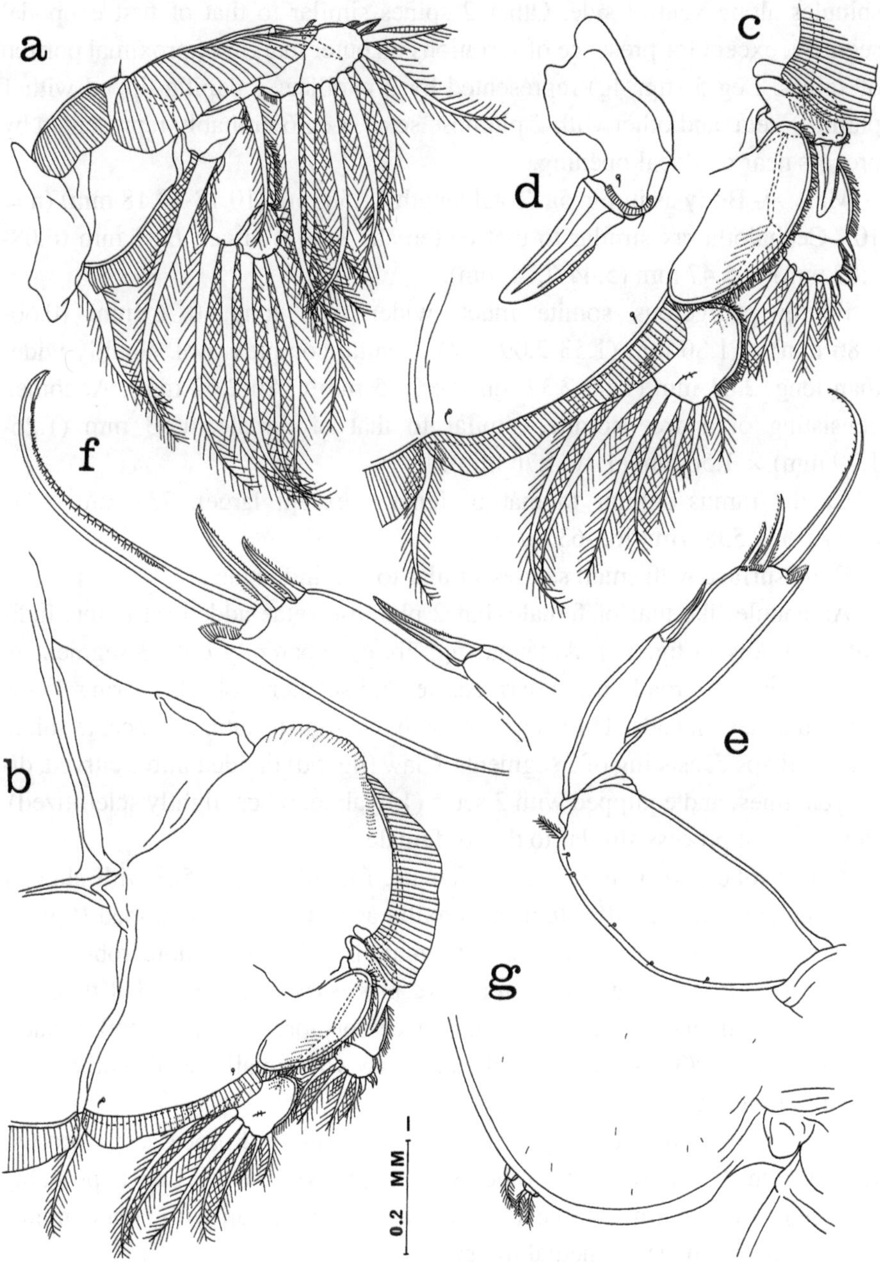

Fig. 4. *Caligus curtus* Müller, 1785, female. a, leg 2, dorsal (scale H); b, leg 3, ventral (H); c, exopod and endopod of leg 3, ventral (C); d, spine and basal swelling of first exopodal segment of leg 3, ventral (G); e, leg 4, ventral (C); f, distal portion of exopod of leg 4, dorsal (I); g, leg 5 and genital area, ventral (C).

spinules along ventral side. Other 2 spines similar to that of first exopodal segment, except for presence of serrations on outer margin of proximal portion of spines. Leg 5 (fig. 4g) represented by 2 papillae; 1 papilla tipped with 1 plumose seta, and other with 2 plumose setae. Leg 6 probably represented by process near oviducal opening.

Male. — Body as in fig. 5a. Total length 11.58 mm (10.13-13.18 mm) (n = 10). Cephalothorax similar to that of female, except larger, 6.81 mm (6.08-7.87 mm) × 6.47 mm (5.44-7.74 mm).

Fourth pedigerous somite much wider than long, 0.76 mm (0.66-0.86 mm) × 1.80 mm (1.53-2.09 mm). Genital complex oval, slightly wider than long, 2.01 mm (1.69-2.32 mm) × 2.25 mm (1.86-2.62 mm). Abdomen consisting of 1 free somite, similar to that of female, 1.53 mm (1.26-1.79 mm) × 1.50 mm (1.16-1.76 mm).

Caudal ramus similar to that of female, except larger, 737 μm (631-830 μm) × 538 μm (465-631 μm).

Body surface with small setules similar to that in female.

Antennule like that of female, but 2 plumose setae added (at points indicated by dots in fig. 1d). Antenna (fig. 5b, c) apparently only 3-segmented. First segment a broad base, unarmed. Second segment robust, bearing 3 corrugated adhesion pads. Third segment with incomplete suture; hence, terminal claw perhaps consisting of 2 segments. Claw (fig. 5d) divided into 2 curved, divergent tines, and equipped with 2 setae (1 hyaline, other slightly sclerotized). Postantennal process similar to that of female.

Mouth tube and mandible as in female. Maxillule (fig. 5e) with slightly more bulbous base than in female. Maxilla as in female. Maxilliped (fig. 5f) different from that of female. Corpus maxillipedis much more robust than in female. Myxa possessing 2 spinelike processes and 1 rounded tubercle. Proximal spinelike process with small accessory process on ventral surface. Shaft and claw 830 μm long combined. Claw with small dorsal seta. Sternal furca as in female.

Legs 1-4 as in female. Leg 5 (fig. 5g) consisting of 2 processes; 1 process tipped with 1 plumose seta, other with 2 plumose setae. Leg 6 probably represented by ventral flap near genital area. Ventral flap with small spinule at apex and 3 setules on medial margin.

Remarks. — Parker et al. (1968) provided a thorough account of *Caligus curtus* including a detailed description, historical review, and complete host and locality records. However, since *C. curtus* is the type-species of the type-genus of the family Caligidae, it was deemed necessary, for the sake of a

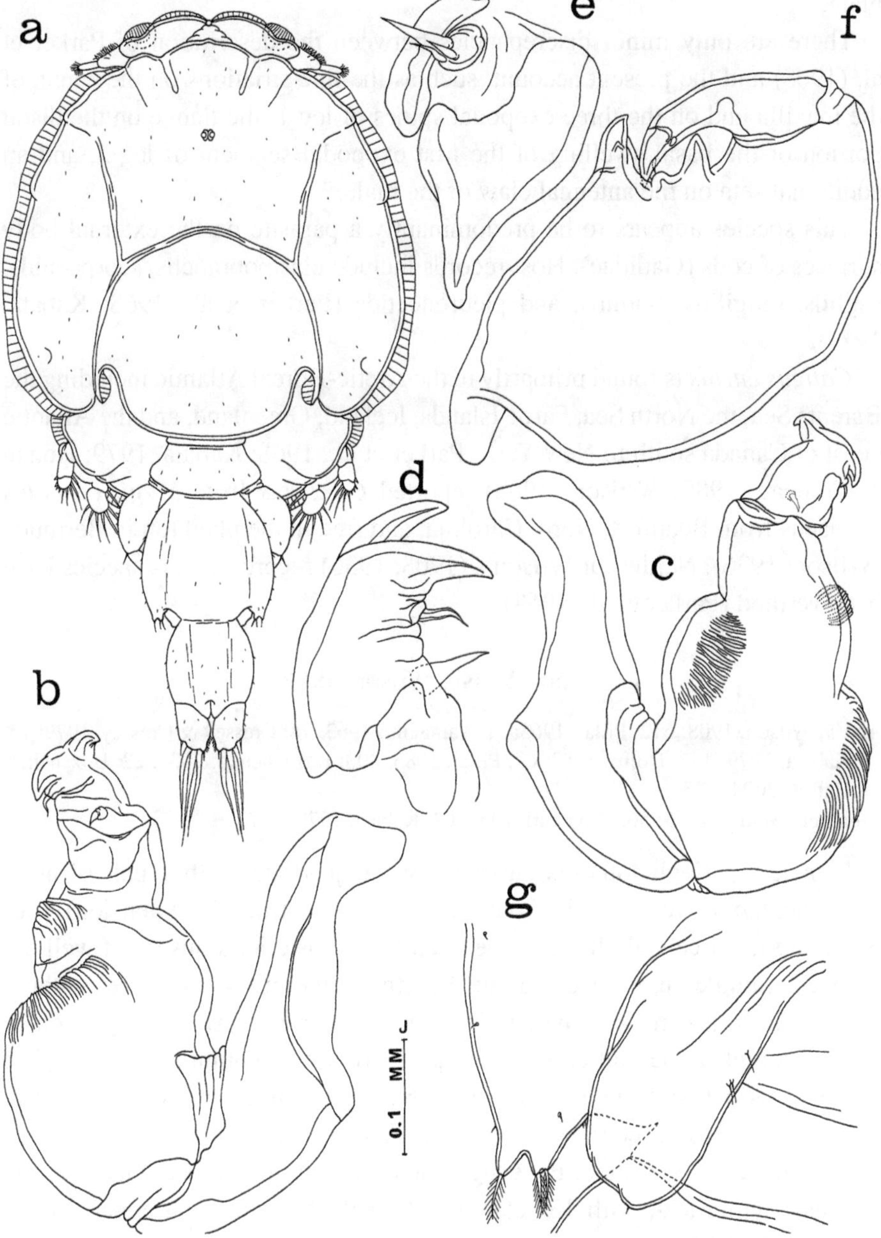

Fig. 5. *Caligus curtus* Müller, 1785, male. a, body, dorsal (scale A); b, antenna, ventro-inner (I); c, same, outer (I); d, claw of antenna, dorsal (J); e, maxillule, ventral (I); f, maxilliped, ventral (H); g, leg 5 and genital area, ventral (F).

complete generic revision, to include a full description of both the female and male.

There are only minor discrepancies between the description of Parker et al. (1968) and the present account, such as the fine striations on the canna of the maxilla and on the three exopodal spines of leg 1, the flange on the distal portion of the basal swelling of the first exopodal segment of leg 3, and an additional seta on the antennal claw of the male.

This species appears to be predominantly a parasite on the external body surfaces of cods (Gadidae). Host records include elasmobranchs, scorpaenids, triglids, mugilids, bothids, and pleuronectids (Parker et al., 1968; Kabata, 1979).

Caligus curtus is found primarily in the Arctic-Boreal Atlantic including the Barents Sea, the North Sea, Faroe Islands, Iceland, Greenland, and the Atlantic coast of Canada south to New York (Parker et al., 1968; Kabata, 1979; Hogan & Trudeau, 1989). Wilson (1908) reported *C. curtus* from *Mugil cephalus* Linnaeus from Beaufort, North Carolina, and free in the plankton in Bermuda (Wilson, 1936). Neither of Wilson's (1908, 1936) records of this species have been verified (Parker et al., 1968).

Genus **Abasia** Wilson, 1908

Abasia Wilson, 1908: 612; Pillai, 1963a: 1; Yamaguti, 1963: 78; Cressey & Cressey, 1979: 15; Kabata, 1979: 157; Dojiri, 1983: 67; Pillai, 1985: 403; Kazachenko, 2001: 22; Boxshall & Halsey, 2004: 725.

Alicaligus Shiino, 1955: 56; Yamaguti, 1963: 62; Kabata, 1979: 157.

Female. — Cephalothorax subovate or subquadrate, with 1 pair of anterior indentations resulting in frontal protrusion (rostrumlike extension). Lateral margins of cephalothorax folded ventrally. Posterior sinuses not well developed. Lunules present or absent. Fourth pedigerous somite free, without dorsal plates, and may be hidden in dorsal view by anteriormost portion of genital complex. Genital complex inflated. Abdomen comprising 1 or 2 free somites, with first somite greatly inflated (except in *A. pusilla* Cressey & Cressey, 1979), and second somite small. Caudal ramus with 6 setae.

Antennule 2-segmented. First segment with several plumose setae. Second segment cylindrical, with 13 setae and 1 aesthetasc. Antenna apparently 3-segmented (although claw may represent 2 segments). Terminal segment a recurved claw (except in *A. pseudorostris* Wilson, 1908 spinelike). Postantennal process absent. Mouth tube with strigil and intrabuccal stylet. Mandible comprising 4 sections, with inner margin of terminal section dentiferous. Maxillule a setiferous papilla with or without associated process. Maxilla brachiform. Maxilliped prehensile; corpus maxillipedis stout. Sternal furca absent.

Leg 1 with 2-segmented exopod possessing 3 terminal spines, and varying number of small setae (completely absent in *A. pusilla* and *A. inflata* Cressey & Cressey, 1979) on inner margin of second exopodal segment. Endopod reduced to small lobe. Leg 2 biramous, with 2-or 3-segmented rami. Leg 3 forming ventral shield (apron). Exopod with large spine on first segment (reduced to very small spine in *A. inflata*). Endopod exhibiting varying degress of reduction from 2 segments to none. Leg 4 with 2-segmented exopod (leg 4 absent in *A. inflata*). Leg 5 represented by few setae or completely absent.

Male. — Cephalothorax similar to that in female, except not folded ventrally. Lunules present. Genital complex suboval in outline. Abdomen cylindrical, not inflated. Antenna chelate. Postantennal process present in some species [i.e., *A. platyrostris* Pillai, 1963 and *A. tripartita* (Shiino)]. Legs 5 and 6 represented by few setae.

Type-species. — *Abasia pseudorostris* Wilson, 1908. Cressey & Cressey (1979) inadvertently cited "*A. platyrostris* Wilson, 1908" as the type-species of this genus.

Remarks. — Wilson (1908) first established the genus *Abasia* based on a single species, *A. pseudorostris*, collected from the roof of the buccal cavity of a lizardfish, *Synodus foetens* Linnaeus, from North American waters. According to Wilson, this genus lacks lunules, a thin membrane bordering the cephalothorax, and the postantennal process (referred to as the maxillule by Wilson). Later Pillai (1963a) redescribed the type-species and a new species, *A. platyrostris*, from India, and discovered that the lunules and membranous flange are present in this genus. He also found that the postantennal process is present in the male of *A. platyrostris*.

In their study of copepods parasitic on Indo-Pacific lizardfishes, Cressey & Cressey (1979) described three new species and one previously known species of *Abasia*, and also provided a revised generic diagnosis for the genus and a key to the six described species. Their discovery of *A. pusilla* with indistinct lunules in the female and *A. inflata* without lunules in the female suggest that the presence or absence of this structure may not be as taxonomically important as once thought.

Cressey & Cressey (1979) synonymized *Alicaligus* with *Abasia* (see remarks section of *Abasia tripartita*), and considered *Abasia* a "plastic genus". They also considered the members of *Abasia* specific for synodontid fishes, regarding Shiino's (1955) collection of *Abasia tripartita* from a scombrid as representing an accidental infestation.

Caligus grandiabdominalis Yamaguti, 1954 may belong to *Abasia*. The inflated genital complex and abdomen in the female, reduced antenna of the female, and absence of the sternal furca support the transfer. However, the presence of the postantennal process, and structures of legs 1, 2, and 4 do not conform to the generic diagnosis of *Abasia*.

Abasia pseudorostris Wilson, 1908
(figs. 6-9)

Abasia pseudorostris Wilson, 1908: 612; Bere, 1936: 587; Pillai, 1963a: 2; Yamaguti, 1963: 79; Cressey & Cressey, 1979: 1, 15; Dojiri, 1983: 71.

Material examined. — Type-material containing 3 females and 5 males (USNM 32811) from roof of mouth of lizardfish, *Synodus foetens* (Linnaeus), from Beaufort, North Carolina.

Female. — Body as in fig. 6a. Total length (excluding setae on caudal ramus) 3.66 mm (2.72-5.18 mm) (n = 3). Cephalothorax, 1.25 mm (1.03-1.63 mm) × 1.10 mm (0.93-1.33 mm), with anterolateral notches resulting in separation of anterior portion of lateral zone from cephalic zone. Anterior portion of cephalic zone extended to form large frontal area (fig. 6b). This frontal area possessing transparent membrane on anterior margin. Lateral zones of cephalothorax folded ventrally and also rimmed with transparent membrane hidden in dorsal view. Dorsal transverse rib shifted posteriorly; consequently, thoracic zone smaller than cephalic area. Posterior sinuses very shallow. Free margin of thoracic zone not extending to posterior tip of lateral zone. Lunules apparently absent (Pillai, 1963a described indistinct lunules for this species, but no traces of these structures observed in these specimens). Tip of antennule well within limit of cephalothorax.

Fourth pedigerous somite distinct, wider than long, 0.22 mm (0.16-0.30 mm) × 0.38 mm (0.30-0.50 mm). Genital complex narrower anteriorly, longer than wide 1.28 mm (1.00-1.79 mm) × 1.05 mm (0.93-1.29 mm). Abdomen with 2 free somites. First somite expanded, 0.77 mm (0.56-1.06 mm) × 0.61 mm (0.46-0.83 mm). Second somite smaller than first, 0.20 mm (0.17-0.27 mm) × 0.32 mm (0.27-0.40 mm).

Caudal ramus (fig. 6c) 68 × 65 μm and carrying 3 large, terminal plumose setae and 3 smaller, subterminal plumose setae.

Egg sacs (fig. 6a) 1.76 mm × 0.30 mm, containing 24 eggs (n = 1).

Antennule (fig. 6d) 2-segmented. First segment 100 μm long (measured along nonsetiferous margin) and equipped with 15 smooth setae. Second segment 86 μm long carrying, usual 13 + 1 aesthetasc. Antenna (fig. 6b, e) highly reduced in size, apparently tripartite. Base relatively small and unarmed. Sec-

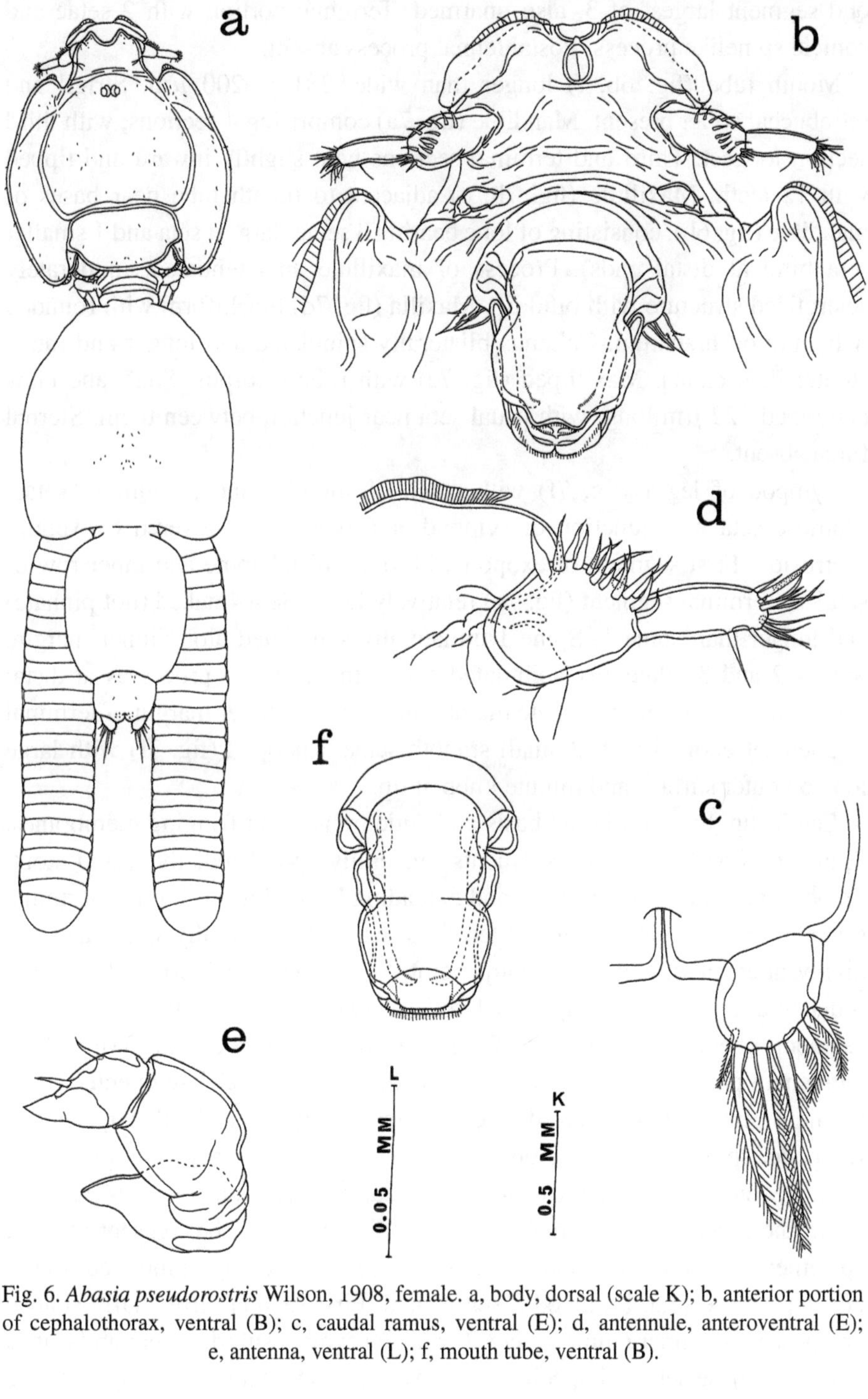

Fig. 6. *Abasia pseudorostris* Wilson, 1908, female. a, body, dorsal (scale K); b, anterior portion of cephalothorax, ventral (B); c, caudal ramus, ventral (E); d, antennule, anteroventral (E); e, antenna, ventral (L); f, mouth tube, ventral (B).

ond segment largest of 3, also unarmed. Terminal portion with 2 setae and conical spinelike process. Postantennal process absent.

Mouth tube (fig. 6b, f) longer than wide, 281 × 200 μm. Strigil and intrabuccal stylet present. Mandible (fig. 7a) comprising 4 sections, with third section longest. Third and terminal sections bent slightly inward and tipped with 12 teeth. Maxillule (fig. 7b, c) adjacent to mouth tube near bases of mandible (fig. 6b), consisting of lobe bearing 3 setae (largest seta and 1 smaller seta bifid at distal ends). Process of maxillule an attenuated, moderately sclerotized structure with bifid tip. Maxilla (fig. 7d) brachiform with rounded flabellum on brachium. Calamus bilaterally spinulated and longer and much stouter than canna. Maxilliped (fig. 7e) with robust corpus. Shaft and claw combined 322 μm long, with usual seta near junction between them. Sternal furca absent.

Sympod of leg 1 (fig. 7f) with patch of spinules and 2 plumose setae. Plumose seta near junction of sympod and exopod not covered by ventral protrusion. First segment of exopod with outer distal spine and inner row of setules. Terminal segment (fig. 7g) relatively large. Seta 4 naked (not pinnate) and longer than spine 1. Spine 1 unilaterally spinulated along inner margin. Spines 2 and 3 bilaterally spinulated and with accessory processes at distal ends. All 3 claws with pectinate membranes at bases. Inner margin of terminal segment of exopod with 2 small smooth setae. Endopod (fig. 7f) with large lobe on outer surface and minute knob at tip.

Leg 2 (fig. 8a) with broad basis and wide transparent fringing membranes. Spines of first 2 exopodal segments bilaterally spinulated, long, and stout. Spine of second segment of exopod extending beyond distal limit of terminal segment. First exopodal spine of terminal segment unilaterally spinulated and highly attenuated, ending in sharp point. Second spine of terminal segment rounded at apex and not highly sclerotized. Outer margin of first and second segments of endopod with rows of minute spinules. Sympod of leg 3 (fig. 8b, c) with small patch of spinules near intercoxal plate. Exopod 3-segmented. First segment represented by basal swelling, terminally armed with large spine. Exopodal spine spinulated along outer margin and with rows of spinules at base. Second segment largest of 3 exopodal segments and possessing outer spine and inner seta (presumably broken off in fig. 8b, c, but present in other specimen). Terminal exopodal segment with 3 bilaterally spinulated spines and 1 plumose seta. Endopod 2-segmented. First segment irregularly shaped and possessing inner plumose seta. Terminal segment subtriangular and tipped with 1 plumose seta in 1 specimen (fig. 8b) or quadrangular and tipped with 2

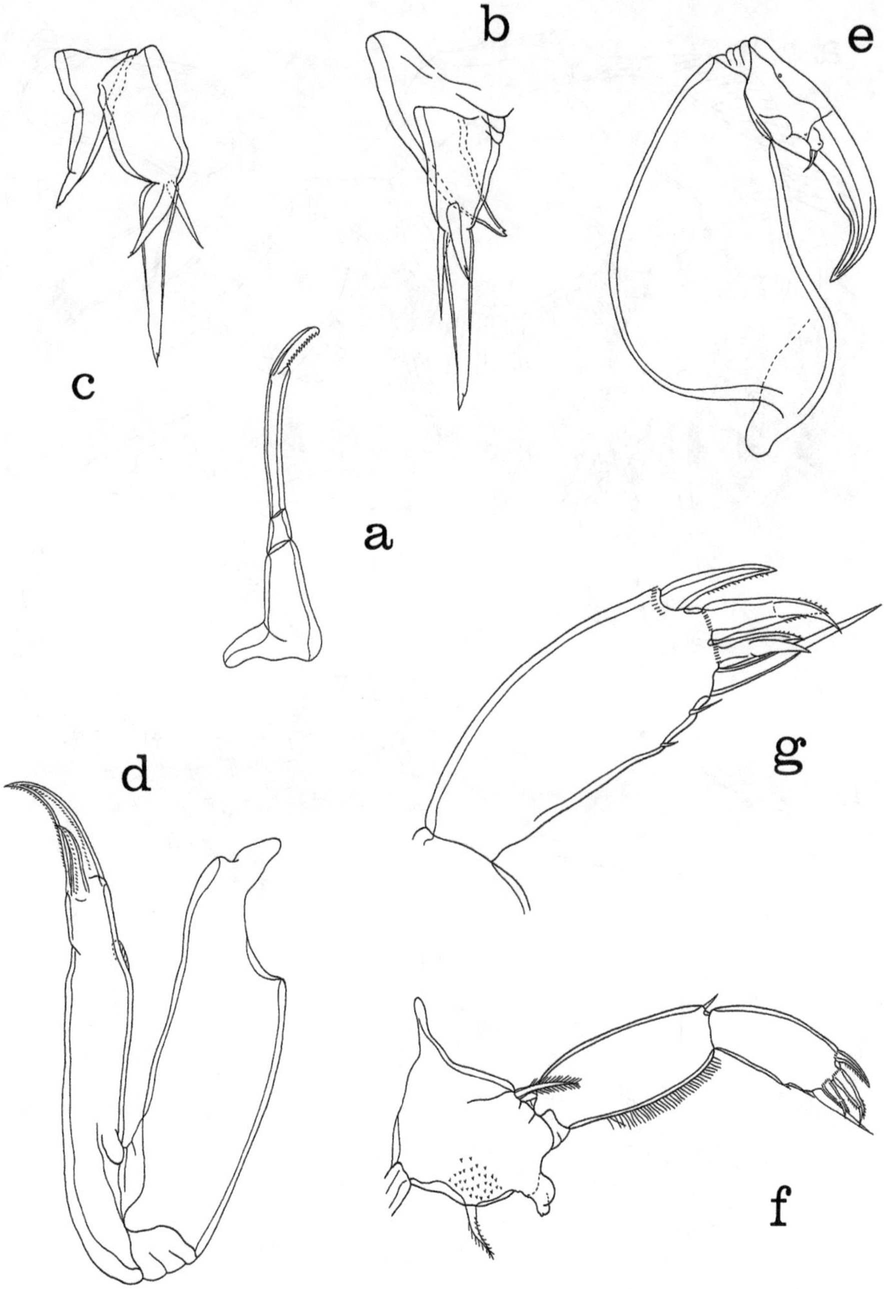

Fig. 7. *Abasia pseudorostris* Wilson, 1908, female. a, mandible, ventro-inner (scale E); b, maxillule, outer (L); c, same, antero-outer (L); d, maxilla, dorso-inner (E); e, maxilliped, ventral (B); f, leg 1, ventral (E); g, terminal exopodal segment of leg 1, ventral (L).

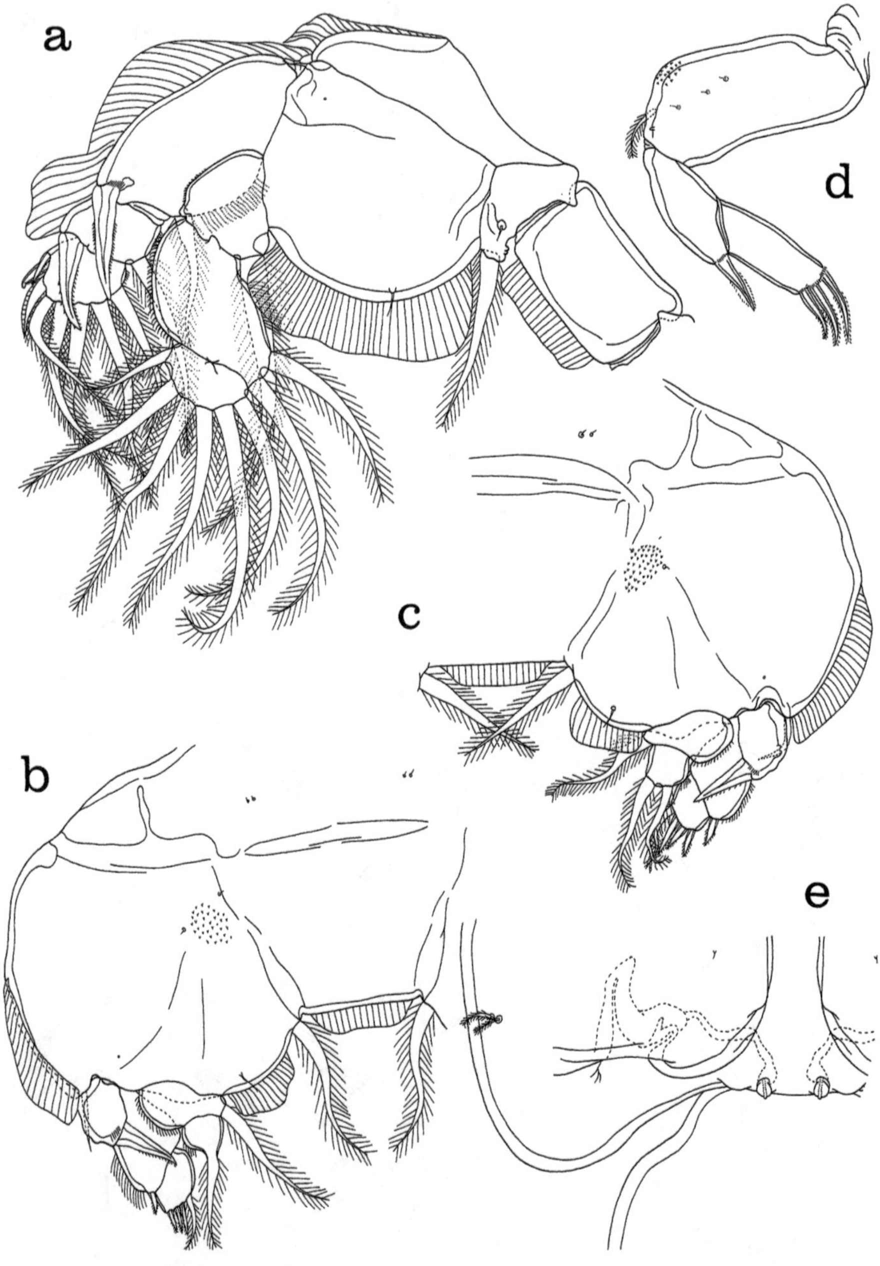

Fig. 8. *Abasia pseudorostris* Wilson, 1908, female. a, leg 2 and intercoxal plate, ventral (scale E); b, leg 3, right side, ventral (E); c, same, left side, ventral (E); d, leg 4, ventral (E); e, leg 5 and genital area; ventral (B).

plumose setae in other specimen (fig. 8c). Leg 4 (fig. 8d) sympod with outer patch of spinules near base of plumose seta. Exopod 2-segmented, with 3 terminal spines almost equal in length. All 4 exopodal spines with pectinate membranes at bases. Leg 5 (fig. 8e) represented by 3 small plumose setae (2 smaller setae arising from same sclerite). Leg 6 possibly represented by posteroventral lobe medial to opening of egg-laying apparatus.

Male. — Body as in fig. 9a, b. Total length 1.84 mm (1.60-2.01 mm) (n = 5). Cephalothorax similar in shape to that of female, but smaller, 1.02 mm (0.90-1.13 mm) long. Width of cephalothorax variable depending upon flexion of lateral zone (compare fig. 9a and b). Lunules also absent in male.

Fourth pedigerous somite wider than long, 103 μm (66-130 μm) × 264 μm (200-299 μm). Genital complex quadrangular to oval, 412 μm (365-432 μm) × 431 μm (362-465 μm). Abdomen with 2 free somites. First somite small, 113 μm (86-146 μm) × 246 μm (211-281 μm). Second somite larger than first, 167 μm (124-200 μm) × 258 μm (216-292 μm).

Caudal ramus similar in shape to that of female, but smaller, only 25 μm long.

Antennule as in female. Antenna (fig. 9c, d) different from that of female. Second segment robust and carrying prominent myxa on inner margin. Myxa with approximately 5 spinules on ventral surface. Terminal claw with 2 smooth setae and 1 minute ventral knob near midlength.

Mouth tube, mandible, maxillule, maxilla, and maxilliped as in female.

Terminal segment of exopod of leg 1 with spines 2 and 3 slightly different from those in female (compare figs. 7g and 9e). Seta 4 pinnate with very short plumosities. Endopod of leg 1 (fig. 9f) without large outer lobe, and possessing 2 minute knobs at tip. Legs 2-4 as in female. Leg 5 (fig. 9g) represented by 3 plumose setae. Leg 6 consisting of 3 smooth setae near posteroventral flap.

Remarks. — *Abasia pseudorostris* was described as a new genus and species from the lizardfish *Synodus foetens* (Linnaeus) caught at Beaufort, North Carolina, by Wilson (1908). Later, Bere (1936) reported this species from the mouth of *S. foetens* from Lemon Bay, Florida. Pillai (1963a) completely redescribed *A. pseudorostris* based on Bere's specimens and observed lunules on both the female and the male of this species, although he admitted that the structures were indistinct. In the type-specimens, these lunules were not observed in either sex. There are other differences between Pillai's description and the present description. For instance, Pillai described oblique grooves that divide the lateral zone of the cephalothorax into three regions. Only the anterior groove is present in the type-specimens. Pillai described 12 or 13 setae

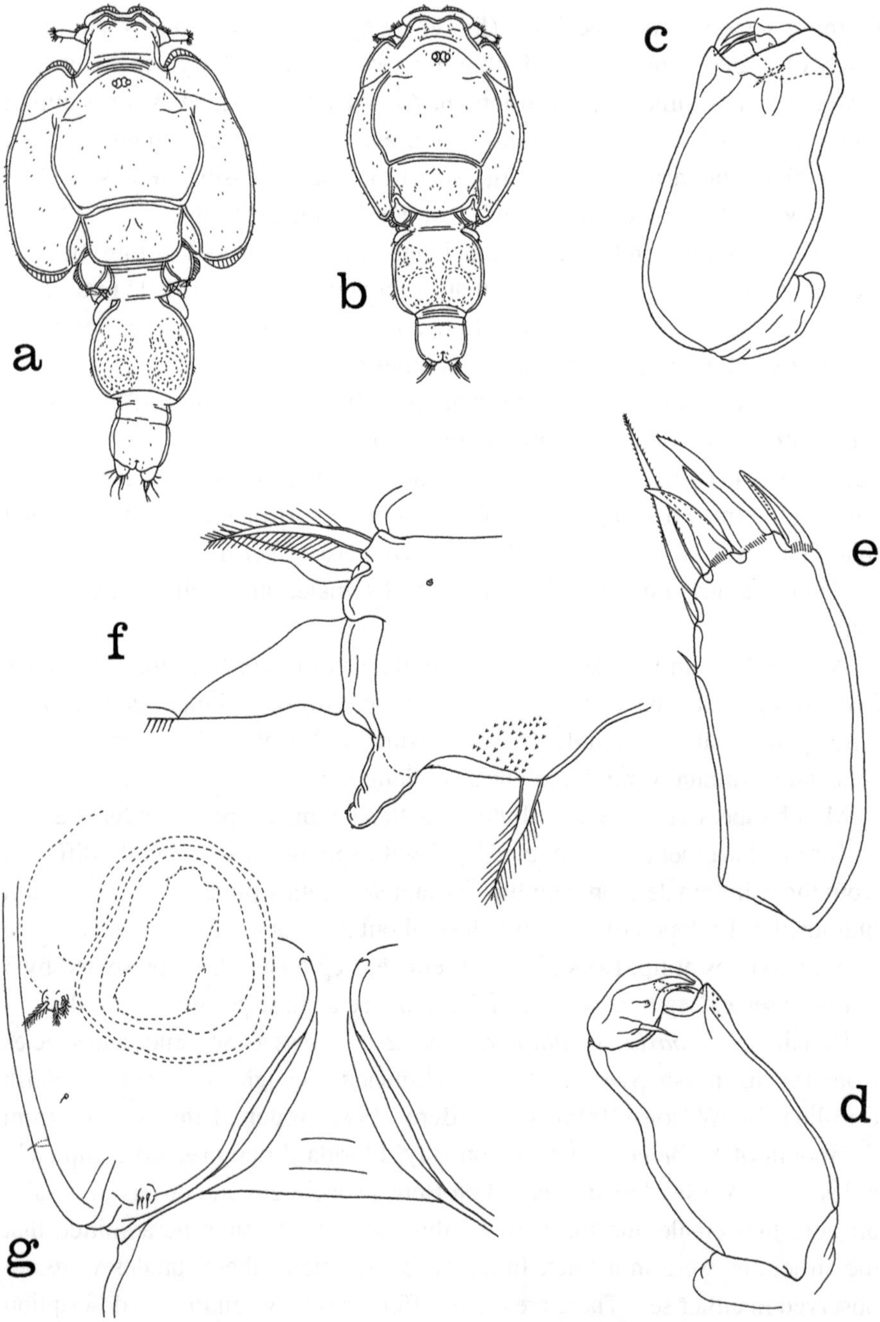

Fig. 9. *Abasia pseudorostris* Wilson, 1908, male. a, body, dorsal (scale K); b, same (K); c, antenna, dorsal (E); d, same, ventral (E); e, terminal portion of leg 1, ventral (L); f, sympod-exopod joint and endopod of leg 1, ventral (L); g, legs 5 and 6, and genital area, ventral (E).

on the first segment and 11 or 12 setae on the second segment of the antennule. He also apparently missed the accessory processes on spines 2 and 3 of the terminal segment of leg 1. Leg 5 of the female and leg 6 of the male were not observed by Pillai, but are certainly present.

Although originally reported from the roof of the buccal cavity of *S. foetens* by Wilson (1908), Pillai (1963a) and Cressey & Cressey (1979) expressed some doubt concerning the recorded attachment site. The basis for this doubt rests on the morphology of the anterior end of the females of *A. pseudorostris*. It has been suggested (Pillai, 1963a) that the presence of lunules (not observed in the type-specimens) and the membranous flange on the anterior margin of the frontal area suggests a predilection for an attachment site different from the roof of the buccal cavity. The folding of the lateral areas of the cephalothorax ventrally reflects an adaptation toward attachment to gill filaments. This is supported by members of other copepod genera (e.g., *Hermilius* and *Lernanthropus*) that exhibit this ventral folding of the cephalothorax and display a preference for gill filaments. In fact, R. Cressey apparently has only collected this species of copepod from the gill filaments and never from the roof of the buccal cavity (Cressey & Cressey, 1979). However, these authors failed to mention Bere's (1936) record of *A. pseudorostris* from the "mouth" of *S. foetens*. Pillai (1963a) himself described a congener, *A. platyrostris*, from a "groove on the roof of the buccal cavity". It is possible that Wilson (1908) was not in error. However, the attachment site preference for the described species of *Abasia*, except *A. platyrostris*, does in fact appear to be the gill filaments.

Abasia tripartita (Shiino, 1955)
(figs. 10-13)

Alicaligus tripartitus Shiino, 1955: 56; Yamaguti, 1963: 63; Kabata, 1979: 162.
Abasia tripartita Cressey & Cressey, 1979: 1, 18; Dojiri, 1983: 77.

Material examined. — Six paratypes (4 females, 2 males) attached lengthwise with anterior end directed toward proximal end of gill filament of scombrid, *Sarda orientalis* (Temminck & Schlegel), from Owase, Mie Prefecture, Japan, on 8 December 1952, collected by Dr. Sueo M. Shiino. This material on loan from Mie University with the help of Dr. Kunihiko Izawa.

Female. — Body as in fig. 10a, b, and c. Total length (excluding setae on caudal ramus) 3.13 mm (2.79-3.32 mm) (n = 4). Cephalothorax quadrangular in dorsal view, wider than long, 0.82 mm (0.80-0.83 mm) × 0.97 mm (0.93-1.00 mm), convex dorsally, with lateral zone folded ventrally. Anterolateral clefts present on either side of frontal area, but not as prominent as in *Abasia*

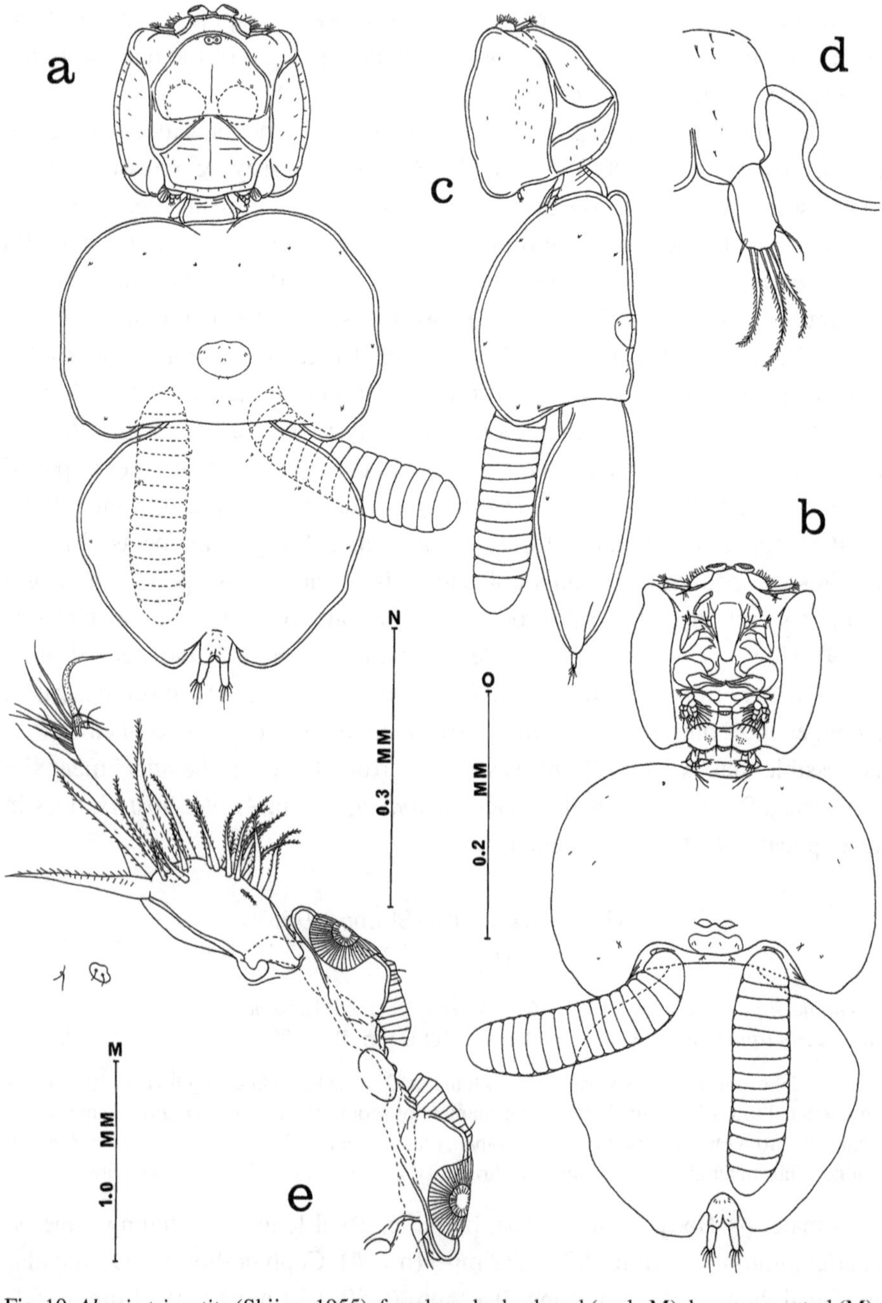

Fig. 10. *Abasia tripartita* (Shiino, 1955), female. a, body, dorsal (scale M); b, same, ventral (M); c, same, lateral (M); d, caudal ramus, ventral (N); e, frontal plate and antennule, ventral (O).

pseudorostris. This frontal area (fig. 10e) carrying membranous flange along anterior margin. Dorsal transverse rib upside down V-shape. Posterior sinuses extremely shallow, but with distinct membranous flange. Free margin of thoracic zone at same level as posterior tip of lateral zone. Lunules (fig. 10a, b and e) close together, and located anteroventrally on frontal area. Tip of antennule well within lateral limit of cephalothorax.

Fourth pedigerous somite small, twice as wide as long, 0.10 × 0.28 mm (0.27-0.30 mm). Genital complex extremely large, longer and wider than cephalothorax, 1.13 mm (1.06-1.20 mm) × 1.64 mm (1.56-1.66 mm), and with large winglike lobes curved ventrally. Abdomen comprising 2 free somites, combined somites 1.30 mm (1.00-1.49 mm) × 1.43 mm (1.06-1.59 mm). First somite almost as long and wide as genital complex, widest at anterior third, flattened dorsoventrally, and with posterior lobes flanking second somite. Second somite (anal somite) small, 0.17 × 0.18 mm.

Caudal ramus (fig. 10d) longer than wide, 92 × 49 μm, carrying 3 long, plumose terminal setae and 3 smaller, smooth subterminal setae.

Egg sacs (fig. 10a-c) 1.45 mm (1.16-1.46 mm) × 0.32 mm (0.30-0.33 mm), containing 15-16 eggs.

Antennule (fig. 10e) 2-segmented. First segment 122 μm long (measured along nonsetiferous margin) and carrying 19 plumose setae. Seta on posterodistal corner long and stout. Second segment 135 μm long, with usual armament of 13 + 1 aesthetasc. Antenna (fig. 11a) with small basal portion. Next segment relatively large and unornamented. Terminal portion a claw 84 μm long, with 2 setae and 2 minute spinules. Postantennal area (fig. 10e) comprised of sclerite with 2 setules in addition to 1 setule slightly removed from sclerite.

Mouth tube (fig. 11b) slender, 211 × 146 μm. Strigil and intrabuccal stylet present. Mandible (fig. 11c) with 4 sections, and tipped with 12 teeth. Maxillule (fig. 11d) a lobe, 43 × 22 μm, bearing 2 small smooth setae and 1 large plumose seta, and situated adjacent to bases of mandible (fig. 11b). Maxilla (fig. 11e) brachiform. Large flabellum near distal end of brachium. Calamus with 3 serrated membranes, longer than canna. Canna with serrated membrane extending from dorsal to ventral surface. Maxilliped (fig. 11f) with robust, unornamented corpus. Shaft and claw with usual seta and 1 setule. At junction of shaft and claw (fig. 11g) a depression (cavity) present on dorsal surface. Tip of claw broken off. Sternal furca absent.

Sympod of leg 1 (fig. 11h) with 2 smooth setae (plumosities may have broken off?). Seta near sympod-exopod junction not covered by ventral

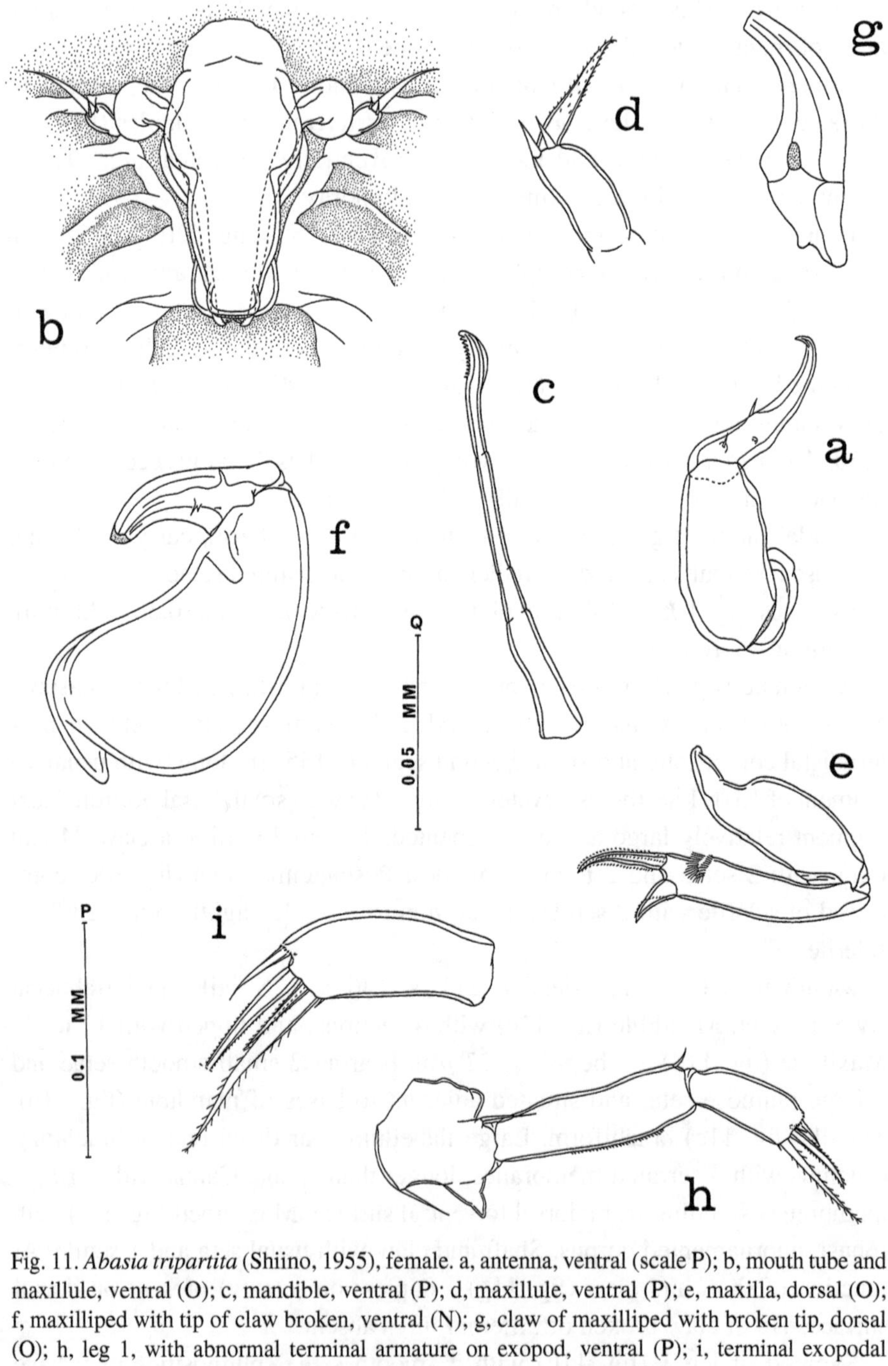

Fig. 11. *Abasia tripartita* (Shiino, 1955), female. a, antenna, ventral (scale P); b, mouth tube and maxillule, ventral (O); c, mandible, ventral (P); d, maxillule, ventral (P); e, maxilla, dorsal (O); f, maxilliped with tip of claw broken, ventral (N); g, claw of maxilliped with broken tip, dorsal (O); h, leg 1, with abnormal terminal armature on exopod, ventral (P); i, terminal exopodal segment of leg 1 with normal armature, ventral (Q).

protrusion. First segment of exopod with smooth seta at outer distal corner and row of small spinules along inner margin. Normal armament of terminal exopodal segment (fig. 11i) with 3 spines unequal in length, middle one (spine 2) being longest. Spine 1 with row of spinules at its base. Spines 2 and 3 bilaterally spinulated and without accessory processes at distal ends. Pinnate seta 4 more than twice as long as outermost spine. One small, smooth seta on inner margin near base of pinnate seta 4. Abnormal armament of terminal segment of exopod as in fig. 11h. Endopod a large conical process. Leg 2 (fig. 12a) of usual caligid form. Spines of first and second segments of exopod long and highly attenuate. First spine of terminal exopodal segment smooth and minute; second spine larger and unilaterally spinulated. First and second segments of endopod with rows of spinules along outer border. Leg 3 (fig. 12b) with wide intercoxal plate rimmed along posterior margin with 2 membranous flaps. Sympod with 2 patches of spinules; 1 near base of exopod, other near intercoxal plate. Exopod (fig. 12 b, c) 2-segmented. First segment represented by basal swelling with terminally situated, bilaterally spinulated spine. Second segment with 2 outer, bilaterally spinulated spines, 1 smaller, smooth seta, and 1 inner, plumose seta. Abnormal exopod (fig. 12d) observed in 1 specimen. Endopod a rounded lobe with outer row of spinules, and carrying 1 plumose seta. Sympod of leg 4 (fig. 12e) with row of minute spinules on ventral surface near long smooth seta on outer distal end. Exopod 2-segmented, with spines of terminal segment decreasing in length from inner to outer margin. All 4 exopodal spines bilaterally spinulated, and with pectinate membranes at bases. Leg 5 (and leg 6?) (fig. 12f) represented by 2 small lobes; 1 tipped with 2 smooth setae, other equipped with only 1 seta.

Male. — Body of typical caligid form as in fig. 13a. Total length 1.59 mm (1.53-1.66 mm) (n = 2). Cephalothorax orbicular, 0.80 mm (0.76-0.83 mm) × 0.73 mm (0.70-0.76 mm), with extremely shallow posterior sinuses rimmed with large membranous flanges. Free margin of thoracic zone extending slightly beyond posterior tips of lateral zone. Tips of antennule almost reaching lateral limits of cephalothorax. Lunules present.

Fourth pedigerous somite wider than long, 0.11 mm (0.10-0.12 mm) × 0.18 mm. Genital complex widest at midlength, longer than wide, 0.32 mm (0.31-0.33 mm) × 0.28 mm (0.28-0.29 mm). Abdomen presumably comprising 2 free somites. First somite short and very weakly sclerotized. Second somite longer than wide.

Antennule as in female. Antenna (fig. 13b) 3-segmented. Base rather long and unornamented. Second segment largest and with row of spinules on inner

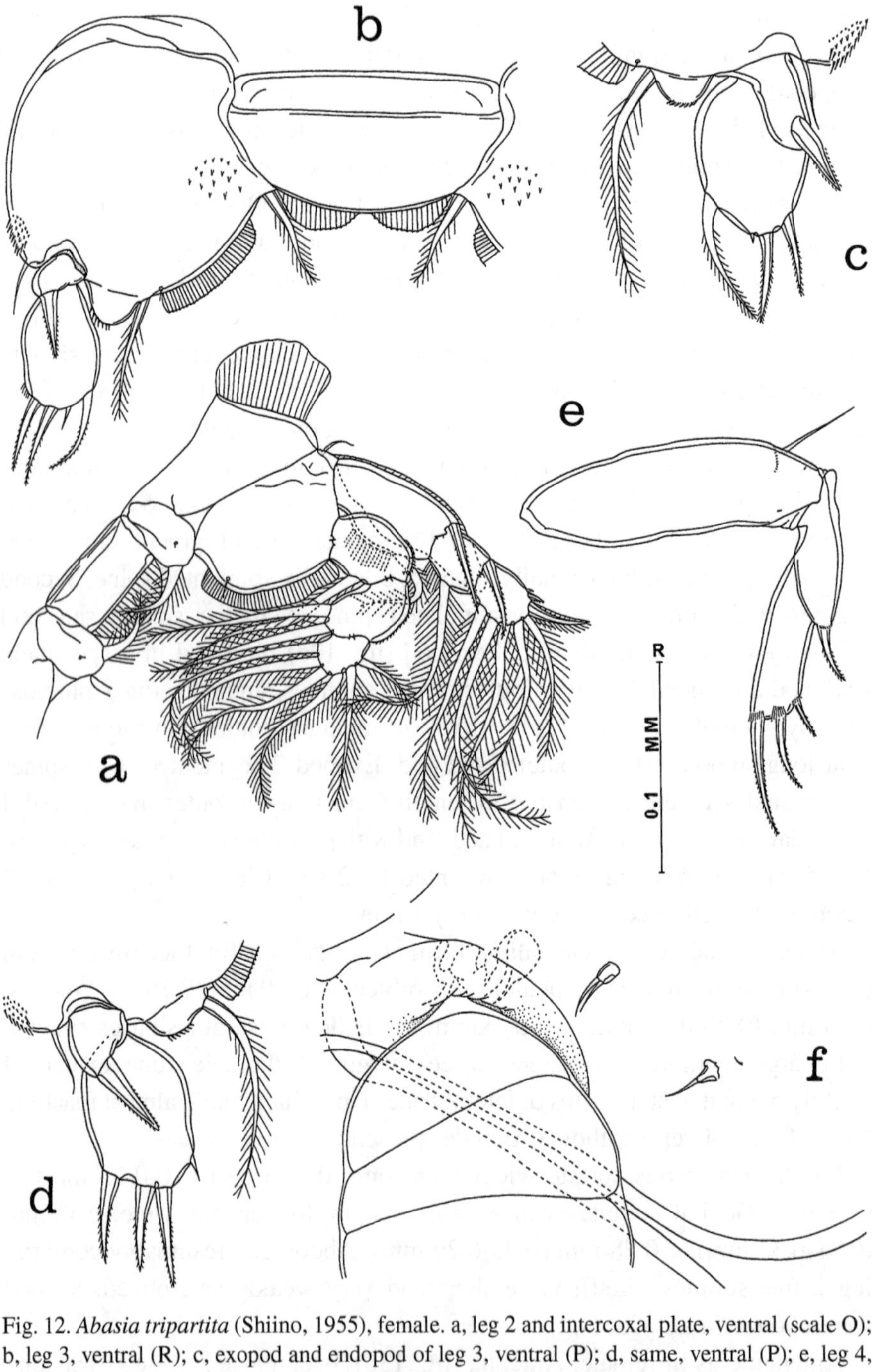

Fig. 12. *Abasia tripartita* (Shiino, 1955), female. a, leg 2 and intercoxal plate, ventral (scale O); b, leg 3, ventral (R); c, exopod and endopod of leg 3, ventral (P); d, same, ventral (P); e, leg 4, ventral (R); f, leg 5 and genital area, ventral (N).

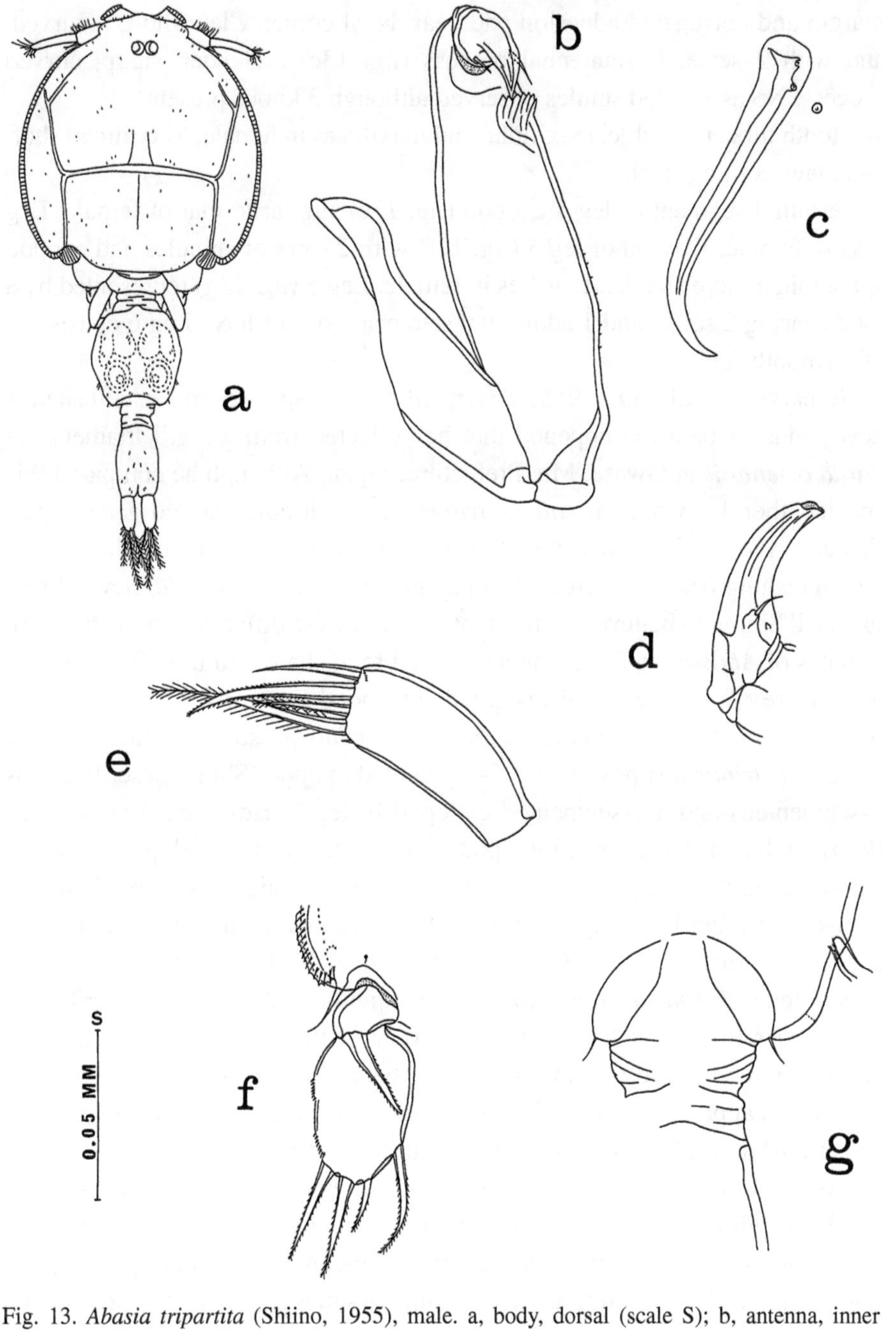

Fig. 13. *Abasia tripartita* (Shiino, 1955), male. a, body, dorsal (scale S); b, antenna, inner (P); c, postantennal process, ventral (P); d, claw of maxilliped with broken tip, ventral (O); e, terminal exopodal segment of leg 1, ventral (Q); f, exopod of leg 3, ventral (P); g, legs 5 and 6, and genital area, ventral (O).

margin and corrugated adhesion pad near distal corner. Claw short, recurved, and with 2 setae. Postantennal process (fig. 13c) a slender, sharp, curved process. No associated setules observed although 3 knobs present.

Mouth tube, mandible, maxillule, and maxilla as in female. Maxilliped shaft and claw as in fig. 13d.

Terminal segment of leg 1 exopod (fig. 13e) similar to that of female. Leg 2 as in female. Exopod of leg 3 (fig. 13f) with 2 rows of spinules and exopod spine much more slender. Leg 4 as in female. Leg 5 (fig. 13g) represented by a lobe bearing 2 setae, and 1 additional seta near base of lobe. Leg 6 consisting of 1 smooth seta.

Remarks. — Shiino (1955) described a new species and established a new genus of parasitic copepod that he collected from the gill filaments of *Sarda orientalis* at Owase, Mie Prefecture, Japan. Although he compared this species, then known as *Alicaligus tripartitus*, with other caligid genera, particularly *Abasia*, there was, at that time, significant morphological differences distinguishing *Alicaligus* from the one known species of *Abasia*. Several features led Shiino to believe that these two genera were different. For instance, in females of *Abasia* the frontal area (referred to as the rostrum by Pillai, 1963a and Cressey & Cressey, 1979) is greatly extended anteriorly. *Abasia* was then thought to be devoid of lunules while *A. tripartitus* possesses distinct lunules. *Alicaligus tripartitus* possesses a 2-segmented exopod (Shiino described it as 1-segmented) and a 1-segmented endopod in leg 3. However, *Abasia pseudorostris* has a 3-segmented exopod and a 2-segmented endopod in leg 3. These features collectively appeared to separate Shiino's species from the genus *Abasia* until Cressey & Cressey (1979), in their comprehensive study of copepods parasitic on lizardfishes from the Indo-West Pacific, discovered three new species of *Abasia* (*A. inflata*, *A. pusilla*, and *A. pillaii*) and redescribed *A. platyrostris* Pillai, 1963. With the discovery of these three species, Cressey & Cressey concluded that the distinction between *Alicaligus* and *Abasia* no longer was apparent and transferred *Alicaligus tripartitus* to the genus *Abasia*. This transfer is fully justified, especially in view of the fact that the species of *Abasia* display a graduated reduction in the number of segments of the rami of leg 3. In addition, the lunule of the female, once thought to carry a great deal of taxonomic weight, is distinct in *A. pillaii*, inconspicuous in *A. pusilla*, and absent in *A. inflata*. Therefore, even within the same genus, the lunules may be either present or absent.

The above description is very similar to that of Shiino's (1955) account, but differs in minor points from that of Cressey & Cressey (1979). The basal segment of the antennule described by these two authors lacks the large stout seta on the posterodistal corner and possesses 26 plumose setae. The antennal claw of the female is much shorter and more robust than that of our description. Leg 5 (and leg 6?) of the female are figured for the first time. Cressey & Cressey believed that legs 5 and 6 were absent in the female. The antenna of our male specimens possesses only one adhesion pad while their specimens possess two.

Although the type-specimens of *Abasia tripartita* were collected from *S. orientalis*, a scombrid, subsequent collections were from lizardfishes (Synodontidae). Since other species of *Abasia* have been collected from synodontids, Shiino's record must represent an accidental infestation as noted by Cressey & Cressey (1979).

Genus **Alanlewisia** Boxshall, 2008

Alanlewisia Boxshall, 2008: 231.

Female. — Cephalothorax subcircular with posterior sinuses. Cephalothoracic shield without transverse rib (suture). Paired frontal plates with paired, ventrally situated anteromedial flaps (lunule-like structures). Nauplius eye present. Fourth pedigerous somite free, without dorsal plates. Genital complex subrectangular. Abdomen short, with 1 free somite. Caudal ramus with 6 setae.

Antennule 2-segmented; proximal segment stout, with 27 plumose setae; distal segment elongate, with usual 13 setae. Antenna 3-segmented. Postantennal process present. Mouth tube formed by tapering labrum and labium. Mandible of usual form. Maxillule bilobed; anterior lobe a papilla bearing 3 setae and posterior lobe an unarmed spiniform process. Maxilla 2-segmented, brachiform. Maxilliped subchelate, with massive protopod (corpus) and distal subchela comprising fused endopodal segment and terminal claw. Sternal furca present and situated in front of paired spinous processes.

Leg 1 with 2-segmented exopod possessing distally 3 spines and 1 seta and 3 vestigial setae on medial margin; endopod small, with 1-segmented ramus tipped with 2 setae and 1 short spine. Leg 2 biramous, with 3-segmented rami. Leg 3 biramous with 2-segmented rami; large, curved, clawlike spine present on basl swelling of first exopodal segment; velum and apron present. Leg 4 uniramous; coxa and basis fused; exopod 2-segmented and tipped with 4 spines. Leg 5 represented by 4 setae on posterolateral margin of genital complex lateral to J-shaped process. Leg 6 represented by paired, unarmed lobes closing off genital apertures.

Male. — Cephalothorax similar to that in female, except lacking lunule-like structures on ventral surface of paired frontal plates. Genital complex comparatively smaller than in female. Antenna with large corrugated pad on ventral surface of second segment and terminal subchela, carrying large digitiform accessory process, resulting in bifid appearance. Postantennal process longer than that of female and ornamented with fine surface striations. No J-shaped process. Both legs 5 and 6 with 3 plumose setae.

Type-species. — *Alanlewisia fallolunulus* (Lewis, 1967).

Remarks. — The type-species was originally reported and described by Lewis (1967) as *Lepeophtheirus fallolunulus* Lewis, 1967 from the bluespine unicornfish [*Naso unicornis* (Forsskål)] caught at Oahu, Hawaii. However, due to the absence of lunules, presence of a well-developed 1-segmented leg 1 endopod, and a reduced 2-segmented exopod of leg 3, Boxshall (2008) considered it to represent a new genus of the Caligidae. After providing a detailed redescription, he established *Alanlewisia* as a new genus named in honor of the discoverer of this species and designated *A. fallolunulus* as its type-species. We agree with Boxshall (2008) that the possession of a modified frontal plate in the female, pair of spiniform processes on the ventral cephalothoracic wall immediately posterior to the base of the sternal furca, a reduced leg 1 endopod bearing sparsely plumose setae, and a hooked (J-shaped) process on the posterolateral corner of the genital complex of the female are very unusual or unique features of the type-species that warrant recognition of this genus.

In addition to Oahu, Hawaii, *A. fallolunulus* is also known from New Caledonia in the South Pacific (Boxshall, 2008).

Alanlewisia fallolunulus (Lewis, 1967)
(figs. 14-16)

Lepeophtheirus? *fallolunulus* Lewis, 1967: 84.
Anuretes fallolunulus Ho & Lin, 2000: 232.

Material examined. — None. Description provided below is modified from Boxshall (2008).

Female. — Body as in fig. 14a. Total length (not including setae on caudal ramus) 2.90 mm (2.69-3.05 mm) (n = 10). Cephalothorax subcircular, with distinct posterior sinuses. Frontal plates produced into rounded expansions laterally on dorsal side (fig. 14c). Lunules absent, but frontal plates with flap of integument on ventral side (see fig. 3A of Boxshall, 2008). Cephalothoracic shield lacking cross groove between two longitudinal grooves (fig. 14a).

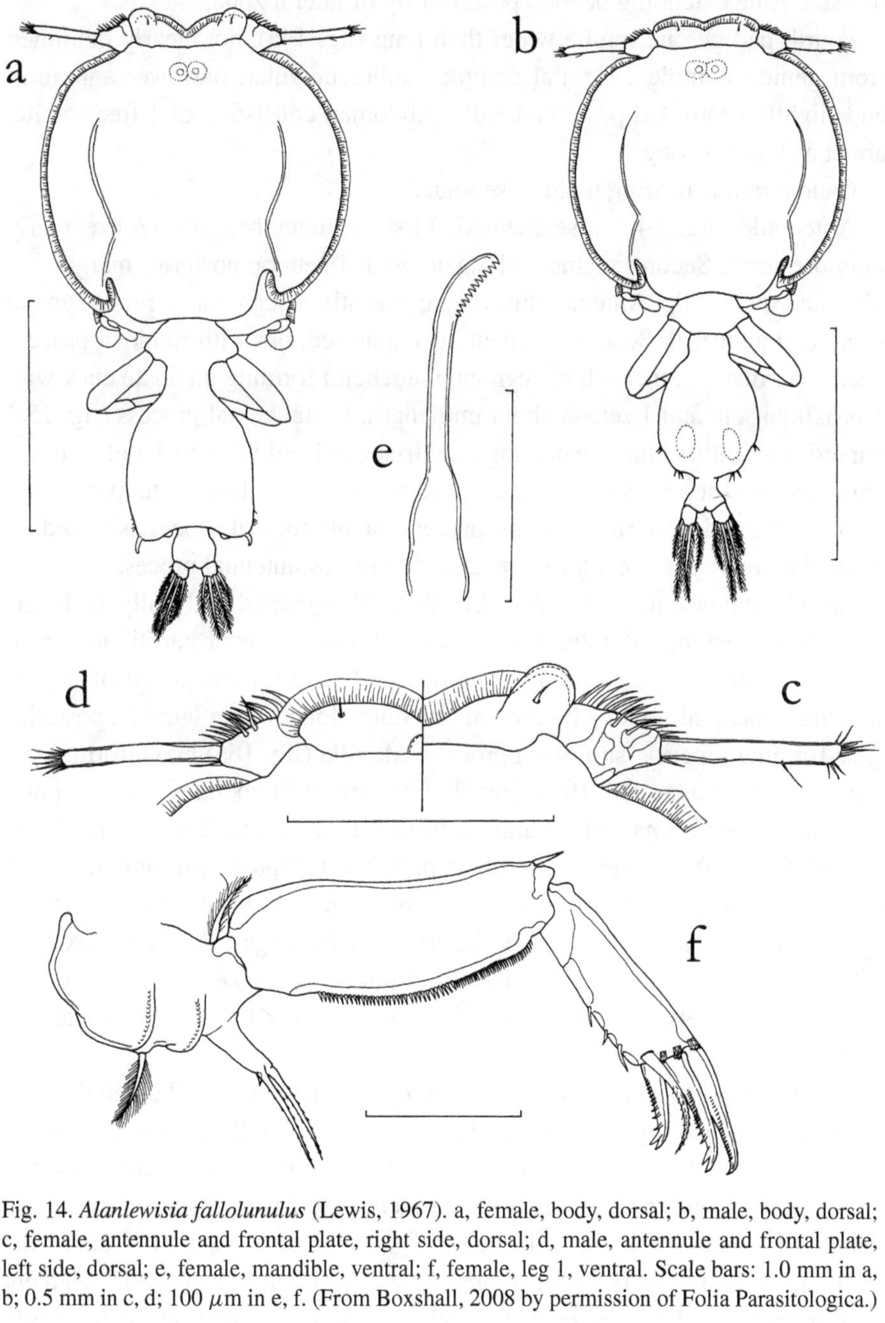

Fig. 14. *Alanlewisia fallolunulus* (Lewis, 1967). a, female, body, dorsal; b, male, body, dorsal; c, female, antennule and frontal plate, right side, dorsal; d, male, antennule and frontal plate, left side, dorsal; e, female, mandible, ventral; f, female, leg 1, ventral. Scale bars: 1.0 mm in a, b; 0.5 mm in c, d; 100 μm in e, f. (From Boxshall, 2008 by permission of Folia Parasitologica.)

Antennule extending beyond lateral limit of cephalothorax. Free margin of thoracic zone extending beyond posterior tip of lateral zone.

Fourth pedigerous somite wider than long (fig. 14a), not clearly delimited from genital complex. Genital complex subrectangular, narrower anteriorly and slightly protruded posterolaterally. Abdomen consisting of 1 free somite, about as wide as long.

Caudal ramus bearing 6 plumose setae.

Antennule (fig. 14c) 2-segmented. First segment bearing 27 irregularly plumose setae. Second segment elongate, with 1 seta on posterior margin and 12 naked setae at tip. Antenna (fig. 15a) apparently 3-segmented. First segment small and unarmed. Second segment also unarmed, but with rounded process near inner distal corner. Third segment (subchela) forming uncinate claw with 1 proximal seta and 1 seta at about midlength. Postantennal process (fig. 15a) curved, with rather blunt tip bearing fine striations in addition to 2 multisetulate papillae present on basal part of process and 1 multisetulate papilla on sternum located posterior to small process; small conical process situated on cephalothoracic surface adjacent and medial to postantennal process.

Mouth tube (see figs. 2A, 3A of Boxshall, 2008) tapering distally, with narrow apical opening. Mandible (fig. 14e) styliform, without partitions, armed with 12 teeth medially near apex. Maxillule (fig. 15a) comprised of papilla bearing 3 unequal setae, with medial one much longer than laterals; posterior lobe forming tapering spiniform process. Maxilla (fig. 16a) brachiform. Syncoxa (lacertus) unarmed. Basis (brachium) carrying flabellum near midpoint and tipped with canna and calamus; latter 2.5 times longer than former. Maxilliped (fig. 15b) with relatively robust protopod (corpus) bearing basal patch of corrugations on posterior surface. Subchela (shaft and claw) with 1 seta and rounded process at level of suture. Sternal furca (see figs. 2B, 5A of Boxshall, 2008) appearing double, with usual pair of tines and another pair of sharp processes immediately posterior to box of furca. Tines of furca simple and parallel to each other.

Leg 1 biramous; sympod unusual in being ornamented with 2 parallel ridges on ventral surface (fig. 14f). Exopod 2-segmented, with 3 vestigial inner setae on distal segment; spine 1 simple, spines 2 and 3 unipectinate and bifid distally; seta 4 adjacent to spine 3. Endopod slender, tipped with 2 sparsely plumose setae and minute spine (fig. 14f). Leg 2 (fig. 15c) biramous, with 3-segmented rami and armed as in typical caligids, except spine of first exopodal segment extremely large, flexed over distal segments, and equipped on outer margin with single row of conspicuous spinules. Apron of leg 3 (fig. 16b) bearing

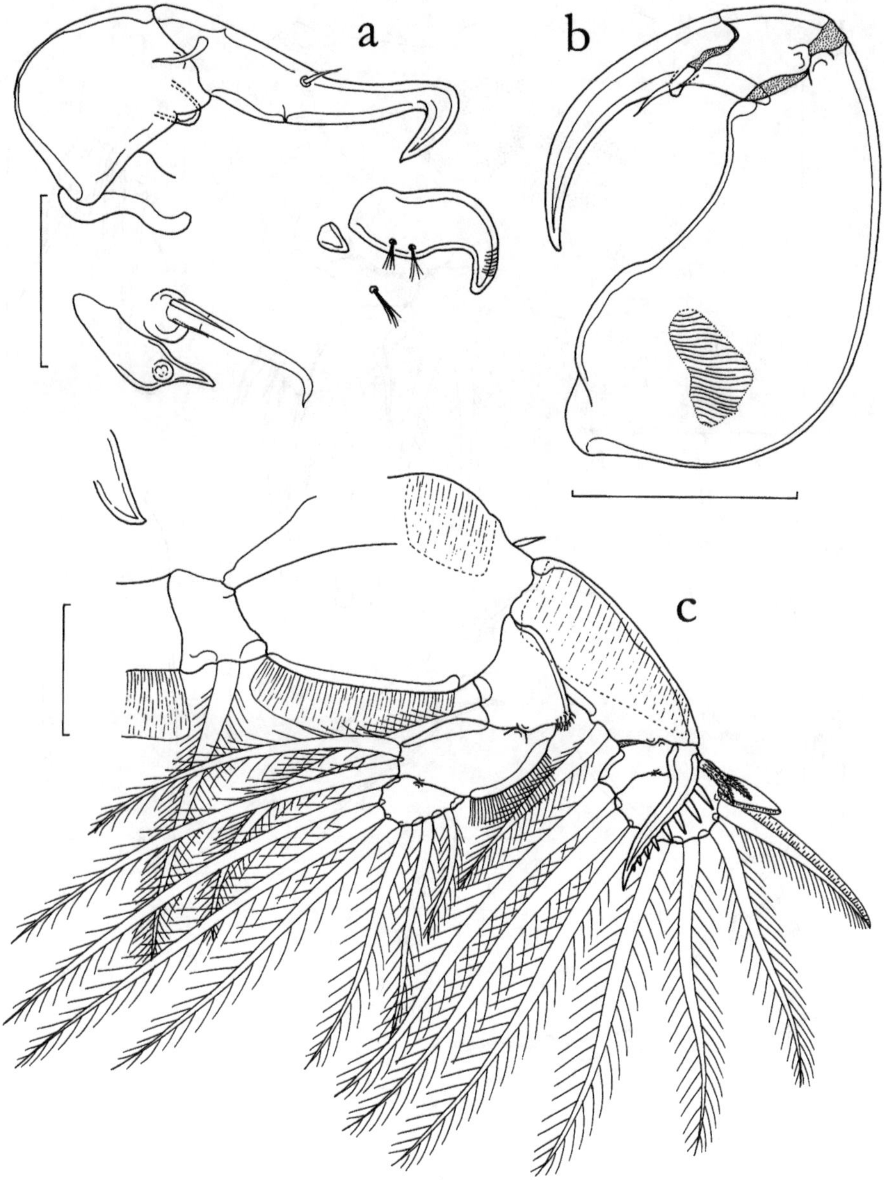

Fig. 15. *Alanlewisia fallolunulus* (Lewis, 1967), female. a, antenna, postantennal process, maxillule, and post-oral process, ventral; b, maxilliped, posterior; c, leg 2, ventral. Scale bars: 100 μm in a, c; 200 μm in b. (From Boxshall, 2008 by permission of Folia Parasitologica.)

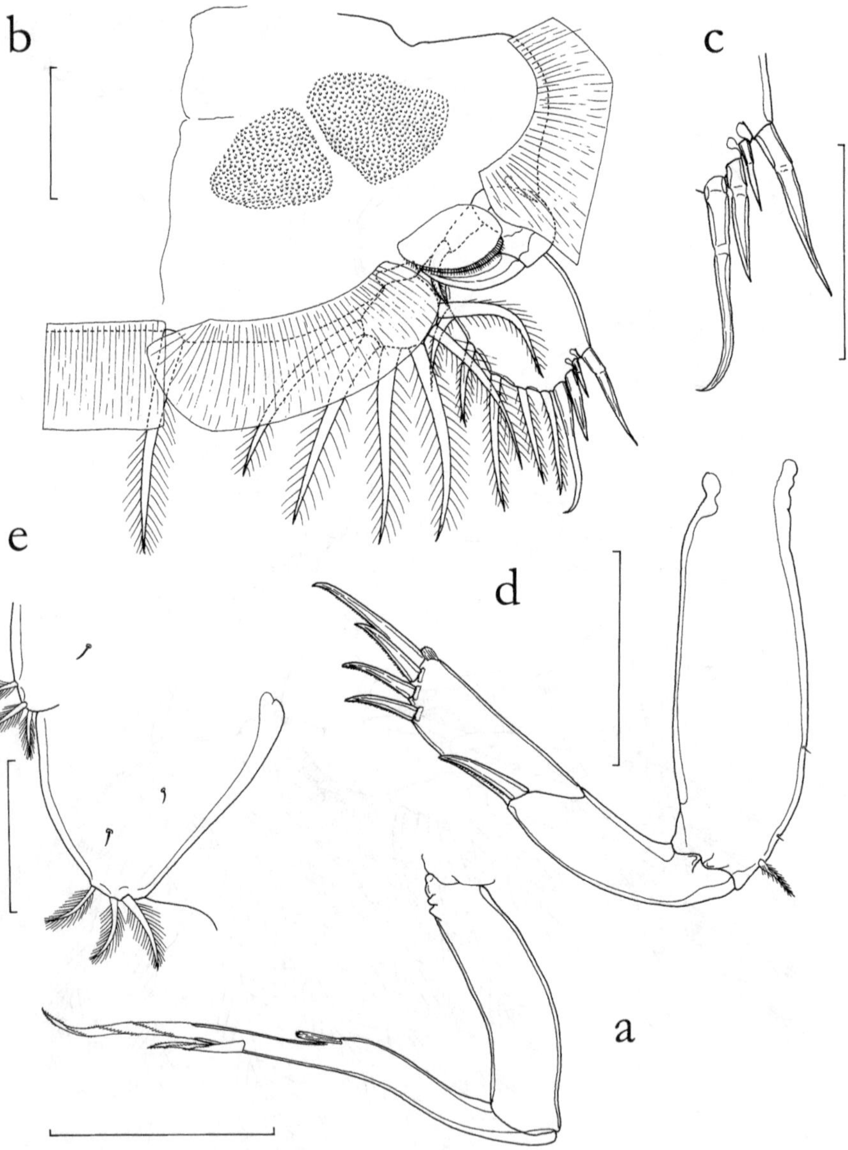

Fig. 16. *Alanlewisia fallolunulus* (Lewis, 1967). a, female, maxilla, ventral; b, female, leg 3, ventral; c, female, detail of exopodal spines on outer margin of leg 3, ventral; d, female, leg 4, ventral; e, male, legs 5 and 6, ventral. Scale bars: 200 μm in a, d; 100 μm in b, c, e. (From Boxshall, 2008 by permission of Folia Parasitologica.)

2 large patches of denticles on ventral surface in addition to usual marginal membranes and velum. Both rami 2-segmented, with exopod distinctly larger (broader) than endopod; having 4 spines on distal exopodal segment (fig. 16c) indicating incomplete fusion of second and third segments. Leg 4 exopod (fig. 16d) 2-segmented, with distal segment bearing 4 spines resulting from fusion of second and third segments. Leg 5 (see fig. 3C of Boxshall, 2008), located at posteroventral surface of genital complex, comprising isolated outer seta and 3 medial plumose setae. Curved J-shaped process present medial to leg 5 (see fig. 3C of Boxshall, 2008).

Male. — Body as in fig. 14b. Total length 2.45 mm (2.27-2.63 mm) (n = 3). Cephalothorax similar in shape to that of female. Frontal plates (fig. 14d) without rounded lateral expansion on dorsal side and lacking ventral integument flap.

Fourth pedigerous somite wider than long. Genital complex ovoid and longer than wide. Abdomen consisting of 1 free somite, about as wide as long.

Antenna (see fig. 5B of Boxshall, 2008) with large, weakly corrugated ridged pad on ventral surface of middle segment (basis). Terminal claw recurved, equipped with large digitiform accessory process and 2 naked setae. Postantennal process (see fig. 5C of Boxshall, 2008) longer than that of female, ornamented with fine surface striations.

Legs 1-4 as in female. Leg 5 (fig. 16e) positioned ventrolaterally on genital complex and armed with 3 plumose setae. Leg 6 (fig. 16e) a swelling also with 3 plumose setae.

Remarks. — The species description provided above was modified from Boxshall (2008). For a list of features that distinguishes *Alanlewisia fallolunulus* from all other species of caligids, see Remarks section for this genus.

Genus **Alebion** Krøyer, 1863

Alebion Krøyer, 1863: 239; Brady, 1883: 135; Bassett-Smith, 1899: 462; Wilson, 1907a: 702; Wilson, 1932: 418; Barnard, 1955a: 254; Vaissière, 1959: 535; Yamaguti, 1963: 99; Cressey, 1972: 1; Kabata, 1979: 200; Dojiri, 1983: 84; Boxshall & Montú, 1997: 66; Pillai, 1985: 501; Ho, 1998: 253; Kazachenko, 2001: 34; Boxshall & Halsey, 2004: 725.
Caligera Beneden, 1892: 258; Bassett-Smith, 1899: 462.

Female. — Cephalothorax suborbicular, with distinct posterior sinuses. Frontal plate distinct, without lunules. Fourth pedigerous somite small, free, and bearing 1 pair of dorsal aliform plates. Genital complex of various shapes (subrectangular, subtriangular, to irregular), with majority of species (except *A. glaber* Wilson, 1905) bearing elongate, spikelike projections representing leg 5. Abdomen with 2 free somites. First somite with lateral aliform processes, either short flaps or long digitiform processes. Caudal ramus with 6 setae.

Antennule 2-segmented. First segment robust, with numerous short setae. Second segment, very short, cylindrical, and armed with numerous setae. Postantennal process absent. Mouth tube long and slender, with intrabuccal stylet. Mandible comprising 4 sections, with third section extremely elongate, and bearing 12 teeth on inner margin of terminal section. Maxillule consisting of large rounded process and adjacent setiferous papilla. Maxilla brachiform. Maxilliped subchelate and prehensile. Corpus maxillipedis with patches of denticles or corrugated pads. Sternal furca absent.

Leg 1 biramous, with 2-segmented rami. Terminal exopodal segment with 3 spines (middle spine modified to broad, striated paddlelike structure), 1 seta on inner distal corner, and 3 inner plumose setae. Endopod with 3 plumose setae on terminal segment. Leg 2 biramous, with 3-segmented rami. Second and third exopodal spines modified as in middle terminal spine of leg 1. Leg 3 with large ventral apron, biramous with 3-segmented rami. Exopod with formula: I-1; I-1; II, 6. First segment of exopod without basal swelling, and exopodal spine of first segment not enlarged to clawlike structure. Next 3 exopodal spines modified as in legs 1 and 2. Endopod with formula: 0-0; 0-2; 4. First 2 segments of endopod with enlarged outer margins. Leg 4 reduced, either indistinctly 2-segmented (sympod and exopod) or a lobe. Leg 5 an elongate spikelike projection bearing spines and setae or a small setiferous papilla. Leg 6 absent.

Male. — Cephalothorax similar to that in female. Fourth pedigerous somite with dorsal aliform plates greatly reduced in size, not as prominent as in female. Genital complex suboval. Abdomen consisting of 2 free somites, without lateral processes. Antenna different from female, bearing corrugated adhesion pads on second segment and inner accessory process on terminal claw. Corpus maxillipedis with spinulose myxa. Second and third exopodal spines of leg 2 modified (frequently papillate) from those of female. Leg 5 consisting of setiferous papilla and adjacent seta. Leg 6 represented by setae.

Type-species. — *Alebion carchariae* Krøyer, 1863.

Remarks. — Cressey (1972) revised *Alebion* Krøyer, 1863, a genus whose members are parasitic predominantly on sharks and occasionally on rays. Cressey recognized eight species of this genus as valid. They are: *A. carchariae* Krøyer, 1863; *A. glaber* Wilson, 1905; *A. gracilis* Wilson, 1905; *A. crassus* Wilson, 1932; *A. maculatus* Wilson, 1932; *A. elegans* Capart, 1953; *A. lobatus* Cressey, 1970; and *A. pacificus* Cressey, 1972. A key to both the females and males of the species of *Alebion* is provided by Cressey (1972).

Alebion can easily be distinguished from all other caligid genera by the modified exopodal spines of legs 1-3. Other features that help differentiate this genus from others include the presence of the lateral abdominal processes in the female, the structure of the antennule (robust first segment and very short second segment), the broad rounded process of the maxillule, the 3-segmented endopod of leg 3 with lateral expansions on first and second segments, and a reduced leg 4. Both *A. carchariae* and *A. glaber* possess a corrugated area in the postantennal region. In addition, the dentiferous strigil found in all other caligid genera was not found in either of these two species examined. Unfortunately, it is not known if their congeners exhibit these two features. Consequently, they cannot be included yet as taxonomic characters of this genus. *Alebion* appears to be closely related to *Euryphorus* based on four selected characteristics (table II).

Alebion carchariae Krøyer, 1863
(figs. 17-23)

Alebion carchariae Krøyer, 1863: 239; Brady, 1883: 135; Bassett-Smith, 1898a: 366 (probably *Alebion maculatus* Wilson, 1932); Bassett-Smith, 1899: 462; Wilson, 1907a: 704, 708; Brian, 1908: 3; Brian, 1912: 11; Wilson, 1932: 422; Leigh-Sharpe, 1934: 27; Wilson, 1935a: 1; Bere, 1936: 593; Gnanamuthu, 1951: 1244; Capart, 1953: 655; Rose & Vaissière, 1953: 85; Barnard, 1955a: 254; Heegaard, 1955: 49; Vaissière, 1959: 535; Heegaard, 1962: 175; Yamaguti, 1963: 99; Cressey, 1970: 4; Cressey, 1972: 3; Dojiri, 1983: 87; Pillai, 1985: 512; Benz, 1984: 219; Boxshall & Montú, 1997: 66.

Alebion difficilis Beneden, 1892: 258; Wilson, 1907a: 703, 707; Capart, 1953: 658; Rose & Vaissière, 1953: 85; Yamaguti, 1963: 100.

Caligera difficilis Beneden, 1892: 258.

Alebion difficile Bassett-Smith, 1899: 462.

Alebion fuscus Wilson, 1921: 2; Wilson, 1932: 421; Carvalho, 1940: 275; Carvalho, 1951: 137; Vaissière, 1959: 538, 549; Yamaguti, 1963: 100.

Alebion gracilis Lewis, 1966b: 136; Cressey, 1967b: 5.

Material examined. — Holotype female (although museum label marked male) borrowed from Zoologisk Museum, Copenhagen, Denmark. Eleven females, 4 males, 1 immature female, and 1 immature male (USNM 149677) from *Carcharhinus* sp. from Indian Ocean (south of Madagascar), 33°13′00″S 43°51′00″E, at "Anton Bruun" station 381, on 30 August 1964. These specimens on loan from National Museum of Natural History, Smithsonian Institution, Washington, D.C.

Female. — Body as in fig. 17a. Total length (not including setae on caudal ramus) 9.49 mm (9.10-9.96 mm) ($n = 10$). Cephalothorax subcircular in outline, narrower anteriorly, longer than wide, 5.03 mm (4.81-5.41 mm) × 4.71 mm (4.42-5.15 mm). Posterior sinus deep, with conspicuous transparent membrane on inner edge. Frontal plate without lunules. Tip of antennule

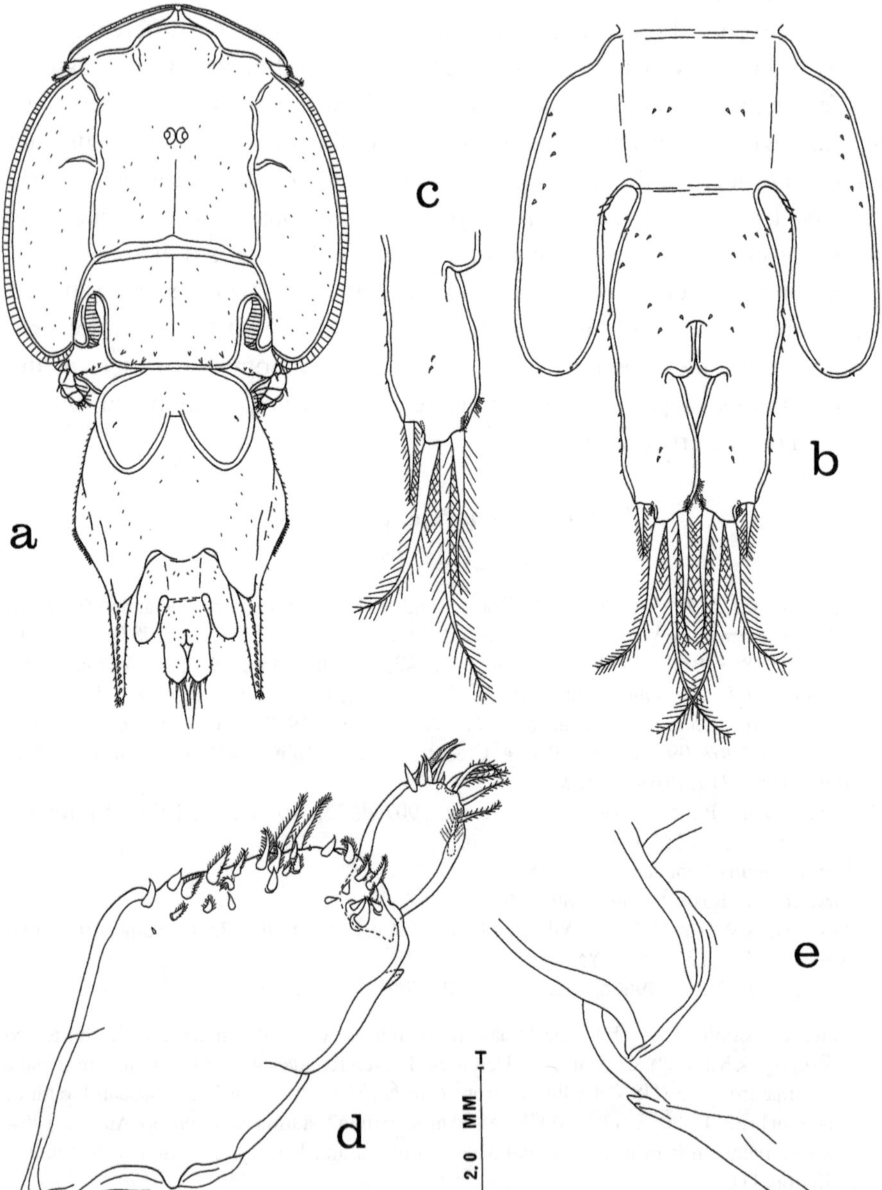

Fig. 17. *Alebion carchariae* Krøyer, 1863, female. a, body, dorsal (scale T); b, abdomen and caudal ramus, dorsal (H); c, caudal ramus, dorsal (F); d, antennule, ventral (G); e, junction of first and second segments of antennule, dorsal (D).

well within lateral limit of cephalothorax. Free margin of thoracic zone of cephalothorax extending slightly beyond posterior limit of lateral zone.

Fourth pedigerous somite width more than twice length, 0.51 mm (0.46-0.63 mm) × 1.22 mm (1.10-1.39 mm), and possessing pair of dorsal aliform plates. Dorsal plates large, longer than wide, 1.42 mm (1.29-1.54 mm) × 1.03 mm (1.00-1.10 mm). Genital complex wider than long, 2.19 mm (2.09-2.26 mm) × 3.01 mm (2.62-3.19 mm), widest near midlength, widest part with conspicuous spines along lateral margin, and bearing long spikelike leg 5 on posterolateral corner. Abdomen (fig. 17b) with 2 free somites. First somite 0.58 mm (0.50-0.63 mm) × 0.50 mm (0.40-0.60 mm), with large posteriorly directed lateral processes. These processes longer than wide, 1.28 mm (1.16-1.46 mm) × 0.34 mm (0.30-0.43 mm), and extending to posterior margin of anal somite. Second abdominal somite (anal somite) approximately as wide as long, 0.70 mm (0.63-0.73 mm) × 0.68 mm (0.63-0.73 mm).

Caudal ramus (fig. 17 b, c) fused to anal somite, 610 × 322 μm, with 4 relatively large plumose setae and 2 minute plumose setae.

Body covered with setules (sensilla) and spines as in fig. 17a.

Antennule (fig. 17d) 2-segmented. First segment robust, 380 μm, with 27 setae (17 plumose and 10 smooth) and 2 minute setae, and bifid process (fig. 17e) on posterodistal corner. Second segment very short, 189 μm, and bearing 6 plumose posterior setae (2 sharing common base), 7 smooth setae, and 1 aesthetasc. Antenna (fig. 18a) apparently 4-segmented. First segment unornamented. Second segment with posteriorly directed, rugose platelike process. Third segment robust and carrying dorsal adhesion pad. Fourth segment a curved claw (fig. 18b), with usual 2 setae. Postantennal area (fig. 18c) a corrugated, rounded protrusion with 3 groups of setules.

Mouth tube (fig. 18d) long and slender, 552 × 357 μm. Intrabuccal stylet (fig. 18e) present on labrum. Dentiferous strigil absent on labium (fig. 18f). Mandible (fig. 18g) comprising 4 sections, with extremely long third section. Sections 124, 51, 221, and 62 μm long from proximal to distal end. Fourth section with 12 teeth. Maxillule (fig. 18h) consisting of papilla tipped with 3 short setae, and quadrangular corrugated process. Corrugation with irregular pattern (fig. 18i). Maxilla (fig. 19a) brachiform and slender. Flabellum consisting of many long setules arranged in semicircular row near distal third of brachium. Calamus much longer than canna, and bearing 3 serrated membranes. Canna bilaterally serrate. Corpus maxillipedis (fig. 19b) moderately slender with 1 distal outer and 1 distal inner corrugated protrusions. Shaft and claw (fig. 19c) separated by suture in dorsal view, but

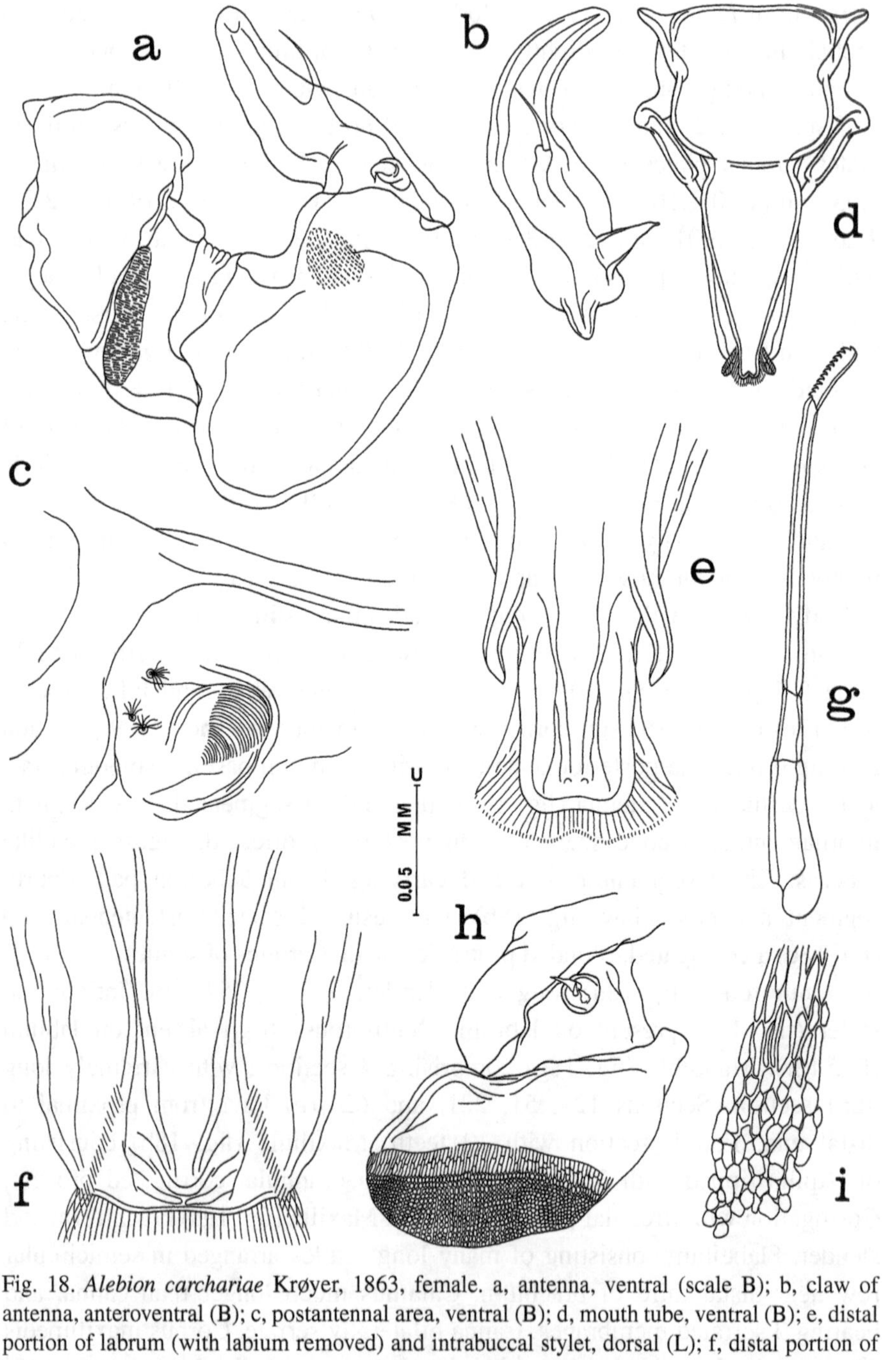

Fig. 18. *Alebion carchariae* Krøyer, 1863, female. a, antenna, ventral (scale B); b, claw of antenna, anteroventral (B); c, postantennal area, ventral (B); d, mouth tube, ventral (B); e, distal portion of labrum (with labium removed) and intrabuccal stylet, dorsal (L); f, distal portion of labium (with labrum removed), ventral (L); g, mandible, ventral (E); h, maxillule, ventral (I); i, surface feature of maxillule, ventral (U).

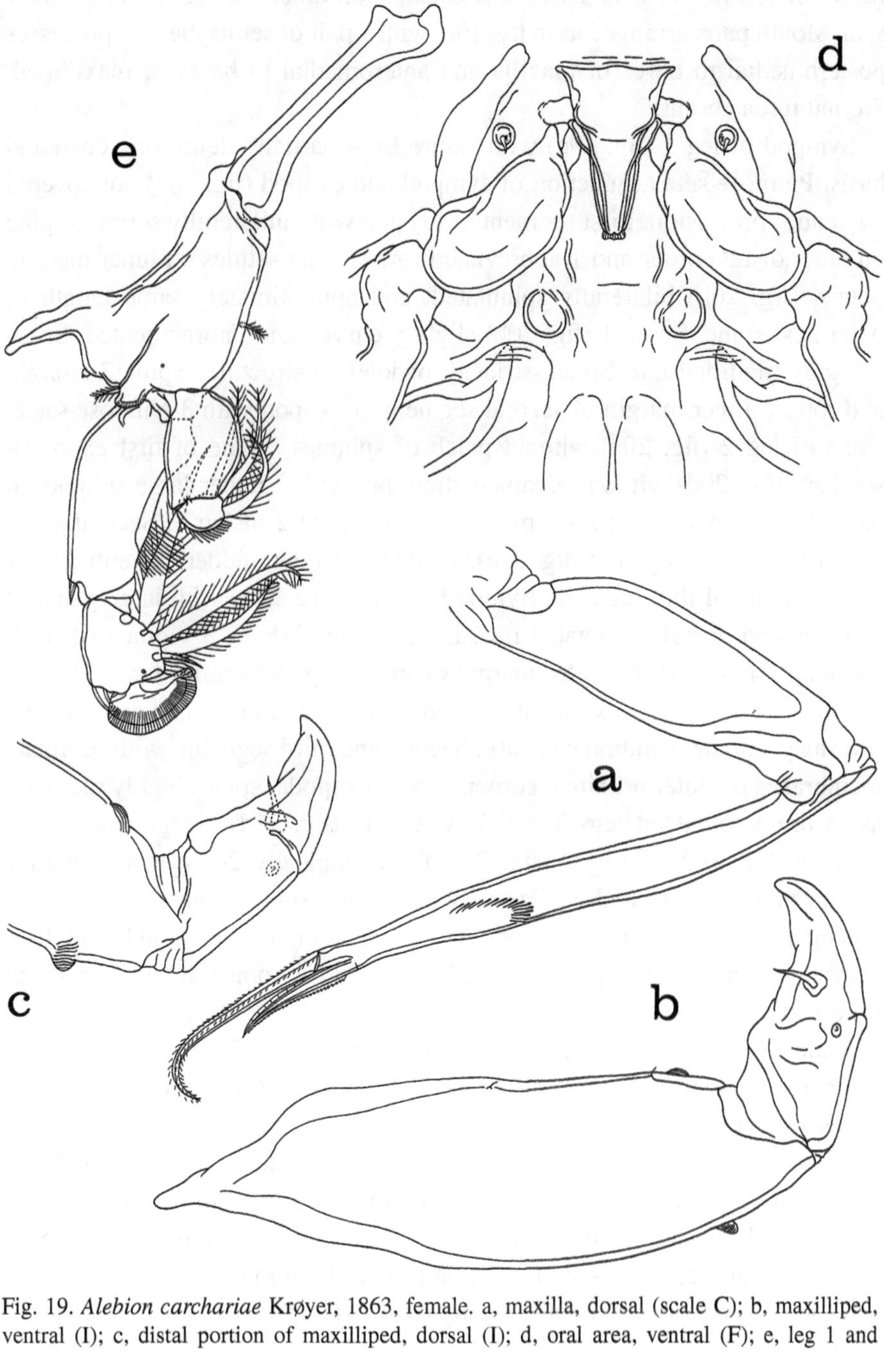

Fig. 19. *Alebion carchariae* Krøyer, 1863, female. a, maxilla, dorsal (scale C); b, maxilliped, ventral (I); c, distal portion of maxilliped, dorsal (I); d, oral area, ventral (F); e, leg 1 and intercoxal plate, ventral (C).

fused on ventral surface. These 2 segments combined 529 μm. Claw with 1 seta. Mouth parts arranged as in fig. 19d, with 1 pair of semispherical processes posteromedial to bases of maxilla and anteromedial to bases of maxilliped. Sternal furca absent.

Sympod of leg 1 (fig. 19e) with suture lines partially delimiting coxa and basis. Plumose seta at junction of sympod and exopod (fig. 20a) not covered by ventral protrusion. First segment of exopod with unilaterally serrated spine on outer distal corner and 1 abbreviated row of long setules on inner margin. Seta 4 (fig. 20b) bilaterally spinulated, and approximately same length as outermost spine. Spine 1 (fig. 20b) slightly curved and unornamented. Spine 2 highly modified into broad, striated, paddlelike structure. Spine 3 straight and robust. Inner margin of second segment of exopod with 3 plumose setae. Coxa of leg 2 (fig. 20c) without patch of spinules. Spine of first exopodal segment (fig. 20d) without ornamentation, but with 2 rather large spinules at base. Next 2 exopodal spines modified as in spine 2 of leg 1. Second spine of third exopodal segment (fig. 20e) small and slender, hidden in ventral view by first spine of this segment. Sympod of leg 3 (fig. 21a) with large patch of spinules and dorsal corrugated pad. Exopod (fig. 21b, c) 3-segmented, with formula: I-1; I-1; II, 6. Outer margin of first exopodal segment with 4 large spinules. Spine of first exopodal segment slender, attenuate, and smooth, and bearing pectinate membrane at base. Second and third segments with pectinate membranes on outer proximal corner. Next 3 exopodal spines highly modified as in those found on legs 1 and 2. Endopod (fig. 21d) 3-segmented, with formula: 0-0; 0-2; 4. Leg 4 (fig. 21e, f) incompletely 2-segmented, with 1 dorsal, plumose seta and 3 bilaterally spinulated spines. Leg 5 (fig. 21g, h) arising from posterolateral corner of genital complex, long and spikelike, bearing dorsal row of spinules, and 3 setae and 4 spines at tip. Leg 6 not observed.

Male. — Body as in fig. 22a. Total length 6.61 mm (5.74-7.10 mm) (n = 4). Cephalothorax similar to that of female, 3.39 mm (3.02-3.65 mm) × 3.07 mm (2.69-3.22 mm).

Fourth pedigerous somite 0.47 mm (0.33-0.53 mm) × 1.02 mm (0.93-1.10 mm) (including dorsal plates). Genital complex longer than wide, 1.33 mm (1.06-1.46 mm) × 1.02 mm (0.93-1.06 mm). Abdomen comprising 2 free somites. First somite 0.46 mm (0.43-0.46 mm) × 0.43 mm (0.40-0.46 mm). Second somite 0.55 mm (0.50-0.60 mm) × 0.42 mm (0.40-0.43 mm).

Caudal ramus similar to that of female, 432 × 227 μm.

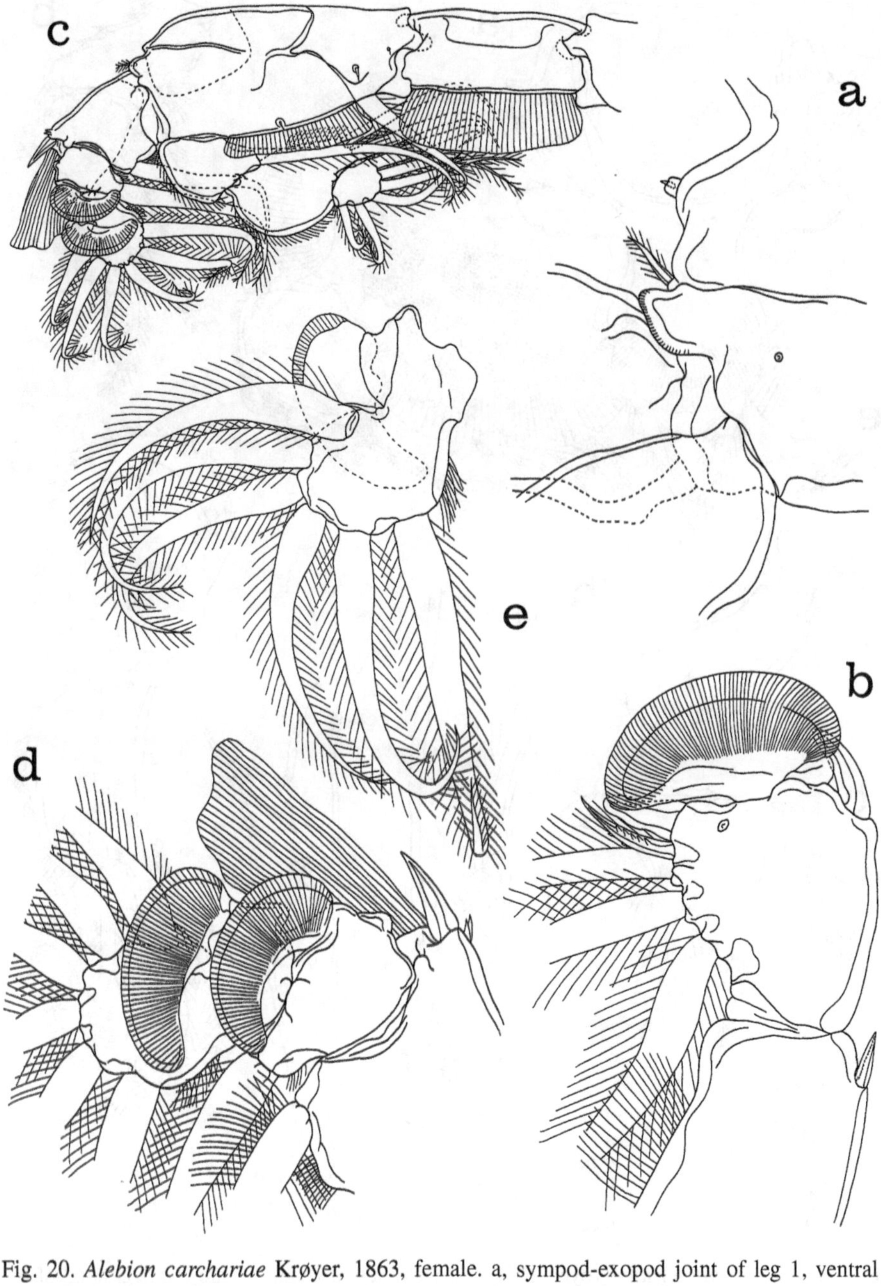

Fig. 20. *Alebion carchariae* Krøyer, 1863, female. a, sympod-exopod joint of leg 1, ventral (scale J); b, distal portion of exopod of leg 1, ventral (J); c, leg 2 and intercoxal plate, ventral (F); d, exopod of leg 2, ventral (G); e, terminal exopodal segment of leg 2, dorsal (G).

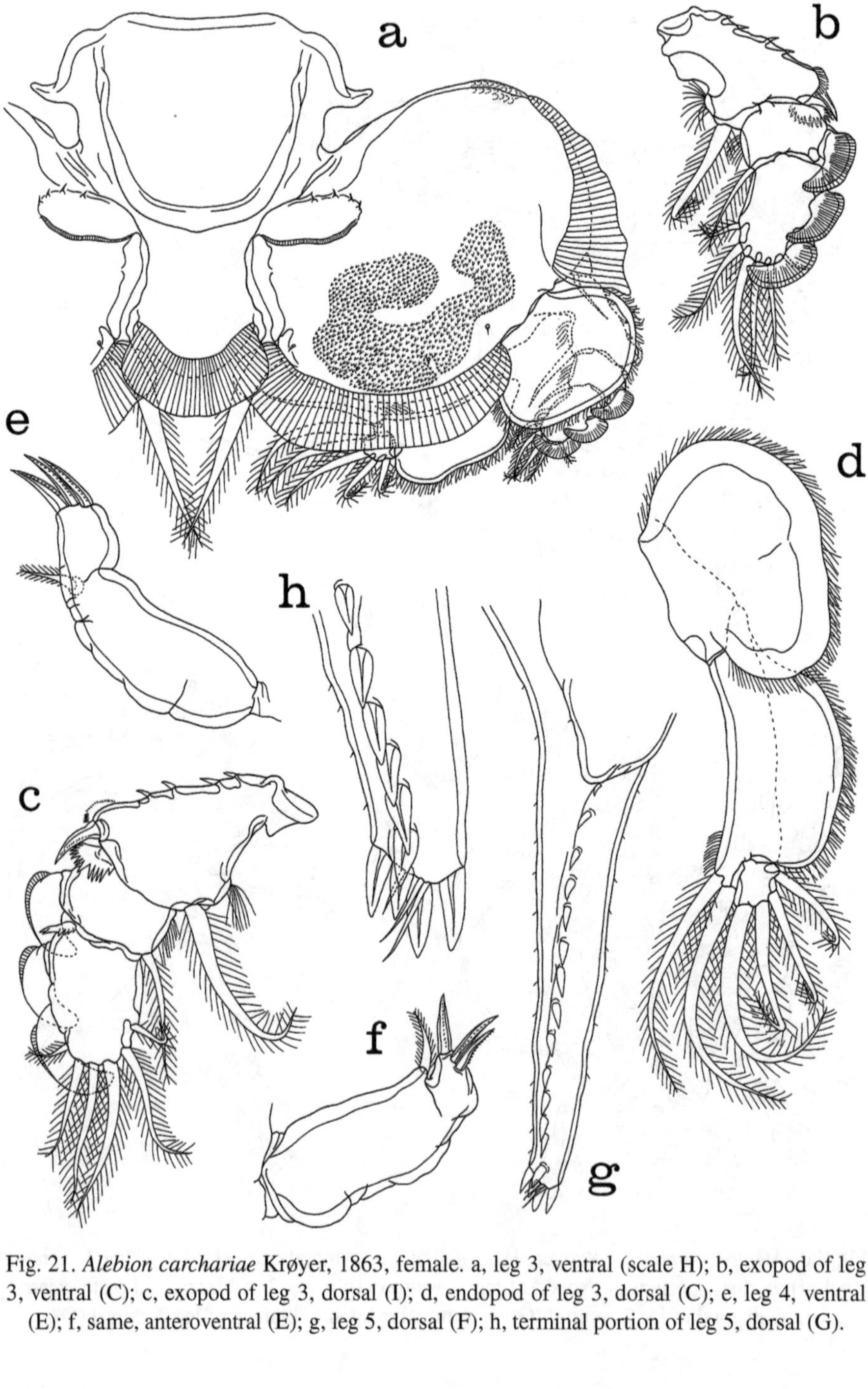

Fig. 21. *Alebion carchariae* Krøyer, 1863, female. a, leg 3, ventral (scale H); b, exopod of leg 3, ventral (C); c, exopod of leg 3, dorsal (I); d, endopod of leg 3, dorsal (C); e, leg 4, ventral (E); f, same, anteroventral (E); g, leg 5, dorsal (F); h, terminal portion of leg 5, dorsal (G).

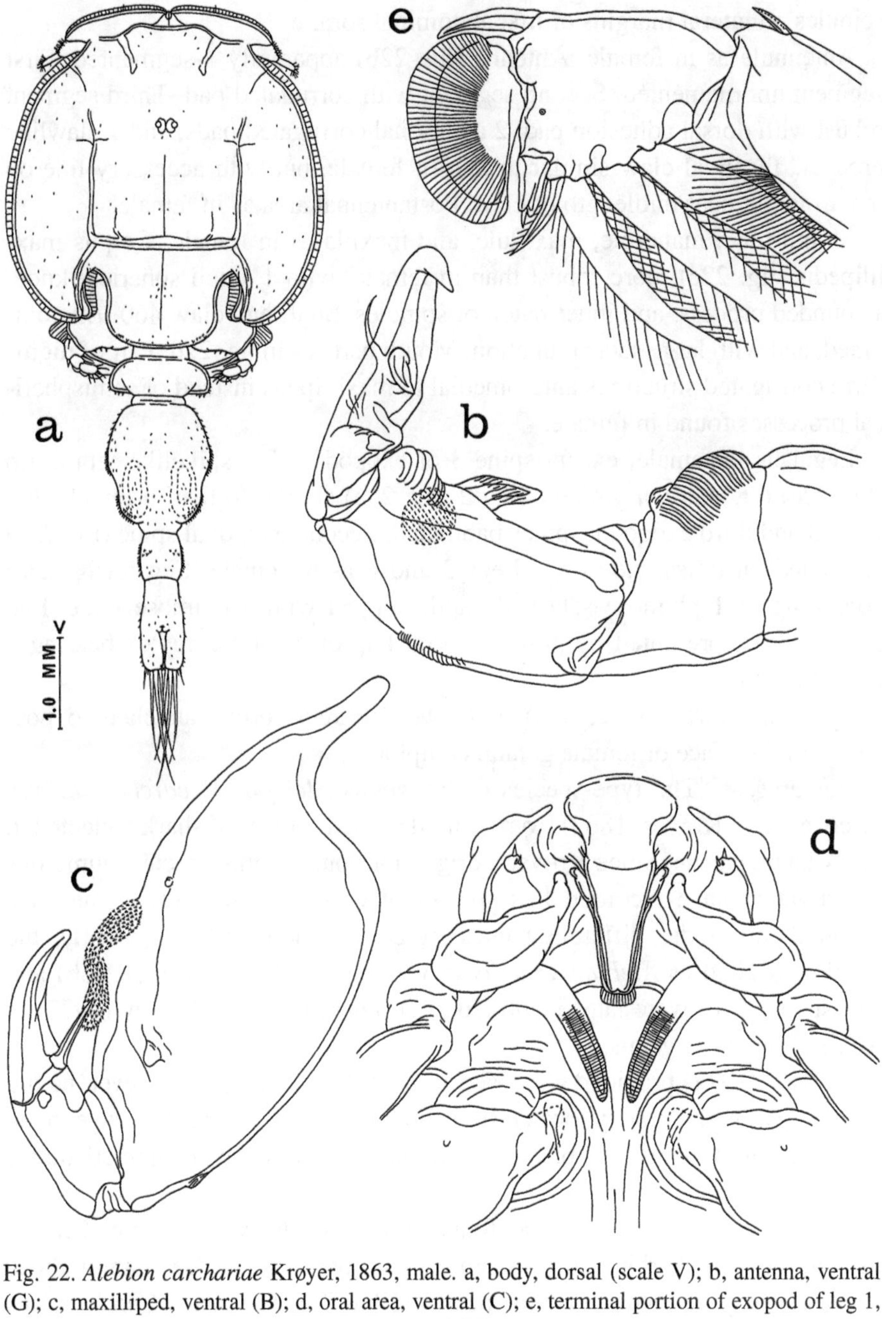

Fig. 22. *Alebion carchariae* Krøyer, 1863, male. a, body, dorsal (scale V); b, antenna, ventral (G); c, maxilliped, ventral (B); d, oral area, ventral (C); e, terminal portion of exopod of leg 1, ventral (E).

Body surface with small setules similar to that in female, except patches of spinules on lateral margins of first abdominal somite.

Antennule as in female. Antenna (fig. 22b) apparently 4-segmented. First segment unornamented. Second segment with corrugated pad. Third segment robust with dorsal adhesion pad, 2 additional corrugated pads, and 1 clawlike process. Terminal claw similar to that of female, but with accessory tine on inner margin near midlength of claw. Postantennal area as in female.

Mouth tube, mandible, maxillule, and maxilla as in female. Corpus maxillipedis (fig. 22c) more robust than in female, with 1 small spherical knob, 1 rounded process, and inner patch of spinules. Shaft and claw 460 μm combined, and with large seta at junction. Mouth parts as in fig. 22d. Pair of digitiform corrugated structures anteromedial to maxillipeds instead of semispherical processes found in female.

Leg 1 as in female, except spine 3 of exopod with a spikelike seta at tip (fig. 22e) (cf. *Alebion glaber*). Leg 2 (fig. 23a) similar to that in female, but second and third exopodal spines papillated. Second exopodal spine (fig. 23b) papillated on dorsal side only. Legs 3 and 4 as in female. Leg 5 (fig. 23c) consisting of 1 plumose seta and papilla tipped with 3 plumose setae. Leg 6 (fig. 23c) represented by posteroventral flap of genital complex bearing 3 plumose setae.

Spermatophore (fig. 23d, e) with two different forms, attached to posteroventral surface of female genital complex.

Remarks. — The type-species of the genus *Alebion*, *A. carchariae*, was discovered by Krøyer (1863) from an unidentified species of shark collected in the Atlantic Ocean. Since Krøyer's original account of this species, numerous investigators have recorded this species of copepod. However, some have reported this under different names (refer to synonymy). Apparently, the specimens cited as *Alebion gracilis* (a valid species) by Lewis (1966b) and Cressey (1967c) were later identified as *A. carchariae* by Cressey (1972) in his revision of the genus.

Cressey (1972) provided a redescription of *A. carchariae*. However, he did not figure a small dorsally located spine on the terminal segment of the exopod of leg 2. This spine is also present in *A. glaber*, and may be characteristic for all its congeners.

Alebion carchariae has been found almost exclusively on members of the families Carcharhinidae (requiem sharks) and Sphyrnidae (hammerhead sharks) with a predilection for the former (table IV). The genus *Carcharhinus* contains seven species that have been reported to harbor *A. carchariae*. Although primarily found on sharks, there is one report of this parasitic copepod

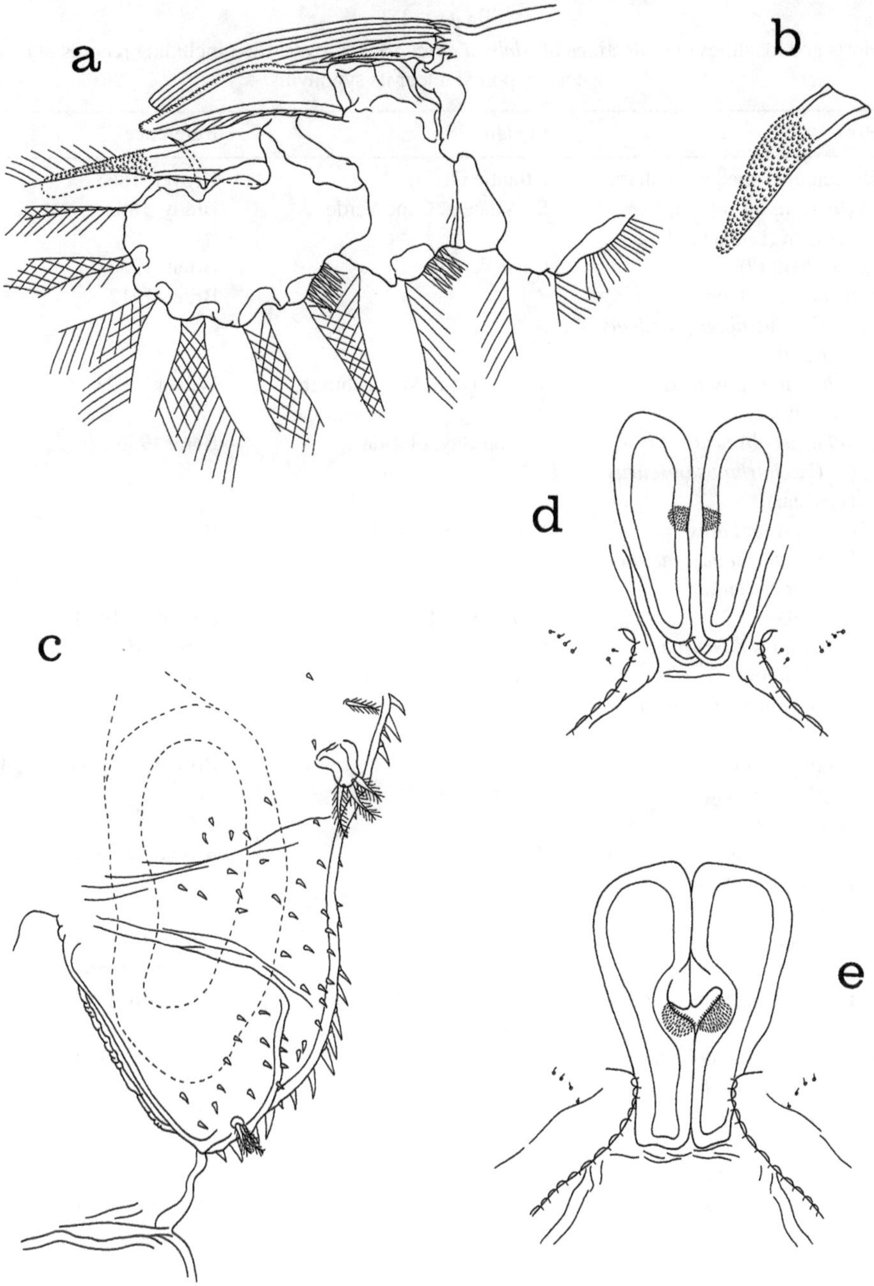

Fig. 23. *Alebion carchariae* Krøyer, 1863, male. a, exopod of leg 2, ventral (scale E); b, spine of second exopodal segment of leg 2, dorsal (E); c, legs 5 and 6, and genital area, ventral (B); d, spermatophore attached to genital region of female, ventral (F); e, same (different morphological form), ventral (F).

TABLE IV

Hosts and localities of collections of *Alebion carchariae* Krøyer, 1863 including records of this species reported under its synonyms

Host	Locality	Reference
Unidentified species of shark	Atlantic Ocean	Krøyer, 1863
Zygaena malleus [= *Sphyrna zygaena* (Linnaeus)]	St. Vincent, Cape Verde	Brady, 1883
Large shark (?)	Cape Verde	Brian, 1908
Carcharias milberti [= *Carcharhinus plumbeus* (Nardo)]	?	Brian, 1912
Carcharhinus obscurus (Lesueur)	Woods Hole, Massachusetts	Wilson, 1921
Carcharias obscurus [= *Carcharhinus obscurus* (Lesueur)]	Lemon Bay, Florida	Bere, 1936
Carcharias limbatus [= *Carcharhinus limbatus* (Müller & Henle)]	Lemon Bay, Florida	Bere, 1936
Not recorded	São Paulo, Brazil	Carvalho, 1940
Carcharhinus obscurus	Gorée, Senegal	Capart, 1953
Hypoprion brevirostris [= *Negaprion brevirostris* (Poey)]	Gorée, Senegal	Capart, 1953
Carcharias glaucus [= *Prionace glauca* (Linnaeus)]	?	Rose & Vaissière, 1953
Zygaena malleus [= *Sphyrna zygaena* (Linnaeus)]	?	Rose & Vaissière, 1953
Unidentified species of shark	Ifafa, Natal	Barnard, 1955a
Carcharhinus sp.	Durban	Barnard, 1955a
Unidentified species of shark	09°23′N 15°07′W (West Africa)	Heegaard, 1955
Carcharhinus maculipinnis (Poey) [= *Carcharhinus brevipinna* (Müller & Henle)]	Senegal	Vaissière, 1959
Coryphaena equiselis Linnaeus	Senegal	Vaissière, 1959
Sphyrna diplana Springer [= *Sphyrna lewini* (Griffith & Smith)]	Senegal	Vaissière, 1959
Sphyrna tudes (Valenciennes)	Senegal	Vaissière, 1959
Carcharhinus obscurus	Senegal	Vaissière, 1959
Galeocerdo cuvieri (Peron & Lesueur)	Peron Peninsula, Shark Bay, Western Australia	Heegaard, 1962

TABLE IV
(Continued)

Host	Locality	Reference
Unidentified species of shark	Oahu, Hawaii	Lewis, 1966b
Carcharhinus albimarginatus (Rüppell)	Station 400 (Indian Ocean)	Cressey, 1967b
Carcharhinus leucas (Valenciennes)	Station 409 (near Madagascar)	Cressey, 1967b
Carcharhinus sp.	Stations 381, 412 (Indian Ocean)	Cressey, 1967b
Galeocerdo cuvieri	Comores Island	Cressey, 1967b
Carcharhinus longimanus (Poey)	Station 291 (Indian Ocean)	Cressey, 1967b
Carcharhinus obscurus	Durban	Cressey, 1967b
Carcharhinus milberti (Valenciennes)	West coast of Florida	Cressey, 1970
Carcharhinus leucas	Near Sarasota	Cressey, 1970
Carcharhinus obscurus	Near Sarasota	Cressey, 1970
Carcharhinus limbatus	Near Sarasota	Cressey, 1970
Carcharhinus maculipinnis [= *Carcharhinus brevipinna* (Müller & Henle)]	Near Sarasota	Cressey, 1970
Negaprion brevirostris (Poey)	Near Sarasota	Cressey, 1970
Galeocerdo cuvieri	Near Sarasota	Cressey, 1970
Sphyrna mokarran (Rüppell)	Near Sarasota	Cressey, 1970
Sphyrna lewini (Griffith & Smith)	Near Sarasota	Cressey, 1970
Various species of sharks	Off Trivandrum, India	Pillai, 1985
Bignose shark, tiger shark, and scalloped hammerhead	Western North Atlantic	Benz, 1984
Unidentified species of shark	Brazil	Boxshall & Montú, 1997

from a pompano dolphin, *Coryphaena equisetis* Linnaeus, by Vaissière (1959). As noted by Cressey (1972), *A. carchariae* has been found in all major oceans, but has not yet been found in the polar regions or the south Pacific.

Alebion glaber Wilson, 1905
(figs. 24-28)

Alebion glabrum Wilson, 1905b: 129; Rathbun, 1905: 93.
Alebion glaber Wilson, 1907a: 708; Fowler, 1912: 481; Wilson, 1932: 419; Vaissière, 1959: 548; Yamaguti, 1963: 100; Cressey, 1972: 6; Dojiri, 1983: 93.

Material examined. — Four females (TC 2293) from skin of bronze whaler shark, *Carcharhinus greyi greyi* (Owen) caught at Marion Bay, South Australia on 1 January 1977. Five females (TC 2290) from tail of *Carcharhinus greyi greyi* caught at same locality on 29 January

1979. Both collections obtained by Dr. Wolfgang Zeidler and on loan from South Australian Museum.

Female. — Body as in fig. 24a. Total length (excluding setae on caudal ramus) 10.50 mm (9.79-11.79 mm) (n = 5). Cephalothorax subcircular, longer than wide, 4.83 mm (4.52-5.35 mm) × 4.64 mm (4.05-5.28 mm), with relatively deep posterior sinuses. Free margin of thoracic zone extending beyond tip of lateral zone and partially fused to fourth pedigerous somite. Tip of antennule well within lateral limit of cephalothorax.

Fourth pedigerous somite with pair of dorsolateral plates (alae or wings), 0.90 mm (0.83-1.00 mm) × 1.94 mm (1.83-2.16 mm) (including plates). Genital complex rectangular, with rounded corners, little longer than wide, 2.76 mm (2.49-3.12 mm) × 2.51 mm (2.32-2.82 mm). Abdomen presumably comprising 2 free somites. First somite with lateral expansions (alae), wider than long, 1.07 mm (0.93-1.29 mm) × 1.27 mm (1.13-1.49 mm) (including alae). Second somite longer than wide, 0.88 mm (0.83-1.00 mm) × 0.73 mm (0.66-0.83 mm).

Caudal ramus (fig. 24b) longer than wide, 460 × 242 μm, and bearing 4 large plumose setae and 2 smaller plumose setae.

Body with small setules (sensilla) as in fig. 24a.

Antennule (fig. 24c) 2-segmented. First segment robust, 403 μm long (measured along nonsetiferous margin), and carrying 27 setae (18 plumose, 9 smooth) and 1 minute knob. Second segment comparatively short, 161 μm long, bearing 6 plumose posterior setae (2 sharing common base), 7 smooth setae, and 1 aesthetasc. Antenna (fig. 24d) apparently 4-segmented. First segment unornamented. Second segment (fig. 24d, e) with posteriorly directed, corrugated platelike process. Third segment (fig. 25a) with dorsal corrugated pad. Claw (fig. 25b) curved, with usual 2 setae. Postantennal area (figs. 24d, 25c) a corrugated, raised area with 3 groups of setules.

Mouth tube (fig. 25d) long and slender, 2 times as long as wide, 697 × 334 μm. Near midlength of labium a small, anteromedially directed, conical projection. Intrabuccal stylet (fig. 25e) present near distal end of labrum. Labrum tipped with usual fringing membrane. Dentiferous strigil not present on labium (fig. 25f). Mandible (fig. 25g) comprising 4 sections, similar to *A. carchariae*. Sections 140, 70, 211, and 54 μm long from proximal to distal ends. Maxillule (fig. 25h) similar to *A. carchariae*, except quadrangular process carrying small irregularly shaped process on its distal, medial corner. Maxilla (fig. 25i) as in *A. carchariae*, but with calamus (fig. 25j) carrying small distally directed barb. Canna bilaterally serrated as in fig. 25j. Corpus

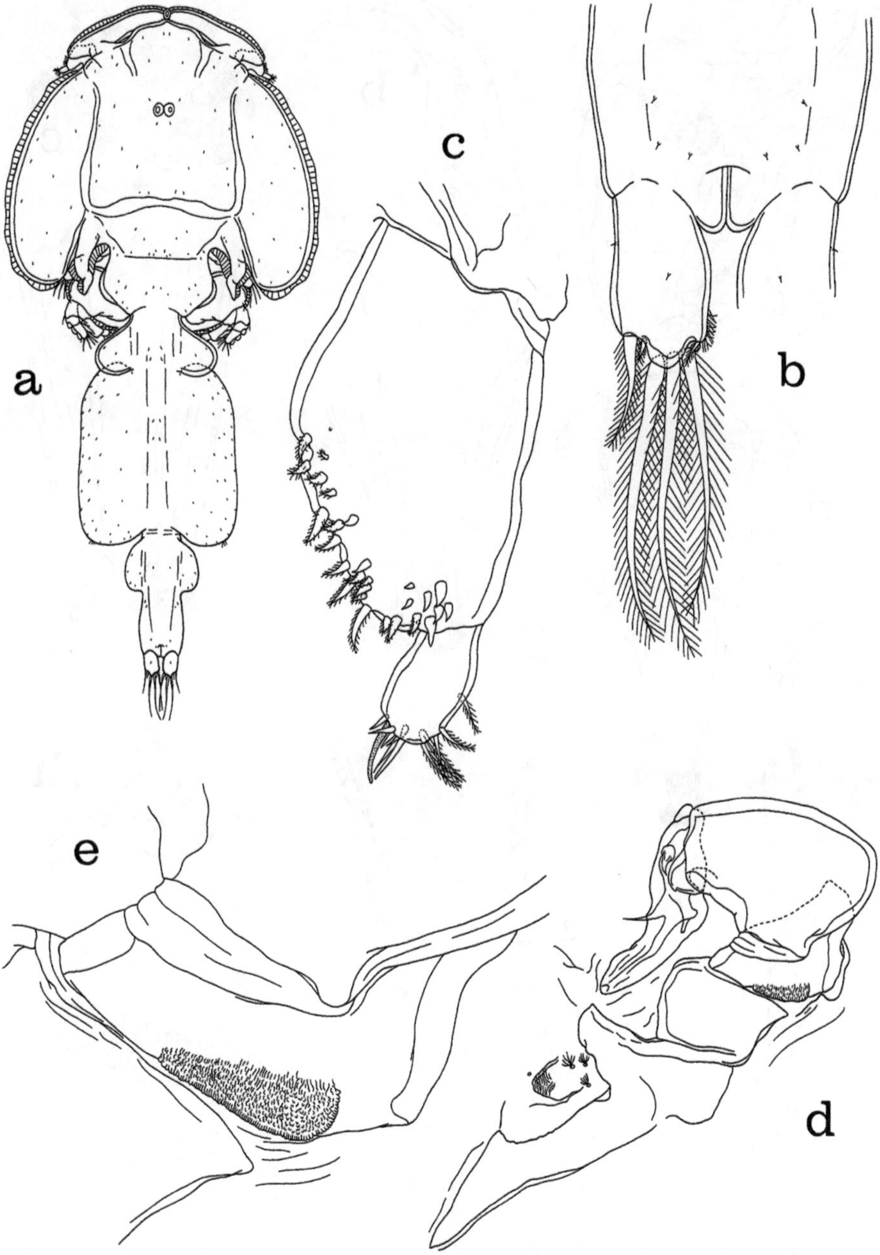

Fig. 24. *Alebion glaber* Wilson, 1905, female. a, body, dorsal (scale A); b, anal somite and caudal ramus, dorsal (C); c, antennule, ventral (G); d, antenna and postantennal area, ventral (C); e, base of antenna, ventral (E).

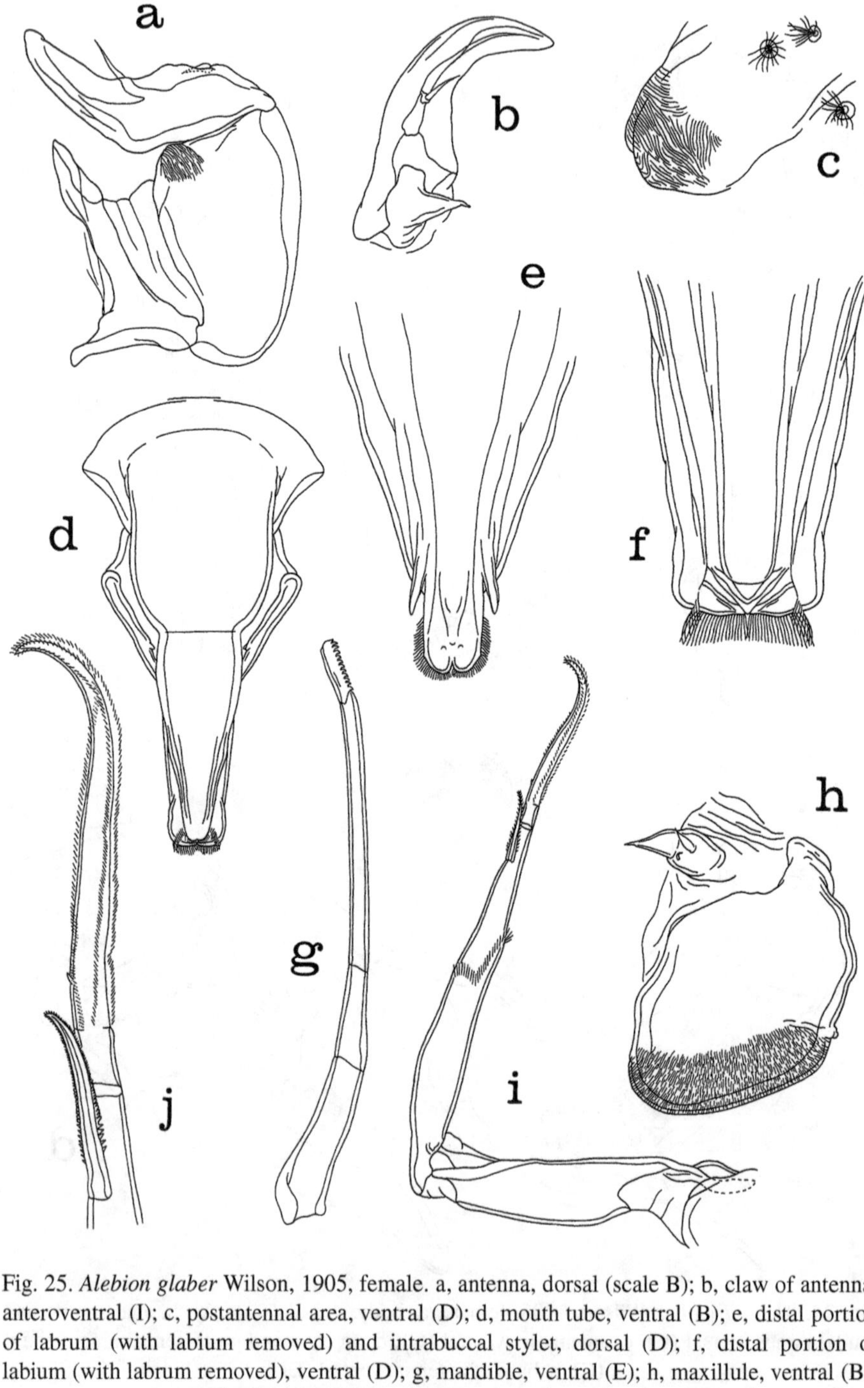

Fig. 25. *Alebion glaber* Wilson, 1905, female. a, antenna, dorsal (scale B); b, claw of antenna, anteroventral (I); c, postantennal area, ventral (D); d, mouth tube, ventral (B); e, distal portion of labrum (with labium removed) and intrabuccal stylet, dorsal (D); f, distal portion of labium (with labrum removed), ventral (D); g, mandible, ventral (E); h, maxillule, ventral (B); i, maxilla, dorsal (C); j, terminal portion of maxilla, dorsal (J).

maxillipedis (fig. 26a, b) slender, with 3 patches of denticles. Shaft and claw separated by suture in dorsal view, but fused on ventral surface. These 2 segments combined 506 μm long. Claw with 1 seta.

Sympod of leg 1 (fig. 26c) with suture lines partially delimiting coxa from basis. Plumose seta at junction of sympod and exopod (fig. 26d) not covered by ventral protrusion. First segment of exopod with unilaterally serrated spine on outer distal corner and row of long setules on inner margin. Pinnate seta 4 slightly shorter than outermost spine. Spines 1 and 2 (fig. 26e) as in *A. carchariae*. Spine 3 (fig. 26e-g) variable, but always with spikelike seta at tip. Coxa of leg 2 (fig. 27a) with patch of spinules near intercoxal plate. Spine of first segment of exopod (fig. 27b) short and with 1 row of spinules at its base and 1 along its axis. Second and third exopodal segments similar to *A. carchariae*. Second spine of third exopodal segment (fig. 27c) small and slender, hidden in ventral view by first spine of this segment. Sympod of leg 3 (fig. 27d) with 2 patches of spinules and dorsal corrugated pad. Exopod and endopod (fig. 28a) 3-segmented, with endopod overlapping exopod. Exopod (fig. 28b) and endopod (fig. 28c) similar to that in *A. carchariae*. Spine of first exopodal segment slender, attenuate, and bilaterally spinulated. Leg 4 (fig. 28d, e) an irregular lobe bearing 1 plumose seta, 2 bilaterally spinulated spines, and 1 unornamented spine. Leg 5 (fig. 28f, g) a lobe bearing 3 plumose setae. Leg 6 (fig. 28f) possibly represented by irregularly shaped lobe medial to leg 5, near origin of egg sacs.

Male. — As described by Cressey (1972).

Remarks. — A preliminary account of *Alebion glaber* was given by Wilson (1905b). It was not accompanied by illustrations at the time, but only a very superficial description. Believing that the generic name of this copepod was neuter in gender, he named the new species *Alebion glabrum*. However, Wilson (1907a) realized that the generic name is indeed masculine and consequently corrected the specific name to *glaber*. In the same paper, he described more fully and illustrated for the first time this species. Although Wilson's description lacks detail, it was sufficient for subsequent identification of this parasitic copepod.

Cressey (1972) revised the entire genus and provided a description of this species. As pointed out originally by Wilson and later by Cressey, the most conspicuous difference between *Alebion glaber* and all its seven congeners is the absence of the long spikelike leg 5 in the female. With *Alebion glaber* as the only described member of this genus without the long spikelike projections of the fifth leg, the argument similar to the *Lepeophtheirus-Dentigryps* conflict applies here also (see discussion on *Dentigryps*).

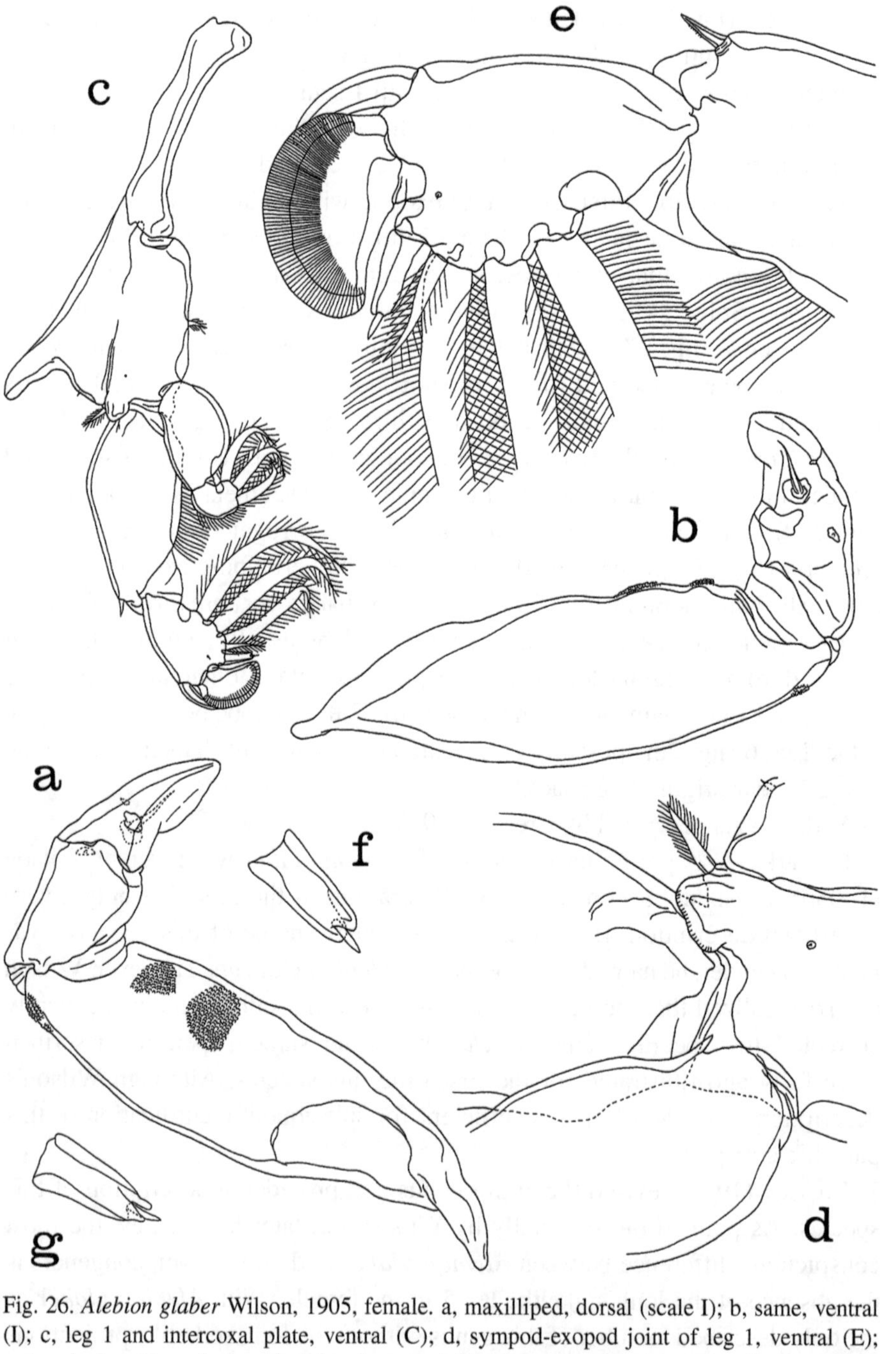

Fig. 26. *Alebion glaber* Wilson, 1905, female. a, maxilliped, dorsal (scale I); b, same, ventral (I); c, leg 1 and intercoxal plate, ventral (C); d, sympod-exopod joint of leg 1, ventral (E); e, terminal portion of exopod of leg 1, ventral (E); f, innermost spine (spine 3) of terminal exopodal segment of leg 1, ventral (D); g, same, ventral (D).

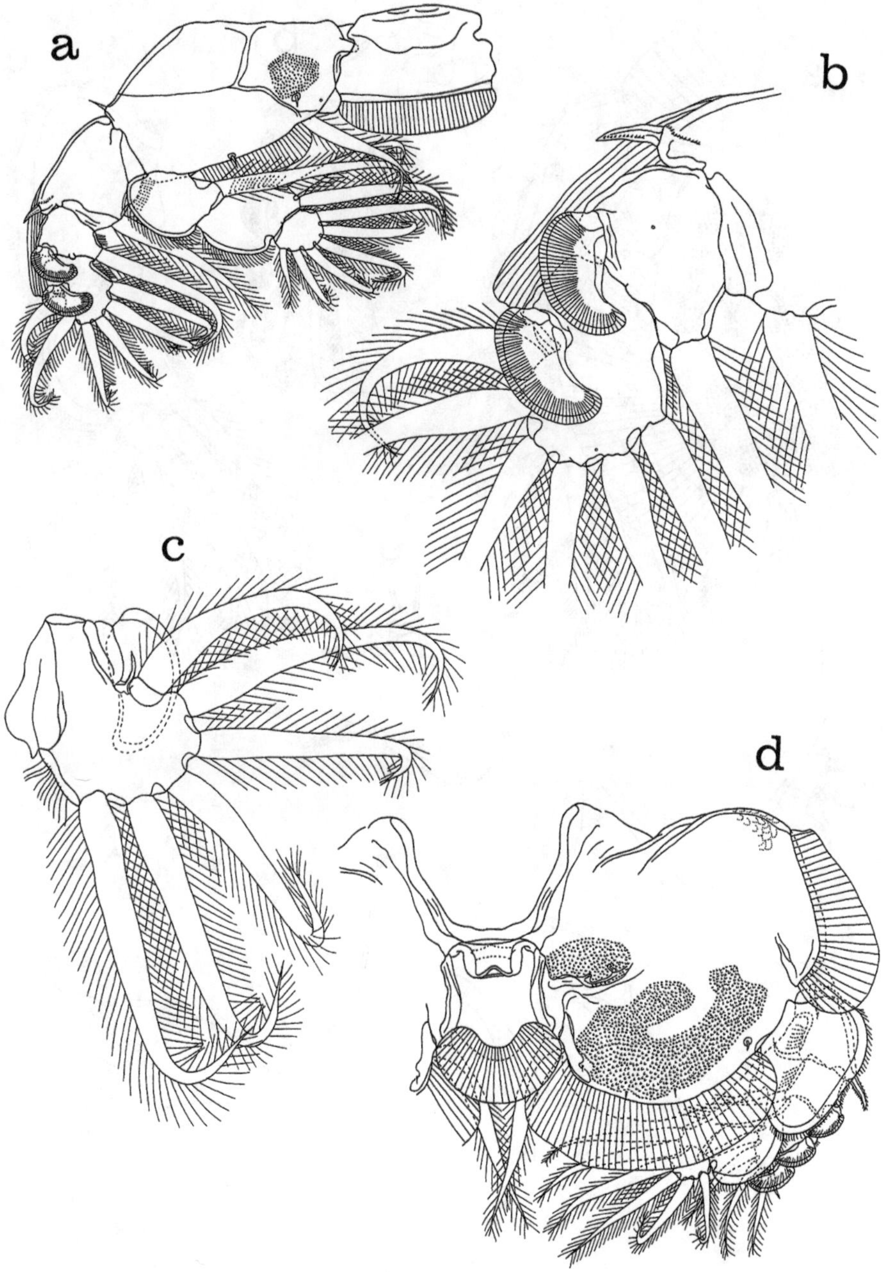

Fig. 27. *Alebion glaber* Wilson, 1905, female. a, leg 2 and intercoxal plate, ventral (scale F); b, exopod of leg 2, ventral (G); c, terminal exopodal segment of leg 2, dorsal (G); d, leg 3, ventral (H).

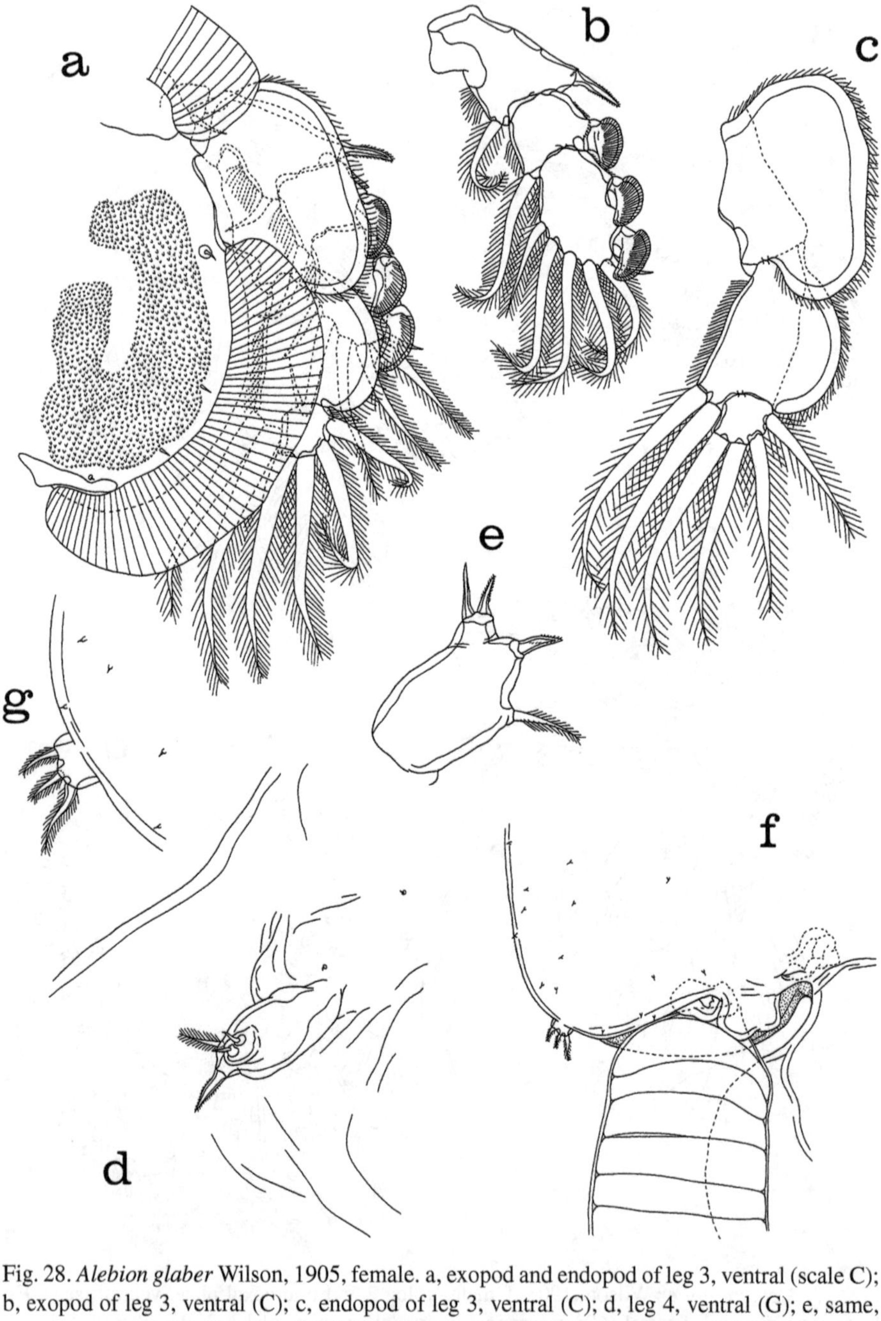

Fig. 28. *Alebion glaber* Wilson, 1905, female. a, exopod and endopod of leg 3, ventral (scale C); b, exopod of leg 3, ventral (C); c, endopod of leg 3, ventral (C); d, leg 4, ventral (G); e, same, posterior (J); f, leg 5 and genital area, ventral (F); g, leg 5, ventral (G).

This species is predominantly parasitic on small inshore species of sharks (Cressey, 1972), such as the smooth dogfish *Mustelus canis* (Mitchill) and the spiny dogfish *Squalus acanthias* Linnaeus, from the east coast of the United States. Wilson (1907a, 1932) also recorded the sand shark, barn-door skate, bonito, and the brown shark as hosts for this parasite.

Genus **Anchicaligus** Stebbing, 1900

Anchicaligus Stebbing, 1900: 667; Thomsen, 1949: 6; Kabata, 1979: 168; Dojiri, 1983: 101; Prabha, 1983: 50; Kazachenko, 2001: 22; Ho & Lin, 2002: 1369; Boxshall & Halsey, 2004: 725; Ho & Lin, 2004: 27.
Caligulina Heegaard, 1972: 303; Ho, 1980: 164; Kabata, 1979: 156.

Female. — Cephalothorax suborbicular, narrower anteriorly, and possessing sensory pit near posterolateral corner. Pair of large lenses present in median eyes. Posterior sinuses conspicuous, but shallow. Frontal plate large, with lunules. Fourth pedigerous somite free, without dorsal plates. Genital complex globular or somewhat triangular. Abdomen consisting of 1 free somite, short and broad, and with postanal protrusion. Caudal ramus small, with 6 setae.

Antennule 2-segmented. First segment stout, with many plumose setae. Second segment cylindrical, with usual 13 setae and 1 aesthetasc. Antenna apparently 4-segmented as in other caligids. Postantennal process present. Mouth tube with strigil and intrabuccal stylet. Mandible comprising 4 sections, with inner margin of terminal section dentiferous. Maxillule consisting of small process and 3 setae. Maxilla brachiform. Maxilliped subchelate. Sternal furca absent, but sternal area with posteriorly directed, sclerotized protrusions.

Leg 1 with 2-segmented exopod possessing 3 terminal spines, 1 inner distal seta, and 3 setae along inner margin. Endopod reduced to small lobe. Leg 2 biramous with 3-segmented rami. Leg 3 with large spine on first segment of exopod. Velum present. Exopod of leg 4 2-segmented, with total of 4 terminal spines. Leg 5 represented by 2 setiferous papillae.

Male. — Cephalothorax similar to that in female. Genital complex comparatively smaller. Antenna with corrugated adhesion pads on second segment and terminating in short claw. Postantennal process hamate. Legs 5 and 6 represented by protrusions bearing setae.

Type-species. — *Anchicaligus nautili* (Willey, 1896).

Remarks. — *Anchicaligus* is a monotypic genus containing the species *A. nautili*. Willey (1896) first mentioned this species as parasitizing *Nautilus* from New Britain (South Pacific). Then Stebbing (1900) described and illustrated this species under the binomen "*Anchicaligus nautili* (Willey)".

According to Ho (1980), the two most diagnostic features of this genus are the pair of large lenses in the eyes and the presence of postanal protrusions.

This genus is of particular interest because it contains the only caligid species known that infests an invertebrate. All other species of the Caligidae parasitize fish (elasmobranchs and teleosts). Another copepod family, Taeniacanthidae (Poecilostomatoida), includes members parasitic on fishes and invertebrates. According to Huys et al. (2012) 14 genera (*Anchistrotos* Brian, 1906; *Biacanthus* Tang & Izawa, 2005; *Caudacanthus* Tang & Johnston, 2005; *Cirracanthus* Dojiri & Cressey, 1987; *Metataeniacanthus* Pillai, 1963; *Nudisodalis* Dojiri & Cressey, 1987; *Parataeniacanthus* Yamaguti, 1939; *Pseudotaeniacanthus* Yamaguti & Yamasu, 1959; *Scolecicara* Ho, 1969; *Taeniacanthodes* Wilson, 1936; *Taeniacanthus* Sumpf, 1871; *Taeniastrotos* Cressey, 1969; *Tucca* Krøyer, 1837; and *Umazuracola* Ho, Ohtsuka & Nakadachi, 2006) are parasitic on fishes. However, members of three taeniacanthid genera (*Clavisodalis* Humes, 1970; *Echinirus* Humes & Cressey, 1961; and *Echinosocius* Humes & Cressey, 1961) are symbiotically associated with sea urchins (Echinoidea). Species associated with invertebrates are not morphologically very different from their counterparts parasitic on fishes (Dojiri & Humes, 1982; Humes & Dojiri, 1984). This is also the case with *Anchicaligus* and *Caligus*; in fact, the only major morphological difference between these genera is the presence of large lenses in the former and its absence in the latter.

Anchicaligus nautili (Willey, 1896)
(figs. 29-33)

Anchicaligus nautili Willey, 1896: 145; Stebbing, 1900: 668; Monod & Dollfus, 1932: 175; Capart, 1941: 177; Yamaguti, 1963: 47; Haven, 1972: 78; Ho, 1980: 157; Dojiri, 1983: 103.
Caligulina ocularis Heegaard, 1972: 304; Kabata, 1979: 164; Ho, 1980: 164.

Material examined. — Holotype male of "*Caligulina ocularis*" Heegaard on loan from Zoologisk Museum, Copenhagen, Denmark. Five females and 1 male *Anchicaligus nautili* from mantle chamber of *Nautilus pompilius* Linnaeus collected at Palau by Mr. Bruce A. Carlson and in collection of second author (J.-S. Ho).

Female. — Body as in fig. 29a, b. Total length (not including setae on caudal ramus) 4.16 mm (3.92-4.48 mm) (n = 5). Cephalothorax narrower anteriorly, 2.36 mm (2.22-2.49 mm) × 2.13 mm (1.96-2.26 mm), with sensory pit on posterolateral corner, and shallow posterior sinuses. Antennule not extending beyond lateral limit of cephalothorax. Lunules present. Pair of large lenses located in anterior region of cephalic area. Free margin of thoracic zone extending beyond posterior tip of lateral zone.

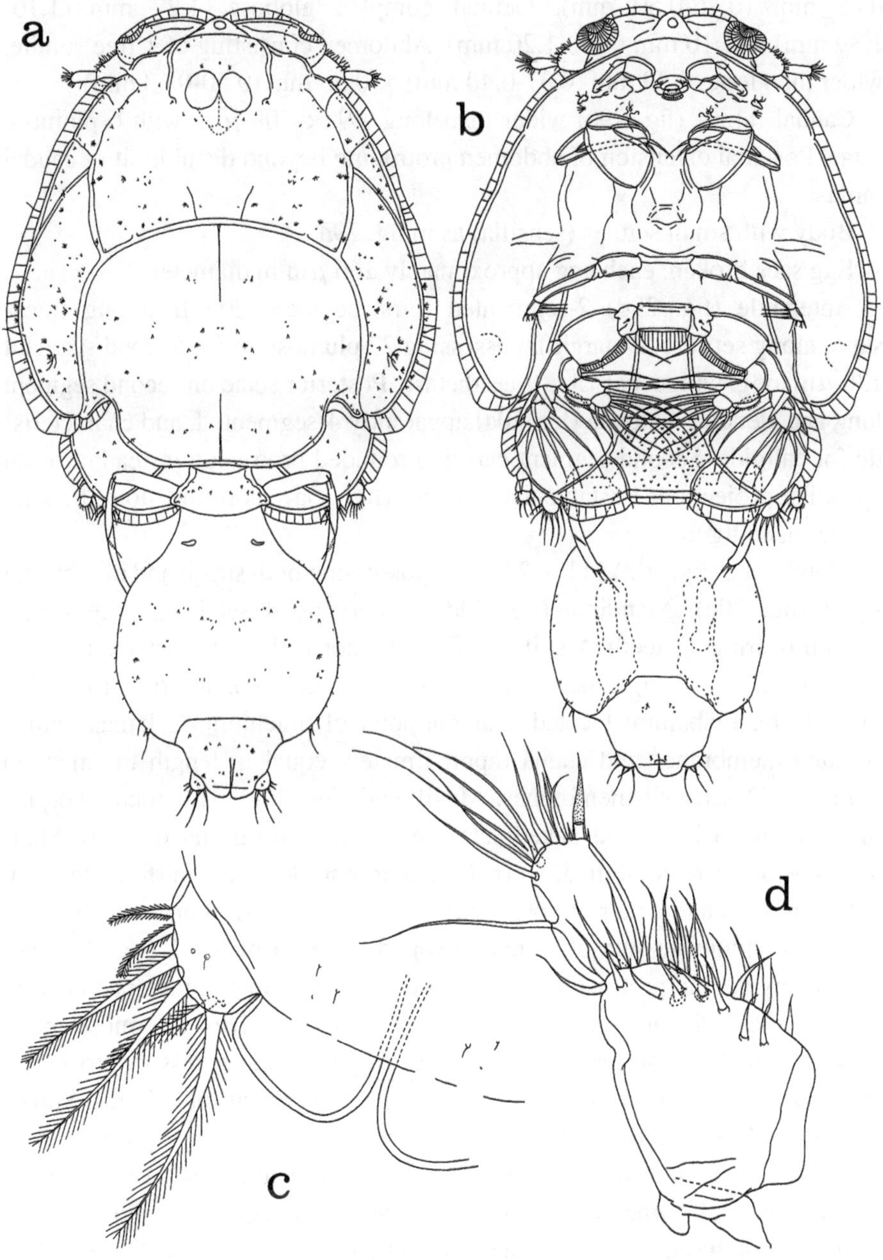

Fig. 29. *Anchicaligus nautili* (Willey, 1896), female. a, body, dorsal (scale M); b, same, ventral (M); c, caudal ramus, ventral (O); d, antennule (plumosities not drawn), ventral (O).

Fourth pedigerous somite wider than long, 0.24 mm (0.20-0.27 mm) × 0.48 mm (0.46-0.50 mm). Genital complex globose, 1.27 mm (1.10-1.39 mm) × 1.10 mm (1.00-1.20 mm). Abdomen consisting of 1 free somite, wider than long, 0.39 mm (0.37-0.40 mm) × 0.55 mm (0.50-0.63 mm).

Caudal ramus (fig. 29c) wider than long, 49 × 108 μm, with 6 plumose setae. Postanal extension of abdomen protruding beyond distal limit of caudal ramus.

Body with small setules (sensilla) as in fig. 29a.

Egg sacs broken, each egg approximately 266 μm in diameter.

Antennule (fig. 29d) 2-segmented. First segment 205 μm long (measured along setiferous margin), possessing 27 plumose setae. Second segment 130 μm long, with usual 13 + 1 aesthetasc. Posterior setae on second segment long and slender. Antenna (fig. 30a) apparently 4-segmented, and characteristic for caligids. Second segment carrying rounded protrusion instead of usual spinelike projection. Postantennal process (fig. 30a) short and stout, bearing setules as in figure.

Mouth tube (fig. 30b) 311 × 221 μm, possessing both strigil and intrabuccal stylet (fig. 30c). Mandible (fig. 30d) comprising 4 sections, with fourth section bearing 12 teeth. Maxillule (fig. 30e) consisting of 3 setae and small sclerotized dentiform process (not shown in fig. 30e). Maxilla (fig. 30f) brachiform. Flabellum located near midpoint of brachium. Calamus with 3 serrated membranes and canna (approximately equal in length to calamus) with only 2 serrated membranes. Maxilliped (fig. 30g) with robust corpus. Small rounded knob and naked seta present near distal end of shaft. Shaft and claw 380 μm combined. Sternal furca absent. Sternal area (fig. 30h) with cuticular folds and 1 pair of posteriorly directed, sclerotized protrusions.

Base of plumose seta at sympod-exopod junction of leg 1 (fig. 31a) not covered by ventral protrusion. Inner border of first segment of exopod bearing short setules. Pinnate seta 4 (fig. 31b) much longer than outermost spine. Spines 2 and 3 with serrated outer borders and possessing accessory processes. Endopod a rounded protrusion. Leg 2 (fig. 31c, d) typical for caligids. Spines of first and second exopodal segments serrated (fig. 31d). Apron of third leg (fig. 32a) as in other caligids, except usual dorsal adhesion pad located on anterolateral corner absent. Exopodal spine of leg 3 (fig. 32b) located terminally on basal swelling. Inner seta on first exopodal segment absent. Leg 4 (fig. 32c) very slender, with 2-segmented exopod; terminal spines decreasing in length from inner to outer margin. Exopod much longer than sympod. A minute spine located on lateral side of outermost spine of second segment

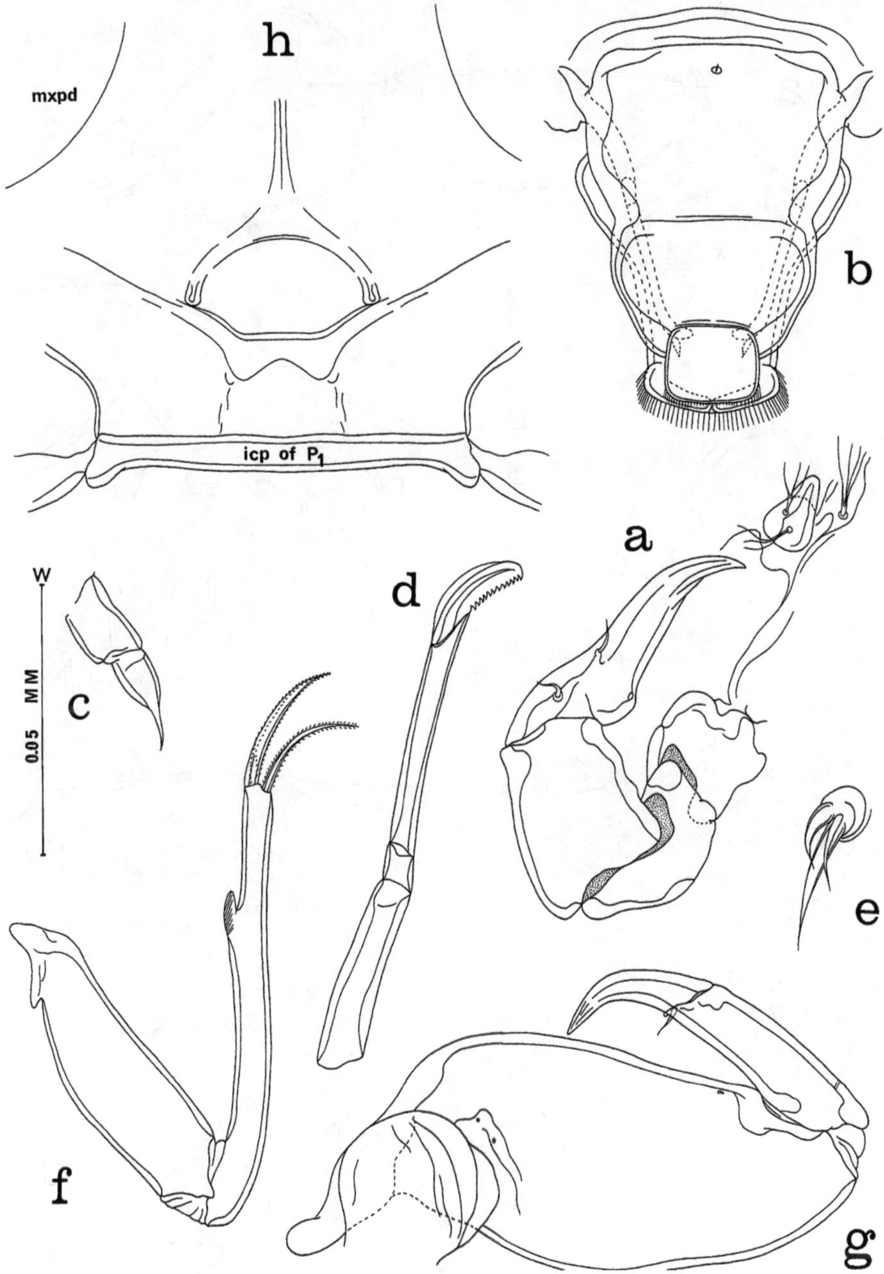

Fig. 30. *Anchicaligus nautili* (Willey, 1896), female. a, antenna and postantennal process, ventral (scale O); b, mouth tube, ventral (O); c, intrabuccal stylet, ventral (W); d, mandible, ventral (W); e, maxillule (small dentiform process not drawn), ventral (P); f, maxilla, dorsal (N); g, maxilliped, dorsal (N); h, area between maxillipeds and first pair of legs, ventral (I).

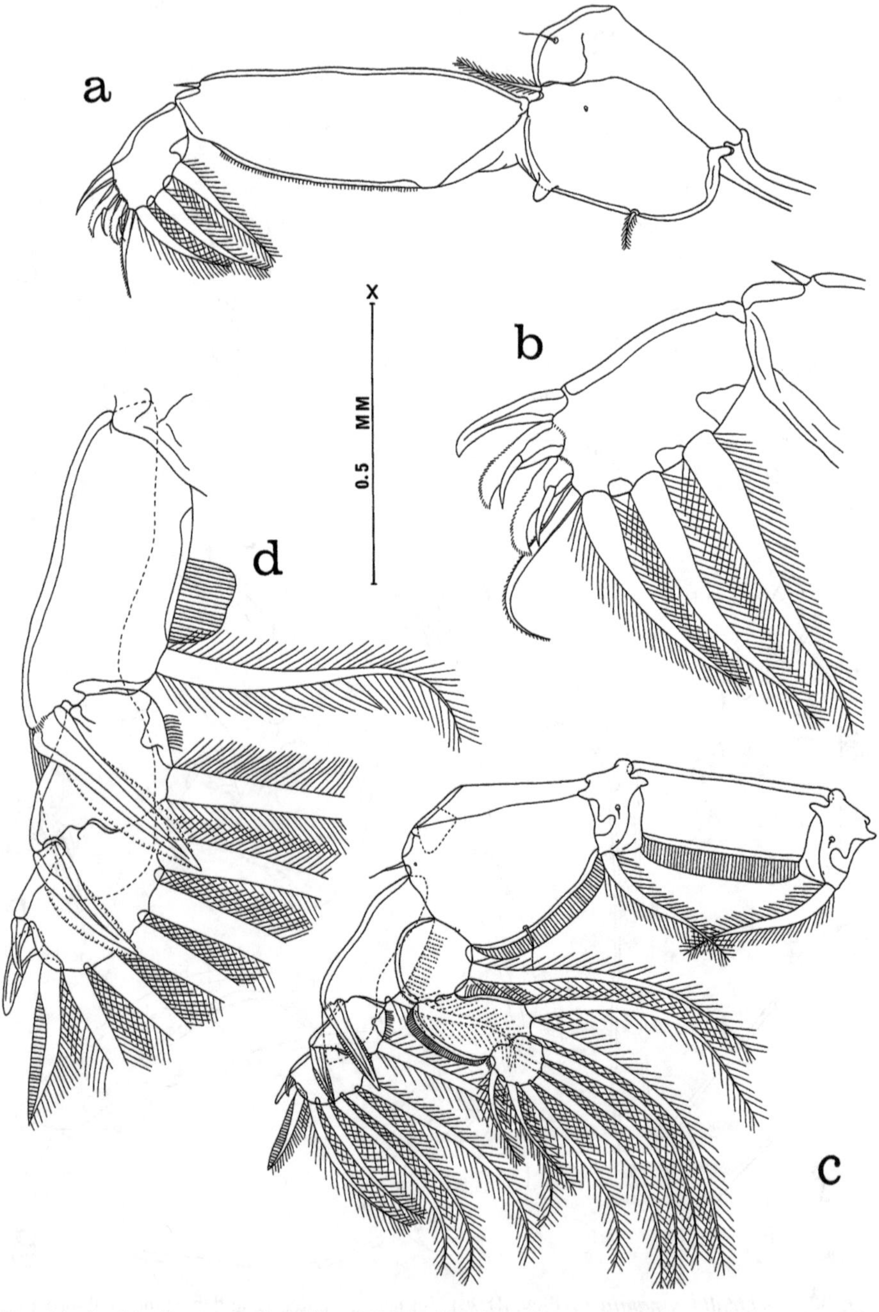

Fig. 31. *Anchicaligus nautili* (Willey, 1896), female. a, leg 1, ventral (scale N); b, terminal portion of exopod of leg 1, ventral (R); c, leg 2 and intercoxal plate, ventral (X); d, exopod of leg 2, ventral (O).

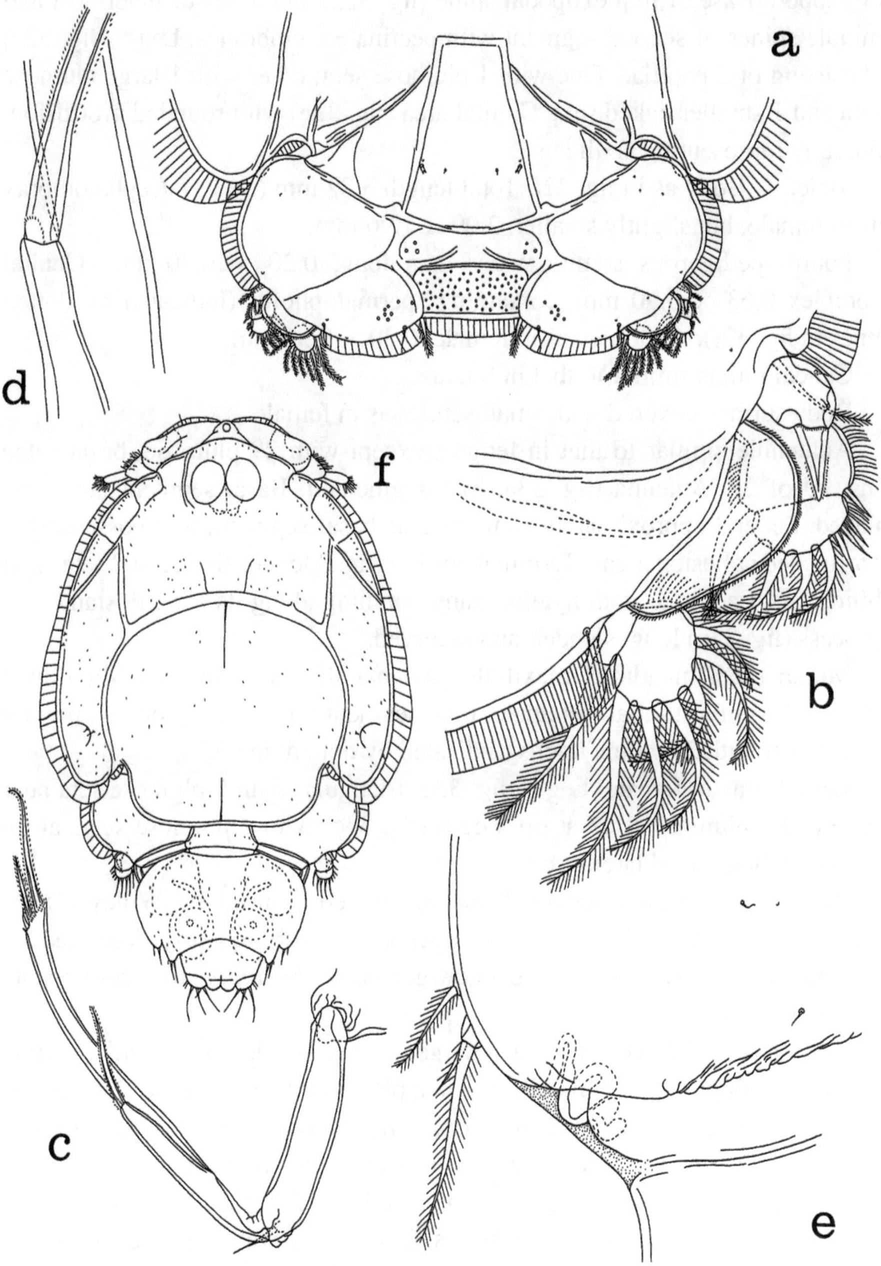

Fig. 32. *Anchicaligus nautili* (Willey, 1896). Female. a, leg 3, ventral (scale S); b, exopod and endopod of leg 3, ventral (O); c, leg 4, ventral (X); d, base of spine of first exopodal segment of leg 4, ventral (P); e, leg 5 and genital area, ventral (O). Male. f, body, dorsal (M).

of exopod. Base of first exopodal spine (fig. 32d) and bases of innermost and middle spines of second segment with pectinate membranes. Leg 5 (fig. 32e) consisting of 2 papillae. One with 1 plumose seta; other with 1 large plumose seta and 1 smaller naked seta. Genital area (fig. 32e) with rounded protrusion, possibly representing sixth leg.

Male. — Body as in fig. 32f. Total length 3.02 mm (n = 1). Cephalothorax as in female, but slightly smaller, 2.09 × 1.96 mm.

Fourth pedigerous somite wider than long, 0.20 × 0.40 mm. Genital complex 0.63 × 0.90 mm, containing spermatophores (indicated by dotted lines in fig. 32f). Abdomen as in female, 0.30 × 0.50 mm.

Caudal ramus similar to that in female.

Body surface covered with small setules as in female.

Antennule similar to that in female, except with 29 plumose, basal setae instead of 27. Antenna (fig. 33a, b) 3-segmented. Basal segment large and naked. Second approximately same size as basal segment, and possessing 2 corrugated adhesion areas. Terminal portion (fig. 33c) consisting of 2 setae and bifid claw. Each tine with hyaline flange running along its axis. Postantennal process (fig. 33a) long, slender, and recurved.

Mouth tube, mandible, maxillule, and maxilla as in female. Maxilliped (fig. 33d) similar to that in female. First segment with small process on inner side. Shaft with 2 hyaline knobs (indicated by dotted lines).

Legs 1-4 as in female. Leg 5 (fig. 33e) triangular with 1 plumose seta near base and 2 plumose setae at tip. Leg 6 (fig. 33e) with 2 plumose setae at tip and a smaller, naked lateral seta.

Remarks. — First discovered and mentioned in print by Willey (1896) as *Caligus nautili*, this species of copepod was subsequently described by Stebbing (1900) who established a new genus, *Anchicaligus*, to accommodate Willey's species.

Heegaard (1972) described a new genus and species of caligid copepod based on a single specimen collected in a plankton net during the Dana Expedition from 1928-1930. This new taxon, *Caligulina ocularis*, was, according to Heegaard, distinctive in possessing large lenses and an unusual segmentation and armature of leg 3 (among other characteristics listed).

The holotype (male) of Heegaard's *C. ocularis* was borrowed from the Zoologisk Museum, Copenhagen, Denmark. Based on a detailed comparison of this specimen with *Anchicaligus nautili*, *C. ocularis* must be placed into synonymy with *A. nautili* as suspected by Ho (1980). According to Heegaard (1972), the diagnostic features of leg 3 of *C. ocularis* were the absence of the

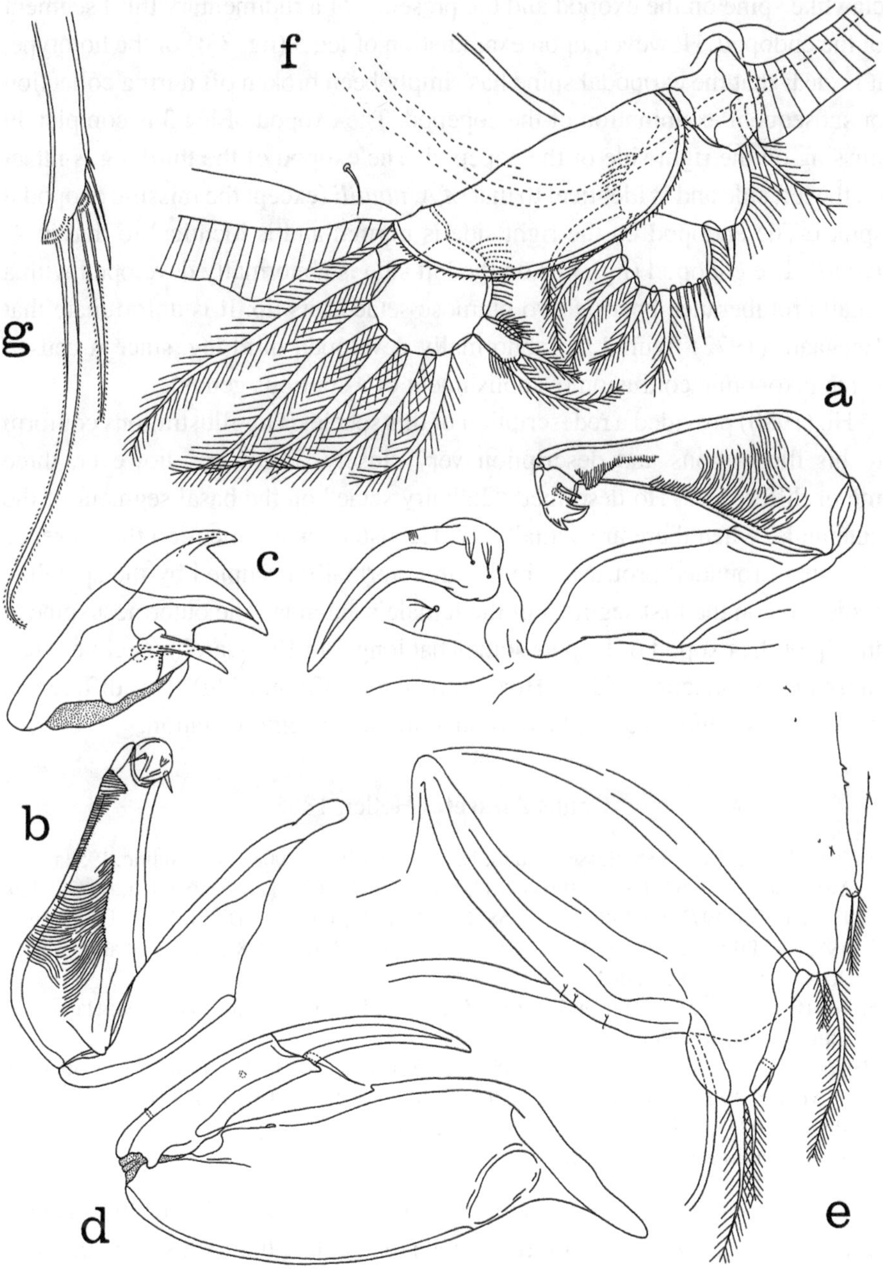

Fig. 33. *Anchicaligus nautili* (Willey, 1896), male. a, antenna and postantennal process, ventral (scale O); b, antenna, inner (O); c, claw of antenna, ventrolateral (Q); d, maxilliped, dorsal (N); e, legs 5 and 6, and genital area, ventral (O); f, exopod and endopod of leg 3, ventral (R); g, terminal portion of exopod of leg 4, ventral (R).

clawlike spine on the exopod and the presence of a rudimentary third segment of the endopod. However, upon examination of leg 3 (fig. 33f) of the holotype, it is clear that the exopodal spine has simply been broken off during collection or subsequent examination of the copepod. The exopod of leg 3 is completely missing on the right side of the copepod. The exopod of the third leg is intact on the left side and is identical to that of *A. nautili* (except the missing exopodal spine). The endopod on the right side is normal and is identical to that in *A. nautili*. The endopod (fig. 33f) on the left side is abnormally developed with a small protuberance bearing two plumose setae at its tip. It is unfortunate that Heegaard (1972) figured the abnormally developed third leg, since it caused some taxonomic confusion (establishment of a spurious genus).

Ho (1980) provided a redescription of this species. Our illustrations conform to his illustrations and description very closely. There are, however, three minor differences. Ho described "26 hairy setae" on the basal segment of the antennule, when there are actually 27. He also does not mention the presence of a small rounded protrusion in the area normally occupied by the spinelike projection on the first segment of the female's antenna. The outermost spine at the tip of the exopod of leg 4 is somewhat longer in Heegaard's specimen than those in the collection of J.-S. Ho (compare figs. 32c and 33g). This difference, of course, is minor and can be attributed to intraspecific variation.

Genus **Anuretes** Heller, 1865

Anuretes Heller, 1865: 186; Bassett-Smith, 1899: 457; Wilson, 1905a: 647; Shiino, 1954a: 260; Yamaguti, 1963: 67; Lewis, 1964a: 187; Pillai, 1967: 362; Hameed & Pillai, 1973a: 409; Ho & Dojiri, 1977: 95; Pillai, 1977: 58; Kabata, 1979: 160, 169; Dojiri, 1983: 108; Prabha, 1983: 56; Pillai, 1985: 453; Boxshall & Montú, 1997: 52; Ho, 1998: 251; Boxshall & Halsey, 2004: 725; Ho & Lin, 2004: 89.

Eirgos Bere, 1936: 593; Yamaguti, 1963: 95; Kabata, 1964a: 649; Ho, 1970: 108; Hameed & Pillai, 1973a: 408; Pillai, 1977: 58.

Heniochophilus Yamaguti & Yamasu, 1959: 124; Yamaguti, 1963: 81; Kabata, 1965a: 30; Pillai & Mohan, 1965: 273; Pillai, 1967: 362; Hameed & Pillai, 1973a: 409; Pillai, 1977: 58; Prabha, 1983: 57; Pillai, 1985: 458; Kazachenko, 2001: 31.

Haeniochophilus Kabata, 1979: 158.

Female. — Cephalothorax suborbicular, as large as or larger than genital complex. Free margin of thoracic zone of cephalothorax usually partially covering fourth pedigerous somite and occasionally anteriormost portion of genital complex. Frontal plate without lunules. Posterior sinuses present, occasionally very shallow. Fourth pedigerous somite small, without dorsal aliform plates. Genital complex quadrangular or globose. Abdomen vestigial and fused with genital complex or absent with only postanal protrusions remaining. Caudal ramus small, with setae.

Antennule and antenna as in *Lepeophtheirus*. Postantennal process and dentiform process of maxillule present. Sternal furca present; rarely absent. Maxillary whip occasionally present near base of maxilla. Other oral appendages as in *Lepeophtheirus*.

Leg 1 as in *Lepeophtheirus*, except seta 4 frequently positioned between spines 2 and 3 instead of usual position on inner distal corner of terminal exopodal segment. Leg 2 as in *Lepeophtheirus*. Leg 3 with ventral apron, biramous, and bearing 2-segmented rami. First segment represented by exopodal spine and occasionally 1 inner seta; second segment with varying numbers of setae (7-9). Basal swelling of first exopodal segment either absent or fused to clawlike spine. First segment of endopod without inner plumose seta; second segment with varying number of setae (2-6). Leg 4 with sympod and 2-segmented exopod with spinal formula of I-0; III or I-0; IV. Leg 5 a setiferous protrusion. Leg 6 absent.

Male. — Generally as in female. Genital complex very small in relation to cephalothorax. Abdomen vestigial as in female. Antenna differing from that of female by possessing corrugated adhesion pads and frequently with accessory tines on terminal claw. Legs 5 and 6 represented by setiferous protrusions.

Type-species. — *Anuretes heckeli* (Krøyer, 1863).

Remarks. — *Anuretes* was established by Heller (1865) for "*Lepeophtheirus heckelii*" described by Krøyer (1863) from "*Ephippus gigas*" (= *Ephippus goreensis* Cuvier) from the Brazilian coast and *Chaetodipterus faber* (Broussonet) from New Orleans (table V). *Eirgos anurus* Bere, 1936 is synonymous with *A. heckeli* as suspected by Ho (1970) (see remarks for *A. heckeli*). Since the establishment of *Anuretes*, 23 additional species (*A. perplexus* Bassett-Smith, 1898; *A. parvulus* Wilson, 1913; *A. plectorhynchi* Yamaguti, 1936; *A. renalis* Heegaard, 1945; *A. brevis* Pearse, 1951; *A. branchialis* Rangnekar, 1953; *A. furcatus* Capart, 1953; *A. quadrilaterus* Shiino, 1954; *A. serratus* Shiino, 1954; *A. menehune* Lewis, 1964; *A. anomalus* Pillai, 1967; *A. fedderini* Price, 1966; *A. rotundigenitalis* Hameed, 1976; *A. rotundus* Prabha & Pillai, 1983; *A. shiinoi* Prabha & Pillai, 1983; *A. hoi* Prabha & Pillai, 1983; *A. plataxi* Prabha & Pillai, 1986; *A. yamagutii* Prabha & Pillai, 1986; *A. chelatus* Prabha & Pillai, 1986; *A. grandis* Ho & Lin, 2000; *A. occultus* Ho & Lin, 2000; *A. similis* Ho & Lin, 2000; and *A. justinei* Venmathi Maran, Ohtsuka & Boxshall, 2008) have been described. A list of the hosts and localities of collections for the species of *Anuretes* is provided (table VI).

TABLE V

Hosts and localities of collections of *Anuretes heckeli* (Krøyer, 1863)

Species	Host	Locality	Reference
Anuretes heckeli	"*Ephippus gigas*" (= *Ephippus goreensis* Cuvier)	Brazilian coast	Krøyer, 1863
	Chaetodipterus faber (Broussonet)	New Orleans	Krøyer, 1863
	Chaetodipterus faber	Louisiana	Causey, 1953a
	Lobotes surinamensis (Bloch)	Louisiana	Causey, 1953a
	Chaetodipterus faber	Texas Coast	Causey, 1953b
	filefish	Texas Coast	Causey, 1953b
	Chaetodipterus faber	Louisiana	Causey, 1955
		Mississippi	Causey, 1955
		Texas	Causey, 1955
	Vomer setapinnis [= *Selene setapinnis* (Mitchill)]	Mississippi	Causey, 1955
	Caranx hippos (Linnaeus)	Tuxpan, Mexico	Causey, 1955
	Chaetodipterus faber	Tuxpan, Mexico	Causey, 1960
Reported under the binomen *Eirgos anurus*	*Chaetodipterus faber*	Gulf of Mexico	Bere, 1936
	Chaetodipterus faber	Beaufort, North Carolina	Pearse, 1947
	Chaetodipterus faber	Texas coast	Pearse, 1952
	Scomberomorus maculatus (Mitchill)	Texas coast	Pearse, 1952
	Chaetodipterus faber	Apalachee Bay, Florida	Ho, 1970

The main distinction between members of *Anuretes* and *Lepeophtheirus* is the absence or great reduction of the abdomen in the species of *Anuretes* (Heller, 1865; Heegaard, 1945b; Capart, 1953; and Lewis, 1964a). However, there is a great number of species of *Lepeophtheirus* that exhibit a highly reduced abdomen (e.g., *L. cossyphi* Krøyer, 1863; *L. dissimulatus* Wilson, 1905; and *L. plotosi* Barnard, 1948). In addition, two species of *Anuretes* (*A. serratus* Shiino, 1954 and *A. quadrilaterus* Shiino, 1964) possess a distinct abdomen (though small and fused to genital complex). It appears that this feature alone is not a good generic discriminant as noted by several researchers (Pillai, 1967; Ho & Dojiri, 1977; Kabata, 1979; and Ho & Lin, 2000).

Another feature used to distinguish *Anuretes* from *Lepeophtheirus* is the elongation of the free margin of the thoracic zone of the cephalothorax to dorsally cover the fourth pedigerous somite in the females of *Anuretes*. However, the female of *A. rotundigenitalis* Hameed, 1976 has its fourth pedigerous somite almost fully exposed.

TABLE VI

Hosts and localities of collections of species of *Anuretes* Heller, 1865

Species	Host	Locality	Reference
A. anomalus Pillai, 1967	*Platax teira* (Forsskål)	Trivandrum, India	Pillai, 1967
	Spilotichthys pictus [= *Diagramma picta* (Thunberg)]	Heron Island, Australia	Ho & Dojiri, 1977 (reported under the binomen *Lepeophtheirus anomalus*)
A. branchialis Rangnekar, 1953	See table VII		
A. brevis Pearse, 1951	*Archosargus probatocephalus* (Walbaum)	Bahamas	Pearse, 1951
A. chelatus Prabha & Pillai, 1986 (transferred to *Pseudanuretes*)	*Pomacanthodes imperator* [= *Pomacanthus imperator* (Bloch)]	Trivandrum, India	Prabha & Pillai, 1986
A. fedderini Price, 1966 (transferred to *Pseudanuretes*)	*Holacanthus ciliaris* (Linnaeus)	Caribbean	Price, 1966
A. furcatus Capart, 1953 (transferred to *Lepeophtheirus*)	*Mobula rochebrunei* (Vaillanti)	Senegal	Capart, 1953
A. grandis Ho & Lin, 2000	*Diagramma picta*	Taiwan	Ho & Lin, 2000
A. heckeli (Krøyer, 1863)	See table V		
A. hoi Prabha & Pillai, 1986	*Diagramma pictum* [= *Diagramma picta*]	Trivandrum, India	Prabha & Pillai, 1986
A. indicus (Pillai, 1977)	*Platax teira*	India	Pillai, 1977
A. justinei Venmathi Maran, Ohtsuka & Boxshall, 2008	*Plectorhinchus lineatus* (Linnaeus)	New Caledonia	Venmathi Maran, Ohtsuka & Boxshall, 2008
A. menehune Lewis, 1964	*Naso hexacanthus* (Bleeker)	Hawaii	Lewis, 1964a

TABLE VI
(Continued)

Species	Host	Locality	Reference
A. occultus Ho & Lin, 2000	*Gaterin schotaf* [= *Plectorhinchus schotaf* (Forsskål)]	Trivandrum, India	Prabha & Pillai, 1983 (reported under the binomen of *Lepeophtheirus rotundigenitalis*)
A. parvulus Wilson, 1913 (transferred to *Pseudanuretes*)	*Pomacanthus arcuatus* (Linnaeus)	Dry Tortugas, Florida	Wilson, 1913, 1935a
	Angelichthys bermudensis (= *Holacanthus bermudensis* Jordan & Rutter)	Dry Tortugas, Florida	Wilson, 1935a
A. perplexus Bassett-Smith, 1898	*Lutjanus* sp.	Sri Lanka	Bassett-Smith, 1898b (refer to Pillai, 1969)
A. plataxi Prabha & Pillai, 1986	*Platax teira*	Trivandrum, India	Prabha & Pillai, 1986
A. plectorhynchi Yamaguti, 1936	*Plectorhynchus pictus* (Thunberg)	Pacific Ocean	Yamaguti, 1936a
A. quadrilaterus Shiino, 1954	*Zenopsis nebulosa* (Temminck & Schlegel)	Japan	Shiino, 1954a
A. renalis Heegaard, 1945 (transferred to *Lepeophtheirus*)	*Diodon* sp.	Japan	Heegaard, 1945b
A. rotundigenitalis Hameed, 1976	*Diagramma crassispinum* [= *Plectorhinchus gibbosus* (Lacépède)]	Cape Comorin, India	Hameed, 1976
A. rotundus Prabha & Pillai, 1983	*Pomacanthodes imperator* [= *Pomacanthus imperator* (Bloch)]	Trivandrum, India	Prabha & Pillai, 1983
A. serratus Shiino, 1954	*Xesurus scalprum* (= *Prionurus scalprum* Valenciennes)	Wakayama, Japan	Shiino, 1954a
	Naso hexacanthus (Bleeker)	Hawaii	Lewis, 1964a

TABLE VI
(Continued)

Species	Host	Locality	Reference
A. shiinoi Prabha & Pillai, 1983	*Naso* sp.	Trivandrum, India	Prabha & Pillai, 1983
A. similis Ho & Lin, 2000	*Spilotichthys pictus* [= *Diagramma picta*]	Trivandrum, India	Prabha & Pillai, 1986 (reported under the binomen *Anuretes plectorhynchi*)
A. yamagutii Prabha & Pillai, 1986 (relegated to synonymy with *A. anomalus*)	*Spilotichthys pictus* [= *Diagramma picta* (Thunberg)]	Trivandrum, India	Prabha & Pillai, 1986
	Plectorhinchus cinctus (Temminck & Schlegel)	Kuwait	Ho & Sey, 1996

According to Shiino (1954a), *Anuretes* could be distinguished from *Lepeophtheirus* by the segmentation of leg 4. Species of *Anuretes* have a 2-segmented exopod of leg 4. However, there are several species of *Lepeophtheirus* that share this feature with the members of *Anuretes*, e.g., *L. appendiculatus* Krøyer, 1863; *L. grohmanni* Krøyer, 1863; *L. intercurrens* Krøyer, 1863; *L. marceps* Wilson, 1944; *L. pectoralis* (Müller, 1776); *L. unispinosus* Pearse, 1952; and *L. watanabei* Shiino, 1954, among others (see list of Ho & Dojiri, 1977). The spinal formula is either I-0; III or I-0; IV.

The description of *A. brevis* Pearse, 1951 (based on a male specimen) and *A. fedderini* Price, 1966 were not detailed enough for a comparative study; therefore, they have not been taken into account for the generic diagnosis. *Anuretes fedderini* and *Anuretes parvulus* Wilson, 1913, have been transferred to *Pseudanuretes* (see discussion on this genus).

Anuretes furcatus Capart, 1953, and *A. renalis* Heegaard, 1945, exhibit a small, but distinct abdomen, and a 3-segmented exopod of leg 4. Furthermore, *A. renalis* possesses a basal swelling at the origin of the clawlike spine of leg 3. According to the above diagnosis, these two species can be better accommodated in *Lepeophtheirus*.

Because of the considerable morphological overlap among the members of *Anuretes* and *Lepeophtheirus*, these two genera were difficult to distinguish. The generic boundary that distinguished them was ill-defined, which prompted Ho & Dojiri (1977) to synonymize the two genera.

Although these two genera appear to be very closely related, *Anuretes* can be distinguished from *Lepeophtheirus* by a combination of: (1) vestigial abdomen, (2) 2-segmented exopod of leg 3, (3) absence of basal swelling or fusion of it with spine, (4) absence of inner plumose seta on the first endopodal segment of leg 3, and (5) 2-segmented exopod of leg 4. The structure of leg 3, particularly in respect to the basal swelling of the first exopodal segment, has not been reported in species of *Lepeophtheirus*. The unusual position of pinnate seta 4 between spines 2 and 3 on the terminal exopodal segment of leg 1 is not found in species of *Lepeophtheirus*. This was first noted by Pillai (1967). However, not all species of *Anuretes* exhibit this feature. Moreover, this characteristic is also shared with species of *Pseudanuretes*.

Although the generic diagnosis provided above easily distinguishes members of the two genera, *Anuretes* remains a somewhat heterogeneous assemblage of species, whose validity is still questionable.

There are a few caligid genera that are so morphologically similar to *Anuretes* that their validity has been questioned. These genera are *Eirgos*, *Heniochophilus*, *Mappates*, and *Pseudanuretes*. A detailed discussion of this taxonomic problem has been conducted by Pillai (1967, 1977) and Hameed & Pillai (1973a). These genera were distinguished from each other by the extent of the elongation of the free margin of the thoracic zone of the cephalothorax in the female, and the presence or absence of sclerotized projections (i.e., postantennal process, dentiform process of maxillule, and sternal furca). These characteristics, however, have been considered features of dubious taxonomic merit (Kabata, 1965a; Pillai, 1967).

Yamaguti (1963) divided the *Anuretes-Eirgos-Heniochophilus-Mappates-Pseudanuretes* complex into two different families and three subfamilies. He established a new family Eirgidae to accommodate the genus *Eirgos* according to Bere's (1936) interpretation of the posterior extension of the cephalothorax of *E. anurus*. Bere believed that this posteromedian extension originated from a fusion of dorsal plates from the third pedigerous somite. This unusual explanation is not warranted; the cephalothoracic extension in the female is simply a result of the lengthening of the thoracic zone of the cephalothorax in a posterior direction. An extreme condition of this phenomenon is exhibited by females of *Mappates*. The diagnosis that *Eirgos* exhibits only the first two pedigerous somites (thoracic segments of Bere, 1936 and Yamaguti, 1963) fused with the cephalosome to form the cephalothorax with the third and fourth pedigerous somites remaining free cannot be accepted. As suspected by Ho (1970) and Hameed & Pillai (1973a), *Eirgos anurus* is synonymous with *Anuretes heckeli* (refer to remarks on *A. heckeli*). Since *A. heckeli* does indeed have three pedigerous somites fused with the cephalosome, it conforms to the family diagnosis for the Caligidae.

Yamaguti (1963) also established a separate subfamily for *Mappates*, calling it the Mappatinae. The two most significant characteristics of this subfamily, according to Yamaguti, are the posterior extension of the thoracic region of the dorsal shield and the abdomen fused with the genital complex. However, these characteristics are shared by *Anuretes*, *Heniochophilus* (here considered a synonym of *Anuretes*), and *Pseudanuretes*, which were not included in this subfamily. In fact, *Anuretes* and *Pseudanuretes* were placed in the subfamily Anuretinae, and *Heniochophilus* was incorporated into the Lepeophtheirinae despite the similarity among the four genera. Because of the morphological similarity of these genera, Yamaguti's subfamilies cannot be recognized (Kabata, 1979).

Since *Eirgos* and *Heniochophilus* are treated as synonyms of *Anuretes* (their synonymies are discussed later in this revision), *Anuretes*, *Mappates*, and *Pseudanuretes* are the only three genera of this complex recognized as valid.

Anuretes heckeli (Krøyer, 1863)
(figs. 34-38)

Lepeophtheirus heckelii Krøyer, 1863: 184.

Anuretes heckeli Heller, 1865: 186; Bassett-Smith, 1899: 457; Wilson, 1905a: 647; Heegaard, 1945b: 12, 14; Causey, 1953a: 2; Causey, 1953b: 10; Rangnekar, 1953: 241; Causey, 1955: 4; Causey, 1960: 328; Yamaguti, 1963: 68; Ho, 1970: 108; Dojiri, 1983: 118; Boxshall & Montú, 1997: 52; Morales-Serna et al., 2012: 50.

Eirgos anurus Bere, 1936: 593; Pearse, 1947: 9; Pearse, 1952: 26; Yamaguti, 1963: 96; Kirtisinghe, 1964: 77; Ho, 1970: 107.

Material examined. — Sixteen females from "*Ephippus gigas*" (= *Ephippus goreensis* Cuvier) on loan from Naturhistorisches Museum, Wien, Austria. According to Dr. Gerhard Pretzmann, curator at this museum, these specimens probably represent syntypes. Lectotype and paralectotypes designated by first author (M. Dojiri) on 2 August 1982. Twenty females from branchial cavity of spadefish, *Chaetodipterus faber* (Broussonet), caught at Apalachee Bay, Florida on 20 July 1965 by second author (J.-S. Ho). Fourteen females, 4 immature females, and 1 male from *Chaetodipterus faber* collected at Charlotte Harbor, Florida on 25 June 1975. These specimens on loan from Dr. Roger F. Cressey, National Museum of Natural History, Smithsonian Institution, Washington, D.C.

Female. — Body as in fig. 34a, b. Total length (excluding setae on caudal ramus) 2.27 mm (2.09-2.46 mm) (n = 10). Cephalothorax more than 1/2 total length, 1.54 mm (1.43-1.66 mm) × 1.29 mm (1.23-1.33 mm), with relatively shallow posterior sinuses. Free margin of thoracic zone extending far beyond posterior tip of lateral zone, and covering anteriormost portion of genital complex. Dorsal transverse rib of cephalothorax absent, although medial suture present. Antennule within lateral limit of cephalothorax.

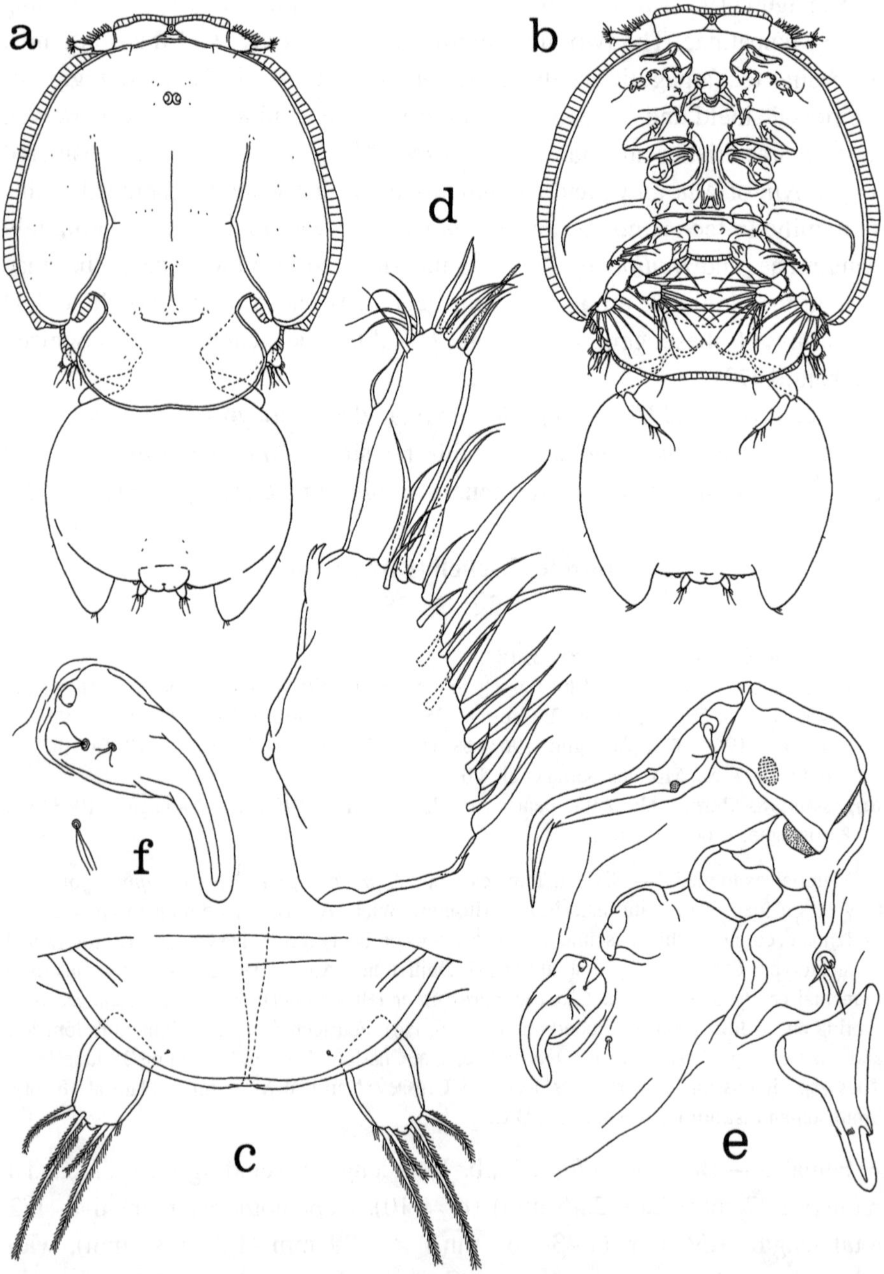

Fig. 34. *Anuretes heckeli* (Krøyer, 1863), female. a, body, dorsal (scale H); b, same, ventral (H); c, abdomen and caudal ramus, dorsal (D); d, antennule (plumosities not drawn), ventral (U); e, antenna, postantennal process, and maxillule, ventral (D); f, postantennal process, ventral (U).

Fourth pedigerous somite covered dorsally by posterior extension of thoracic zone of cephalothorax (fig. 34a) and ventrally by apron of leg 3 (fig. 34b). Genital complex ovoid, wider than long, 0.80 mm (0.73-0.86 mm) × 1.10 mm (0.93-1.20 mm) (excluding posterolateral processes). Processes approximately 0.19 mm (0.17-0.23 mm) × 0.23 mm (0.20-0.30 mm). Abdomen (fig. 34c) much wider than long, 0.09 mm (0.08-0.11 mm) × 0.23 mm (0.21-0.25 mm).

Caudal ramus (fig. 34c) somewhat quadrangular, 65 × 49 μm, and carrying 5 plumose setae and 1 small protrusion.

Egg sac 2.53 mm (2.22-3.09 mm) × 0.23 mm, containing 28 eggs (22-33 eggs) (n = 4).

Antennule (fig. 34d) 2-segmented. First segment 130 μm long (measured along setiferous margin), with 25 plumose setae along anterior margin, 1 spinelike process near base of proximalmost seta, and bifid process on posterodistal corner. Second segment cylindrical, 92 μm long, with usual 13 + 1 aesthetasc. Antenna (fig. 34e) apparently 4-segmented. Second segment with posteriorly directed spinelike process. Third segment with usual dorsal adhesion pad. Claw recurved, sharp, and with characteristic 2 setae. Postantennal process (fig. 34e, f) with curved, blunt tine, 3 groups of setules and 1 small hyaline knob (arising from a depression near its base).

Mouth tube (fig. 35a) longer than wide, 194 × 130 μm. Intrabuccal stylet present on labrum. Strigil present on distal end of labium. Mandible (fig. 35b) comprising 4 sections, with first and third sections nearly equal in length and much longer than other 2. Terminal section with 12 teeth. Maxillule (fig. 34e) consisting of papilla tipped with 3 setae, and 1 straight process with inner, minute accessory process. Maxilla (fig. 35c) with usual flabellum. Calamus approximately 3 times as long as canna. Maxilliped (fig. 35d) with robust corpus. Shaft and claw combined 254 μm long. Sternal furca (fig. 35e) with straight, parallel, bluntly rounded tines. Pair of posteriorly directed spinelike projections (fig. 35f) situated medial to bases of maxilla.

Plumose seta at junction of sympod and exopod of leg 1 (fig. 35g) not covered by ventral protrusion. First exopodal segment with spine on outer distal corner and setules on inner margin. Pinnate seta 4 (fig. 36a) slightly shorter than outermost spine, and originating near base of spine 3. Spines 1, 2, and 3 spinulated, with spines 1 and 2 possessing spinules near their bases. Spines 2 and 3 with accessory processes at distal ends. Endopod ending in attenuate process. Area between legs 1 and 2 (fig. 35g) with pair of sclerotized processes projecting posteriorly. Leg 2 (fig. 36b) of usual caligid form. Spine of first exopodal segment (fig. 36c) bilaterally spinulated and with pectinate

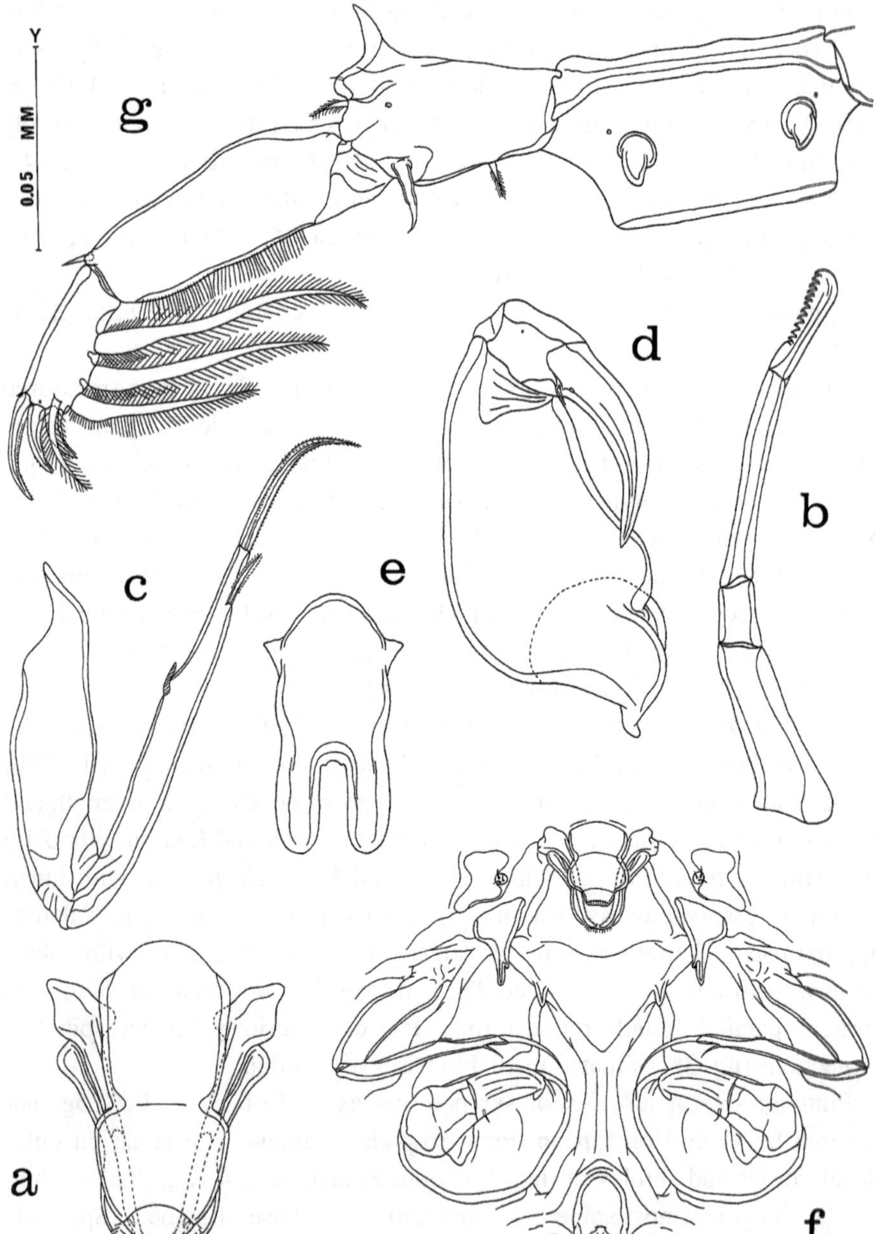

Fig. 35. *Anuretes heckeli* (Krøyer, 1863), female. a, mouth tube, ventral (scale D); b, mandible, ventral (Y); c, maxilla, dorsal (D); d, maxilliped, ventral (J); e, sternal furca, ventral (D); f, oral area, ventral (G); g, leg 1, intercoxal plate, and sclerotized area between legs 1 and 2, ventral (D).

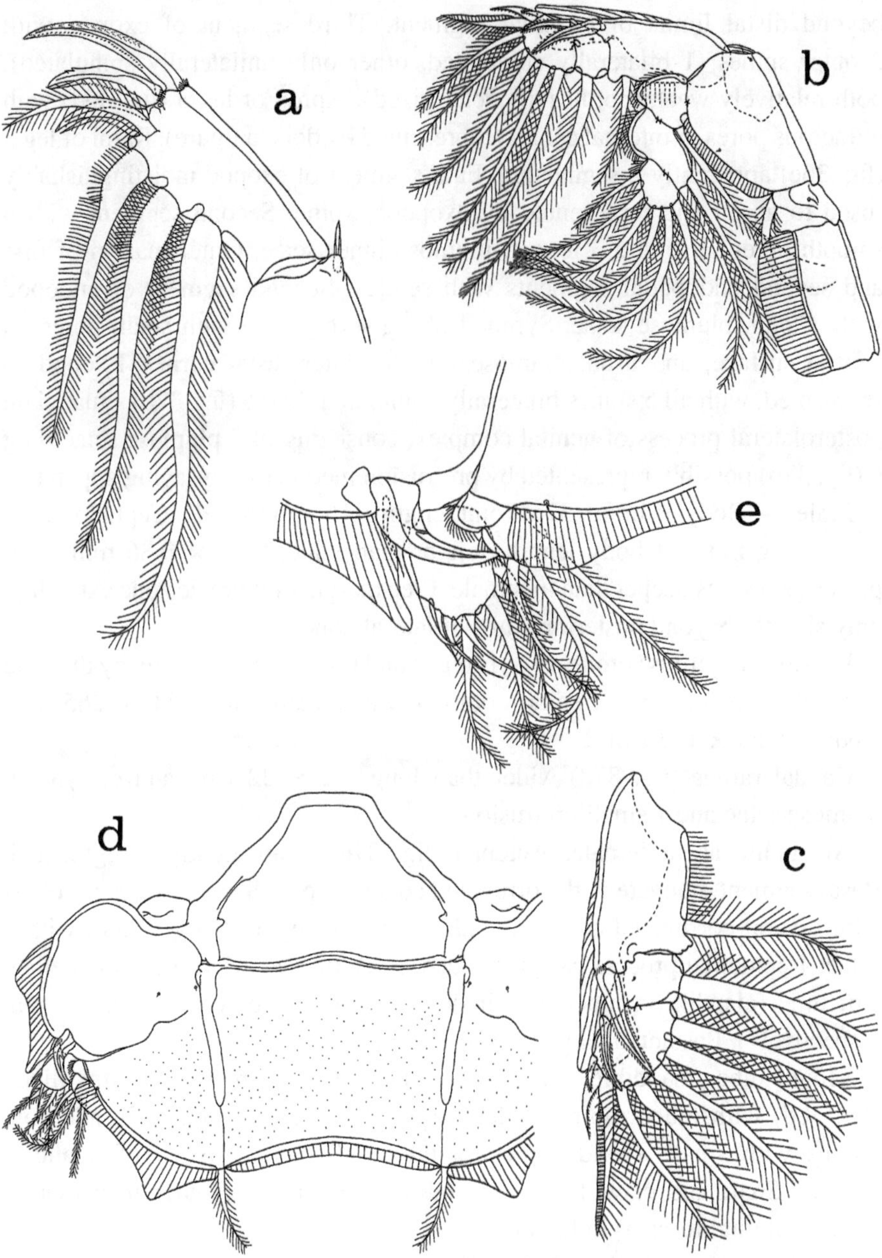

Fig. 36. *Anuretes heckeli* (Krøyer, 1863), female. a, terminal portion of exopod of leg 1, ventral (scale L); b, leg 2 and intercoxal plate, ventral (G); c, exopod of leg 2, ventral (E); d, leg 3, ventral (B); e, exopod and endopod of leg 3, ventral (D).

membrane at its base. Spines of second and third exopodal segments extending beyond distal limits of terminal segment. Third segment of exopod with 2 outer spines (1 bilaterally spinulated, other only unilaterally spinulated); both relatively weak (not heavily sclerotized). Apron of leg 3 (fig. 36d) with numerous pores or internal canals (represented by dots in figure). Rami of leg 3 (fig. 36e) apparently 2-segmented. First segment of exopod indistinguishably fused to apron and represented by exopodal spine. Second segment with 4 smooth outer setae and 3 longer, plumose inner setae. Outer margin of first and second endopodal segments with setules. Second segment of endopod with 3 long, plumose setae. Sympod of leg 4 (fig. 37a) with small pores on ventral surface, and usual plumose seta on outer distal corner. Exopod 2-segmented, with all 5 spines bilaterally spinulated. Leg 5 (fig. 37b) situated on posterolateral process of genital complex, consisting of 3 plumose setae. Leg 6 (fig. 37b) possibly represented by protrusion medial to egg-laying apparatus.

Male. — Body as in fig. 37c. Total length 1.16 mm (n = 1). Cephalothorax comprising most of body length, longer than wide, 0.92 × 0.86 mm, with posterior sinuses deeper than in female. Free margin of thoracic zone extending only slightly beyond posterior limits of lateral zone.

Fourth pedigerous somite 49 × 146 μm and not covered dorsally by thoracic zone of cephalothorax. Genital complex wider than long, 151 × 265 μm. Abdomen 54 × 113 μm.

Caudal ramus (fig. 37d) wider than long, 22 × 32 μm, and carrying 6 plumose setae and 1 small protrusion.

Antennule as in female. Antenna (fig. 37e, f) apparently 3-segmented. Basal segment elongate with corrugated adhesion pad. Second segment robust with several corrugated adhesion pads. Terminal segment (fig. 38a-c) with 2 setae, 1 spinelike process carrying 1 accessory tine, and 2 recurved lamellate structures (clawlike in appearance in dorsal and ventro-outer views, fig. 38a, b, respectively). Postantennal process as in female.

Mouth tube, mandible, maxillule, and maxilla as in female. Maxilliped (fig. 38d) with much more slender corpus than that of female. In addition, corpus with small, rounded process proximally and 1 patch of denticles distally. Shaft and claw 135 μm combined and more slender than that of female. Sternal furca as in female.

Leg 1 as in female. Leg 2 (fig. 38e) as in female, except first and second exopodal spines not reaching distal limit of terminal segment of exopod of leg 2 (compare figs. 36c and 38e). Leg 3 as in female, except male carrying dorsal corrugated pad on anterolateral corner of ventral apron. Leg 4 (fig. 38f)

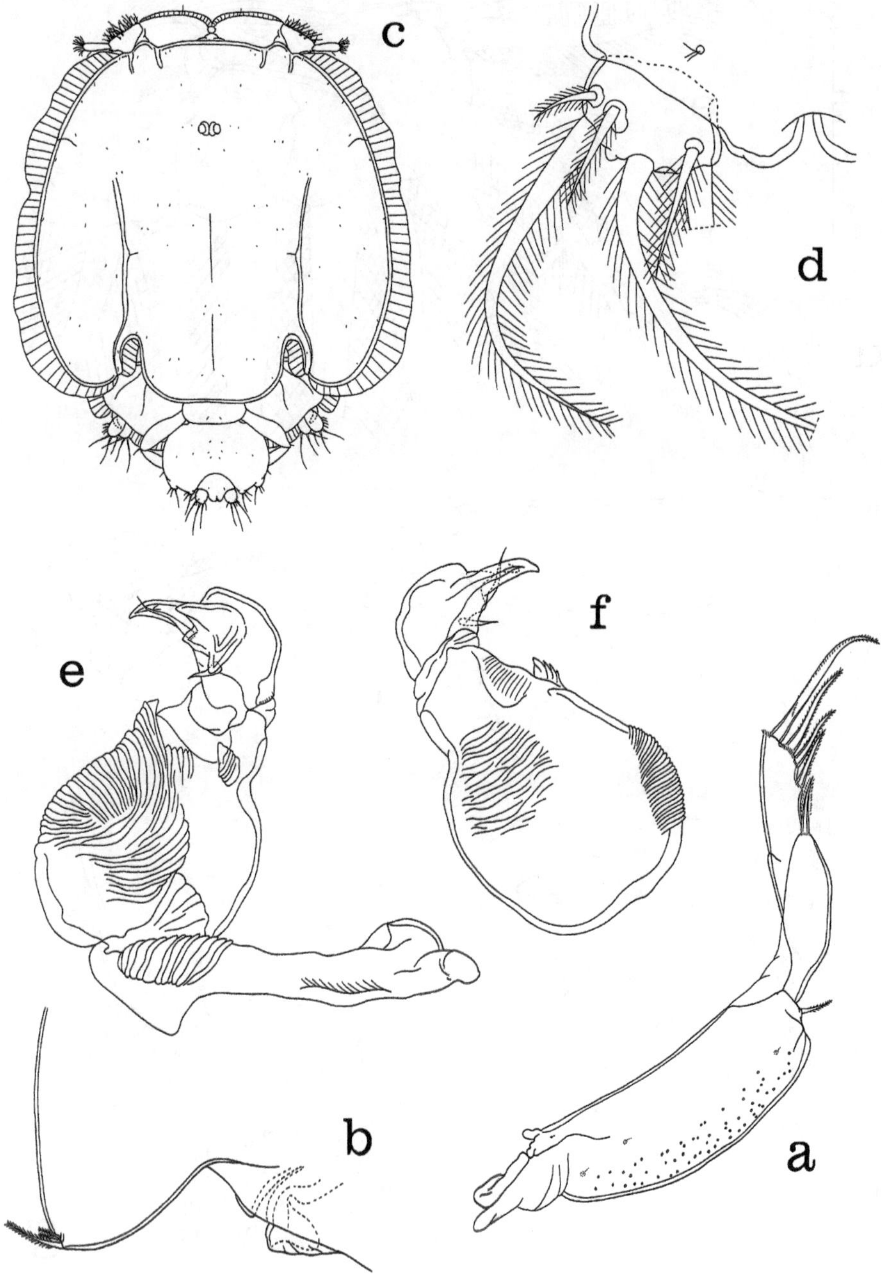

Fig. 37. *Anuretes heckeli* (Krøyer, 1863). Female. a, leg 4, ventral (scale J); b, leg 5 and genital area, ventral (J). Male. c, body, dorsal (C); d, caudal ramus, dorsal (Y); e, antenna, ventro-inner (U); f, distal two segments of antenna, dorsal (U).

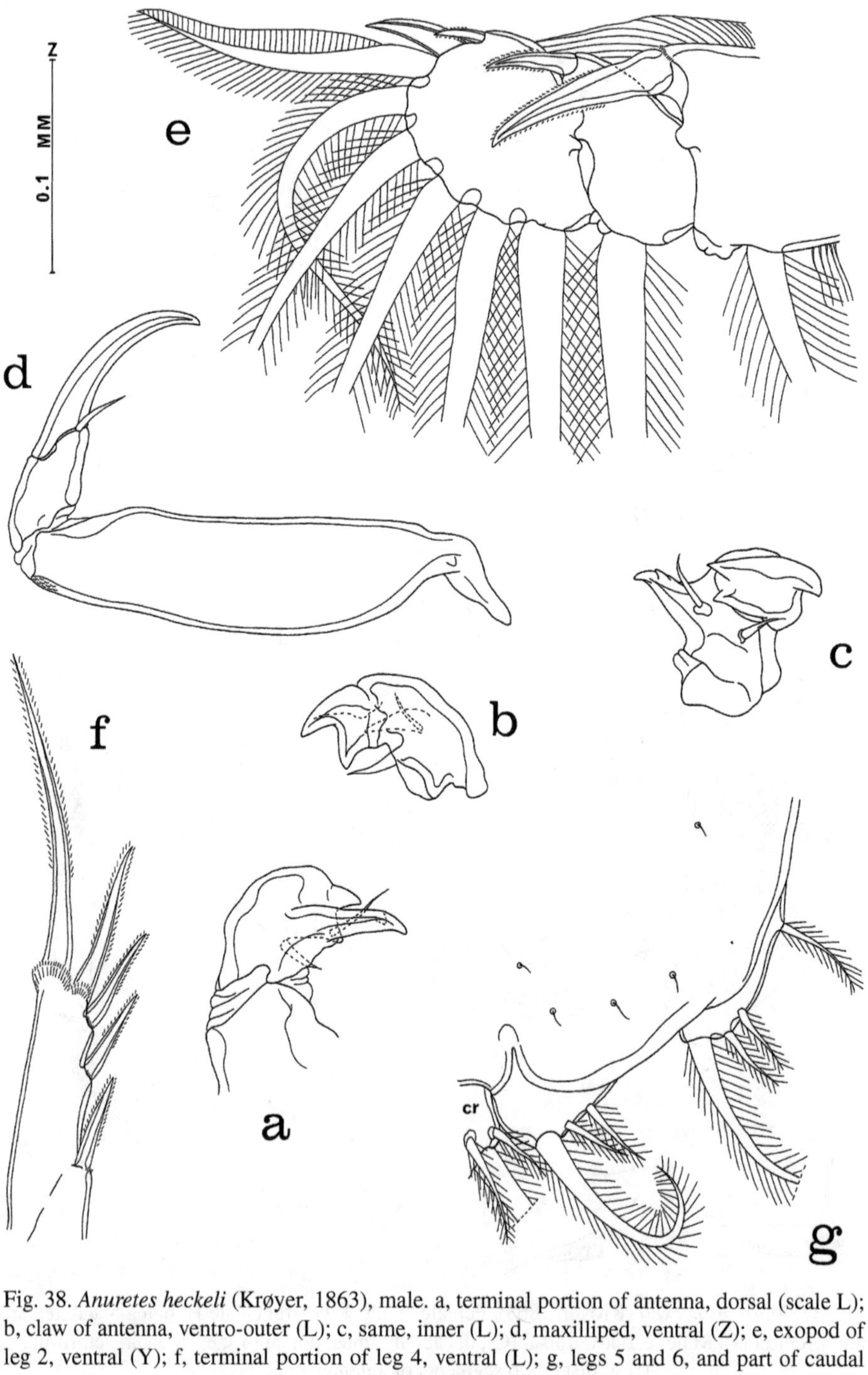

Fig. 38. *Anuretes heckeli* (Krøyer, 1863), male. a, terminal portion of antenna, dorsal (scale L); b, claw of antenna, ventro-outer (L); c, same, inner (L); d, maxilliped, ventral (Z); e, exopod of leg 2, ventral (Y); f, terminal portion of leg 4, ventral (L); g, legs 5 and 6, and part of caudal ramus, dorsal (Y).

similar to that of female. Legs 5 and 6 (fig. 38g) on posterolateral corner of genital complex. Leg 5 represented by small process carrying 4 plumose setae. Leg 6 consisting of process bearing 3 plumose setae.

Remarks. — Bere (1936) originally described *Eirgos anurus* as a new species from the mouth cavity of the spadefish *Chaetodipterus faber* caught in the Gulf of Mexico. Bere provided an incorrect generic diagnosis (see discussion on *Anuretes*) and phylogenetic position for *Eirgos*. The first three pedigerous somites are fused to the cephalosome to form the cephalothorax as in other caligids, not just the first and second as stated by Bere. Therefore, this species cannot be considered as a "link" between the "Trebinae and the Euryphorinae and Pandarinae".

Ho (1970) suggested that *E. anurus* was perhaps synonymous with *Anuretes heckeli* (Krøyer, 1863). According to Wilson's (1905a) account (apparently taken from Krøyer's original description), there were presumably several important differences between the descriptions of these two species. The greatest difference seemed to be that in the female of *A. heckeli* the thoracic zone of the dorsal shield extends only slightly beyond the posterior limits of the lateral lobes. The abdomen is supposedly absent in *A. heckeli*, and there appeared to be some differences in legs 3 and 4. The reported hosts for *E. anurus* and *A. heckeli* overlap a great deal as do the localities of collections (table V). Most of the records come from the spadefish *Chaetodipterus faber* from the Gulf of Mexico. In addition, the genital complex (especially the posterolateral processes) of the female, the maxillule, and sternal furca are almost identical. The differences mentioned above are attributable to errors in observation by Krøyer. The two species are indeed synonymous.

Krøyer (1863) attributed *A. heckeli* to Kollar. Dr. Gerhard Pretzmann, curator of the museum where these specimens were housed, sent me the presumed type-material of *A. heckeli*. The museum label also attributes the species to Kollar, who was once curator at the museum. Whether Kollar actually published a description or simply named the specimens on a museum label remains a mystery, since Krøyer gave no literature citation that referenced Kollar.

Anuretes branchialis Rangnekar, 1953
(figs. 39-41)

Anuretes branchialis Rangnekar, 1953: 239; Rangnekar, 1957: 12; Yamaguti, 1963: 68; Dojiri, 1983: 126; Lin & Ho, 2002a: 46; Ho & Lin, 2004: 90; Samotylova, 2010: 99.

Heniochophilus japonicus Yamaguti & Yamasu, 1959: 121; Yamaguti, 1963: 82; Kabata, 1965a: 22; Pillai, 1967: 358.

Heniochophilus branchialis Pillai & Mohan, 1965: 270; Pillai, 1985: 459.

Material examined. — Three females from *Platax teira* (Forsskål) (USNM 169893) collected at "Philippine Islands, Wharf Sulu, by "Albatross", "Archipelago, Sitanki" on 27 February 1908; 1 female from *Platax teira* (original field number LT 11761) collected at "Iloilo Market", Philippine Islands by the "Albatross" on 2 June 1908; 1 female from *Platax teira* (USNM 169894) from "Celebes, Dodepo, Pasejogo" by "Albatross" on 16 November 1909; 17 females from *Platax orbicularis* (Forsskål) (USNM 207099) from St. John's Fish Market, Colombo, Sri Lanka (landed at Kalpitiya) and collected by T. Iwamoto on 12 May 1970. This material obtained from fishes housed in fish collection by first author (M. Dojiri) at National Museum of Natural History, Smithsonian Institution, Washington, D.C.

Female. — Body as in fig. 39a. Total length (excluding setae on caudal ramus) 1.61 mm (1.53-1.69 mm) (n = 7). Cephalothorax somewhat rectangular, with rounded corners (compare fig. 39a, b), much longer than wide, 1.13 mm (1.03-1.29 mm) × 0.69 mm (0.63-0.76 mm), with posterior sinuses small, either partially or completely hidden by thoracic zone in dorsal view. Antennule not extending beyond lateral limit of cephalothorax. Free margin of thoracic zone extending beyond posterior tip of lateral zone.

Fourth pedigerous somite covered in dorsal view by thoracic zone of dorsal shield. Genital complex quadrangular, wider than long, 0.55 mm (0.50-0.60 mm) × 0.65 mm (0.56-0.70 mm). Anterior end partially covered by free margin of thoracic zone. Abdomen (fig. 39c) consisting of 1 free somite, greatly reduced, and fused to genital complex.

Caudal ramus (fig. 39c) longer than wide, 30 × 24 μm, and bearing 5 plumose setae.

Body with small setules (sensilla) as in fig. 39a, b.

Egg sacs 1.46 mm (1.33-1.56 mm) × 0.23 mm (0.22-0.23 mm), containing 19 eggs (17-21 eggs) (n = 3).

Antennule (fig. 39d) 2-segmented. First segment 54 μm long (measured along nonsetiferous margin), with 26 setae (8 of which are small and smooth), and 1 bifid process on posterodistal corner. Second segment 84 μm long with usual armature of 13 + 1 aesthetasc. Antenna (fig. 39e, f) apparently 4-segmented. Second segment with slightly blunt spinelike process. Third segment with dorsal adhesion pad. Claw (or hamulus) with characteristic 2 setae. Postantennal area (fig. 39e) with 1 relatively large, sclerotized seta and 3 groups of setules. Postantennal process (fig. 39f) present only in some specimens. Out of 11 specimens examined, 1 with postantennal process on both sides, 2 with process only on 1 side, and 8 with process absent on both sides.

Mouth tube (fig. 40a) long and slender, 211 × 92 μm, with intrabuccal stylet present on inner surface of labrum. Strigil present on distal end of labium.

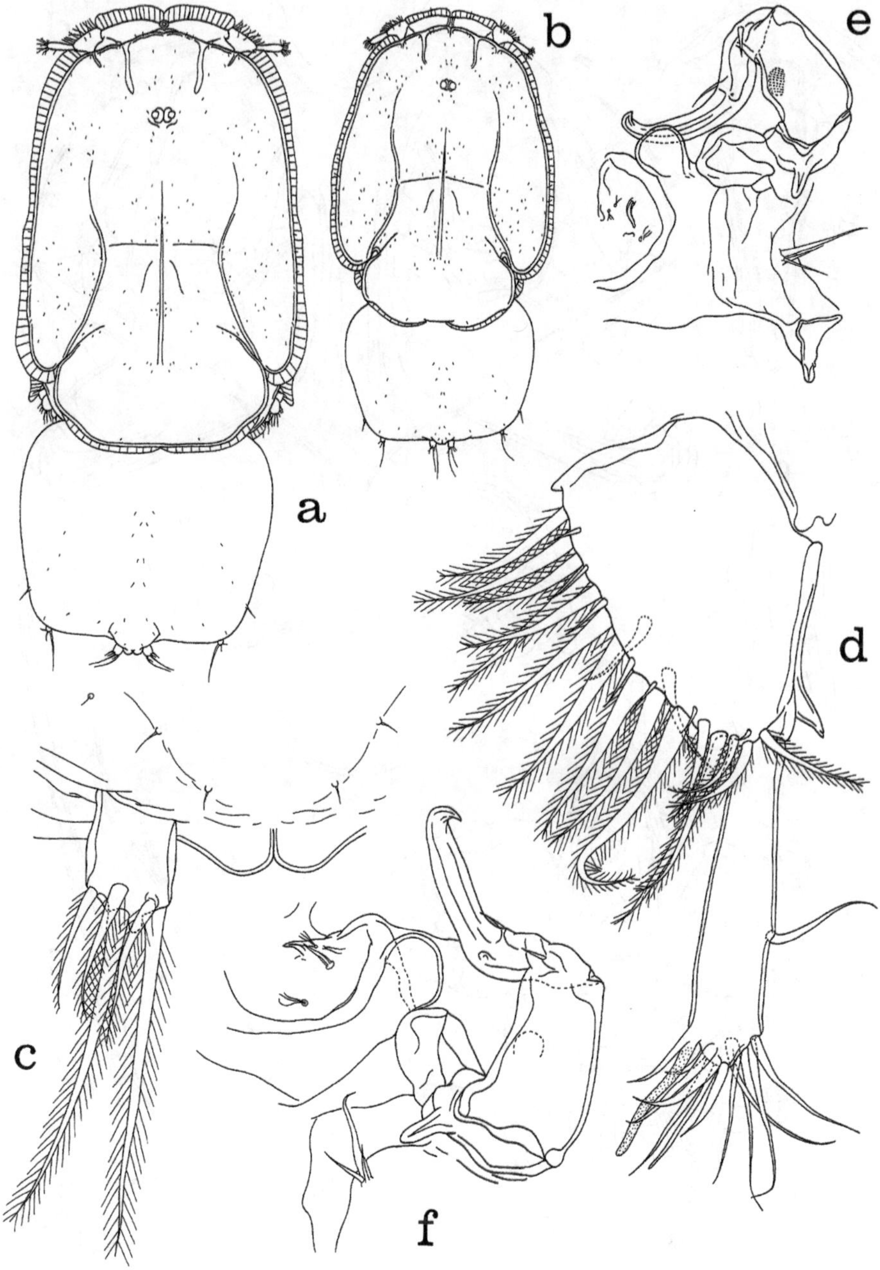

Fig. 39. *Anuretes branchialis* Rangnekar, 1953, female. a, body, dorsal (scale F); b, same (K); c, caudal ramus, dorsal (L); d, antennule, ventral (L); e, antenna, postantennal area, and maxillule, ventral (E); f, antenna, postantennal process, and setae of maxillule, ventral (E).

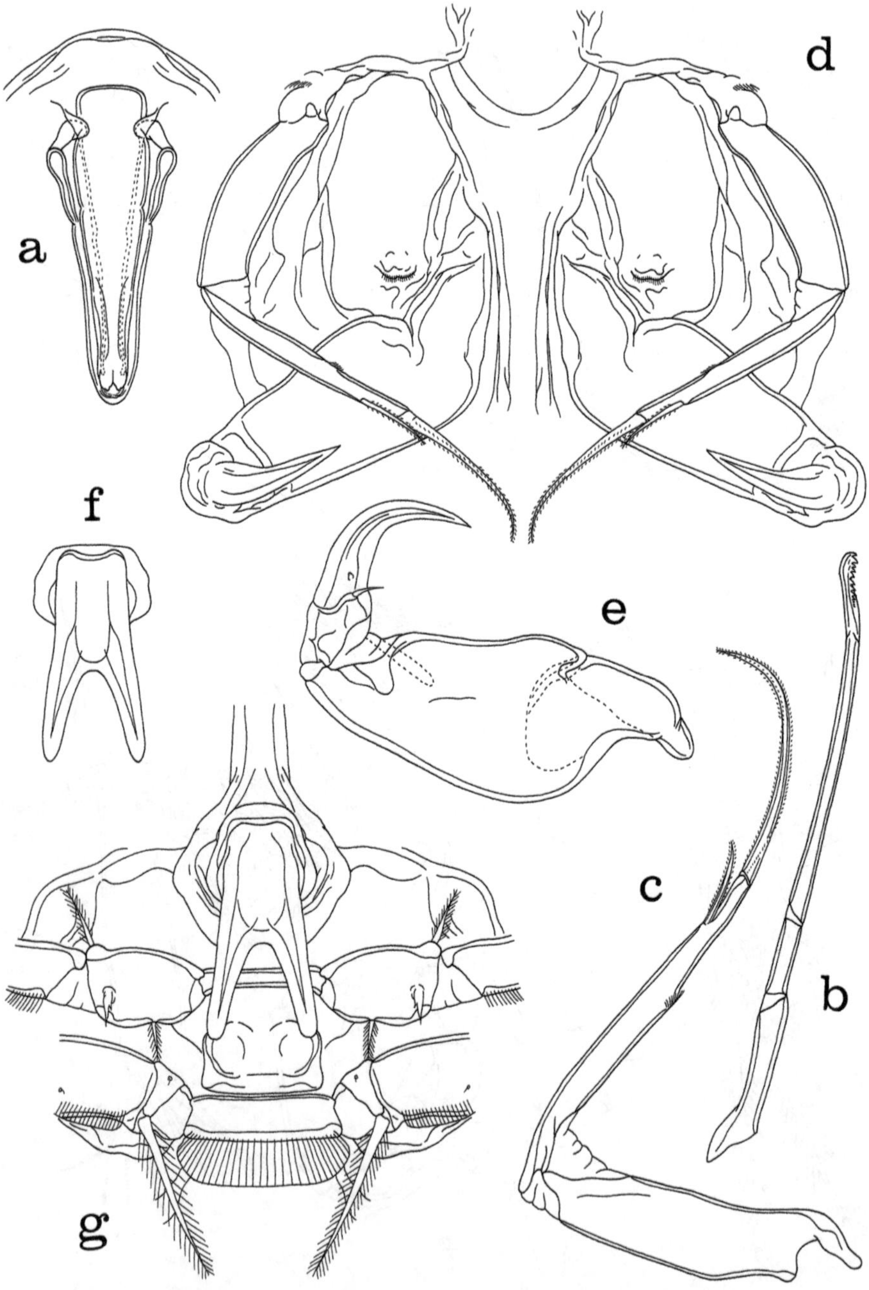

Fig. 40. *Anuretes branchialis* Rangnekar, 1953, female. a, mouth tube, ventral (scale E); b, mandible, ventral (L); c, maxilla, dorsal (Z); d, maxilla and maxilliped, ventral (E); e, maxilliped, ventral (E); f, sternal furca, ventral (E); g, sternal furca and sclerotized area between legs 1 and 2, ventral (E).

Mandible (fig. 40b) comprising 4 sections, with elongate third section. Distal section with 12 teeth (3 proximalmost teeth highly reduced and difficult to see). Maxillule (fig. 39e) consisting of 3 setae, and 1 process bearing small hyaline lobe on medial side. Maxilla (fig. 40c) brachiform with flabellum near midlength of brachium. Calamus more than 2.5 times as long as canna, and bearing 3 serrated membranes. Canna with usual 2 serrated membranes. Posteromedial to bases of maxilla, a pair of sclerotized ridges with setules (fig. 40d). Maxilliped (fig. 40e) of usual form, with shaft and claw 151 μm long. Sternal furca (fig. 40f, g) with straight, diverging tines.

Base of plumose seta at junction of sympod and exopod of leg 1 (fig. 41a) not covered by ventral protrusion. Seta on outer distal corner of first exopodal segment sclerotized and spinelike. Pinnate seta 4 shorter than outermost spine, and located between bases of spines 2 and 3. Spines 1 and 2 with pectinate membranes at bases. All 3 spines with spinules, but spines 2 and 3 with accessory processes at distal ends. Endopod with tapering end and small process on inner margin. Area between legs 1 and 2 with 1 pair of rounded protrusions (fig. 40g). Leg 2 (fig. 41b) of usual form, with 4 exopodal spines bilaterally spinulated. Spine of first segment of exopod with pectinate membrane (or spinules) at base. Leg 3 (fig. 41c) biramous. Exopod 2-segmented, with first segment small. Exopodal spine terminally located on basal swelling and indistinctly fused to it. Second segment with 4 smooth outer setae and 3 plumose inner setae. Endopod 2-segmented. First segment unarmed, except setules on outer margin. Second segment with 2 terminal plumose setae. Leg 4 (fig. 41d) with 2-segmented exopod. Innermost spine of distal segment more than twice length of other 2. Spines decreasing in length from inner to outer margin. Leg 5 (fig. 41e) consisting of papilla tipped with 1 smooth seta and 1 quadrangular lobe with 3 plumose setae.

Male. — Unknown.

Remarks. — *Anuretes branchialis* was first described by Rangnekar (1953), who collected the species from the gill filaments of the scombrid "*Thynnus pelamys*" [= *Katsuwonus pelamis* (Linnaeus)] off the coast of Bombay. Rangnekar (1953) did notice that in the female the free margin of the thoracic zone extended far beyond the tip of the lateral zone and covered the free fourth pedigerous somite and the anterior portion of the genital complex, but did not consider it of generic value. Ho & Lin (2004) provided a very detailed redescription of this species, which they collected from *Monodactylus argenteus* (Linnaeus) and *Platax orbicularis* (Forsskål) from Taiwan.

Apparently unaware of Rangnekar's description, Yamaguti & Yamasu (1959) established a new genus, *Heniochophilus*, to accommodate a species

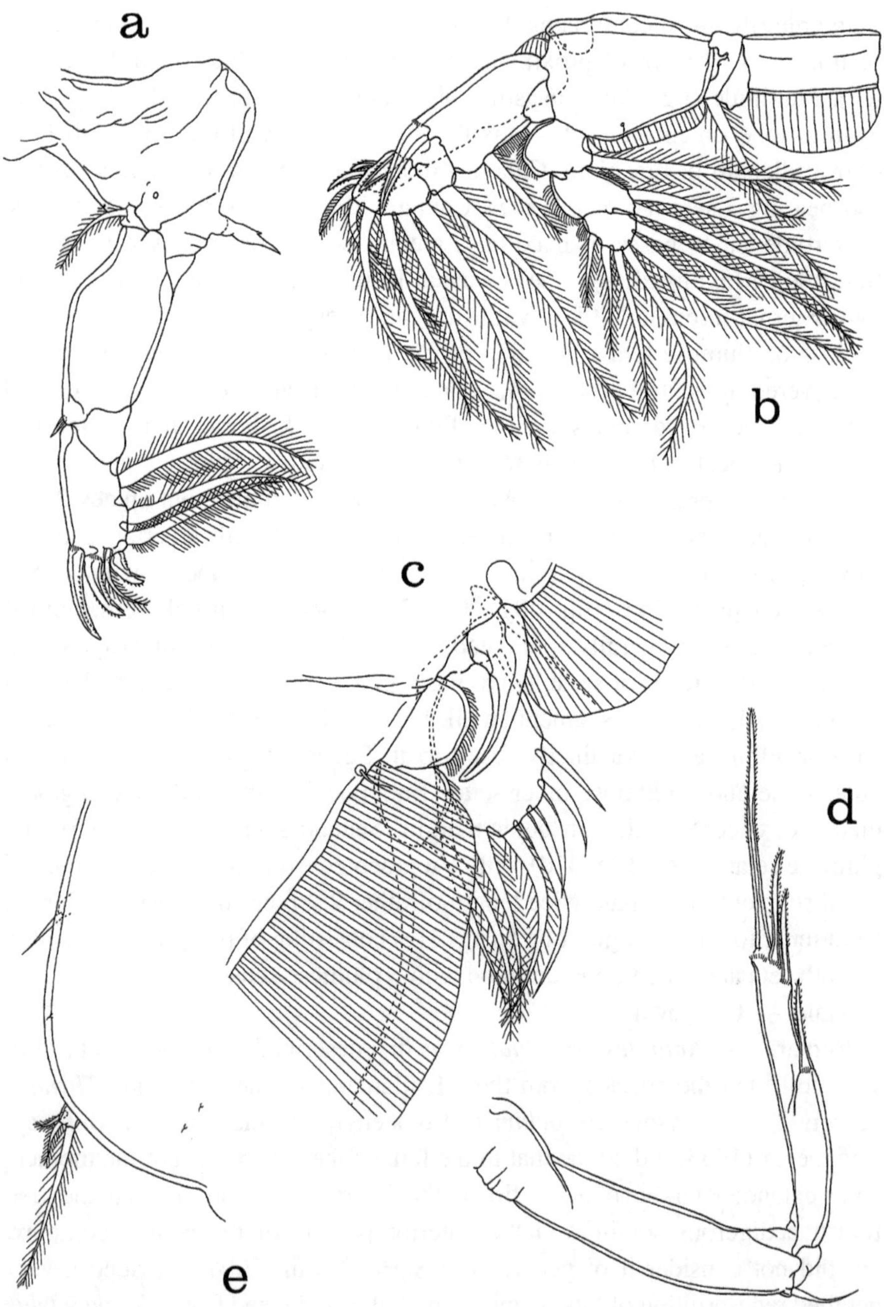

Fig. 41. *Anuretes branchialis* Rangnekar, 1953, female. a, leg 1, ventral (scale Z); b, leg 2 and intercoxal plate, ventral (E); c, exopod and endopod of leg 3, ventral (L); d, leg 4, ventral (Z); e, leg 5, ventral (E).

that exhibited this same feature. This species, *H. japonicus*, was collected from the gills of the chaetodontid *Heniochus acuminatus* (Linnaeus) from Japan. It was later found and redescribed by Kabata (1965a) from the gills of *Platax teira* (Forsskål) at Moreton Bay, Queensland, Australia. Unfortunately, Yamaguti & Yamasu (1959) did not make comparisons with morphologically similar genera. If they had, Yamaguti & Yamasu would have realized that their species was synonymous with Rangnekar's *A. branchialis*. *Anuretes branchialis* was transferred to *Heniochophilus*, and *H. japonicus* relegated to synonymy with *H. branchialis* by Pillai & Mohan (1965). The genus *Heniochophilus* was then considered monotypic until Pillai (1977) described *H. indicus* from *P. teira* from India.

Heniochophilus was distinguished from *Anuretes* by the elongation of the thoracic area of the cephalothorax to dorsally cover the anteriormost portion of the genital complex in the female. In addition, the postantennal process and the process of the maxillule were reported to be reduced in size in *Heniochophilus* (Pillai, 1967). Unfortunately, the postantennal process is not a reliable generic discriminant, particularly since it may be present or absent even within the same species. Therefore, the extension of the thoracic zone of the cephalothorax of the female is the only major distinguishing feature between the two genera. However, there are species of *Anuretes* (e.g., *A. heckeli*, *A. menehune*, and *A. plectorhynchi*) that exhibit this characteristic. Consequently, *Heniochophilus* should be considered a junior synonym of *Anuretes*, as first suggested by Pillai (1967).

Anuretes branchialis has so far been recorded from six species of fishes belonging to five families, and has been found in the Indo-West Pacific (table VII). The specimens that we examined were obtained from the gill filaments of the hosts with the head of the parasite imbedded in the gill tissue, and the anterior end of its cephalothorax oriented toward the base of the gill filament.

Genus **Arrama** Dojiri & Cressey, 1991

Arrama Dojiri & Cressey, 1991: 594; Boxshall & Montú, 1997: 48; Boxshall & Halsey, 2004: 725.

Female. — Cephalothorax caligiform, incorporating pedigerous somites 1-3; its lateral zone expanded into aliform structure and curved ventrad. Frontal plate divided into 2 parts, with distinct gap between them. Nauplius eye present. Posterior sinuses absent. Fourth pedigerous somite forming short neck, without dorsal aliform plates. Genital complex with ventral egg sac

TABLE VII
Hosts and localities of collections of *Anuretes branchialis* Rangnekar, 1953, including records of this species reported under its synonyms

Host family	Host species	Locality	Reference
Scombridae	*Katsuwonus pelamis* (Linnaeus)	Bombay coast	Rangnekar, 1953
Chaetodontidae	*Heniochus acuminatus* (Linnaeus)	Shirahama, Japan	Yamaguti & Yamasu, 1959
Ephippidae	*Platax teira* (Forsskål)	Moreton Bay, Australia	Kabata, 1965a
	Platax teira	Philippines	Present paper
		Celebes	Present paper
	Platax orbicularis (Forsskål)	Ceylon (Sri Lanka)	Present paper
		Taiwan	Lin & Ho, 2002a
Polynemidae	*Eleutheronema tetradactylum* (Shaw)	Arabian Sea	Pillai & Mohan, 1965
Monodactylidae	*Monodactylus argenteus* (Linnaeus)	Taiwan	Lin & Ho, 2002a

attachment area. Abdomen small, wider than long. Caudal ramus partially fused to abdomen and bearing 4 setae. Egg sacs long, uniseriate.

Antennule 2-segmented; proximal segment with 15 pilose setae; distal segment with 1 posterior seta and 12 apical setae plus 1 aesthetasc. Antenna 3-segmented. Postantennal process absent. Mouth tube with protuberant anteroventral surface. Mandible of usual form, bearing 12 teeth on medial margin of distal blade. Maxillule with setiferous lobe and dentiform process with setiform tip. Maxilla 2-segmented, brachiform, bearing flabellum on brachium and relatively stout calamus and canna distally. Maxilliped prehensile, with massive protopod (corpus); distal subchela comprising fused endopodal segment and terminal claw. Sternal furca absent.

Leg 1 biramous, with 2-segmented exopod and greatly reduced endopod; small digitiform process located near origin of endopod. Leg 2 biramous, with 2-segmented rami. Ventral apron absent. Legs 3 and 4 reduced to setiferous lobes. Leg 5 represented by single naked seta near egg sac attachment area. Leg 6 absent.

Male. — Cephalothorax similar to that in female. Genital complex comparatively smaller. All appendages as in female, except for bearing 17 pilose setae on proximal segment of antennules and second segment of antenna with corrugated medial surface.

Type-species. — *Arrama tandani* Dojiri & Cressey, 1991.

Remarks. — This genus can be distinguished from all other caligid genera by a combination of characters that include: (1) only 4 setae on the caudal ramus, (2) only 14 or 15 setae on the first segment of the antennule, (3) the reduced apical armature of the leg 1 exopod, (4) the 2-segmented rami of leg 2, (5) the absence of the ventral apron of leg 3, and (6) the reduction of legs 3 and 4 to setiferous lobes.

A reduced number of setae occurs on the caudal ramus of *Arrama*. Both species of the genus bear only four setae on the caudal ramus, a characteristic shared only with one other caligid genus, *Belizia* Cressey, 1990. Although four genera (i.e., *Anuretes*, *Kabataella*, *Mappates*, and *Pseudanuretes*) have five caudal setae, the majority of the caligid genera possess six, which most likely represents the plesiomorphic condition in the entire order Siphonostomatoida. *Arrama* Dojiri & Cressey, 1991 also has one of the fewest number of setae on the first segment of the antennule, with only 14 or 15 setae. *Abasia* is similar with 15-19, but the predominant condition in caligid genera appears to be 22-27.

The exopodal armature of leg 1 is also reduced in number in *Arrama*. The outer proximal exopodal spine, which is slightly displaced distally, is present and represents the outer distal spine of the first exopodal segment as in other caligid genera. However, the terminal exopodal segment shows a reduced number of elements (setae/spines). Typically, in relatively unmodified siphonostomes such as *Dirivultus*, the terminal exopodal segment of leg 1 exhibits the armature formula III, 4 (three spines and four setae) (Humes & Dojiri, 1980: 146, fig. 2f). This armature is also found in the majority of caligids in which there are three clawlike spines at the tip, one pinnate seta on the inner distal corner of the segment, and three large pinnate setae on the inner margin (Dojiri, 1979: 254, fig. 2c, e). *Arrama cordata* Dojiri & Cressey, 1991 bears only six elements (three spines and three setae); one seta is missing from the plesiomorphic condition of seven. The armature is further reduced to five (four spines and one seta) in *A. tandani* Dojiri & Cressey, 1991. This reduction in the leg 1 armature is not unique to this genus. Other caligids exhibit a reduction in the exopodal armature of leg 1 (Dojiri, 1983).

Although *Arrama* exhibits many characters that are considered apomophic (e.g., the shape of the cephalothorax, the shape of the genital complex, the reduced number of setae discussed above, and the loss of the ventral apron of leg 3 discussed below), one feature stands out as a plesiomorphic character. This is the maxillule, which consists of two parts. Lewis (1969) suggested that the sympod of an originally biramous maxillule may have been suppressed, resulting in a separation of the dentiform process (one ramus) and a setiferous papilla (second ramus). In most caligids, the dentiform structure is a curved triangular process, somewhat removed from a setiferous papilla. However, in genera such as *Abasia*, *Dartevellia*, and *Arrama*, the setiferous papilla is a distinct lobe bearing three setae. The dentiform process, a remnant of a setiferous ramus, in *Arrama* is tipped with a pinnate setiform element. These two structures represent an originally bipartite appendage as suggested by Lewis (1969).

The most distinguishing features of this genus are the folding of the cephalothorax and the absence of the ventral apron of leg 3. Apparently, the predilection of the two species of *Arrama* for the gill filaments has helped to determine the morphology of the cephalothorax. Because of the significance of these morphologic adaptations, they are discussed in more detail below.

Many caligid genera exhibit a dorsoventrally flat, suction-cuplike dorsal shield. These genera, as expected, live on the external body surfaces, buccal cavity walls, and branchial cavity walls of their hosts. These attachment sites are all smooth surfaces that provide excellent substrates for the suction-cuplike cephalothorax to produce a tight seal. The shape and function of the cephalothorax appears to have evolved in response to the caligids' predilection for this type of attachment surface. Smooth surfaces, which are the predominant substrate among members of the Caligidae, most likely represent the plesiomorphic attachment sites.

In members of the Caligidae, the posterior margin of the cephalothorax is completely closed off by the enlargement of the intercoxal plate and sympod of leg 3 to form a broad structure referred to as the ventral apron, thus improving its suction capability (Kabata, 1979). Transparent marginal membranes rim the anterior margins of the frontal plate and lateral margins of the cephalothorax and help seal off the concavity beneath the dorsal shield (Kabata, 1979).

Species of some caligid genera (i.e., *Abasia* and *Hermilius*), however, have become adapted to a more sedentary existence on the gill filaments and in the nasal cavities of their hosts as is the case for *Arrama* and *Kabataella*. The cephalothorax in *Abasia*, *Hermilius*, and *Arrama* is folded longitudinally so that the lateral areas are bent ventrally. This alteration of the cephalothorax results in a median longitudinal groove that envelopes laterally the gill filament of the host. The shape of the cephalothorax of these genera and the corrugated surface of the gill filament does not suit a suction-cuplike attachment. Presumably, the ventral aprons in both *Abasia* and *Hermilius*, although present, have lost their function. The ventral apron in *Kabataella*, which lives in the nasal cavities of its host, is reduced in size, and most likely has lost its function also. Apparently, the most derived state is exhibited by *Arrama*, which lacks the ventral apron. In addition, both legs 3 and 4 are reduced to small setiferous lobes further distinguishing this genus from all other caligid genera.

The type-species is a parasite living on the gill filaments of the plotosid catfish, *Cnidoglanis macrocephalus* (Valenciennes), from Australia and the only known congener, *A. cordata* Dojiri & Cressey, 1991, is also a gill filament parasite of another species of Australian plotosid catfish, *Paraplotosus* sp.

Arrama tandani Dojiri & Cressey, 1991
(figs. 42-45)

Arrama tandani Dojiri & Cressey, 1991: 596.

Material examined. — None. Description given below is modified from Dojiri & Cressey (1991).

Female. — Body as in fig. 42a, b. Total length 3.06 mm (2.93-3.19 mm) and width at widest points 1.48 mm (1.33-1.65 mm) (n = 3). Cephalothorax (fig. 42c) much wider than long, 0.79 mm (0.73-0.90 mm) × 1.48 mm (1.33-1.65 mm); tips of antennule well within lateral limits of cephalothorax; posterior sinuses absent; lateral zone expanded into aliform structure, curved ventrad, and forming midventral longitudinal groove encompassing gill filament of host; tips of lateral zone extending slightly beyond posterior margin of thoracic zone; thoracic zone demarcated anteriorly by partial suture and approximately one-fourth length of cephalic zone. Fourth pedigerous somite (fig. 42c) forming short neck, not well defined; this somite and genital complex together forming trunk, combined measuring 2.00 mm (1.95-2.10 mm) × 0.97 mm (0.87-1.14 mm); trunk greatly constricted and narrower at posterior end near junction with abdomen. Abdomen (fig. 42d, e) small, wider than long, 0.14 mm (0.11-0.19 mm) × 0.23 mm (0.18-0.31 mm). Caudal ramus (fig. 42e) fused to abdomen, bearing 1 naked and 3 pinnate setae. Egg sac attachment area (fig. 42d) at ventrolateral corner.

Frontal plate (fig. 42a, c) in 2 sections, with distinct gap between halves; reduced striated marginal membrane at anterior end; 2 oval pits located in center of anterior margin of cephalothorax, and directly posteromedial to frontal plate. Antennule (fig. 42f) 2-segmented; first segment relatively slender, with 15 plumose setae; second segment cylindrical and bearing 1 subapical seta and 12 setae + 1 aesthetasc (2 posterior setae sharing common base) at tip. Antenna (fig. 43a) 3-segmented; first segment unarmed, relatively small, and without posteriorly directed spiniform process; second segment longest of

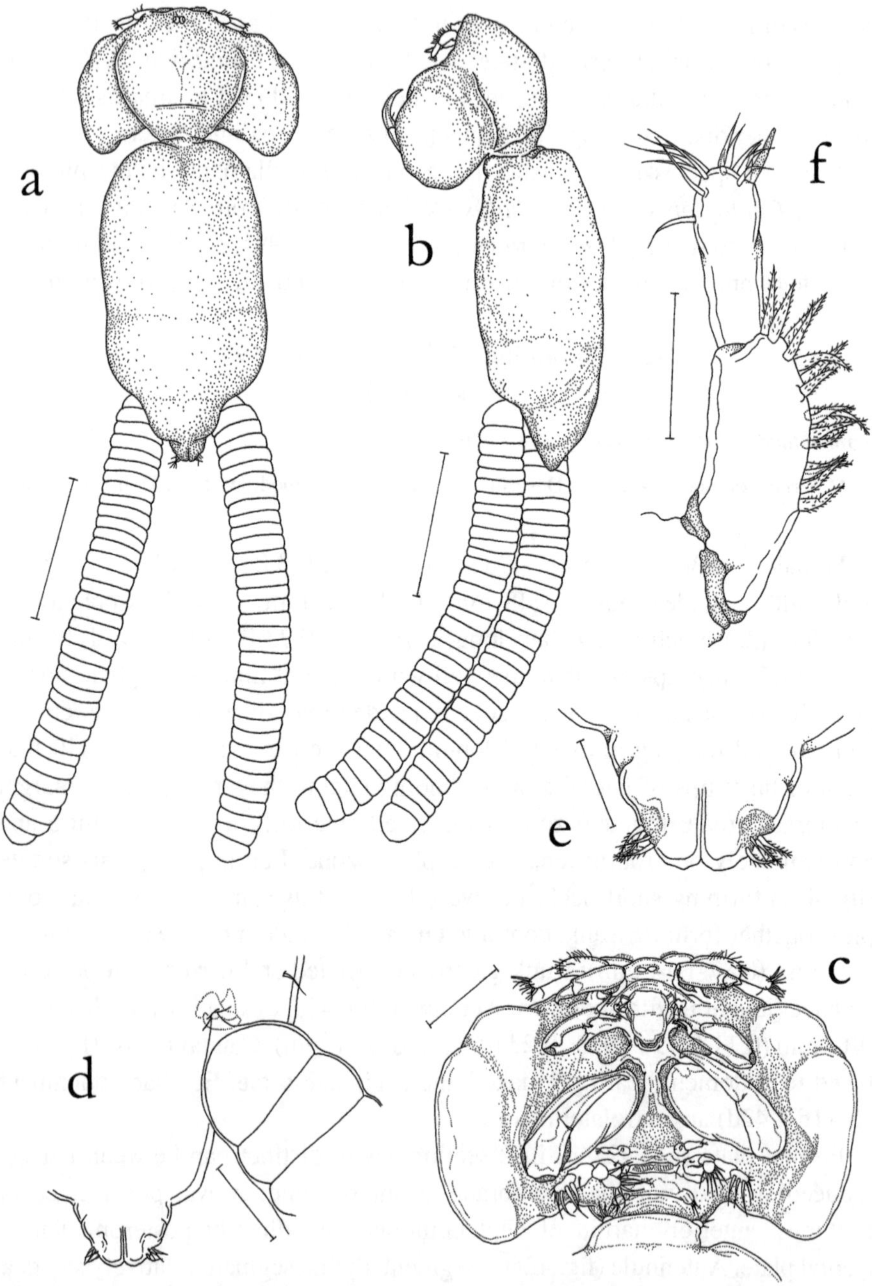

Fig. 42. *Arrama tandani* Dojiri & Cressey, 1991, female. a, body, dorsal; b, same, lateral; c, cephalothorax, ventral; d, posterior portion of urosome with egg sac, ventral; e, anal somite, ventral; f, antennule, ventral. Scale bars: 1.0 mm in a, b; 0.3 mm in c; 0.2 mm in d; 0.1 mm in e, f. (From Dojiri & Cressey, 1991 by permission of the Journal of Crustacean Biology.)

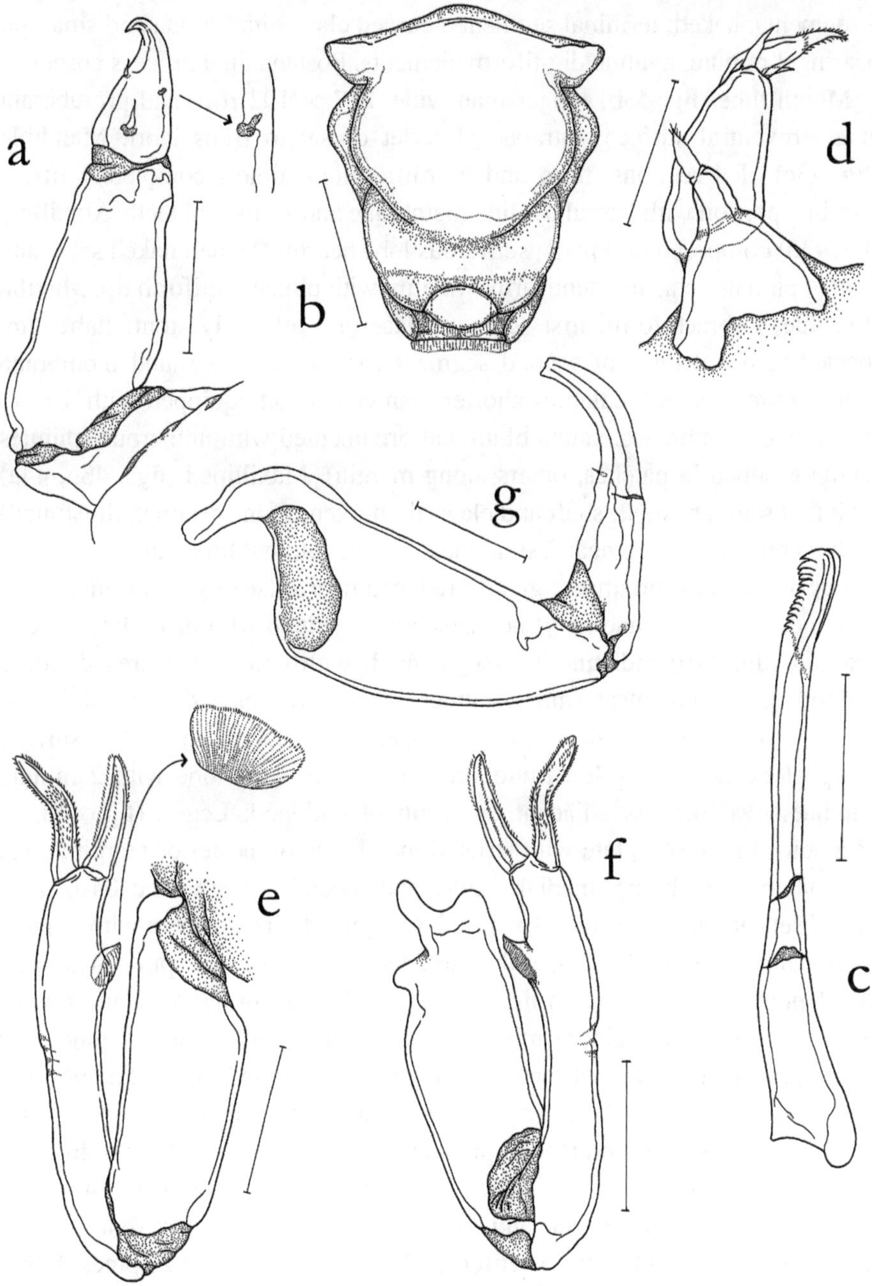

Fig. 43. *Arrama tandani* Dojiri & Cressey, 1991, female. a, antenna, ventral; b, mouth tube, ventral; c, mandible, ventral; d, maxillule, posterior; e, maxilla, ventral; f, same, dorsal; g, maxilliped, dorsal. Scale bars: 0.1 mm in a, b, e-g; 0.05 mm in c, d. (From Dojiri & Cressey, 1991 by permission of the Journal of Crustacean Biology.)

3 segments, naked; terminal segment a curved claw with 2 setae and small pit bearing 2 minute, hyaline digitiform elements. Postantennal process absent.

Mouth tube (fig. 43b) longer than wide, 217 × 192 μm, and protuberant at anteroventral surface; intrabuccal stylet dorsal to frons labri. Mandible (fig. 43c) of 4 sections; third and fourth sections almost completely fused; terminal portion with lateral hyaline membrane and 12 medial teeth. Maxillule (fig. 43d) composed of 2 parts; setiferous lobe bearing 2 small naked setae and 1 large pinnate seta, and dentiform structure with pinnate setiform tip. Maxilla (fig. 43e, f) brachiform; first segment (lacertus) relatively stout; flabellum, located at distal third of second segment (brachium), a striated membrane with serrated margin; calamus shorter than canna and equipped with 3 rows of serrated membranes; canna blunt and ornamented with numerous spinules (some grouped in patches, others along margin). Maxilliped (figs. 43g, 44a) with first segment stout; shaft and claw (tip broken off in specimen illustrated) with incomplete suture near 2 small naked setae. Sternal furca absent.

Leg 1 (fig. 44b) biramous, greatly reduced in size (see fig. 42c); intercoxal plate not sclerotized and very inconspicuous; sympod with distal lateral seta. Exopod (fig. 44b) indistinctly 2-segmented, with constricted area dividing exopod into 2 segments; terminal segment with lateral spine, 4 terminal spines, and 1 terminal seta; truncate process (opening to duct?) on ventral surface at tip of exopod (fig. 44c). Endopod (fig. 44b) a small lobe with 2 minute papillae; oval process adjacent to origin of endopod. Leg 2 (fig. 44d, e) biramous; intercoxal plate with striated membrane on posterior margin; coxa with lateral membrane, medial setule, and medial pinnate seta; basis with lateral seta and medial setule. Exopod 2-segmented; first segment with striated marginal membrane, 1 unilaterally barbed spine with row of spinules at base, and 1 pinnate seta (shown in fig. 44d, but broken off in fig. 44e) and row of setules on medial margin; terminal segment with 2 barbed lateral spines and 1 semipinnate and 4 pinnate setae. Endopod 2-segmented; first segment (with relatively straight lateral margin in fig. 44d, but protuberant margin in fig. 44e) bearing lateral setules and medial pinnate seta; second segment (fig. 44d) with lateral patch of hairs and 6 pinnate setae (7 in fig. 44e). Leg 3 (fig. 44f, g) a setiferous lobe bearing 3 setae and 3 or 4 relatively large setules; ventral apron absent. Leg 4 (fig. 44f) also a setiferous lobe with 1 seta and 2 spines. Leg 5 (fig. 42d) represented by single naked seta near egg sac attachment area.

Male. — Body as in fig. 45a. Total length 1.31 mm (1.23-1.41 mm). Cephalothorax about one-half total length, wider than long, 0.58 mm (0.51-0.65 mm) × 0.83 mm (0.76-0.93 mm), and similar in shape to that of female.

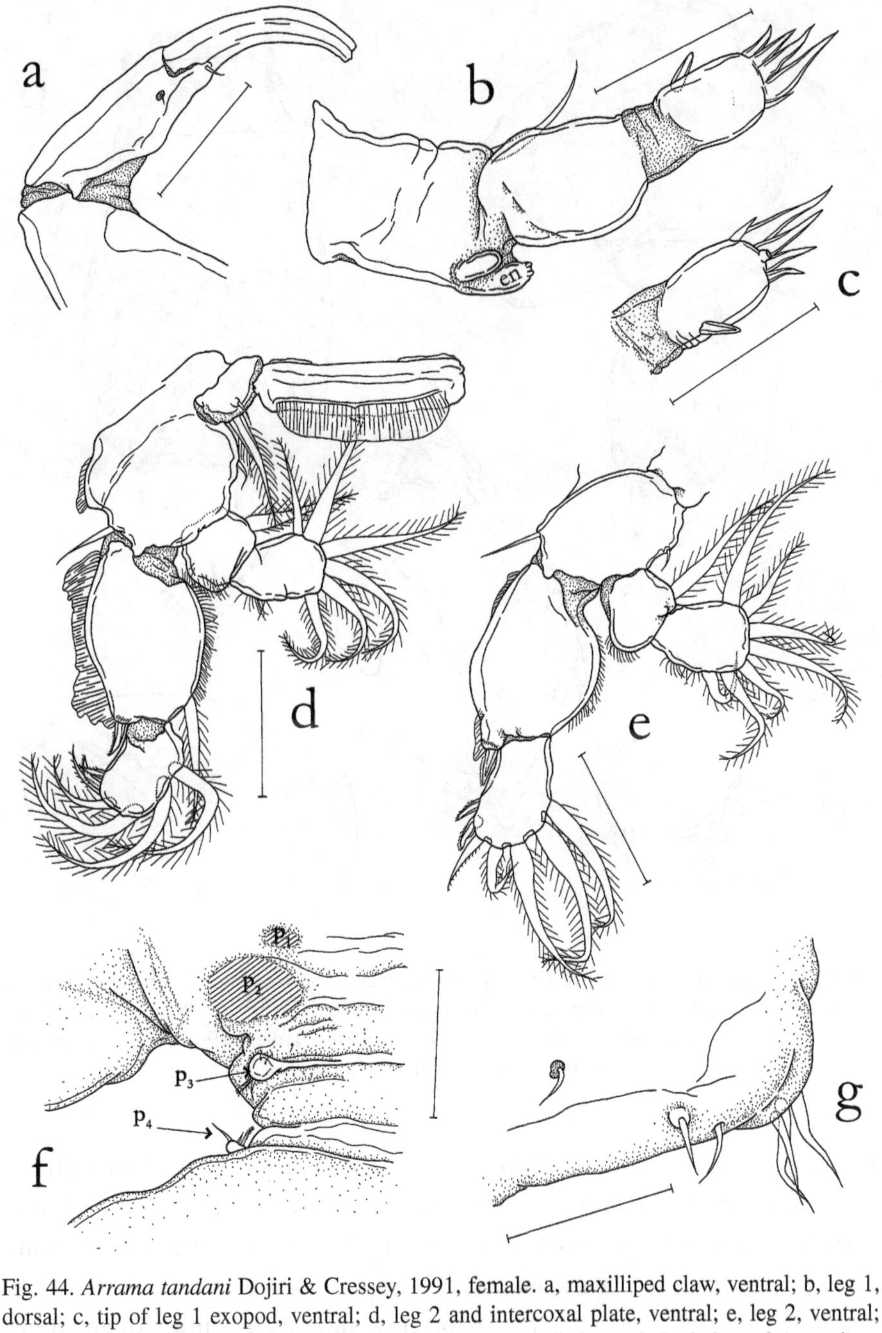

Fig. 44. *Arrama tandani* Dojiri & Cressey, 1991, female. a, maxilliped claw, ventral; b, leg 1, dorsal; c, tip of leg 1 exopod, ventral; d, leg 2 and intercoxal plate, ventral; e, leg 2, ventral; f, legs 3 and 4 (legs 1 and 2 removed), ventral; g, leg 3, anteroventral. Scale bars: 0.1 mm in a, d, e; 0.05 mm in b, c, g; 0.2 mm in f. Symbols: en = endopod; p1-p4 = legs 1-4. (From Dojiri & Cressey, 1991 by permission of the Journal of Crustacean Biology.)

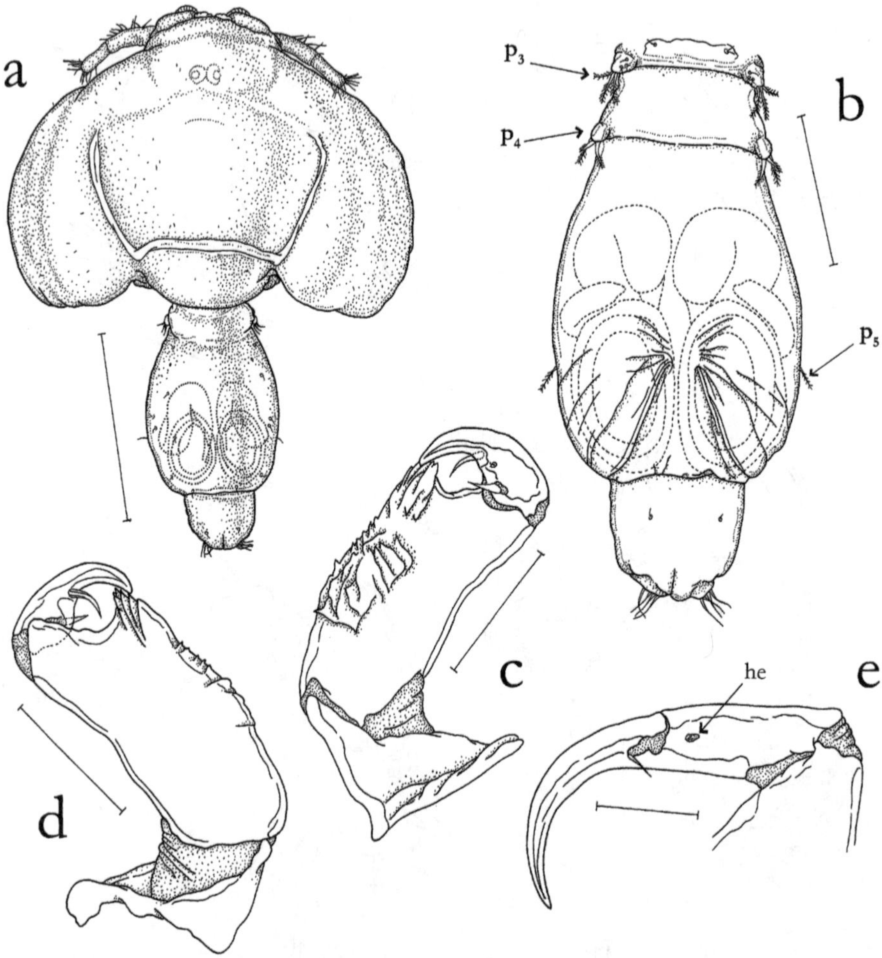

Fig. 45. *Arrama tandani* Dojiri & Cressey, 1991, male. a, body, dorsal; b, urosome, ventral; c, antenna, ventral; d, same, dorsal; e, maxilliped claw, ventral. Scale bars: 0.5 mm in a; 0.2 mm in b; 0.1 mm in c, d. Symbols: p3-p5 = legs 3-5; he = hyaline element. (From Dojiri & Cressey, 1991 by permission of the Journal of Crustacean Biology.)

Trunk subovoid in outline, 0.40 mm (0.35-0.42 mm) × 0.28 mm (0.25-0.32 mm). Abdomen slightly wider than long, quadrangular, and 0.11 mm (0.09-0.12 mm) × 0.15 mm (0.14-0.19 mm). Caudal ramus (fig. 45b) partially fused to abdomen, and bearing 4 naked setae.

All appendages as in female, except those described below. Antennule as in female except 17 plumose setae on first segment. Antenna (fig. 45c, d) different from that of female; second segment with corrugated inner (medial) margin; claw with slightly different shape than in female (compare fig. 43a with fig. 45c). Claw of maxilliped (fig. 45e) more slender than in female.

Remarks. — The type-species *Arrama tandani* Dojiri & Cressey, 1991 may be distinguished from *A. cordata* Dojiri & Cressey, 1991, the only congener described so far, by the shape of the cephalothorax and genital complex, relative sizes of the four caudal setae, pattern of spinules on the canna of the maxilla, armature of the terminal exopodal segment of leg 1, and armature of the terminal exopodal segment of leg 2 (Dojiri & Cressey, 1991).

Both the type-species *A. tandani* and its congener *A. cordata* were collected from the gill filaments of the plotosids *Cnidoglanis macrocephalus* (Valenciennes) and *Paraplotosus* sp., respectively, from Australia. Plotosid fishes are commonly known as eel catfishes, stinging catfishes, coral reef catfishes, and barbel eels, inhabit coastal waters of the Indo-Pacific region, are predominantly estuarine or freshwater (a few species are marine), and are considered edible (Burgess, 1989).

Genus **Avitocaligus** Boxshall & Justine, 2005

Avitocaligus Boxshall & Justine, 2005: 349.

Female. — Cephalothorax caligiform, incorporating pedigerous somites 1-3. Cephalothoracic shield subcircular. Frontal plates without lunules. Nauplius eye present. Free fourth pedigerous somite very short, without dorsal aliform plates. Genital complex distinctly broader than cephalothorax; armed with 2 pairs of (dorsal and ventral) large expanded, lamelliform plates. Abdomen composed of 2 free somites; proximal somite short, with pair of broad, ventral lamelliform plates; distal somite elongate, concealed ventrally by 2 abdominal plates. Caudal ramus dorsoventrally flattened, armed with 6 plumose setae. Egg sacs long, uniseriate, coiled, and concealed in space between ventral genital plates and abdominal plates.

Antennule 2-segmented; proximal segment with 27 plumose setae; distal segment with 1 posterior seta and 12 apical setae, no aesthetasc. Antenna 3-segmented. Postantennal process absent. Mouth tube formed by tapering labrum and labium. Mandible of usual form, bearing 12 teeth on medial margin of distal blade. Maxillule bilobed, anterior lobe a papilla bearing 3 setae and posterior lobe an unarmed spiniform process. Maxilla 2-segmented, brachiform, and bearing calamus and canna distally. Maxilliped subchelate, with massive protopod (corpus) and distal subchela comprising fused endopodal segment and terminal claw. Sternal furca absent.

Leg 1 with 2-segmented exopod and endopod; distal segment of endopod with 3 long plumose setae. Leg 2 biramous, both rami 3-segmented. Leg 3 biramous, both rami 3-segmented; apron present; basal swelling and clawlike spine of first exopodal segment absent. Leg 4 biramous, both rami 2-segmented. Leg 5 and leg 6 unknown.

Male. — Unknown.

Type-species. — *Avitocaligus assurgericola* Boxshall & Justine, 2005.

Remarks. — This is a monotypic genus and known only from one report made by Boxshall & Justine (2005). So far, only one female parasite was found under the operculum of a razorback scabbardfish, *Assurger anzac* (Alexander, 1917), collected off the east coast of Lifou Island, Loyalty Islands, New Caledonia (Boxshall & Justine, 2005).

The most outstanding characteristics of this genus are the expanded lamelliform plates on the genital complex and abdomen, and biramous leg 4. As in species of *Euryphorus*, *A. assurgericola* bears a 2-segmented leg 1 endopod with 3 plumose setae on the terminal segment and also has a biramous leg 4, but unlike *Euryphorus*, this genus is characterized by having a 2-segmented (instead of 3-segmented) exopod.

Avitocaligus shares a 2-segmented leg 1 endopod and 3-segmented leg 3 endopod with four other genera, i.e., *Euryphorus*, *Pupulina*, *Gloiopotes*, and *Alebion*. All five of these genera are paraphyletic and basal to the remaining caligid genera (see Phylogeny of Caligidae section).

Avitocaligus assurgericola Boxshall & Justine, 2005
(figs. 46-49)

Avitocaligus assurgericola Boxshall & Justine, 2005: 352.

Material examined. — None. Description provided below is modified from Boxshall & Justine (2005).

Female. — Body as in fig. 46a-c. Total length (not including setae on caudal ramus) 14.5 mm. Cephalothorax subcircular, distinctly smaller than genital complex. Frontal plate thick, lacking marginal membrane. Lunules absent. Tip of antennule well within lateral limit of cephalothorax.

Fourth pedigerous somite very short, lacking dorsal aliform plate and appearing as 'waist' between cephalothorax and genital complex. Latter with 2 pairs of large plates: dorsal lamellae and middle lamellae (fig. 46a, b). Dorsal lamellae originating along posterior margin of genital complex, produced posteriorly into free plate with its medial half folded upward. Middle lamellae originating anterior to dorsal lamellae, extending posteriorly along margin of

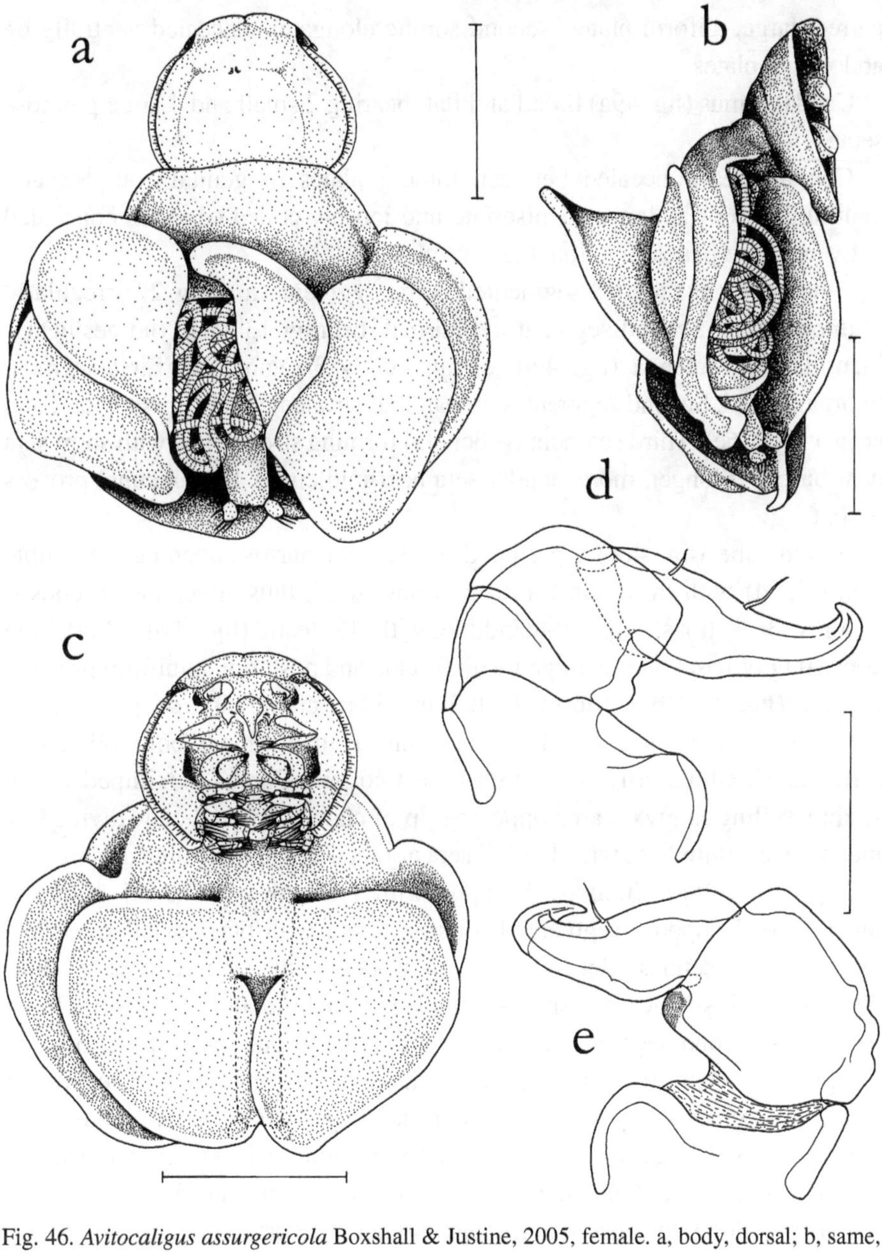

Fig. 46. *Avitocaligus assurgericola* Boxshall & Justine, 2005, female. a, body, dorsal; b, same, lateral; c, same, ventral; d, antenna, ventral; e, antenna, dorsal. Scale bars: 5.0 mm in a-c; 0.5 mm in d, e. (From Boxshall & Justine, 2005 by permission of Folia Parasitologica.)

genital complex forming a pair of horizontal plates (fig. 46b, c). Abdomen (fig. 46c) long and composed of 2 free somites; first somite short, bearing paired, large, aliform plates; second somite elongate, concealed ventrally by abdominal plates.

Caudal ramus (fig. 49a) broad and flat, bearing 2 small and 4 large plumose setae.

Genital area concealed between middle plates of genital complex and ventral plates of abdomen. Uniseriate and loosely coiled egg sacs concealed between these plates (fig. 46a, b).

Antennule (fig. 47a) 2-segmented. First segment carrying 27 irregularly plumose setae. Second segment cylindrical, bearing 13 setae (no aesthetasc identifiable). Antenna (fig. 46d, e) apparently 3-segmented. First segment unornamented. Second segment with patch of striated cuticle dorsally at distal corner (fig. 46e). Third segment (subchela) forming strong claw with short seta near base and longer, more slender seta near midlength. Postantennal process absent.

Mouth tube (fig. 46c) tapering distally, with narrow opening. Mandible (fig. 47c, d) with third and fourth sections fused, thus appearing to consist of only 3 sections. Tip of mandible with 12 teeth (fig. 47d). Maxillule consisting of basal papilla tipped with 3 setae and posterior spiniform process. Maxilla (fig. 47e) brachiform. Flabellum short, comprising deeply serrated membrane. Calamus about 1.5 times longer than canna; both bilaterally serrated. Maxilliped (fig. 47b) with robust corpus. Corpus maxillipedis with slight swelling in myxal area opposing tip of claw. Shaft and claw divided by incomplete suture line, with 1 small seta at junction. Sternal furca absent.

Leg 1 (fig. 48a) biramous. Sympod with 1 naked, outer and 1 plumose, inner setae. Exopod 2-segmented. First segment with outer distal spine and inner row of setules. Second segment tipped with 3 spines and 1 short seta. Spine 1 simple, but spines 2 and 3 with bifid tips (fig. 48b). Inner margin of second segment of exopod with 3 large plumose setae. Endopod 2-segmented, with first segment unarmed. Second segment with outer row of setules and inner group of 3 plumose setae. Leg 2 (fig. 48c) biramous, with small coxa and large basis. Coxa with large plumose inner seta and surface sensilla; basis with short outer distal spine, single sensilla, flap of marginal membrane along posterior margin and another membrane along outer margin extending onto first exopodal segment. Both rami 3-segmented, with usual armature. Outer spines on exopodal segments (fig. 48d) unipectinate. Second endopodal segment elongate, with thickened lateral margin lacking row of pinnules. Intercoxal plate of leg 3 expanded to form broad transverse plate (fig. 49b, c), carrying flap of membrane along posterior margin. Coxa and basis

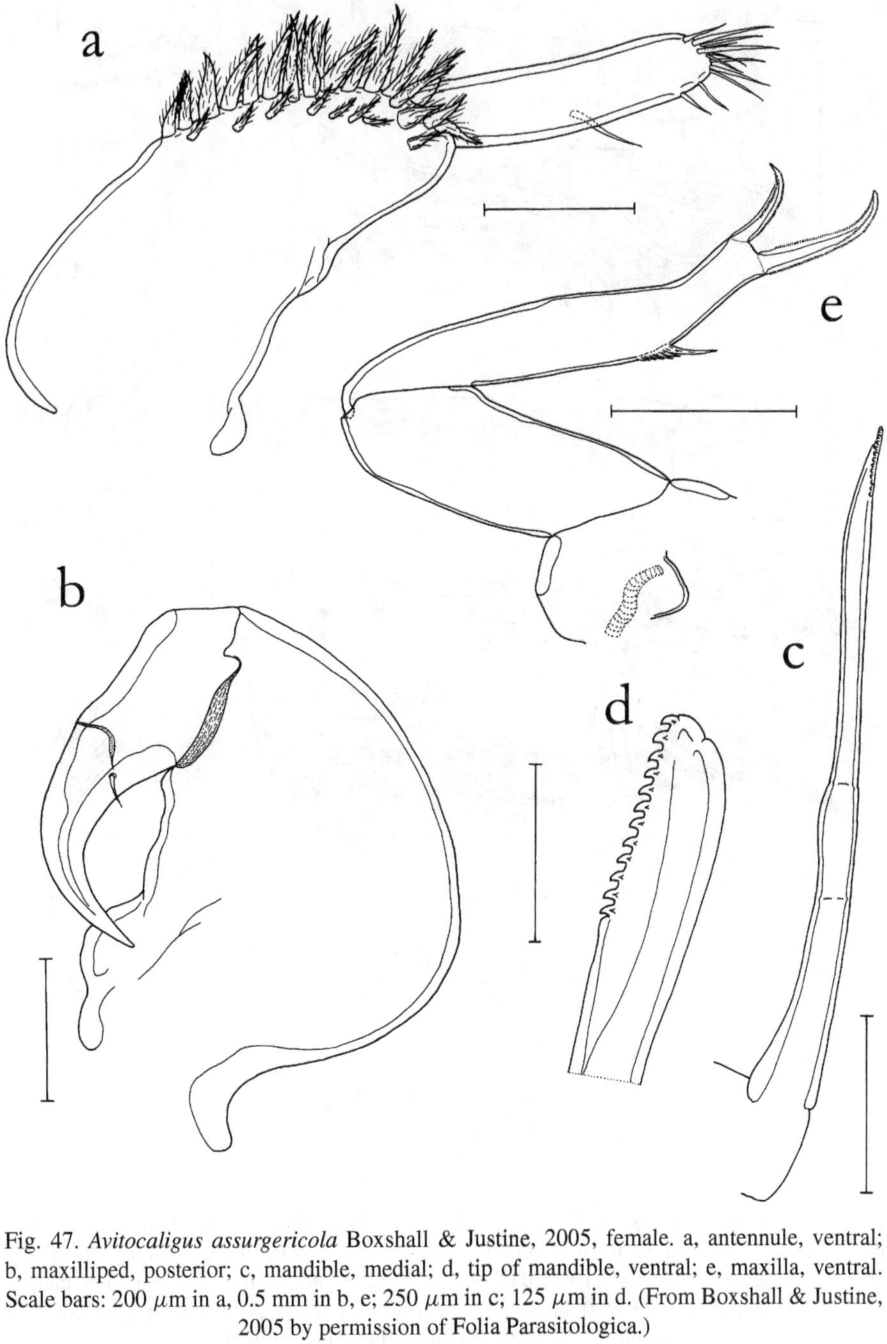

Fig. 47. *Avitocaligus assurgericola* Boxshall & Justine, 2005, female. a, antennule, ventral; b, maxilliped, posterior; c, mandible, medial; d, tip of mandible, ventral; e, maxilla, ventral. Scale bars: 200 μm in a, 0.5 mm in b, e; 250 μm in c; 125 μm in d. (From Boxshall & Justine, 2005 by permission of Folia Parasitologica.)

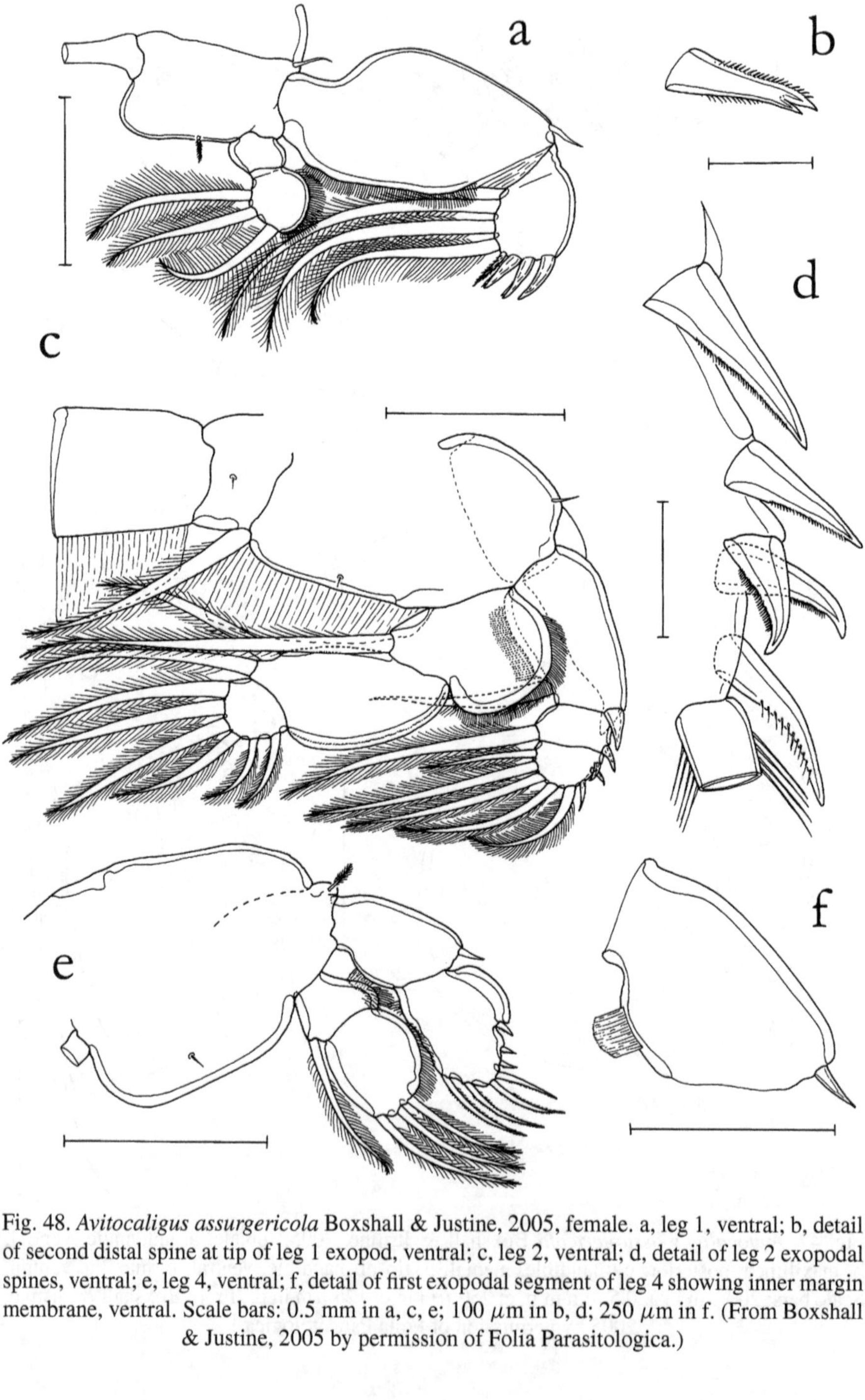

Fig. 48. *Avitocaligus assurgericola* Boxshall & Justine, 2005, female. a, leg 1, ventral; b, detail of second distal spine at tip of leg 1 exopod, ventral; c, leg 2, ventral; d, detail of leg 2 exopodal spines, ventral; e, leg 4, ventral; f, detail of first exopodal segment of leg 4 showing inner margin membrane, ventral. Scale bars: 0.5 mm in a, c, e; 100 μm in b, d; 250 μm in f. (From Boxshall & Justine, 2005 by permission of Folia Parasitologica.)

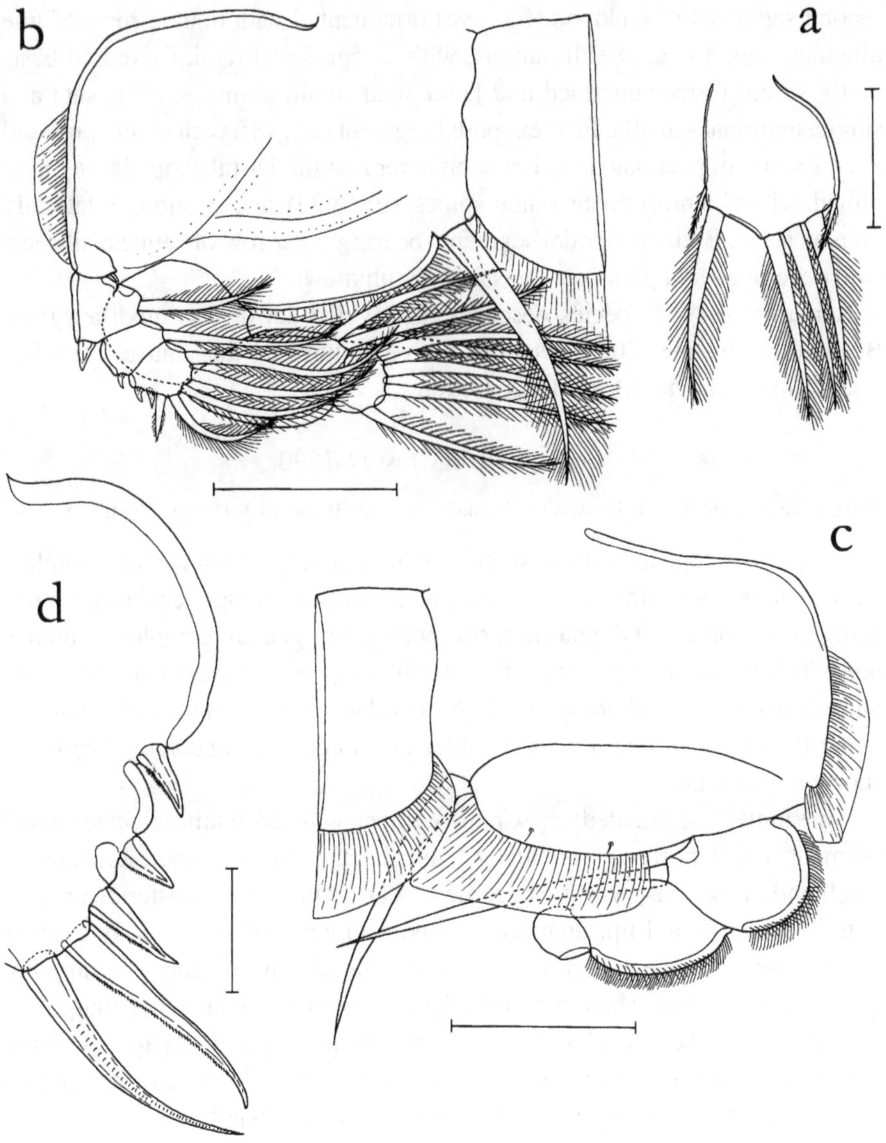

Fig. 49. *Avitocaligus assurgericola* Boxshall & Justine, 2005, female. a, caudal ramus, dorsal; b, leg 3, ventral; c, leg 3 with exopod and most of endopodal seta excluded, dorsal; d, leg 4, detail of outer margin of distal exopodal segment, ventral. Scale bars: 0.5 mm in a-c; 100 μm in d. (From Boxshall & Justine, 2005 by permission of Folia Parasitologica.)

incompletely fused, former with 1 outer marginal membrane and latter with 1 posterior marginal membrane plus 2 sensilla. Both rami 3-segmented. Exopod (fig. 49b) with stout, naked outer spines, but distal spine bipinnate. First and second segments of endopod (fig. 49c) ornamented with double rows of fine pinnules. Leg 4 (fig. 48e) biramous, with 2-segmented rami. Coxa and basis partly fused; former unarmed and latter with small, plumose, outer seta and single, marginal sensilla. First exopodal segment (fig. 48f) with outer spine and short section of pectinate membrane on inner margin. Distal exopodal segment armed with 4 unipectinate outer spines (fig. 49d) and 3 short, bilaterally pinnate setae. Both endopodal segments bearing outer row of setules; all setae well developed and plumose. Legs 5 and 6 unknown.

Remarks. — The species description provided above was modified from Boxshall & Justine (2005). For distinctive features of this unusual species, refer to the *Avitocaligus* Remarks section.

Genus **Belizia** Cressey, 1990

Belizia Cressey, 1990: 151; Boxshall & Montú, 1997: 48; Boxshall & Halsey, 2004: 725.

Female. — Cephalothoracic shield subcircular, larger than genital complex, with shallow posterior sinuses; its posterior free margin covering fourth pedigerous somite and anteriormost portion of genital complex. Lunules present. Nauplius eye present. Free fourth pedigerous somite without dorsal aliform plates. Genital complex subrectangular. Abdomen reduced, distinctly wider than long. Caudal ramus on outer distal corner of abdomen, bearing 4 short pinnate setae.

Antennule 2-segmented; proximal segment with 26 plumose setae; distal segment with 1 posterior seta and 12 apical setae plus 1 aesthetasc. Antenna small and uniramous, with first segment bearing prominent posterior process with spatulate round tip, unarmed second segment and short nearly straight distal claw. Postantennal process present. Mouth tube formed by tapering labrum and labium. Mandible comprising 4 sections, with inner margin of terminal section bearing 12 teeth. Maxillule comprising a small papilla bearing 3 setae and unarmed blunt tip process. Maxilla brachiform and maxilliped subchelate. Sternal furca reduced to pair of sclerotized knobs.

Leg 1 with 2-segmented exopod possessing distally 2 spines, 1 seta, and 3 plumose setae; endopod vestigial. Leg 2 biramous, with 3-segmented rami. Leg 3 biramous, with 3-segmented exopod and 2-segmented endopod; apron present and first endopodal segment inflated along outer margin (velum not separated from basal endopodal segment); endopod of leg 3 with only 4 setae. Leg 4 uniramous; coxa and basis fused; exopod 2-segmented and tipped with 2 spines. Leg 5 represented by setiferous lobes.

Male. — Unknown.

Type-species. — *Belizia brevicauda* Cressey, 1990.

Remarks. — This is a monotypic genus with the only known species reported from Belize in Central America. It was found parasitic in the branchial chambers of pluma porgy, *Calamus pennatula* Guichenot, and creole wrasse *Clepticus parrae* (Bloch & Schneider) (Cressey, 1990).

The type-species is characterized by possessing a combination of unusual features, e.g., the posterior margin of the thoracic zone of the cephalothorax completely covering the fourth pedigerous somite, a general habitus reminiscent of the *Anuretes* group (except for the presence of lunules; see Phylogeny of the Caligidae section), the presence of only four setae on the caudal ramus (majority of caligid genera carry 6 caudal setae), sternal furca being reduced to two sclerotized knobs, only four setae on the terminal segment of the leg 3 endopod, and the velum not distinct (separated) from the first endopodal segment of leg 3 as in most caligid genera, that provide support to Cressey (1990) for *Belizia* being recognized as a valid genus.

Belizia brevicauda Cressey, 1990
(figs. 50-52)

Belizia brevicauda Cressey, 1990: 151.

Material examined. — One female holotype (USNM 241651) and 4 female paratypes (USNM 241652) from branchial cavities of 3 *Calamus pennatula* Guichenot collected at Carrie Bow Cay, Belize by R. F. Cressey and H. B. Cressey (field collection number Cressey + Cressey CB362) on 8 December 1986. Specimens on loan from the National Museum of Natural History, Smithsonian Institution, Washington, D.C.

Female. — Body as in fig. 50a. Total length (excluding setae on caudal ramus) 1.97 mm (1.92-2.05 mm) × 1.23 mm (1.21-1.24 mm) (n = 4). Cephalothorax longer than wide, 1.43 mm (1.39-1.47 mm) × 1.23 mm (1.21-1.24 mm), with tips of antennules well within lateral limits of cephalothorax; posterior sinuses distinct; posterior margin of thoracic zone extending beyond posterior tip of lateral zone and completely covering fourth pedigerous somite. Fourth pedigerous somite wider than long, 74 μm (51-92 μm) × 300 μm (260-320 μm), and demarcation with genital complex difficult to distinguish. Genital complex roughly quadrangular, 0.56 mm (0.54-0.59 mm) × 0.86 mm (0.82-0.88 mm). Abdomen (fig. 50b) about 75 μm (60-95 μm) × 170 μm (140-190 μm). Caudal ramus (fig. 50b) about as long as wide, 30 μm (28-32 μm) × 40 μm (32-45 μm), and bearing 4 plumose setae.

Frontal plate distinct; lunules present. Antennule (fig. 50c) 2-segmened; first segment bearing 26 plumose setae; second segment cylindrical and carrying

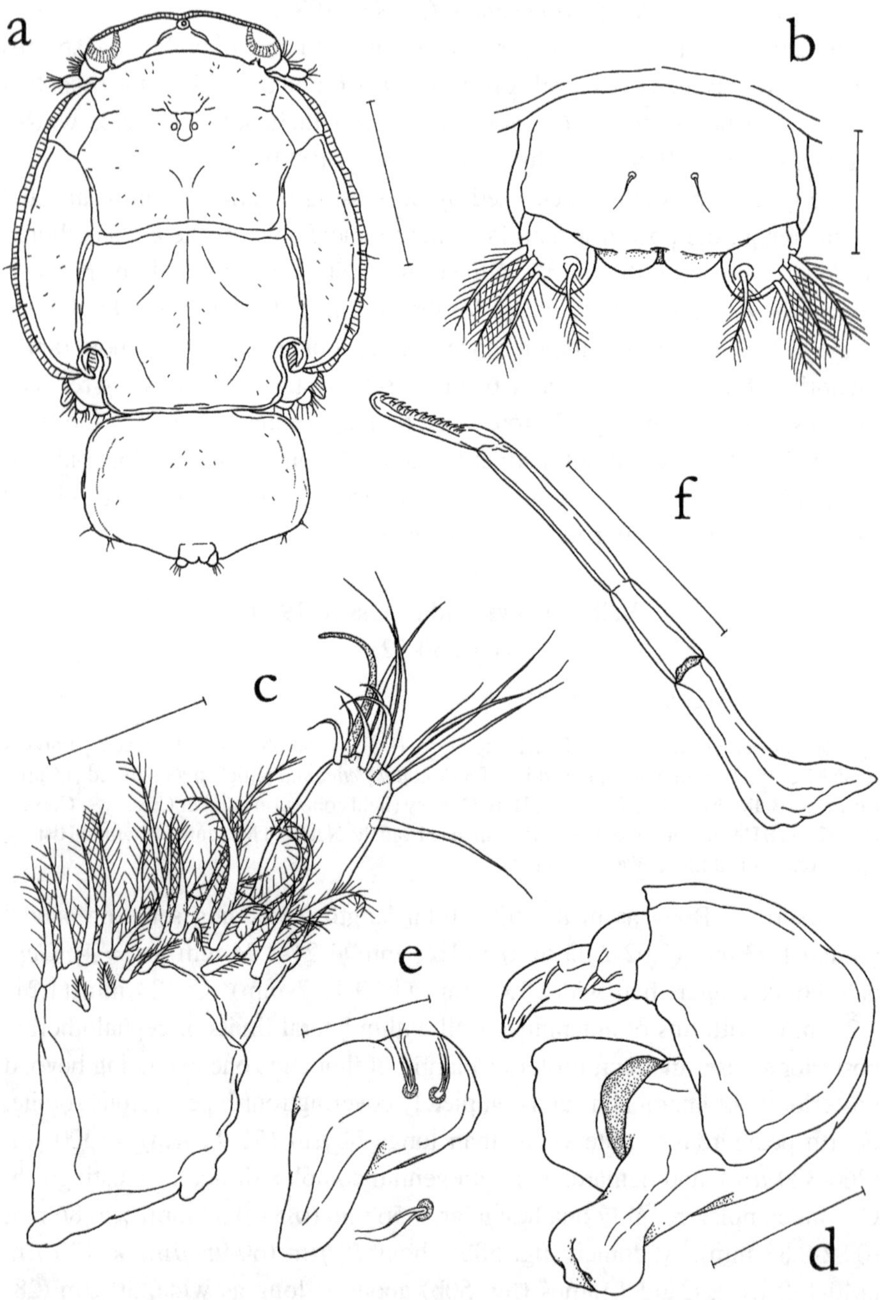

Fig. 50. *Belizia brevicauda* Cressey, 1990, female. a, body, dorsal; b, anal somite and caudal rami, dorsal; c, antennule, ventral; d, antenna, ventral; e, postantennal process, ventral. Scale bars: 0.5 mm in a; 0.05 mm in b-f.

13 + 1 aesthetasc (2 posterior terminal setae sharing common base). Antenna (fig. 50d) 3-segmented; first segment unarmed, small, and with club-shaped posteriorly directed spiniform process; second segment stout, longest of 3 segments, and pointed on outer distal corner; terminal segment a blunt curved claw with two naked setae on inner margin. Postantennal process (fig. 50e) stout, carrying a bifid setule near base and 2 trifid setules at base.

Mouth tube slender, about twice as long as wide, 225 μm × 112 μm, and possessing intrabuccal stylet; dentiferous strigil not observed and apparently absent. Mandible (fig. 50f) comprising 4 sections; terminal section with 12 medial teeth on inner margin and transparent flange on outer margin. Maxillule (fig. 51a) consisting of blunt dentiform process and papilla that bears 3 naked setae. Maxilla (fig. 51b) brachiform and slender; first segment (lacertus) naked; second segment (brachium) extremely slender, with flabellum that consists of striated membrane with serrated margin; calamus with 3 rows of serrated membranes, shorter than bilaterally serrated canna. Maxilliped (fig. 51c) 2-segmented; first segment with rounded knoblike process on outer margin, 2 small conical processes and a small naked seta on inner margin near base, and 1 small naked seta on inner margin closer to base of claw; shaft and claw with small seta on inner margin near incomplete suture. Pair of conical processes present in sternal area (fig. 51d) possibly representing reduced sternal furca; additional similar processes situated on either side of this "sternal furca".

Leg 1 with corrugated intercoxal plate (fig. 51d, e); sympod with outer naked seta; usual inner seta on sympod absent. Exopod 2-segmented (fig. 51e); first segment with small hump on outer margin, 1 small naked seta at outer distal corner, and row of setules along inner margin; second segment with 2 terminal spines armed with hyaline accessory processes on inner margins; third element (possibly representing seta 4) curved and twice as long as spines 1 and 2; 3 inner plumose setae stout, with pinnae along outer margin stouter than those of inner margin; endopod represented by small conical process. Pair of conical cuticular processes (fig. 51d) situated between intercoxal plates of legs 1 and 2. Leg 2 (fig. 52a, b) biramous, with intercoxal plate carrying a membranous flange on posterior margin; coxa with striated membranous flange on dorsal surface and 1 plumose seta and 1 setule near intercoxal plate; basis with 1 small naked seta on outer distal corner and 1 setule and striated membrane along inner margin. Exopod (fig. 52b) 3-segmented; first segment with dorsal striated membrane, 1 outer distal, bilaterally serrated, stout spine that bears a pectinate membrane at its base, and a row of setules and 1 large plumose seta along inner margin; second segment with unilaterally serrated

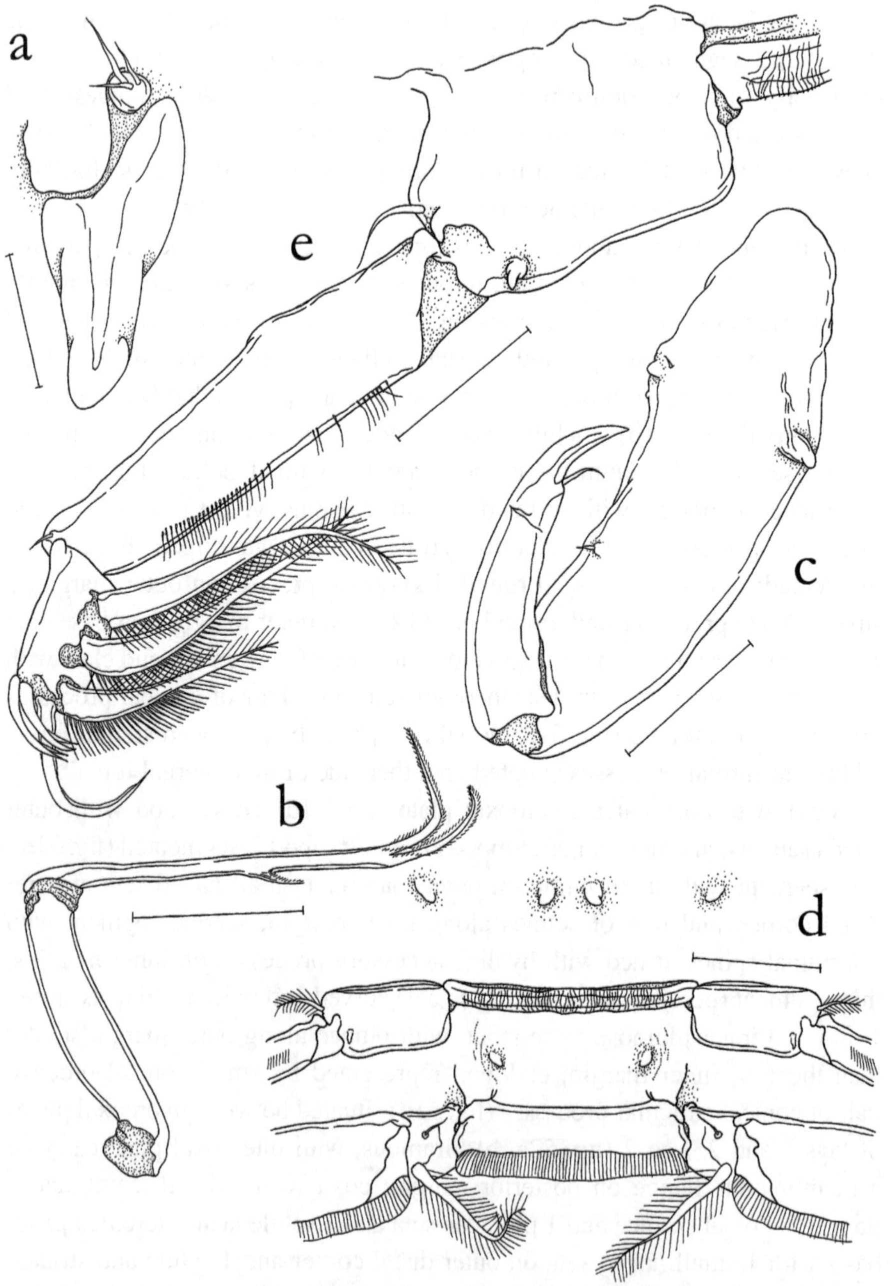

Fig. 51. *Belizia brevicauda* Cressey, 1990, female. a, maxillule, ventral; b, maxilla, ventral; c, maxilliped, ventral; d, sclerotized spiniform process in sternal area and between intercoxal plates of leg 1 and 2, ventral; e, leg 1 ventral. Scale bars: 0.03 mm in a; 0.1 mm in b and d; 0.05 mm in c and e.

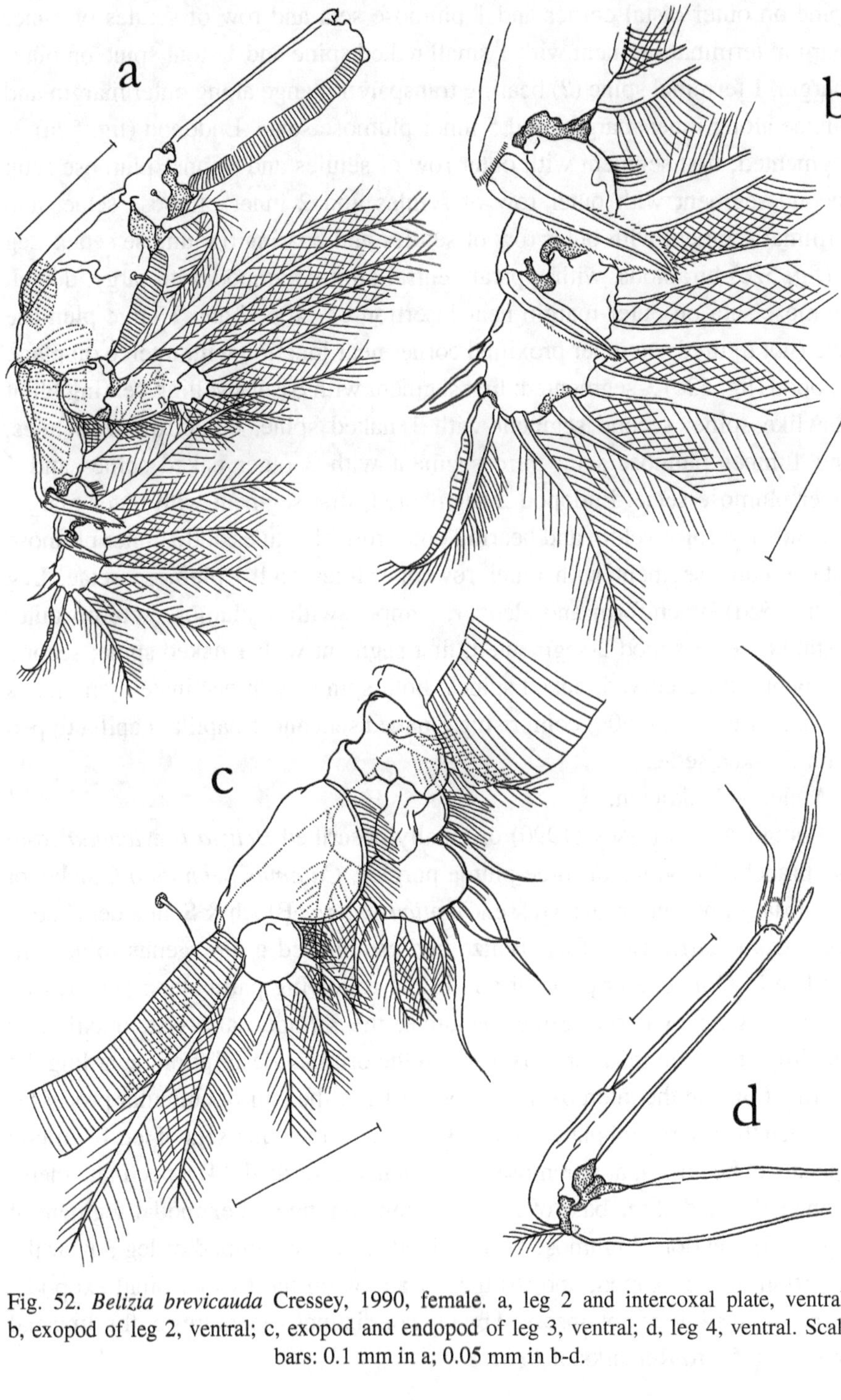

Fig. 52. *Belizia brevicauda* Cressey, 1990, female. a, leg 2 and intercoxal plate, ventral; b, exopod of leg 2, ventral; c, exopod and endopod of leg 3, ventral; d, leg 4, ventral. Scale bars: 0.1 mm in a; 0.05 mm in b-d.

spine on outer distal corner and 1 plumose seta and row of setules on inner margin; terminal segment with 1 small naked spine and 1 stout spine on outer margin, 1 terminal spine (?) bearing transparent flange along outer margin and pinnae along inner margin, and 5 inner plumose setae. Endopod (fig. 52a) 3-segmented; first segment with outer row of setules and 1 inner plumose seta; second segment with outer row of setules and 2 inner plumose setae; and terminal segment with outer row of setules and bearing 6 plumose setae. Leg 3 (fig. 52c) biramous, with typical ventral apron; sympod with large, dorsal, plumose seta on outer margin near insertion of exopod and 1 large plumose seta (not figured) on inner proximal corner near junction with intercoxal plate. Exopod (fig. 52c) 3-segmented; first segment with basal swelling bearing blunt clawlike spine; second segment with 1 naked spine, outer row of setules, and 1 inner plumose seta; third segment with 3 outer naked spines and 4 inner plumose setae. Endopod 2-segmented; first segment with outer margin expanded to form velum and bearing outer row of setules and 1 inner plumose seta; second segment with outer row of setules and 4 plumose setae. Leg 4 (fig. 52d) brachiform and slender; sympod with 1 plumose seta on outer distal corner. Exopod 2-segmented; first segment with 1 naked spine; second segment with 2 curved, naked spines; both spines with pectinate membranes at bases. Leg 5 (fig. 50a) comprising 1 naked seta and 1 papilla; papilla tipped with 2 naked setae.

Male. — Unknown.

Remarks. — Cressey (1990) originally described *Belizia brevicauda* from the branchial cavities of three pluma porgies, *Calamus pennatula* Guichenot (Sparidae), and one creole wrasse, *Clepticus parrae* Bloch & Schneider (Labridae), from Carrie Bow Cay, Belize and established a new genus to accommodate it. Our redescription differs in the following particulars: (1) Cressey did not figure or describe the mandible, the intrabuccal stylet, maxilla, or maxilliped; (2) no mention was made in the original description regarding the corrugations on the intercoxal plate of leg 1 nor the pair of cuticular processes between the intercoxal plates of legs 1 and 2; (3) the outer seta of leg 1 sympod appears to be naked, not plumose as originally described; (4) a pectinate membrane is located at the base of the stout spine on the first exopodal segment of leg 2; (5) one dorsal plumose seta is situated on the sympod of leg 3 near the insertion of the exopod; and (6) the serrations on the two terminal exopodal spines of leg 4 were not observed by us. For distinctive features of this unusual species, refer to Remarks of *Belizia*.

Genus **Caligodes** Heller, 1865

Caligodes Heller, 1865: 180; Bassett-Smith, 1899: 446; Wilson, 1905a: 608; Wilson, 1932: 408; Heegaard, 1945b: 6; Yamaguti, 1963: 63; Kabata, 1979: 159; Dojiri, 1983: 133; Prabha, 1983: 52; Pillai, 1985: 407; Boxshall & Montú, 1997: 52; Ho, 1998: 251; Kazachenko, 2001: 24; Boxshall & Halsey, 2004: 725; Ho & Lin, 2004: 102.

Female. — Cephalothorax orbicular, with shallow posterior sinuses. Frontal plate distinct. Lunules present. Fourth pedigerous somite without dorsal plates, and indistinctly fused to anterior portion of genital complex. Neck formed from fusion of fourth pedigerous somite and narrow, anteriormost region of genital complex. Posterodorsal process longer than genital complex and extending beyond or near posterior limit of abdomen. Abdomen consisting of 1 free somite, longer than genital complex. Caudal ramus small, bearing 6 setae.

Antennule 2-segmented. First segment robust, and bearing numerous setae. Second segment cylindrical, and armed with 12 + 1 aesthetasc. Postantennal process absent. Antenna 4-segmented. Second segment with posteriorly directed spinelike process. Terminal segment a recurved claw. Mouth tube with intrabuccal stylet and strigil. Mandible comprising 4 sections, with 13 teeth on inner margin of terminal section. Maxillule consisting of dentiform process and adjacent setiferous papilla. Maxilla brachiform. Maxilliped subchelate and prehensile. Sternal furca present.

Leg 1 with 2-segmented exopod and vestigial endopod. Leg 2 biramous, with 3-segmented rami; second and third segments of endopod partially fused. Leg 3 biramous, with large ventral apron; rami 2-segmented. Basal swelling of first exopodal segment bearing large terminal spine. Leg 4 uniramous, brachiform. Legs 5 and 6 absent.

Male. — Unknown.

Type-species. — *Caligodes laciniatus* (Krøyer, 1863).

Remarks. — *Caligodes* was first established by Heller (1865) to include the type-species *C. laciniatus* (Krøyer, 1863), described under the binomen *Sciaenophilus laciniatus*. Since that time, four additional species have been included in this genus. They are: *C. carangis* Bassett-Smith, 1898; *C. megacephalus* Wilson, 1905; *C. lamarcki* Quidor, 1913; and *C. alatus* Heegaard, 1945. Among the five nominal species of this genus, the type-species may be the only valid member of *Caligodes* (Dojiri, 1983; Ho & Lin, 2004).

Caligodes megacephalus, which Wilson (1905a) described from the longnose gar *Lepisosteus osseus* (Linnaeus) collected in Woods Hole, Massachusetts, is very similar to *C. laciniatus*. Wilson's main distinctions between these two species were the total body length and sizes of the various tagmata of the females. However, the sizes of *Caligodes* differ greatly depending upon the species of host (Kirtisinghe, 1937; Pillai, 1961; Nunes-Ruivo, 1962; and Cressey & Collette, 1970). Therefore, the total length and lengths and widths of the tagmata cannot be used as specific discriminants. Wilson also based his new species on presumed differences in the claws of the maxillipeds between the two species, and the presence of accessory processes between the sternal furca and the sympod of leg 1 in *C. megacephalus*. However, these structures are very similar in both species. The main distinction appears to be the posteromedian process on the genital complex of the female of *C. megacephalus*. This process may simply represent a misinterpretation by Wilson (compare Wilson's, 1905a fig. 187 and Pillai's, 1961 fig. 15A). Consequently, *C. megacephalus* is probably synonymous with the type-species. However, the type-specimen of *C. megacephalus* must be examined before the synonymy is confirmed.

Caligodes carangis was transferred to a newly established genus *Parechetus* Pillai, 1962 (see remarks on *Parechetus*). It appears that *C. lamarcki* Quidor, 1913 is synonymous with *Parechetus carangis* based on the general habitus and the structure of leg 1.

Caligodes alatus was first described by Heegaard (1945b) from "*Cardus* sp." from Java. Apparently, this species differs significantly from the type-species. In the female of *C. alatus* the genital complex and posterolateral process form a "bell" partially enclosing the inflated abdomen. Legs 1-4 differ considerably based on a comparison between *C. laciniatus* and the illustrations and description of *C. alatus* provided by Heegaard (1945b). A conclusion on the generic placement of *C. alatus* is postponed until an examination of the holotype. Until that time, this species is retained within *Caligodes*.

The type-species exhibits several characteristics that distinguishes it from all other caligids. The long posterodorsal processes on the genital complex in combination with the long cylindrical abdomen in the female gives *C. laciniatus* a unique habitus among the caligid genera. The enlarged outermost spine of the terminal exopodal segment of leg 1, the small, slender exopodal spines of leg 2, and the segmentation and armament of leg 4 help differentiate this species from other members of the family.

Caligodes laciniatus (Krøyer, 1863)
(figs. 53-56)

Chondracanthus laciniatus Kollar?

Sciaenophilus laciniatus Krøyer, 1863: 227.

Caligodes laciniatus Heller, 1865: 180; Bassett-Smith, 1899: 446; Brian, 1902: 8; Poche, 1902: 13; Wilson, 1905a: 608; Brian, 1906: 34; Kirtisinghe, 1937: 439; Heegaard, 1945b: 6, 8, 10; Capart, 1953: 654; Capart, 1959: 88; Rangnekar, 1959: 49; Causey, 1960: 329; Pillai, 1961: 114; Nunes-Ruivo, 1962: 72; Yamaguti, 1963: 63; Kirtisinghe, 1964: 74; Cressey & Collette, 1970: 378; Collette & Parin, 1970: 40, 44, 54; Dojiri, 1983: 136; Pillai, 1985: 407; Luque et al., 1999: 24; Boxshall & Montú, 1997: 52; Lin & Ho, 2002a: 50; Ho & Lin, 2004: 102; Dippenaar, 2005: 1286; Morales-Serna, 2012: 50.

Material examined. — Type-material consisting of 1 intact mature female, 1 decapitated mature female, and 1 intact immature female (labeled "*Chondracanthus laciniatus* Kllr." and "*Caligodes laciniatus* Koll.") on loan from Naturhistorisches Museum, Wien, Austria. No host or site of attachment recorded on label. Eleven intact females (USNM 180573) from *Ablennes hians* (Valenciennes) collected in West Africa. This material on loan from the National Museum of Natural History, Smithsonian Institution, Washington, D.C.

Female. — Body as in fig. 53a. Total length (not including setae on caudal ramus) 7.29 mm (6.31-8.03 mm) (n = 7). Cephalothorax (fig. 53b) circular in outline, almost as wide as long, 1.14 mm (1.03-1.26 mm) × 1.11 mm (1.00-1.23 mm). Dorsal transverse rib situated more posteriorly. Cephalic area much larger than thoracic area of dorsal shield. Posterior sinuses very shallow and possessing transparent membranous flap. Free margin of thoracic zone extending beyond posterior limit of lateral zone. Frontal plate relatively large. Tip of antennule well within lateral limit of cephalothorax. Lunules (fig. 53c) located ventrally on frontal plate. Corrugated area and transparent membrane (fig. 53d) located ventrally along margin of cephalothorax.

Fourth pedigerous somite longer than wide, 0.72 mm (0.66-0.86 mm) × 0.52 mm (0.47-0.60 mm), and not clearly delimited from genital complex (demarcation represented by constriction). Genital complex roughly triangular, narrowing anteriorly, and longer than wide, 2.42 mm (2.06-2.69 mm) × 1.54 mm (1.33-1.66 mm). Dorsal area of posterolateral corners of genital complex with long lamellate processes. These processes (sometimes extending beyond posterior limits of abdomen) wide at proximal end, but tapering posteriorly, with lateral margins curved ventrally, 2.92 mm (2.40-3.39 mm) × 1.00 mm (0.83-1.20 mm). Abdomen longer than genital complex, 3.09 mm (2.72-3.85 mm) × 0.51 mm (0.43-0.56 mm).

Caudal ramus (fig. 53e) 46 × 40 μm, with 4 plumose setae, 2 smooth setae, and inner row of setules.

Cephalothorax with small setules (sensilla) as in fig. 53b.

Egg sac 8.06 × 0.33 mm, containing about 121 eggs (n = 1).

Antennule (fig. 53f) 2-segmented. First segment with distinctive sclerotized area, 49 μm long (measured along nonsetiferous margin), and possessing

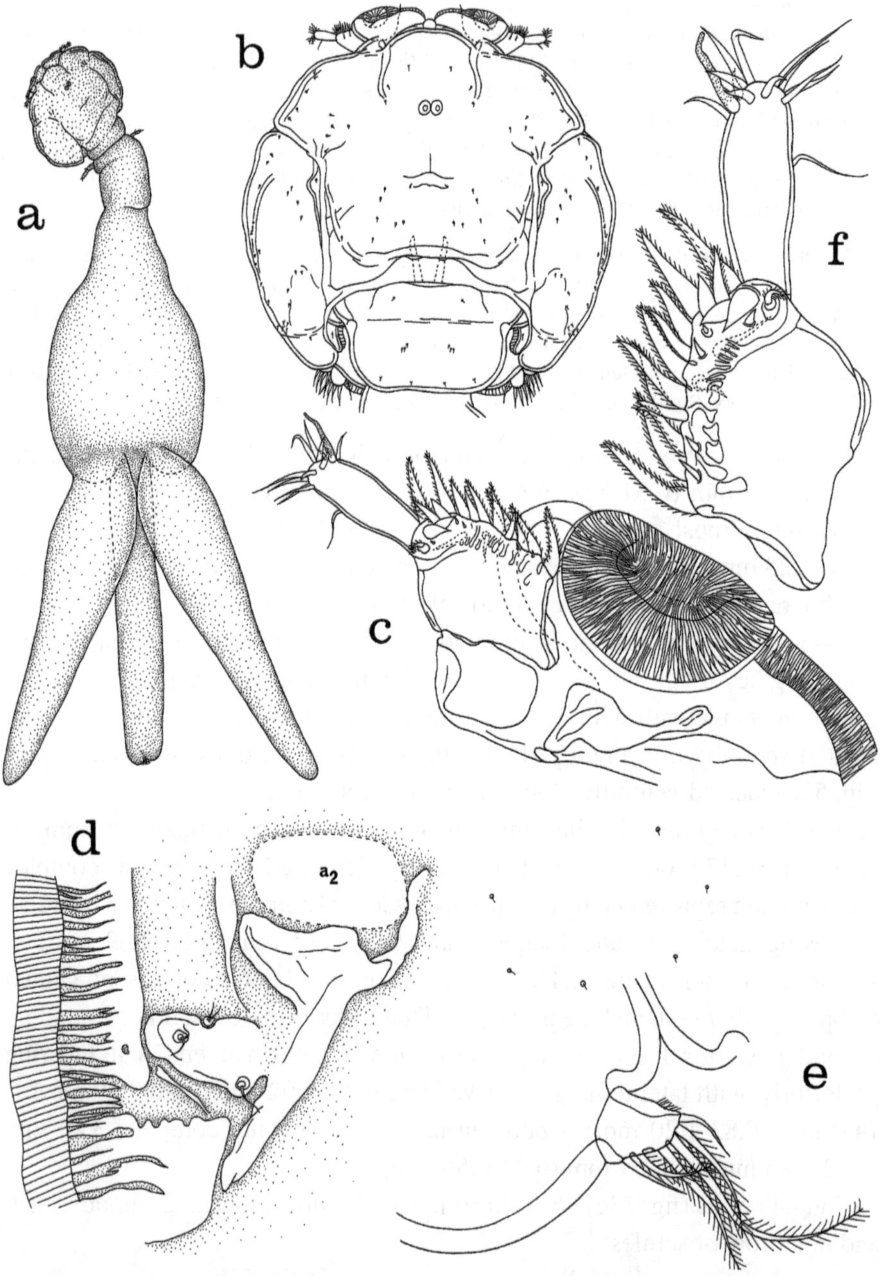

Fig. 53. *Caligodes laciniatus* (Krøyer, 1863), female. a, body, dorsal (scale V); b, cephalothorax, dorsal (F); c, frontal plate with lunule and antennule, ventral (D); d, postantennal area, ventral (D); e, caudal ramus, ventral (D); f, antennule, ventral (D).

17 large setae and 6 small setae (plumosity as in fig. 53f). Second segment 76 μm long, with unusual armature of 12 + 1 aesthetasc. Antenna (fig. 54a) apparently 4-segmented. First segment unornamented. Second segment with posteriorly directed spinelike process. Third segment robust, without dorsal adhesion pad found in many caligids. Claw with 2 setae. Proximal seta, with balloonlike base and highly attenuate tip, located in large circular depression. Second seta slender and situated almost at midlength. Postantennal region (fig. 53d) represented by sclerotized area with 3 associated setules (2 of which are bifid).

Mouth tube (fig. 54b) 248 × 194 μm, with intrabuccal stylet and accessory stylet (fig. 54c) on inner surface of labrum. Labium with strigil. Mandible (fig. 54d) with hyaline membrane on outer margin and 13 teeth on inner margin of fourth section. Maxillule (fig. 54e) comprised of papilla with 3 setae, and sclerotized process with corrugated pad. Maxilla (fig. 54f, g) brachiform. Brachium slender with flabellum near midlength. Flabellum serrated along its edge and striated. Calamus (fig. 54g) longer than canna, curved at distal end, and bearing 2 serrated membranes. Canna (fig. 54g) with spinules along 1 side and 3 larger spinules along other side. Maxilliped (fig. 54h) with relatively robust corpus. Shaft and claw of maxilliped (fig. 54i) 232 μm long, with slender seta near midlength, and small pimples slightly distal to seta. Sternal furca (fig. 54j) with large subcircular base and blunt horseshoe-shaped tines. Sternal furca of holotype (fig. 54k) with rather sharp tines. Area between sternal furca and leg 1 (fig. 55a) with pair of posteriorly directed, sharp spinelike processes.

Leg 1 (fig. 55b) biramous. Sympod with 1 large outer and 1 small inner plumose setae, in addition to large patch of spinules. Exopod 2-segmented. First segment with outer distal spine and inner row of long setules. Outer distal spine (fig. 55c) sparsely serrated and with small spinules at base. Second segment of exopod with unilaterally pinnate seta 4 much shorter than outermost spine. Outermost spine large, triangular, and unilaterally rimmed with transparent membrane. Middle spine (spine 2) with outer row of spinules; inner spine (spine 3) smooth. Inner margin of second segment of exopod with 3 large plumose setae. These setae with large patches of spinules along outer edge. Endopod (fig. 55d, e) 1-segmented, with small knob (fig. 55d) in 1 specimen and 1 long process (fig. 55e) in others examined. Area between intercoxal plates of legs 1 and 2 (fig. 55a) with pair of setules. Leg 2 (fig. 56a) biramous. First exopodal spine (fig. 56b) with striated membrane at its base. All 3 exopodal spines very slender. Second exopodal spine of terminal segment

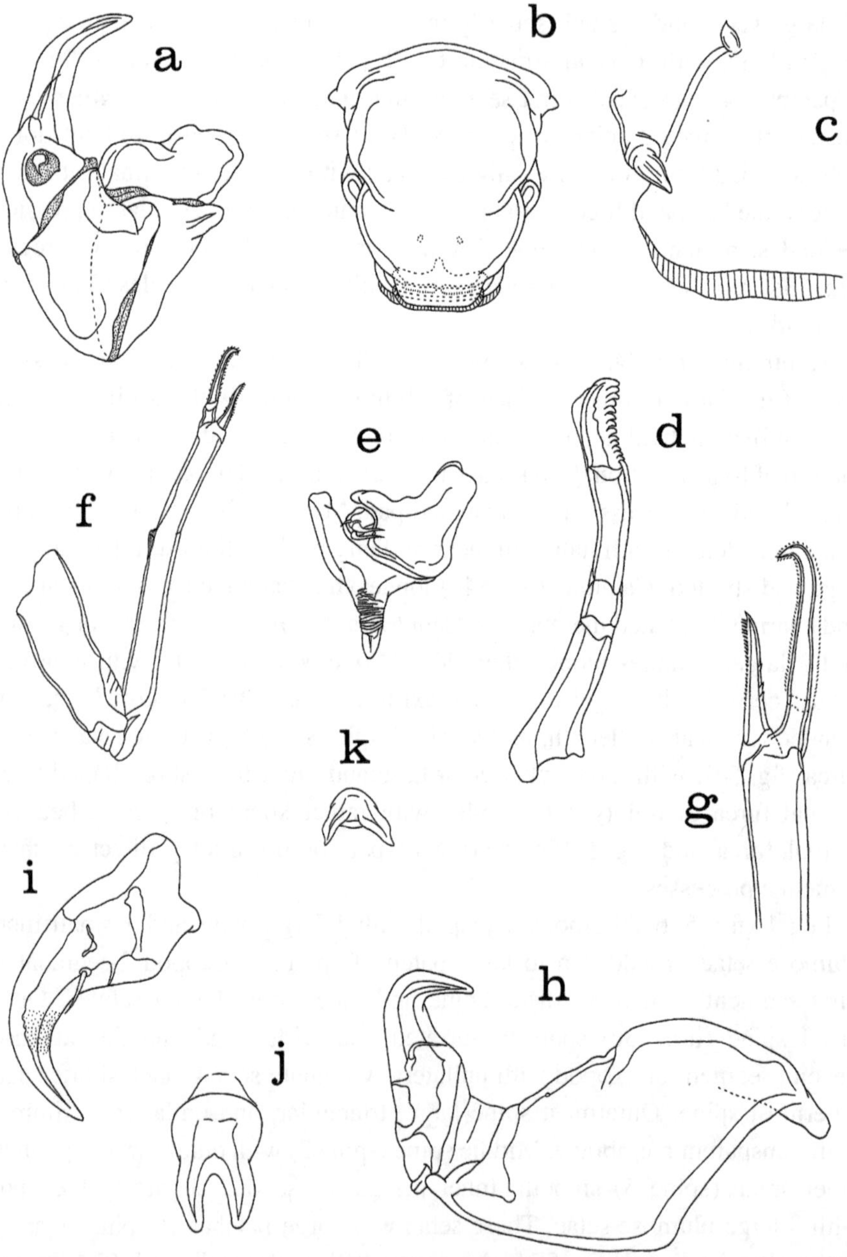

Fig. 54. *Caligodes laciniatus* (Krøyer, 1863), female. a, antenna, ventral (scale D); b, mouth tube, ventral (J); c, intrabuccal stylet, accessory stylet, and part of frons labri, dorsal (Y); d, mandible, ventral (U); e, maxillule, ventral (E); f, maxilla, dorsal (D); g, terminal portion of maxilla, ventral (Y); h, maxilliped, ventral (J); i, shaft and claw of maxilliped, ventral (E); j, sternal furca, ventral (J); k, sternal furca of holotype (USNM 180573), in situ, ventral (D).

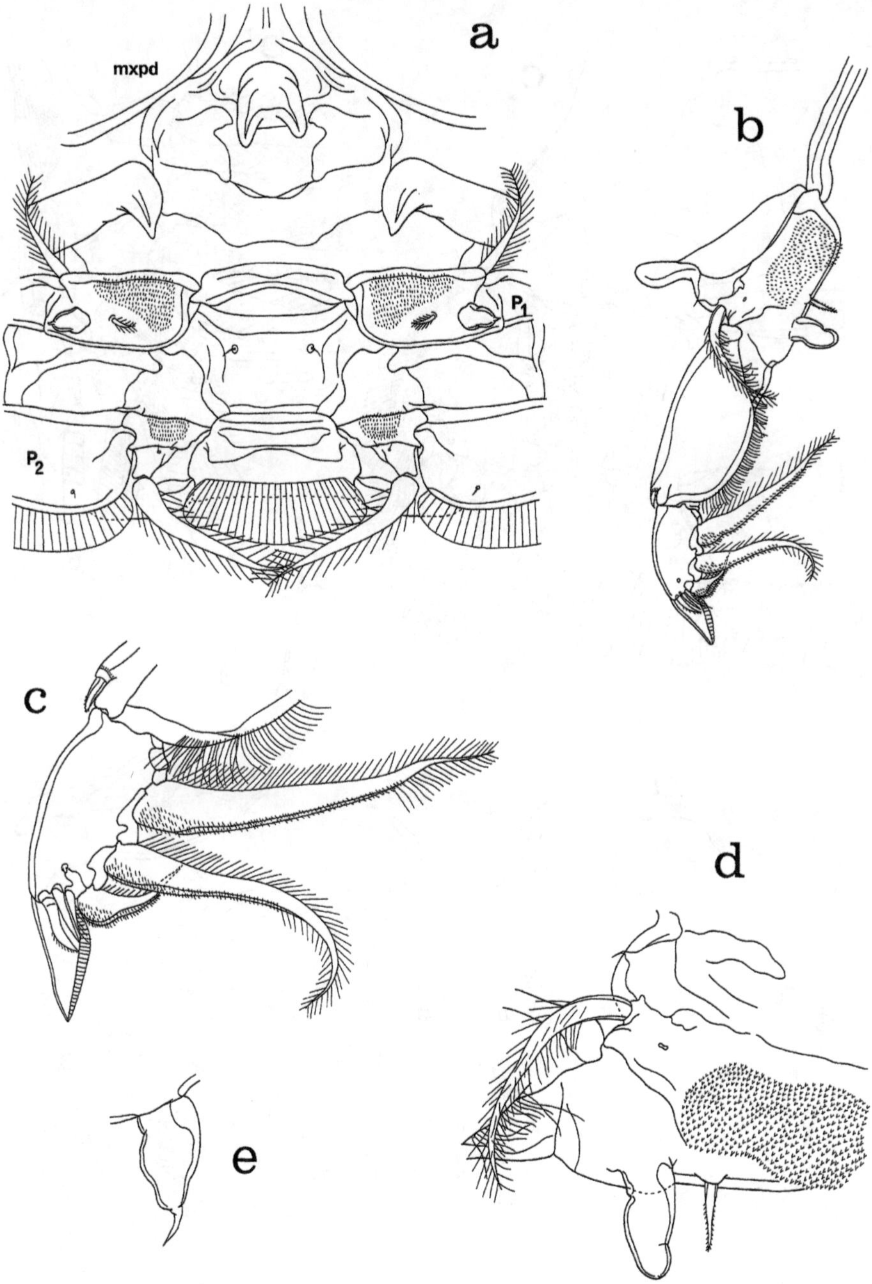

Fig. 55. *Caligodes laciniatus* (Krøyer, 1863), female. a, sternal furca and sclerotized area between legs 1 and 2, ventral (scale E); b, leg 1, ventral (E); c, terminal portion of exopod of leg 1, ventral (U); d, sympod-exopod joint and endopod of leg 1, ventral (U); e, endopod of leg 1, ventral (L).

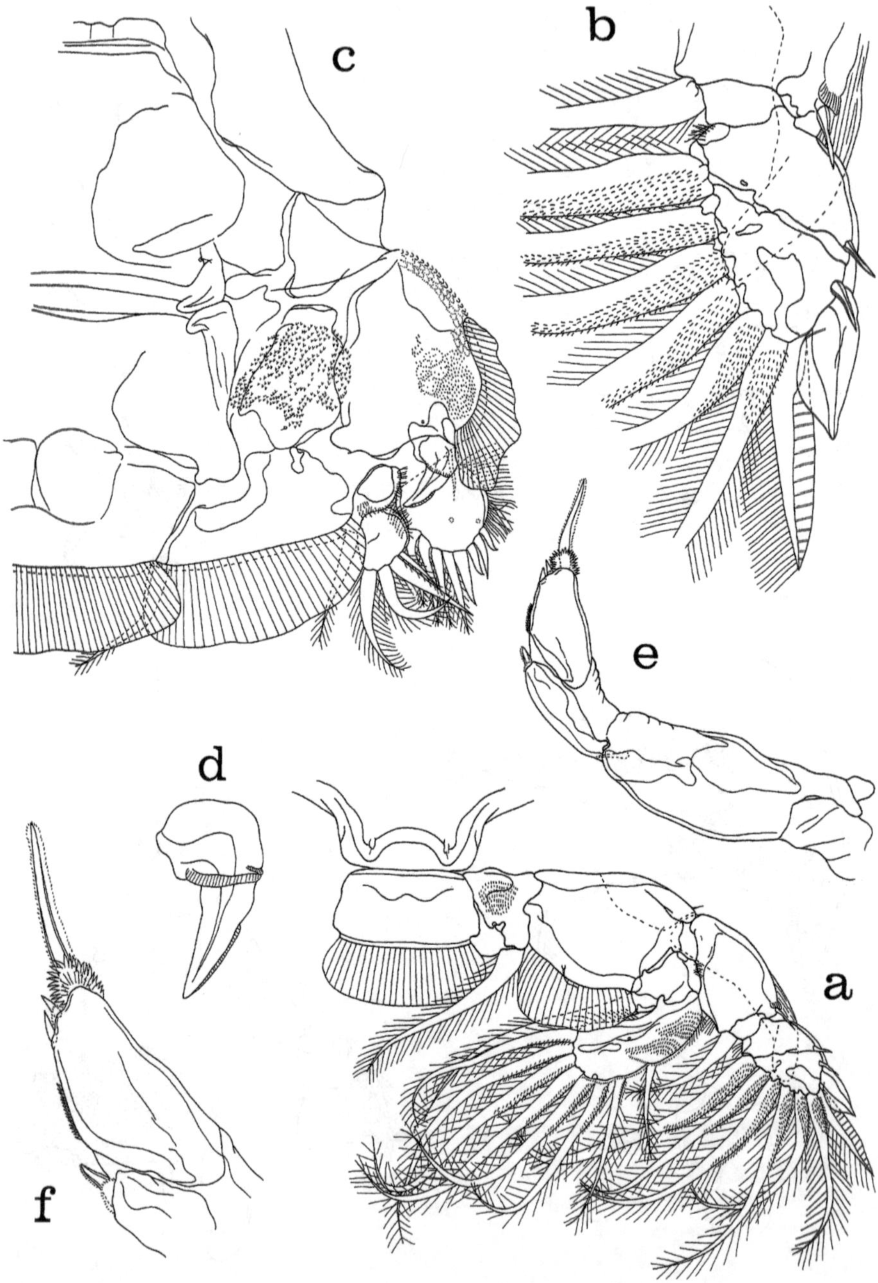

Fig. 56. *Caligodes laciniatus* (Krøyer, 1863), female. a, leg 2 and intercoxal plate, ventral (scale J); b, exopod of leg 2, ventral (U); c, leg 3, ventral (E); d, spine and basal swelling of first exopodal segment leg 3, ventral (L); e, leg 4, ventral (L); f, terminal portion of leg 4, ventral (Y).

not highly sclerotized, but covered with hyaline membrane. Second segment of endopod with patch of spinules on outer portion. Second and third endopodal segments not clearly delimited; otherwise, armature as in other caligids. Setae of leg 2 with rows of spinules on outer margins as in fig. 56a, b. Sympod of leg 3 (fig. 56c) with corrugated pad on dorsal surface of anterolateral corner, and patch of minute spinules near pair of medially directed spinelike processes. Exopod apparently 2-segmented. First segment represented by exopodal spine. Exopodal spine (fig. 56d) fused with basal swelling (membranous flange at base of spine delimiting original boundary), and possessing flange along outer edge. Second exopodal segment with outer and inner row of setules, 4 smooth setae, and 4 plumose setae. Endopod 2-segmented. First segment with outer patch of setules and 1 inner plumose seta. Second endopodal segment with outer patch of setules and 4 plumose setae. Leg 4 (fig. 56e) uniramous, brachiform. Sympod with irregularly shaped sclerotized areas, bearing 1 distal smooth seta. Exopod (fig. 56f) 2-segmented. First segment with unilaterally serrated spine with serrated membrane at its base. Second segment with outer pectinate membrane, 2 small smooth spines (second spine with serrated membrane at base), and 1 large bilaterally serrated spine. Latter spine with huge pectinate membrane at its base. Leg 5 may be represented by posterolateral process of genital complex (see Cressey & Collette, 1970: 381).

Male. — Unknown. Although Rangnekar (1959) stated in her remarks section that Kirtisinghe (1937) collected both females and males from *Strongylura strongylura*, Kirtisinghe actually reported only females.

Remarks. — The type-specimens from the Naturhistorisches Museum, Wien, Austria, have labels that have two names written on them. One is "*Chondracanthus laciniatus* Kllr.", and the other "*Caligodes laciniatus* Koll.". In each case, the discoverer is identified as Kollar. Although *Caligodes laciniatus* is attributed to Krøyer (1863), Krøyer himself stated the original author of the species (under Krøyer's binomen *Sciaenophilus laciniatus*) to be Kollar. Poche (1902) apparently realized that this species was originally identified as *Chondracanthus laciniatus* by Kollar as "...ein Museumname..." then transferred to *Sciaenophilus* by Krøyer (1863). Yamaguti (1963) was also aware of this when he wrote "*C. laciniatus* (Kollar in Krøyer, 1863)". However, since Krøyer never gave a literature citation for Kollar's work, it is not known if Kollar actually described the species. He may simply have given these specimens a name on a museum label, which, of course, does not constitute a valid description. In any case, Heller (1865) proposed the genus *Caligodes* to accommodate this species of copepod, and thus the present binomen *Caligodes laciniatus* was established.

Several redescriptions have been published since the discovery of this species (Kirtisinghe, 1937; Rangnekar, 1959; Pillai, 1961; Cressey & Collette, 1970; and Pillai, 1985), but none with great attention to the details of the appendages, except for those given by Lin & Ho (2002a) and Ho & Lin (2004). In comparing our redescription with those of previous investigators, there are some conspicuous differences. For instance, the shape of the sternal furca of the female appears to vary. The tines of the sternal furca of the specimens Pillai (1961) described are sharp, much like the type-material (refer to fig. 54k of present account). The sternal furca of Cressey & Collette's (1970) specimens is very similar to that of Rangnekar's (1959): blunt, diverging tines, not curved as in the present specimens. Pinnate seta 4 of the second exopodal segment of leg 1 has apparently been overlooked in redescriptions except by Rangnekar (1959), Lin & Ho (2002a), and Ho & Lin (2004). Specimens of *C. laciniatus* from Taiwan differ from the type-material in bearing a central band of spinules around the maxillipedal terminal claw.

Caligodes laciniatus is apparently host specific for needlefishes (family Belonidae). It has been reported from the mouth (under the oral valve) of many species, representing several genera, of needlefishes including *Ablennes hians* (Valenciennes), *Strongylura leiura* (Bleeker), *Platybelone argalus* (Lesueur), *Lhotskia gavialoides* [= *Tylosurus gavialoides* (Castelnau)], *Tylosurus acus* (Lacépède), *Tylosurus crocodilus* (Peron & Lesueuer), and *Tylosurus choram* (Rüppell) (Cressey & Collette, 1970). There is, however, a report of *C. laciniatus* from a nonbelonid host, a barracuda, *Sphyraena argentea* Girard, which was recorded by Causey (1960) from Guaymas, Sonora, Mexico.

Kirtisinghe (1937) noted that the specimens of *C. laciniatus* he obtained from *S. leiura* were conspicuously larger than those from *S. strongylura*. In addition, Pillai (1961) also found that the adult females from *A. hians* were almost twice as large (total length) as those collected from *T. crocodilus*. Cressey & Collette (1970) confirmed that the size of this species of copepod is at least partly dependent upon the species of host. They found that the greatest change occurs in the posterior region of the copepod, the genital complex and the abdomen being affected the most.

Caligodes laciniatus has been reported from most of the world's oceans (see Cressey & Collette, 1970, fig. 181). The majority of the reports have been from tropical and subtropical waters.

Genus **Caritus** Cressey, 1967

Caritus Cressey, 1967c: 1; Kabata, 1979: 158; Dojiri, 1983: 143; Prabha, 1983: 51; Kazachenko, 2001: 24; Boxshall & Halsey, 2004: 725.

Female. — Cephalothorax suborbicular, narrower anteriorly, with wide posterior sinuses. Frontal plate distinct. Lunules present. Fourth pedigerous somite free and without dorsal plates. Genital complex somewhat triangular. Abdomen consisting of 1 free somite, cylindrical. Caudal ramus longer than wide, with 6 short, smooth setae.

Antennule 2-segmented. First segment longer than second, and possessing numerous plumose setae. Second segment cylindrical, with 13 setae and 1 aesthetasc. Antenna apparently 4-segmented. Second segment without posteriorly directed spinelike process. Third segment without dorsal adhesion process. Terminal segment hamate. Postantennal process and sternal furca absent. Mouth tube with intrabuccal stylet and strigil. Mandible comprising 4 sections; terminal section about as long as third section, and armed with 12 teeth on inner margin. Maxillule consisting of spinelike projection and adjacent papilla tipped with 3 setae. Maxilla brachiform. Maxilliped subchelate and prehensile.

Leg 1 with 2-segmented exopod and vestigial endopod. Terminal exopodal segment with 3 spines, 1 smooth seta, and 3 inner plumose setae. Leg 2 biramous; rami 3-segmented. Leg 3 with large ventral apron, and biramous. Exopod indistinctly 3-segmented; first segment represented by basal swelling and large spine. Endopod comprised of 2 lamelliform structures (both velumlike) with outer spinules. Fourth leg uniramous, brachiform, with 3-segmented exopod. Leg 5 a setiferous lobe situated on posterolateral corner of genital complex.

Male. — Cephalothorax and fourth pedigerous somite similar to those in female. Genital complex subovate. Abdomen presumably comprising 3 free somites; third somite longest. Caudal ramus comparatively broader than that in female. Antenna with accessory tine near base of claw. Maxillule different from that of female, with additional tine on inner surface of dentiform process. Maxilliped with papillated myxa on corpus. Other appendages as in female. Legs 1-4 as in female; exopodal spines of leg 2 different from that of female. Legs 5 and 6 represented by protrusions tipped with setae.

Type-species. — *Caritus serratus* Cressey, 1967.

Remarks. — Although based on the general habitus of the females, the presence of the lunules, and the details of the appendages, *Caritus* appears closely allied to *Caligus*. There are, however, some major differences between the members of these two genera. The posteriorly directed spinelike process on the basal segment of the antenna, the postantennal process, and the sternal furca are absent in *Caritus*. The majority of the species of *Caligus* possess all three of these characteristics. Although there are several species of *Caligus* that lack one of these structures, none lacks all three (table III). The structure of the mandible of *Caritus* is markedly different from that of *Caligus* (compare fig. 2e and fig. 58b). The major difference is that in *Caritus* the terminal section of the mandible is as long as the third section (not 1/2 the length of the third section as in *Caligus*).

The most distinctive feature of *Caritus* is the modification of the rami of leg 3. The setae of the exopod are highly reduced, although the large clawlike spine is present on the basal swelling of the first exopodal segment. The greatest modification is displayed by the endopod where this structure has evolved into a lamelliform appendage bearing no armature. The combination of the above features, in addition to the ornamentation of the exopodal spines of leg 2 and the relatively large lobelike leg 5, clearly distinguishes *Caritus* from *Caligus* and all other genera of this family.

Caritus serratus Cressey, 1967
(figs. 57-60)

Caritus serratus Cressey, 1967c: 1; Dojiri, 1983: 145; Prabha & Pillai, 1983; p. 13; Rangnekar, 1984: 347.

Material examined. — One holotype female (USNM 120352), allotype (USNM 120353), and 3 paratypes (all females) (USNM 120354) collected from gills of milkfish, *Chanos chanos* (Forsskål), from Nosy Bé, Madagascar, in April 1964 by Dr. Roger F. Cressey. This material on loan from National Museum of Natural History, Smithsonian Institution, Washington, D.C.

Female. — Body as in fig. 57a. Total length (not including setae on caudal ramus) 4.66 mm (4.61-4.71 mm) (n = 2). Cephalothorax almost half of total body length, subcircular, 2.08 mm (2.06-2.09 mm) × 2.06 mm (2.03-2.09 mm), and with very wide posterior sinuses. Free margin of thoracic zone not clearly delimited, and only slightly extending beyond tip of lateral zone of cephalothorax. Antennule not extending beyond lateral limit of cephalothorax. Lunules present and located ventrally.

Fourth pedigerous somite wider than long, 0.30 mm × 0.61 mm (0.56-0.66 mm), and delimited from genital complex by constriction. Genital complex roughly triangular, 1.29 mm × 1.31 mm (1.29-1.33 mm). Posterolateral corner of genital complex bearing leg 5. Pair of protuberances located posteromedially to leg 5. Abdomen apparently consisting of 1 free somite, 0.78 mm (0.73-0.83 mm) × 0.37 mm (0.33-0.40 mm), with slight constriction at about midlength.

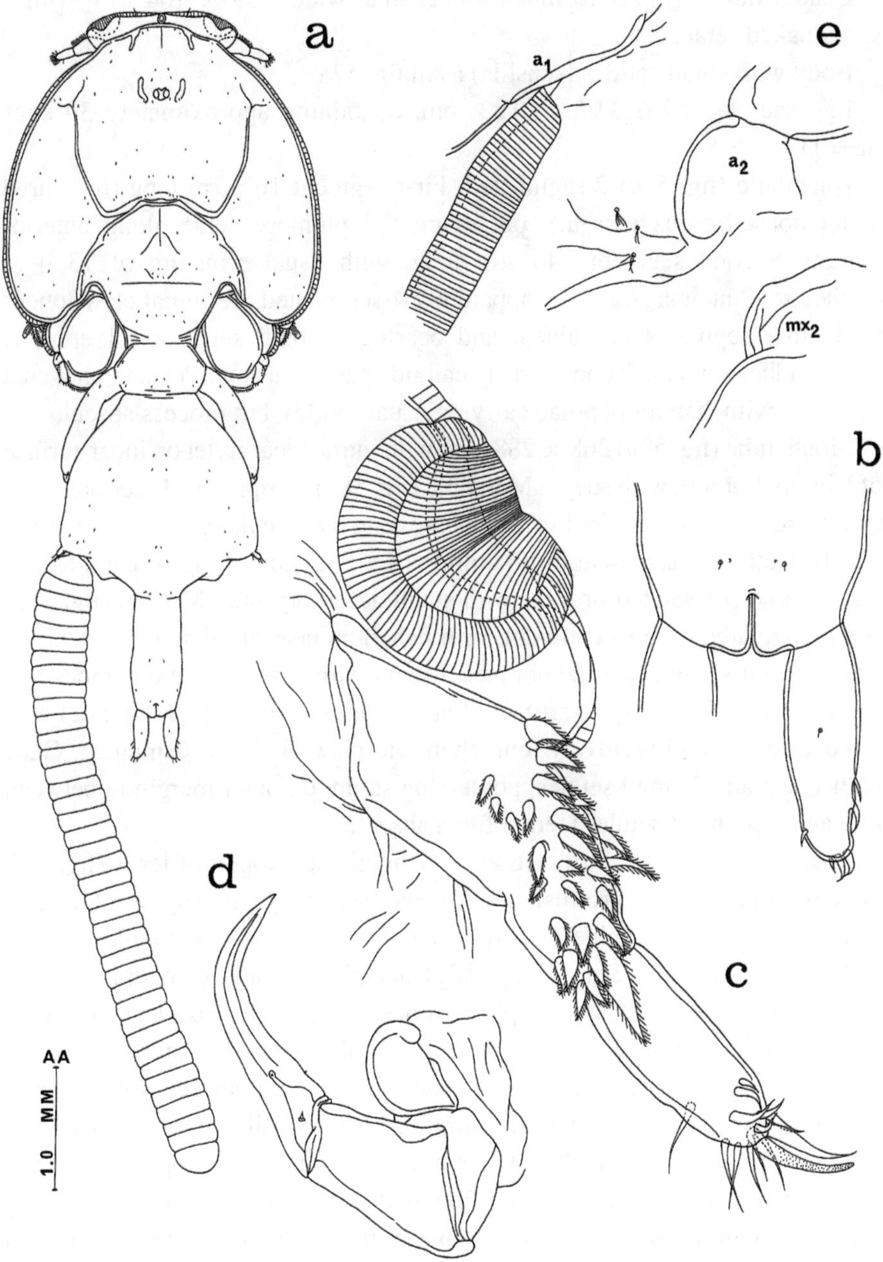

Fig. 57. *Caritus serratus* Cressey, 1967, female. a, body, dorsal (scale AA); b, anal somite and caudal ramus, dorsal (B); c, lunule and antennule, anteroventral (E); d, antenna, ventral (B); e, postantennal area showing bases of antennule and antenna, ventral (B).

Caudal ramus (fig. 57b) much longer than wide, 299 × 104 μm, with 6 short, naked setae.

Body with small setules (sensilla) as in fig. 57a.

Egg sac (fig. 57a) 3.95 × 0.33 mm, containing approximately 39 eggs (n = 1).

Antennule (fig. 57c) 2-segmented. First segment 167 μm long (measured along nonsetiferous margin), possessing 27 plumose setae along anterior margin. Second segment 146 μm long, with usual armature of 13 + 1 aesthetasc. Antenna (fig. 57d) apparently 4-segmented. Terminal claw longer than other segments combined, and bearing 2 small setae and 1 spinule. Dorsal adhesion pad, found in many caligid species, absent. Postantennal area (fig. 57e) with 3 small papillae carrying usual setules, but process absent.

Mouth tube (fig. 58a) 368 × 288 μm, with intrabuccal stylet on inner surface of labrum. Labium with strigil. Mandible (fig. 58b) comprising 4 sections, with comparatively long terminal section possessing hyaline flange on outer margin and 12 teeth on inner margin. Maxillule (fig. 58c) consisting of papilla with 3 setae, and sclerotized process with small accessory tine. Maxilla (fig. 58d) stout (compared to most caligids), with flabellum near distal end of brachium. Calamus with 3 serrated membranes. Canna with 2 serrated membranes and hyaline membrane (may be artifact of preservation). Maxilliped (fig. 58e) with slender corpus and relatively long shaft and claw, 552 μm combined. Claw with 1 seta and 2 small setules, possessing suture on outer margin in between seta and distalmost setule. Sternal furca absent.

Base of plumose seta at junction of sympod and exopod of leg 1 (fig. 58f) not covered by ventral protrusion. Coxa and basis only partially fused to form sympod. Distal outer spine of first segment of exopod with spinules at base and on outer margin. Seta 4 (fig. 58g) not pinnate, and extending beyond distal limit of outermost spine. Spine 1 robust, highly sclerotized, and carrying 4 stout teeth. Pectinate membrane present only at base of spine 1. Spines 2 and 3 longer and less sclerotized than spine 1, without accessory spines, and possessing small spinules on ventral surfaces. Small hyaline knob located near bases of spines 2 and 3. Inner margin of terminal exopodal segment with 3 rather short plumose setae. Endopod with 3 small knobs. Second leg (fig. 58h) with usual caligid form. Spine of first exopodal segment (fig. 59a) bent at almost right angle, and with large serrations. Spine of second exopodal segment only slightly curved with stouter, blunt serrations or teeth. Terminal exopodal segment with 3 outer spines. First spine of terminal exopodal segment small and pointed; second stout and weaker; third with transparent

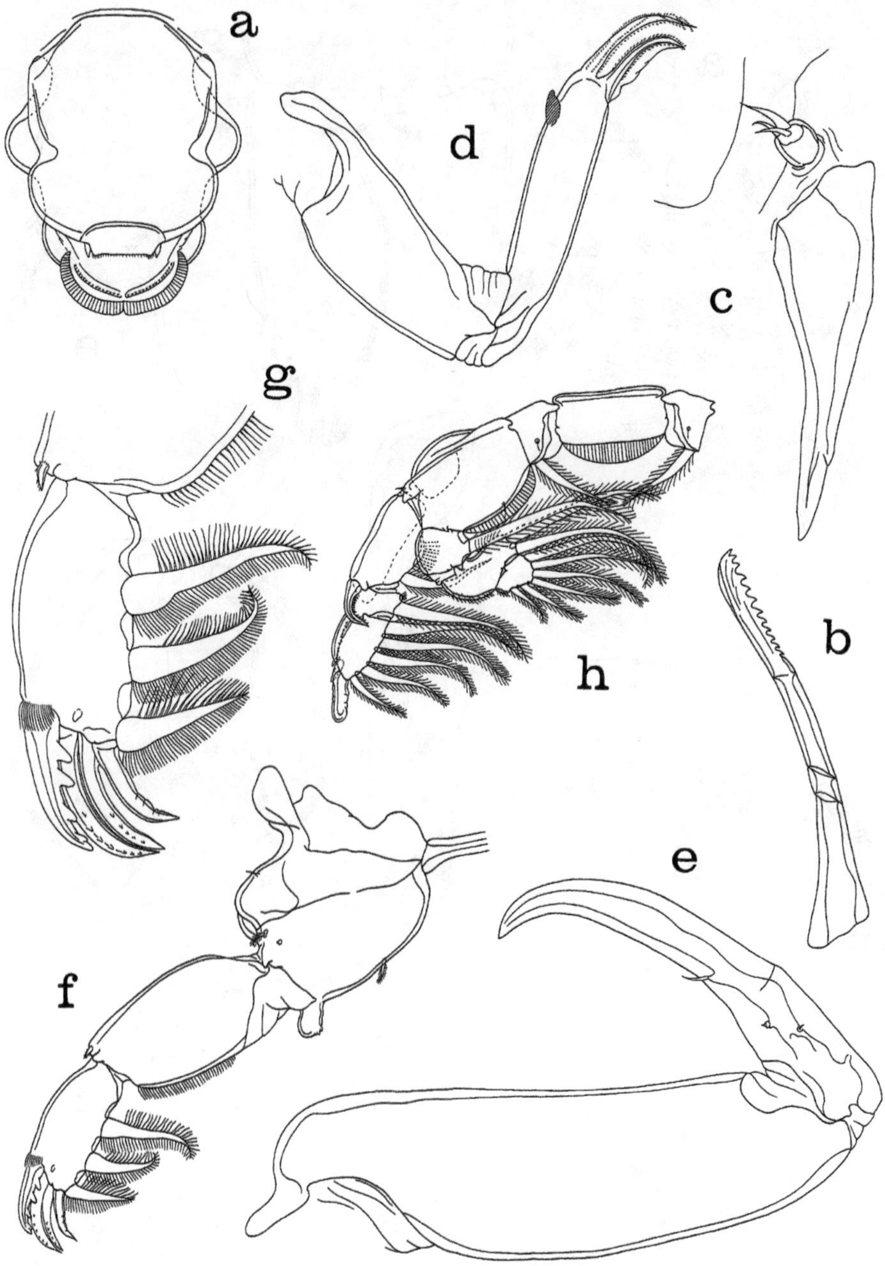

Fig. 58. *Caritus serratus* Cressey, 1967, female. a, mouth tube, posteroventral (scale B); b, mandible, ventral (E); c, maxillule, ventral (E); d, maxilla, dorsal (B); e, maxilliped, ventral (B); f, leg 1, ventral (B); g, terminal portion of exopod of leg 1, ventral (E); h, leg 2 and intercoxal plate, ventral (F).

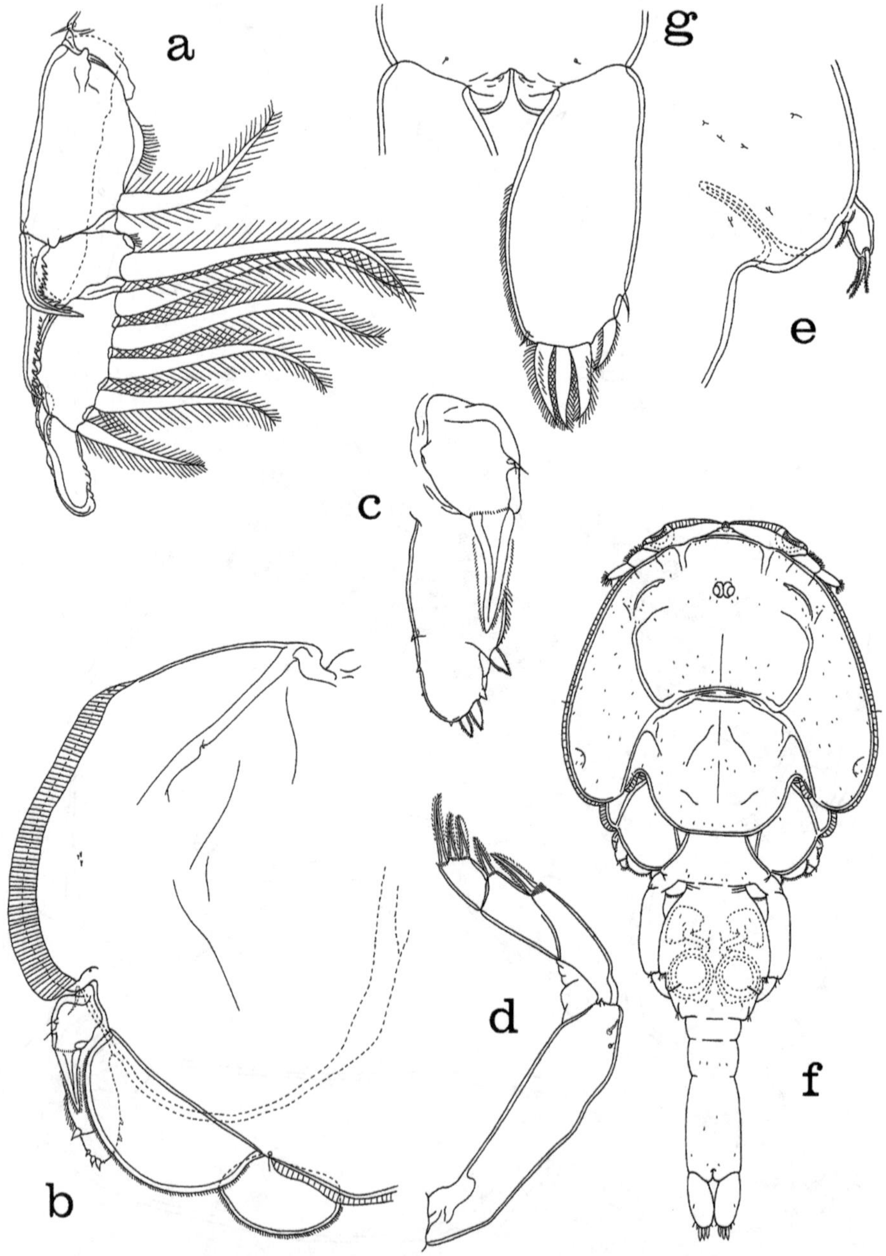

Fig. 59. *Caritus serratus* Cressey, 1967. Female. a, exopod of leg 2, ventral (scale B); b, leg 3, ventral (B); c, exopod of leg 3, ventral (E); d, leg 4, ventral (F); e, leg 5 and genital area, dorsal (B). Male. f, body, dorsal (AA); g, caudal ramus, ventral (B).

membrane and greatly rounded distal end. Apron of third leg (fig. 59b) without usual dorsal adhesion pad. Exopod (fig. 59c) 3-segmented. First segment represented by basal swelling with terminally situated exopodal spine. Spine serrated and equipped with small spinules at base. Second exopodal segment with outer spine and inner small seta. Terminal exopodal segment with 3 outer spines (outermost one only weakly sclerotized) and 4 setae (middle two minute and difficult to see). Endopod 2-segmented (assuming that velum represents first segment). Outer border of endopod with small spinules. Leg 4 exopod (fig. 59d) 3-segmented, with innermost spine markedly longer than other two. Spines 2 and 3 of terminal segment nearly equal in length. Leg 5 (fig. 59e) a lobe bearing 3 plumose setae. Small protrusion with 1 plumose seta located dorsally near base. Leg 6 possibly represented by pair of protuberances located posteromedially to leg 5 (fig. 57a).

Male. — Body as in fig. 59f. Total length 4.48 mm (n = 1) (allotype). Cephalothorax similar in shape to that of female, measuring 1.99 × 1.99 mm. Sensory pit located on posterolateral corner of cephalothorax, not observable in female.

Fourth pedigerous somite wider than long, 0.33 × 0.66 mm. Genital complex 0.83 × 0.70 mm. Abdomen with 3 free somites. First and second somites short; third somite long; all 3 somites combined 1.06 × 0.37 mm.

Caudal ramus (fig. 59g) more than twice as long as wide, 368 × 173 μm, with 6 setae (4 large setae with plumosities).

Body surface with small setules as in female.

Antennule as in female. Antenna (fig. 60a) with long claw sharply bent, and possessing accessory tine and 1 seta near its base. Postantennal area as in female.

Mouth tube and mandible as in female. Maxillule (fig. 60b, c) similar to that of female, but with additional tine on inner surface near tip. This tine not observed on left side (fig. 60c), but perhaps hidden by base of maxilliped. Maxilla as in female. Maxilliped (fig. 60d) with robust corpus carrying several conical projections (myxa) on inner surface. Claw curved and carrying 2 setae.

First leg as in female. Spines of terminal segment of exopod of leg 2 (fig. 60e) different from that of female. First spine comparatively larger and bearing 4 blunt teeth on inner margin; second spine long and sharper; third spine attenuate with membranous flange on outer margin and plumosities on inner margin. Exopod of leg 3 (fig. 60f) with spines and setae more conspicuous. Outer margin of second and third exopodal segments with small spinules. Second exopodal segment with large, plumose inner seta. Spines of

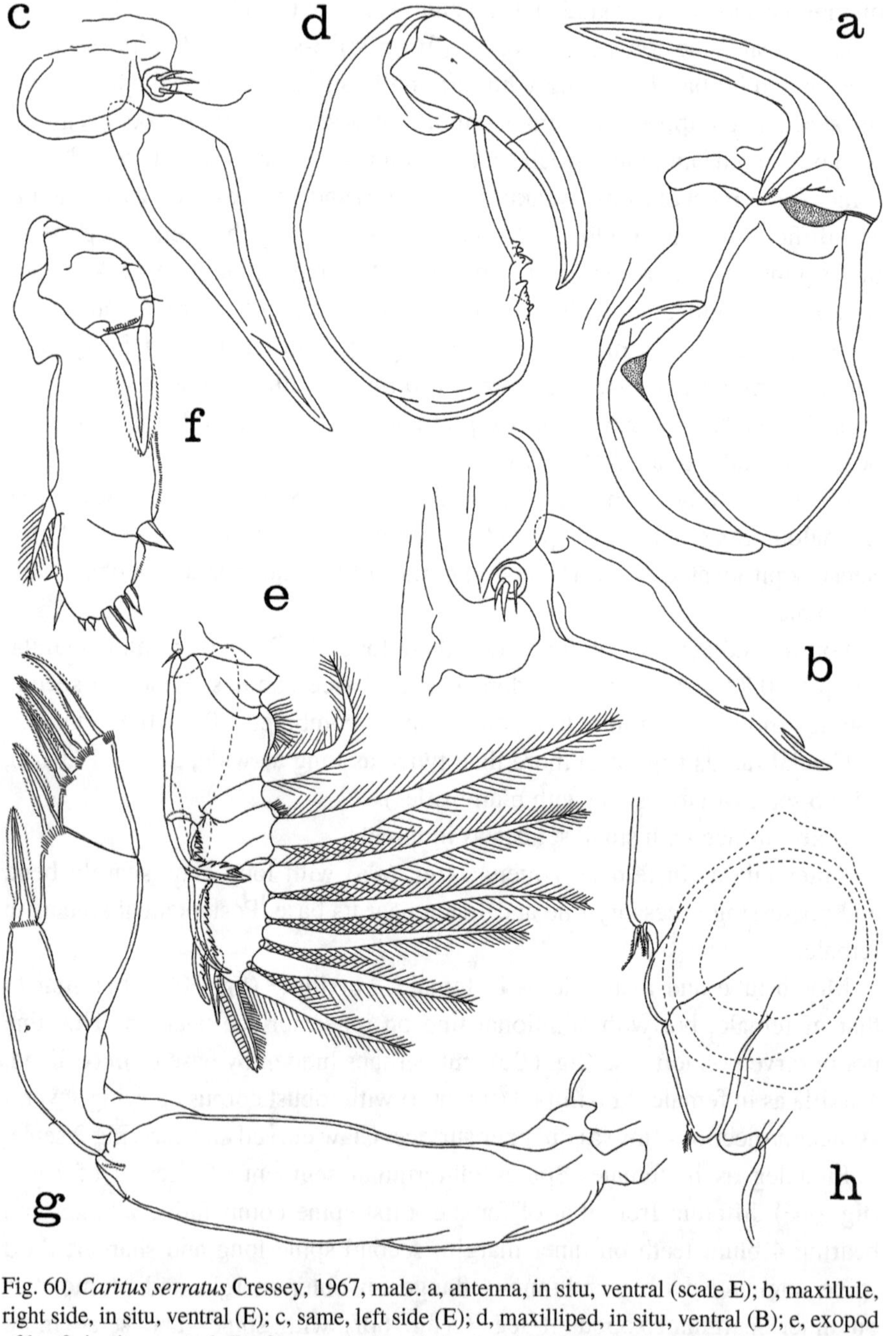

Fig. 60. *Caritus serratus* Cressey, 1967, male. a, antenna, in situ, ventral (scale E); b, maxillule, right side, in situ, ventral (E); c, same, left side (E); d, maxilliped, in situ, ventral (B); e, exopod of leg 2, in situ, ventral (B); f, exopod of leg 3, in situ, ventral (E); g, leg 4, in situ, ventral (B); h, legs 5 and 6, and genital area, ventral (B).

third segment of exopod of leg 4 (fig. 60g) only slightly decreasing in length from inner to outer margin. Serrations of spines minute, very difficult to see. Legs 5 and 6 (fig. 60h) located near posterolateral corner of genital complex. Leg 5 (fig. 60h) represented by 4 small setae. Leg 6 (fig. 60h) consisting of small papilla tipped with 2 small setae.

Remarks. — This species was first described by Cressey (1967d) from the milkfish *Chanos chanos* (Forsskål) collected at Nosy Bé, Madagascar.

There are several differences between the original description and the one given above. Although, admittedly, most of the discrepancies are minor, they are nevertheless discussed below.

The lunules are ventrally located as in all other caligids that possess this structure; Cressey figured them as dorsal structures. They are located ventral to a relatively transparent frontal plate causing them to appear dorsally situated. The basal segment of the antennule bears 27 plumose setae, not 22 as figured by Cressey. In addition, the second segment carries 13 setae + 1 aesthetasc not 12 + 1 aesthetasc. The mandible is not typical for caligids. The terminal section is elongate; the shape of the teeth are slightly different than those in other caligids. This is a very important feature since the mandible is an evolutionarily conservative structure. Consequently, this appendage helps distinguish this genus from all other caligid genera. The maxillule of the female is equipped with an accessory tine on the inner surface, which was not mentioned in the original description. The third segment of the exopod of leg 3 of the female possesses 3 additional minute setae not figured by Cressey. These setae are very small and easily overlooked. However, the armature of the exopod of the female does indeed correspond with that of the male.

Since there is only one male (allotype) in the original collection, it was not dissected by Cressey or by us. Therefore, the above description of the male is based on an undissected specimen. However, a careful examination of the appendages (in situ) revealed that the antenna, maxillule, maxilliped, and the three outer spines of the terminal segment of the exopod of leg 2 are different from those of the female.

Genus **Dartevellia** Brian, 1939

Dartevellia Brian, 1939: 179; Yamaguti, 1963: 80; Kabata, 1979: 159; Dojiri, 1983: 151; Prabha & Pillai, 1983: 55; Kazachenko, 2001: 30; Boxshall & Halsey, 2004: 725.

Female. — Cephalothorax suborbicular, narrower anteriorly, and possessing narrow posterior sinuses. Posteriormost limit of lateral zone of cephalothorax extending well beyond free margin of thoracic zone. Frontal plate small and inconspicuous. Lunules absent. Fourth pedigerous somite free, small, and without dorsal plates. Genital complex quadrangular and shorter and narrower than cephalothorax. Abdomen consisting of 1 free somite, shorter than genital complex, and fused with 1 pair of large, flat alae (winglike structures). Posterior ends of abdominal alae extending well past distal end of abdomen. Caudal ramus absent.

Antennule 2-segmented. First segment cylindrical, longer than second, and bearing short plumose setae; second segment cylindrical, with 13 setae and 1 aesthetasc. Antenna 4-segmented. Second segment with posteriorly directed spinelike process. Terminal segment hamate. Postantennal process and sternal furca absent. Mouth tube as in other caligids. Mandible comprising 4 sections, with 13 teeth on inner margin of terminal section. Maxillule consisting of two papillae. Smaller papilla tipped with bifid process; larger papilla with usual 3 setae. Maxilla brachiform. Maxilliped prehensile with large conical myxa.

Leg 1 with 2-segmented exopod and vestigial endopod. Terminal segment of exopod with usual 3 spines. Seta 4 and usual 3 plumose setae modified to small spiniform setae. Leg 2 biramous, with 3-segmented rami. Leg 3 with large ventral apron, uniramous. Exopod 2-segmented; first segment with basal swelling tipped with large spine. Leg 4 uniramous, brachiform, and indistinctly 2-segmented. Leg 5 represented by seta situated on posterolateral corner of genital complex.

Male. — Unknown.

Type-species. — *Dartevellia bilobata* Brian, 1939.

Remarks. — Since its original description by Brian (1939), *Dartevellia* has remained a monotypic genus and since its establishment, only Capart (1959) has collected *Dartevellia bilobata*. However, Capart did not redescribe the single female that he obtained from the branchial cavity of an unidentified species of *Epinephelus*. Kabata (1979) provided a short description of the general habitus of *Dartevellia*, but erroneously based it on a figure of *Sinocaligus* (referred to as *Pseudopetalus*) not *Dartevellia* (see Kabata, 1979: 160, text fig. 49c).

Since the discovery of the type-species, the validity of *Dartevellia* had not been confirmed. However, the absence of the caudal ramus (or caudal ramus represented by abdominal alae), the shape of the antennule, the structure of the maxillule, the structure of the armature elements of the exopod of leg 1, the nature of leg 3 (armament of exopod and absence of endopod), the structure of leg 4 (first and second exopodal segments indistinctly separated; second and third completely fused), and the extremely flat, large abdominal alae of the female collectively distinguish *Dartevellia* from all other known caligid genera.

Dartevellia bilobata Brian, 1939
(figs. 61-62)

Dartevellia bilobata Brian, 1939: 179; Capart, 1959: 89; Yamaguti, 1963, p. 80; Dojiri, 1983: 153; Kazachenko, 2001: 31.

Material examined. — Eleven females (R. G. Mus. Congo 24048-24056) from branchial cavity of sea catfish, *Arius* sp., from "Banane" (Banana) near mouth of Congo River, Zaire, collected by Dr. E. Dartevelle in January 1938. This material on loan from Koninklijk Museum voor Midden-Afrika, Musée Royal de l'Afrique Centrale, Tervuren, Belgium.

Female. — Body as in fig. 61a. Total length (to posterior end of abdomen) 6.95 mm (6.65-7.32 mm) or 9.60 mm (9.17-10.17 mm) (to distal tip of abdominal alae) (n = 3); remaining specimens too distorted for accurate measurement. Cephalothorax large, 2.90 mm (2.76-2.99 mm) (to posterior end of free margin of thoracic zone) × 3.59 mm (3.42-3.80 mm), with relatively narrow and shallow posterior sinuses. Posterior tip of lateral zone extending well beyond free margin of thoracic zone. Frontal plate small. Lunules absent. Tip of antennule well within lateral limit of cephalothorax.

Fourth pedigerous somite wider than long, 0.51 mm (0.48-0.52 mm) × 1.19 mm (1.14-1.24 mm). Genital complex smaller than cephalothorax, anterior portion narrowest, 2.14 mm (2.09-2.19 mm) × 2.33 mm (2.23-2.47 mm). Abdomen consisting of 1 free somite, 1.38 mm (1.28-1.43 mm) × 0.56 mm (0.48-0.62 mm), with pair of large, flat aliform structures. Abdominal alae 4.10 mm (3.94-4.37 mm) × 1.84 mm (1.71-1.90 mm), with setae as in fig. 61b, appearing to possess system of tubules or canals throughout entire structure.

Caudal ramus apparently absent, but may be represented by the abdominal alae, which bears a few setae.

Egg sacs (fig. 61a) 8.34 mm (6.98-9.69 mm) × 0.33 mm, containing approximately 80 eggs (n = 2).

Antennule (fig. 61c) 2-segmented. First segment relatively slender, longer than second segment, 288 μm long (measured along nonsetiferous margin), with 31 stout plumose setae. Second segment 178 μm long, with 13 + 1 aesthetasc. Antenna (fig. 61d) 3-segmented. First segment small, stout, with blunt process directed posteriorly. Second segment unornamented. Claw 402 μm long with 2 setae; proximal seta short and tear-drop shaped. Postantennal process absent.

Mouth tube (fig. 61e) 472 × 460 μm. Mandible (fig. 61f) comprising 4 sections, with third section longest; terminal section with 13 teeth. Maxillule (fig. 61g) situated next to base of mandible (fig. 61e) and comprised of 2 lobes. Smaller lobe ending in bifid process. Larger lobe tipped with 3 setae. Maxilla

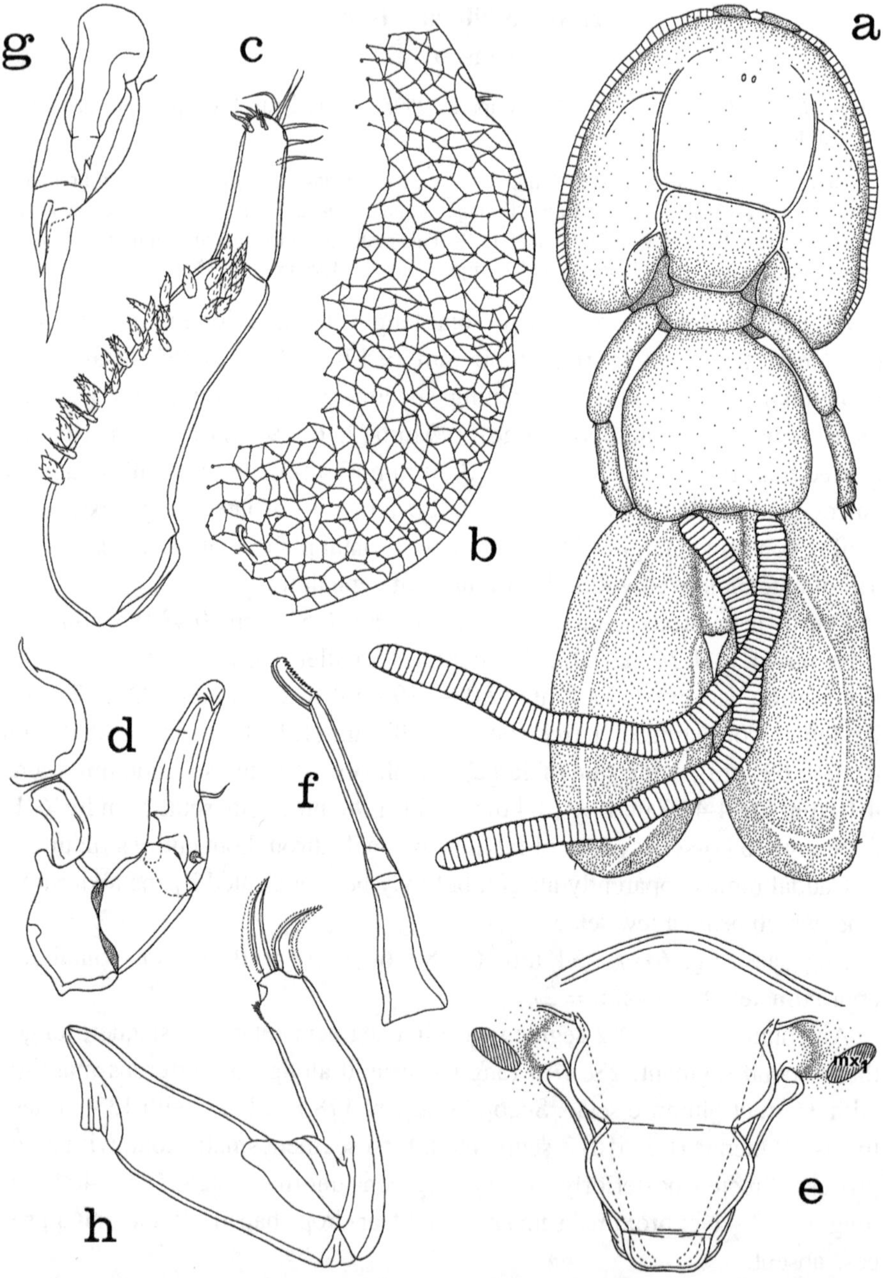

Fig. 61. *Dartevellia bilobata* Brian, 1939, female. a, body, dorsal (scale T); b, terminal portion of abdominal alae, dorsal (X); c, antennule, anteroventral (N); d, antenna, ventral (X); e, mouth tube and position of maxillule, ventral (X); f, mandible, ventral (N); g, maxillule, posterior (R); h, maxilla, ventral (X).

(fig. 61h) brachiform, with flabellum near distal end of brachium. Calamus with 2 serrated membranes, longer than canna. Canna, bent at about midlength, and with 3 serrated membranes. Corpus maxillipedis (fig. 62a) stout with conical myxa near proximal end. Shaft and claw (fig. 62b) 963 μm long, with small seta on inner margin. Sternal furca absent.

Sympod of leg 1 (fig. 62c) with plumose seta on outer distal margin and 1 sclerite (presumably a remnant of seta found at this position in other caligids) on inner margin almost midway between intercoxal plate and endopod. First segment of exopod with small spinule on outer distal corner and row of small spinules along inner margin. Terminal segment of exopod stout, with pinnate seta 4 (spinelike in appearance) much stouter than outermost spine. Spines 1, 2, and 3 unilaterally spinulated and with pectinate membranes at bases. Spines 2 and 3 with accessory processes at distal ends. Usual 3 long plumose setae on inner margin modified and reduced to 3 short spiniform setae. Distalmost spiniform seta of this set with pectinate membrane at base. Endopod with rounded lobe near base on posterior side, and equipped with 3 small setae. Leg 2 (fig. 62d) of usual caligid form. Exopod with 5 spines; 4 spines bilaterally spinulated; fifth spine with spinules on outer margin and long plumosities on inner margin. First and second segments of endopod with usual caligid armature. Terminal segment with 2 serrated membranes (fig. 62e) at bases of 3 outermost setae. Dorsal adhesion pad on sympod of leg 3 absent. Leg 3 (fig. 62f) uniramous. Exopod with usual plumose seta on sympod near basal swelling. Swelling tipped with bilaterally spinulated spine. Second exopodal segment armed with 3 bilaterally spinulated spines of approximately equal size and 1 large spine bilaterally spinulated at its tip and with serrated membrane at its base. Sympod of leg 4 (fig. 62g) longer than exopodal segments combined. Exopod indistinctly 2-segmented. Terminal 3 spines of last exopodal segment of about equal size. All 5 spines bilaterally spinulated. Transparent membrane along outer distal border of exopod. Leg 5 (fig. 62h) situated on posterolateral corner of genital complex consisting of 1 large smooth seta, 86 μm long, and small lobe tipped with minute smooth seta.

Male. — Unknown.

Remarks. — *Dartevellia bilobata* was first discovered and described by Brian (1939) from the branchial cavity of a sea catfish, *Arius* sp. (Ariidae) (also collected from *Pentanemus* and *Polynemus*), from the mouth of the Congo River near Banana on the west coast of Zaire. It is the type-species of its genus which, to this day, remains monotypic. Except for the single specimen collected by Capart (1959), *D. bilobata* has not been collected since Dr. E. Dartevelle obtained this species and gave the specimens to Brian to describe.

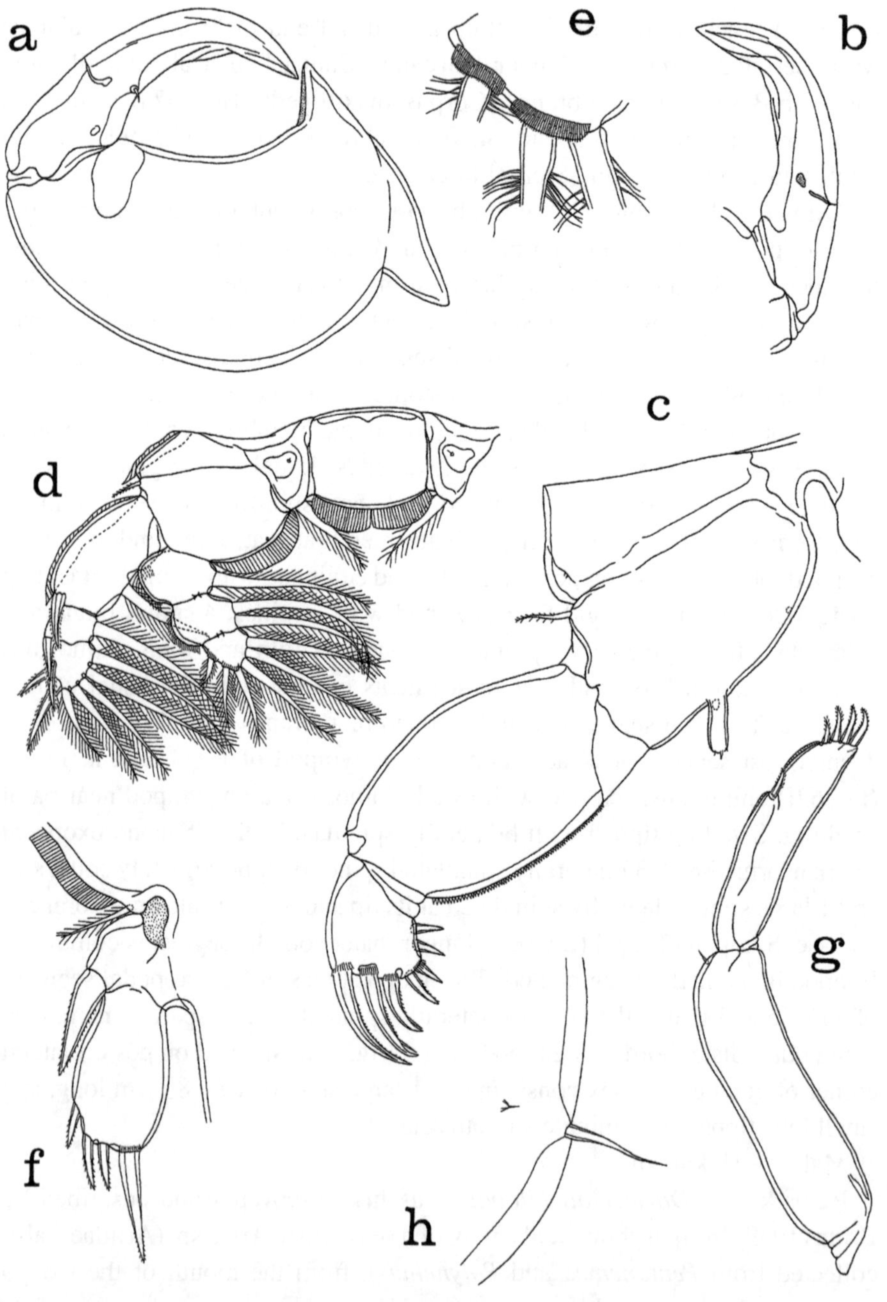

Fig. 62. *Dartevellia bilobata* Brian, 1939, female. a, maxilliped, ventral (scale S); b, shaft and claw of maxilliped, dorsal (S); c, leg 1, ventral (X); d, leg 2 and intercoxal plate, ventral (S); e, outer margin of terminal endopodal segment of leg 2, ventral (O); f, exopod of leg 3, ventral (O); g, leg 4, ventral (M); h, leg 5, dorsal (O).

The museum label is marked "plusieurs spécimens...", but in his paper Brian stated, "Neuf spécimens femelles dans la cavité branchiale d'*Arius* sp.". However, there are 11 specimens (females). Five of them are attached to each other by their maxillipeds into a group. As Brian (1939) stated, "Ces spécimens ne sont pas tous libres mais plusieurs sont entortillés et fixés entre eux formant trois groupes". Two of these groups were separated during my (first author's) examination.

There are several unusual features of this species that are noteworthy. A unique feature is the appearance of the lateral zone of the cephalothorax and the entire leg 4, which appear under low magnification to be covered with surficial scales. However, under closer scrutiny, these "scales" appear to be underneath the cuticle and therefore internal structures. Another unique feature is a system of tubules or canals inside the abdominal alae. These tubules appear as spinules on the surface of the alae when viewed under low magnification, but under higher magnification they appear as a subsurface system of interconnecting tubules. Whether the subsurface appearances of the cephalothorax, leg 4, and abdominal alae are actual structural entities or simply an artifact of preservation and/or clearing is not known.

Although Brian (1939) mentioned that the segmentation of the exopod of leg 2 of *D. bilobata* is not distinct as that in the endopod, our examination has revealed that the segmental boundaries between the three segments of this ramus are very distinct. In addition, he described the exopod of leg 3 as "...petite lamelle ovaliforme, foliacée avec 5 courtes soies et une épine à sa base". However, we found only four spines on the terminal segment and the usual large exopodal spine at the basal swelling.

Genus **Echetus** Krøyer, 1864

Echetus Krøyer, 1864: 389; Wilson, 1905a: 611; Yamaguti, 1963: 69; Kabata, 1979: 157; Dojiri, 1983: 158; Prabha, 1983: 52; Kazachenko, 2001: 25; Boxshall & Halsey, 2004: 726.

Female. — Body divisible into 4 tagmata: cephalothorax, neck, genital complex, and abdomen. Cephalothorax comparatively very small, suborbicular; anterior portion narrower than posterior, with shallow posterior sinuses. Posterior margin of thoracic zone indistinguishably fused to fourth pedigerous somite. Neck long, slender, and presumably formed solely from elongation of fourth pedigerous somite (anteriormost portion of genital complex may possibly contribute to formation of neck). Genital complex globose, with pair of posterolateral lobes and posteromedian protrusion. Abdomen consisting of 1 free somite, much longer than genital complex, spindle-shaped, and tapering posteriorly. Caudal ramus small, armed with 6 setae.

Antennule 2-segmented. First segment with numerous setae. Second segment long and cylindrical, with 13 + 1 aesthetasc. Antenna 4-segmented. Second segment with posteriorly directed spinelike process. Terminal segment hamate. Postantennal process and sternal furca absent. Mouth tube as in other caligids. Mandible comprising 4 sections and equipped with 12 teeth on inner margin of terminal section. Maxillule consisting of weakly sclerotized lobe and adjacent setiferous papilla. Maxilla brachiform. Maxilliped subchelate and prehensile.

Leg 1 with 2-segmented exopod and vestigial endopod. Terminal exopodal segment with outermost spine shortest of 3. Seta 4 and 3 inner plumose setae present. Leg 2 biramous, with 3-segmented rami. Leg 3 with large ventral apron, biramous, and possessing 3-segmented exopod and 2-segmented endopod (velum considered as part of first endopodal segment). Leg 4 uniramous, brachiform, and possessing 2-segmented exopod. Legs 5 and 6 absent.

Male. — Body similar in general habitus to males of *Caligus*. Cephalothorax more slender than in female, with free margin of thoracic zone extending slightly beyond posterior limits of lateral zone. Fourth pedigerous somite small, and without plates. Genital complex subquadrangular. Abdomen comprising 2 free somites. Caudal ramus different from that of female, with relatively larger plumose setae.

Antenna 3-segmented (perhaps 4-segmented), with corrugated adhesion pads and terminal claw. Legs 5 and 6 represented by 2 groups of setae.

Type-species. — *Echetus typicus* Krøyer, 1864.

Remarks. — Krøyer (1864) established the genus *Echetus* for some female specimens that he collected from the red drum *Sciaenops ocellatus* (Linnaeus). The heads were missing on all the specimens. However, due to the degenerate nature of the neck, genital, and abdominal somites, Krøyer placed the new species *Echetus typicus* with the lernaeocerids (Pennellidae). However, Wilson (1905a) transferred the species to the Caligidae and redescribed intact specimens that he discovered in the National Museum of Natural History (then known as the United States National Museum). Since his redescription was not complete and also rather inaccurate, Ho (1966) redescribed both the female and the male of this species.

Apparently, *E. typicus* is species specific, having been found to infest only a single species, *S. ocellatus*. Its site of attachment on the fish also does not vary, always being found in the branchial cavity.

As pointed out by Ho (1966), there are three other genera that exhibit an elongate neck in the female, *Caligodes* Heller, 1865; *Parechetus* Pillai, 1961;

and *Sinocaligus* Shen, 1957. Since the publication of Ho's paper, *Pseudechetus* Prabha & Pillai, 1979 must be added to this list. According to Ho, *Echetus* is the only genus in which the elongate neck of the female is formed solely from the fourth pedigerous somite. In the three genera listed by him, in addition to *Pseudechetus*, the elongation of the anterior portion of the genital complex contributes to the formation of the long, slender neck in the female. It is, however, very difficult to determine the boundary between these two tagmata.

Yamaguti (1963) erected a separate subfamily Echetinae for this genus. However, Ho (1966) discovered that *Echetus* possesses a pair of lunules, and "therefore cannot be separated in the first couplet of Yamaguti's key from the Caliginae". Ho then suggested two possibilities for the subfamilial position of *Echetus*. The first possibility implied by Ho was to place *Echetus* into the subfamily Caliginae. The second possibility was to retain the subfamily Echetinae, and include in it not only *Echetus*, but also *Caligodes*, *Parechetus*, and *Sinocaligus*. The latter possibility is based on the modified nature of the fourth pedigerous somite, genital complex, and the abdomen in the female. *Pseudechetus* would also be included in this group. However, the appendages of *Echetus* are very similar to those described for species of *Caligus*, indicating a relatively close phylogenetic relationship with this genus.

Echetus typicus Krøyer, 1864
(figs. 63-66)

Echetus typicus Krøyer, 1864: 389; Bassett-Smith, 1899: 486; Wilson, 1905a: 611; Fowler, 1912: 480; Bere, 1936: 586; Pearse, 1952: 24; Causey, 1953a: 6; Causey, 1953b: 11; Causey, 1955: 5; Yamaguti, 1963: 70; Ho, 1966: 752; Dojiri, 1983: 161.

Material examined. — Five females and 1 immature female from red drum, *Sciaenops ocellatus*, caught at Panacea, Florida, on 24 July 1965 by one of us (Ju-Shey Ho). Allotype male (USNM 69854) from gills of *Sciaenops ocellatus* (originally labeled as *Sciaenophilus ocellatus*) caught at Englewood, Florida, by Dr. Ruby Bere. Allotype on loan from National Museum of Natural History, Smithsonian Institution, Washington, D.C.

Female. — Body greatly modified as in fig. 63a. Total length (not including setae on caudal ramus) 20.38 mm (18.06-21.21 mm). Cephalothorax (fig. 63a, b) small in comparison to body, longer than wide 0.98 mm (0.90-1.16 mm) × 0.81 mm (0.73-0.86 mm), with relatively wide, but shallow posterior sinuses. Posterior margin of thoracic zone indistinguishable from fourth pedigerous somite. Tip of antennule not extending beyond lateral limit of cephalothorax. Lunules located on ventral surface of frontal plate.

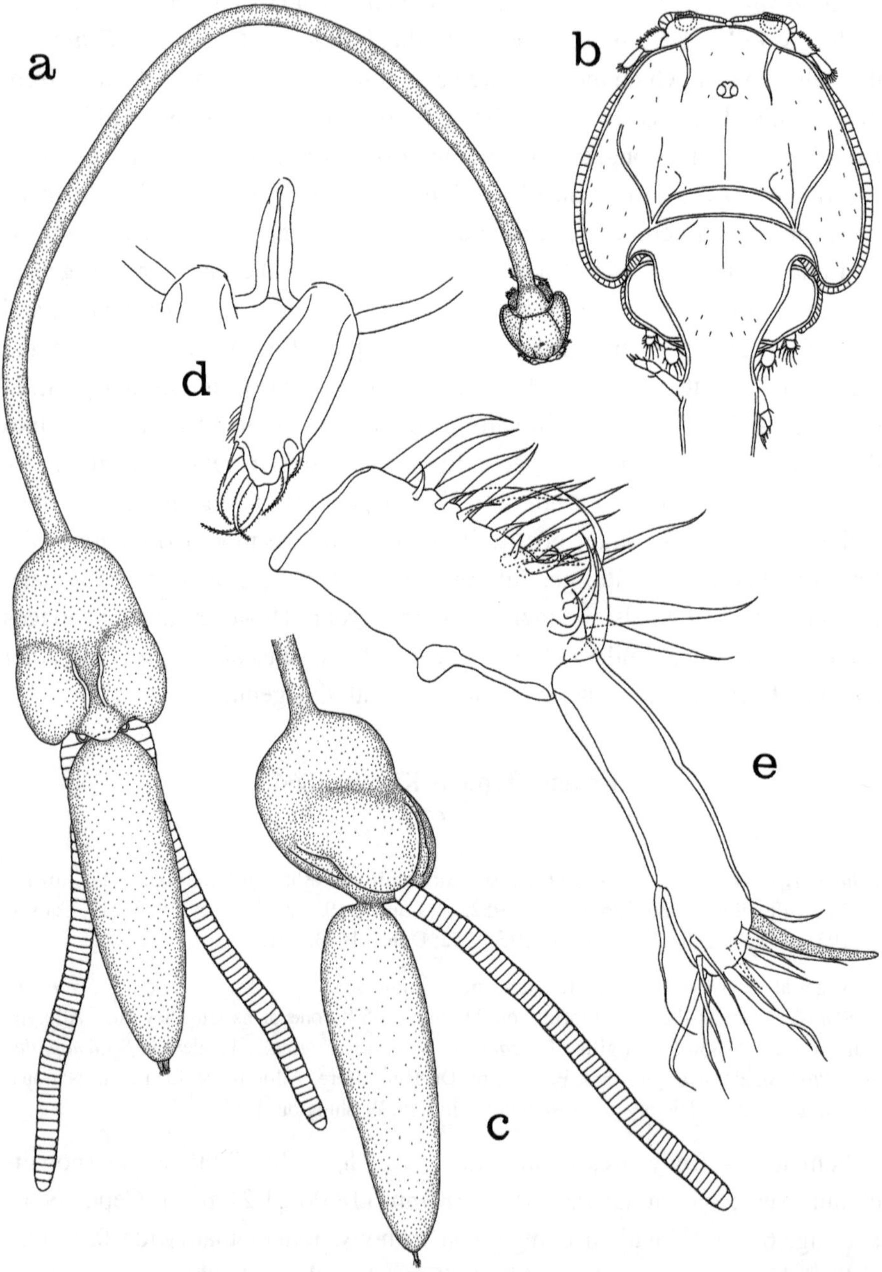

Fig. 63. *Echetus typicus* Krøyer, 1864, female. a, body, dorsal (scale T); b, cephalothorax, dorsal (F); c, posterior part of body (genital complex, abdomen, and caudal ramus) showing egg sacs, lateral (T); d, caudal ramus, ventral (E); e, antennule, ventral (L).

Fourth pedigerous somite (fig. 63a) elongate and slender, in form of long neck, 12.04 mm (10.16-13.18 mm) × 0.28 mm (0.23-0.30 mm). Genital complex (fig. 63a, c) 2.47 mm (2.26-2.66 mm) × 1.91 mm (1.89-1.96 mm), somewhat globose with pair of posterolateral lobes and posteromedian protrusion. Abdomen spindle-shaped, almost 4 times as long as wide, 4.75 mm (4.15-5.71 mm) × 1.29 mm (1.16-1.43 mm), and attached to posteroventral region of genital complex.

Caudal ramus (fig. 63d) much longer than broad, 140 × 59 μm, bearing 6 plumose and 1 smooth seta.

Egg sacs (fig. 63a) long and slender, 5.16 mm (4.98-5.35 mm) × 0.27 mm, containing 51 eggs (49-52 eggs) (n = 2).

Antennule (fig. 63e) 2-segmented. First segment 97 μm long (measured along setiferous margin) and carrying 27 plumose setae. Second segment of equal length (twice length of nonsetiferous margin), slender, and bearing usual armature of 13 + 1 aesthetasc. Antenna (fig. 64a) 4-segmented. Second segment with small, posteriorly directed spinelike projection. Third segment robust, without dorsal adhesion pad. Terminal segment a recurved claw bearing 3 setae. Postantennal area (fig. 64a) with 4 small setules; process absent.

Mouth tube (fig. 64b) longer than wide, 189 × 184 μm. Intrabuccal stylet and strigil present. Mandible (fig. 64c) with 4 sections and bearing 12 teeth. Maxillule (fig. 64d) a blunt, weakly sclerotized process with papilla, bearing 3 setae, fused to base. Maxilla (fig. 64e) brachiform with flabellum on distal half of brachium. Calamus longer than canna; both stout. Maxilliped (fig. 64f) with robust corpus; shaft and claw combined 211 μm long. Sternal furca absent.

Pinnate seta at junction of sympod and exopod of leg 1 (fig. 65a) not covered by ventral protrusion. Patch of spinules located on antero-inner surface of sympod. First segment of exopod with outer distal spine and usual setules on inner margin. Seta 4, with short plumosities on inner margin, at least 3 times as long as outermost spine. All 3 spines with pectinate membranes at bases. Spine 1, with short spinules, much shorter than other 2 spines. Spines 2 and 3 with relatively long plumosities on inner margins. Endopod with patch of setules. Leg 2 (fig. 65b) with usual caligid armature. Neither spine of first and second exopodal segments extend beyond distal limit of terminal exopodal segment. Sympod of leg 3 (fig. 65c) with 2 patches of spinules, 1 on outer margin and other on inner margin. Dorsal corrugated pad located near outer patch of spinules. Exopod 3-segmented, and endopod 2-segmented if velum considered part of first segment. First segment of exopod consisting primarily of basal swelling terminally armed with stout, bilaterally serrated or spinulated spine.

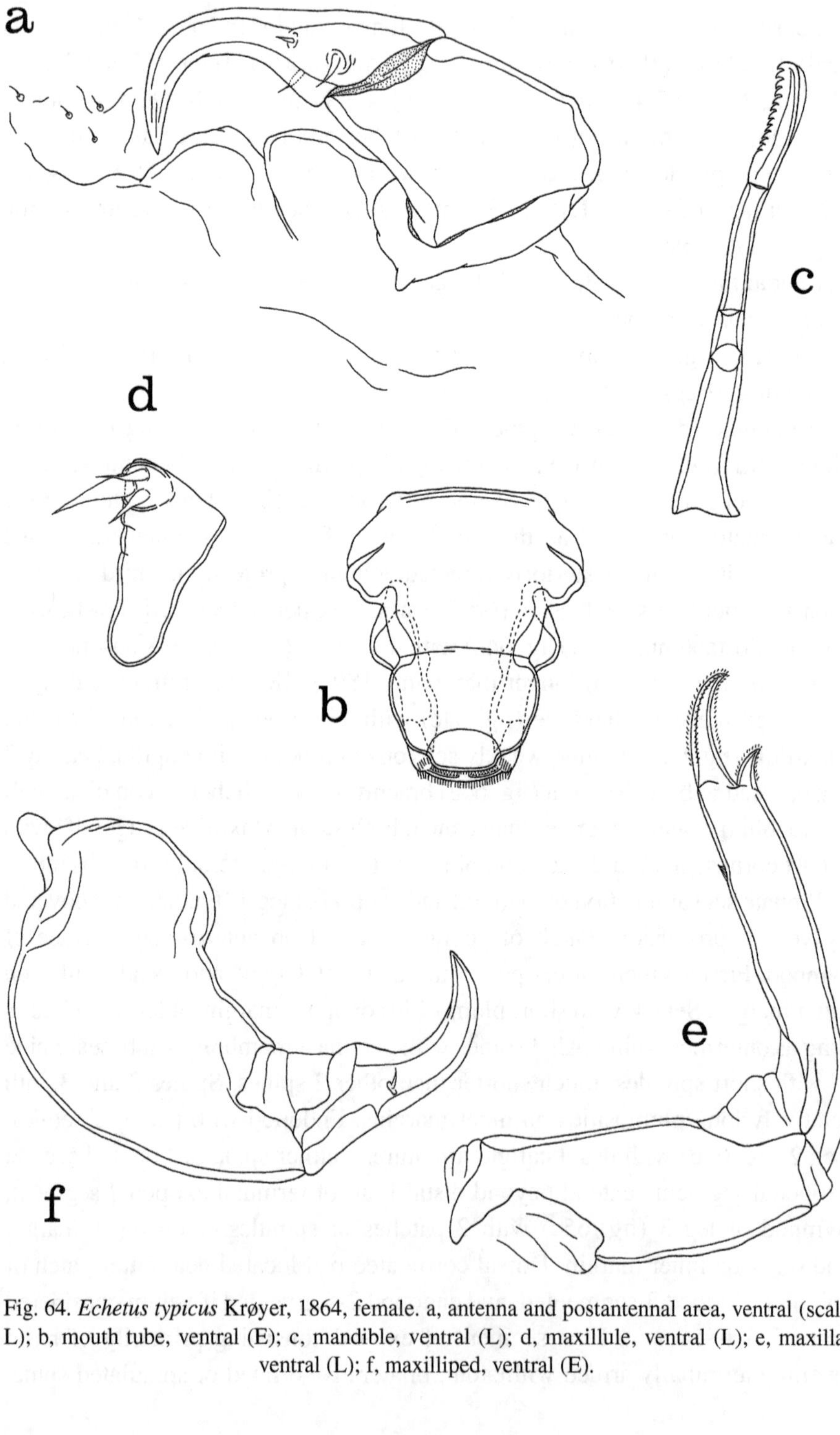

Fig. 64. *Echetus typicus* Krøyer, 1864, female. a, antenna and postantennal area, ventral (scale L); b, mouth tube, ventral (E); c, mandible, ventral (L); d, maxillule, ventral (L); e, maxilla, ventral (L); f, maxilliped, ventral (E).

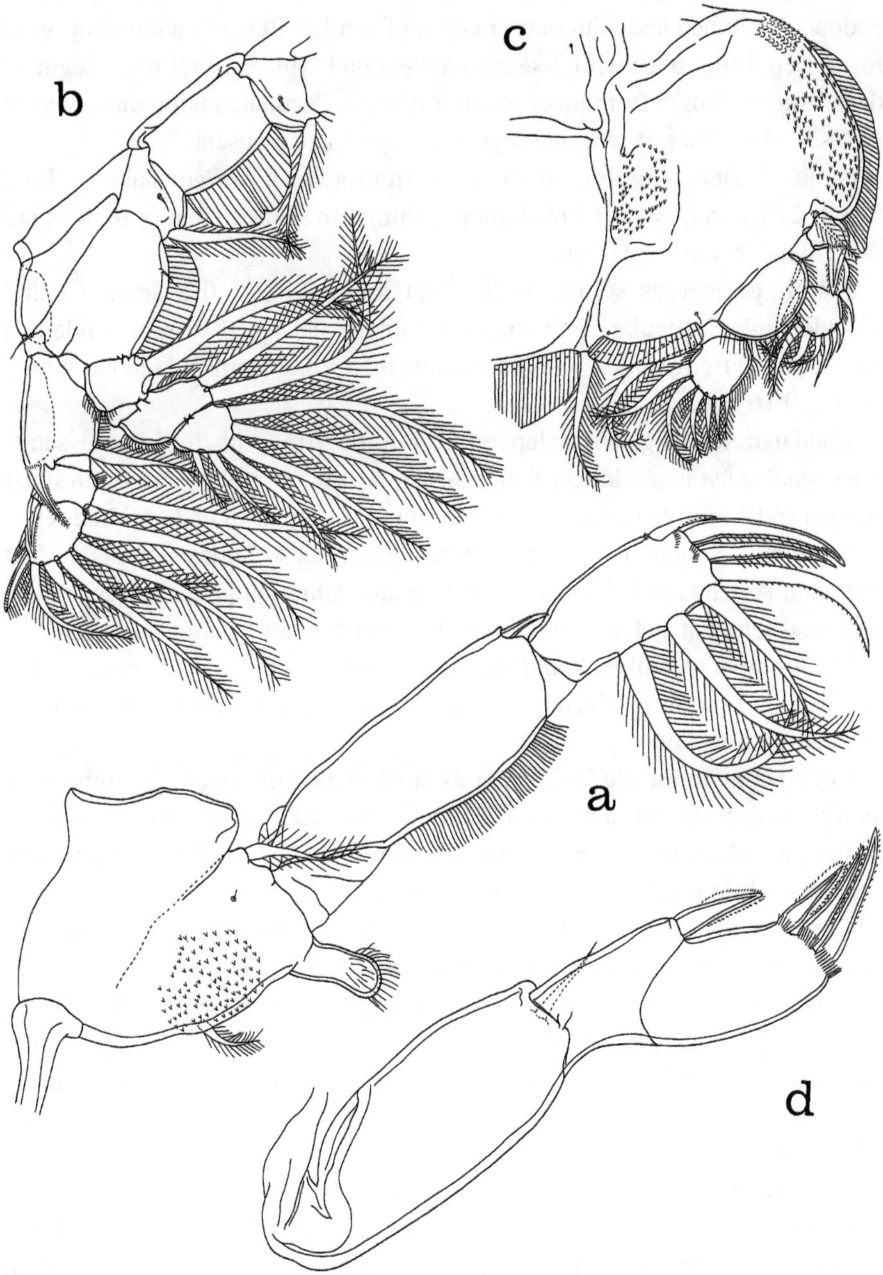

Fig. 65. *Echetus typicus* Krøyer, 1864, female. a, leg 1, anterior (scale L); b, leg 2 and intercoxcal plate, ventral (E); c, leg 3, ventral (E); d, leg 4, ventral (L).

Next 2 exopodal segments with usual caligid armature. Velum (first segment of endopod) rounded and with outer margin of setules. Rest of endopod of usual form. Leg 4 (fig. 65d) with 2-segmented exopod. Spines of terminal segment decreasing in length from inner to outer margin. Pectinate membranes present at bases of 3 spines of terminal segment. Legs 5 and 6 absent.

Male. — Body (fig. 66a) unmodified from general *Caligus* habitus. Total length 2.31 mm (n = 1). Cephalothorax similar in shape to that of female, but more slender, 1.10 × 0.90 mm.

Fourth pedigerous somite wider than long, 0.17 × 0.23 mm. Genital complex subrectangular, with rounded corners, 0.47 × 0.35 mm. Abdomen comprising 2 free somites, with first somite 0.15 × 0.20 mm and second somite 0.24 × 0.16 mm.

Caudal ramus (fig. 66b) slender, 184 × 65 μm, with 6 plumose setae; 3 terminal setae much longer than other 3. A row of setules present on inner margin and 2 transverse rows of minute spinules near bases of 3 terminal setae.

Antennule similar to that of female, except 29 plumose setae on first segment. Antenna (fig. 66c) with 4 corrugated adhesion pads on 2 segments. Terminal segment a claw with 2 setae. Postantennal area as in female.

Mouth tube, mandible, maxillule, and maxilla as in female. Corpus maxillipedis (fig. 66d) more slender than in female. Shaft and claw 211 μm long. Sternal furca absent.

Legs 1-3 as in female. Leg 4 not as stout as that of female, but otherwise similar. Leg 5 (fig. 66e) represented by 2 setae. Leg 6 (fig. 66e) located on posterolateral corner of genital complex, represented by small lobe tipped with 2 setae. A third seta located near base of lobe.

Remarks. — Because of the unusual habitus of *Echetus typicus*, this species has never been incorrectly identified; consequently, it has no synonyms. However, the original female specimen upon which Krøyer (1864) based his description of this species was missing its cephalothorax. Apparently, the head and much of the neck region (elongate fourth pedigerous somite) had been broken off during collection from the host "*Corvina unimaculata*" (*Sciaenops ocellatus*) from New Orleans. Due to the unusual genital complex and abdomen, in addition to the elongate, slender neck, Krøyer felt that the specimens represented a new taxon. Because of the highly modified appearance of the copepod, he placed *E. typicus* within the Pennellidae then known as the "Lernaeocerina".

Wilson (1905a) was the first to collect intact females of *E. typicus*, which were deeply imbedded in the wall on the inner surface of the operculum of

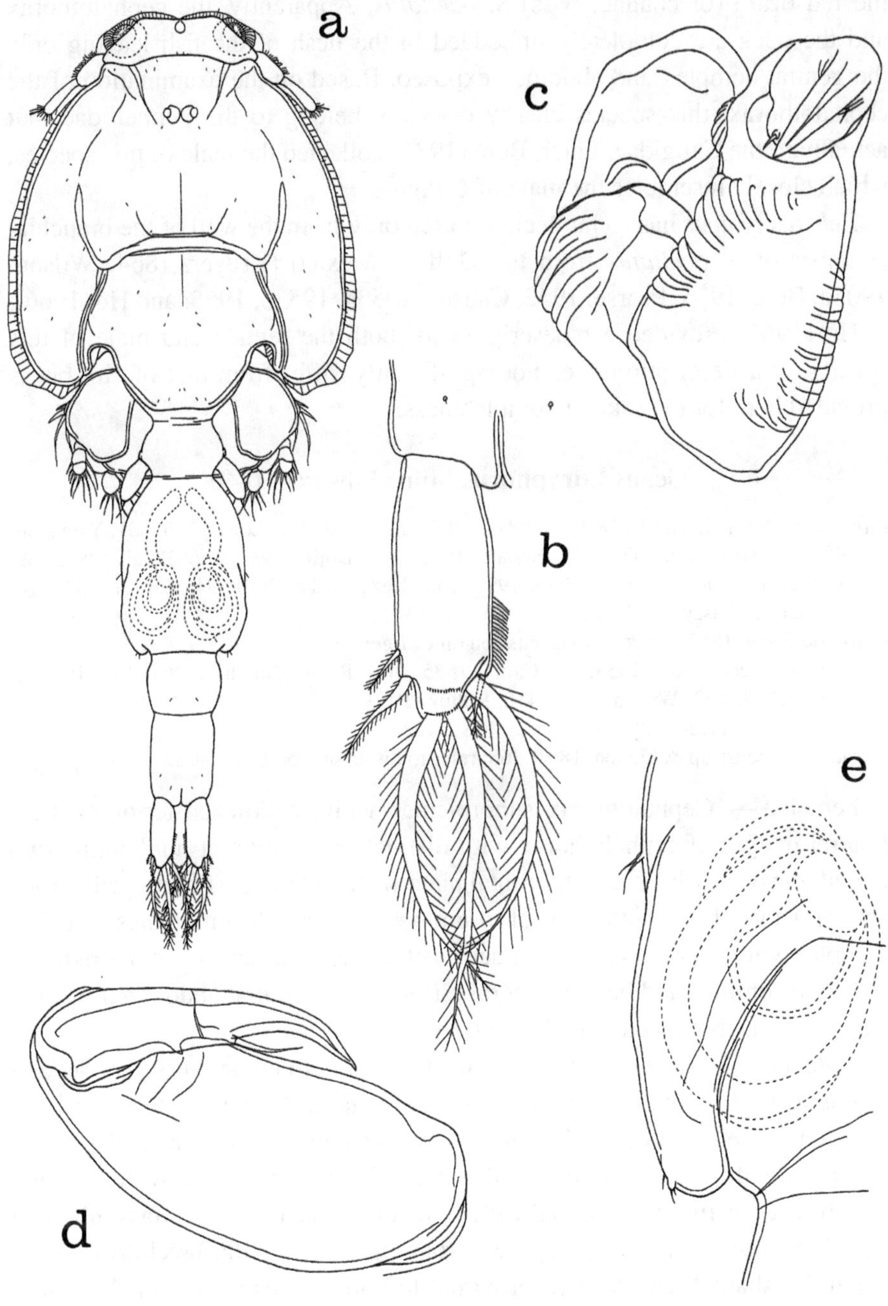

Fig. 66. *Echetus typicus* Krøyer, 1864, male. a, body, dorsal (scale F); b, caudal ramus, ventral (E); c, antenna, in situ, postero-inner (E); d, maxilliped, in situ, ventral (E); e, legs 5 and 6, and genital area, ventral (E).

the red drum (or channel bass) *S. ocellatus*. Apparently, the cephalothorax and the neck are completely imbedded in the flesh of the fish leaving only the genital complex and abdomen exposed. Based on the examination of the cephalothorax, this species clearly does not belong to the Pennellidae, but actually to the Caligidae. Later, Bere (1936) collected the male of this species, which closely resembles the male of *Caligus*.

Echetus typicus has so far been reported only from the wall of the branchial chamber of *S. ocellatus* from the Gulf of Mexico (Krøyer, 1864; Wilson, 1905a; Bere, 1936; Pearse, 1952; Causey, 1953a, 1953b, 1955; and Ho, 1966).

Ho (1966) provided a redescription of both the female and male of this species. Our description does not significantly differ from that of Ho, but is provided here for the sake of completeness.

Genus **Euryphorus** Milne Edwards, 1840

Euryphorus Milne Edwards, 1840: 462; Bassett-Smith, 1899: 461; Lewis, 1967: 31; Yamaguti, 1963: 98; Heegaard, 1972: 315; Kabata, 1979: 202; Dojiri, 1983: 166; Pillai, 1985: 489; Boxshall & Montú, 1997: 67; Ho, 1998: 253; Kazachenko, 2001: 34; Kabata, 2003: 16; Boxshall & Halsey, 2004: 726.

Caligeria Dana, 1852: 57; refer to discussion on *Caligeria*.

Elytrophora Gerstaecker, 1853: 16; Carus, 1885: 360; Bassett-Smith, 1896a: 158; Bassett-Smith, 1899: 462; Wilson, 1932: 416; Lüling, 1953: 85; Markewitsch, 1956: 145; Lewis, 1967: 42; Yamaguti, 1963: 102; Pillai, 1985: 492.

Dysgamus Steenstrup & Lütken, 1861: 368; refer to discussion on *Dysgamus*.

Female. — Cephalothorax suborbicular with shallow posterior sinuses. Lateral margin of cephalothorax with distinct indentation at midlength. Two dorsal transverse ribs present. Frontal plate without lunules. Fourth pedigerous somite short, free, with pair of prominent dorsal aliform plates. Genital complex with pair of short, rounded posterolateral processes, comprising 2 or 3 free somites, and bearing lateral aliform flaps on first abdominal somite. Caudal ramus broad, equipped with 6 setae.

Antennule 2-segmented. First segment robust with numerous setae. Second segment short, cylindrical with 13 + 1 aesthetasc. Antenna with posteriorly directed spinelike process. Terminal segment hamate. Postantennal process absent. Mouth tube with intrabuccal stylet and strigil. Mandible with third and fourth sections fused and armed with 12 teeth on inner margin. Maxillule with dentiform process and adjacent setiferous papilla. Maxilla brachiform, with small fan-shaped lobe located proximal to and oppposite side of flabellum. Maxilliped subchelate and prehensile. Sternal furca present.

Leg 1 conspicuously biramous with 2-segmented rami. Terminal segment of exopod with 3 spines, 1 seta on inner distal corner, and 3 inner plumose setae.

Seta 4 more than twice as long as outermost spine. Spine 3 originating dorsal to and approximately same level as seta 4. Endopod with first segment unarmed, and second segment with 3 plumose setae. Leg 2 biramous with 3-segmented rami. Leg 3 with large ventral apron, biramous, and possessing 3-segmented rami. Exopod with formula: I-1; I-1; III, 5. First exopodal segment largest of 3; basal swelling absent, and exopodal spine of first segment not enlarged to clawlike structure. Endopod with formula: 0-1; 0-2; 4. Leg 4 biramous with 3-segmented exopod and 2-segmented endopod. Formula of exopod: I-0; I-1; III, 4 (or II, I, 4). Endopod with formula: 0-1; 4 (second segment showing partial subdivision). Leg 5 located on posteroventral area of genital complex, and represented by setiferous papilla.

Male. — Cephalothorax and fourth pedigerous somite as in female. Genital complex almost as wide as long, and slightly indented on lateral edge near midlength. Abdomen comprising 2 or 3 free somites. First somite with lateral aliform flaps. Antenna similar to that in female, except third segment with corrugated area near distal inner end, and small curved accessory tine near midlength of claw. Corpus maxillipedis with bilobed protrusion and spinelike process at myxal area. Leg 6 apparently absent. All other appendages similar to those in female.

Type-species. — *Euryphorus nordmanni* Milne Edwards, 1840.

Remarks. — *Euryphorus* Milne Edwards, 1840 is distinguished from all caligid genera, except *Avitocaligus*, by a distinctly biramous fourth leg. This genus appears to be most morphologically similar to *Alebion* (table II). There are four genera (i.e., *Alebion*, *Avitocaligus*, *Gloiopotes*, and *Pupulina*) besides *Euryphorus* that exhibit the combination of a 2-segmented endopod of leg 1 and 3-segmented endopod of leg 3. Although the endopod of leg 1 is 2-segmented in *Gloiopotes*, this ramus is reduced in size. There is a progressive reduction of the endopod from the condition found in *Euryphorus*, *Alebion*, *Avitocaligus*, and *Pupulina* to *Gloiopotes* and finally to the vestigial ramus found in most caligids such as *Caligus* and *Lepeophtheirus*. In addition, there are some caligid species (e.g., *Tuxophorus caligodes* Wilson, 1908) that exhibit a partial division of the terminal endopodal segment of leg 3. The suture suggests a partial fusion of the original second and third endopodal segments found in the third leg of *Euryphorus*. Apparently, there is a trend in the reduction of this ramus from a 2-segmented condition to a vestigial ramus.

The antennule and antenna, mouth parts, and especially legs 1-4 are very similar between *Euryphorus nordmanni* and *E. brachypterus* (Gerstaecker, 1853). The differences that exist are not unusual between congeners, and can be attributed to interspecific variation.

The genital complex and the aliform flaps of the first abdominal somite are conspicuously different between the females of the two species. However, differences in the size and shape of the genital complex are common among species within a single genus. The aliform flaps of the first abdominal somite of the female remain as the single possible generic discriminant. Since the female of *E. brachypterus* possesses a pair of small flaps on the first free somite of the abdomen, the character is not the presence or absence of this structure, but simply the size and shape of it. Evaluated alone, the size and shape of the aliform flaps cannot be considered taxonomic features worthy of generic value. In addition, *E. nordmanni* has been reported from scombrids, which are known hosts for *E. brachypterus*. *Elytrophora* is thus considered a junior synonym of *Euryphorus* as suggested by Heegaard (1972) and supported by Kabata (1979). This action also avoids the recognition of two morphologically similar (almost identical) monotypic genera.

Euryphorus nordmanni Milne Edwards, 1840
(figs. 67-71)

Euryphorus nordmanni Milne Edwards, 1840: 462; Bassett-Smith, 1899: 461; Kirtisinghe, 1937: 445; Capart, 1959: 96; Yamaguti, 1963: 98; Lewis, 1967: 32; Lewis et al., 1969: 416; Heegaard, 1972: 308; Burnett-Herkes, 1974: 101; Kazachenko, 1976: 213; Kazachenko & Avdeev, 1977: 33; Kabata, 1979: 203; Dojiri, 1983: 169; Pillai, 1985: 490; Luque et al., 1999: 13; Lin & Ho, 2002b: 320.

Dysgamus atlanticus Steenstrup & Lütken, 1861: 368; Bassett-Smith, 1899: 461; Heegaard, 1943: 24; T. Scott, 1912: 579; Wilson, 1944: 531; Yamaguti, 1963: 101.

Euryphorus nympha Steenstrup & Lütken, 1861: 365; Heller, 1865: 189; Bassett-Smith, 1899: 461; Shiino, 1954b: 284; Shiino, 1958a: 105; Shiino, 1959b: 20; Shiino, 1959c: 350; Ho, 1963: 83; Yamaguti, 1963: 99; Kirtisinghe, 1964: 88; Pillai, 1964a: 64; Shiino, 1965: 425; Kabata & Gusev, 1966: 164; Carbonell et al., 1999: 347.

Euryphorus coryphaenae Krøyer, 1863: 235; Wilson, 1913: 225; Yamaguti, 1936b: 1; Bonnet, 1948: 7; Causey, 1953a: 7; Causey, 1953b: 11; Causey, 1955: 6.

Nogagus errans Krøyer, 1863: 247; Wilson, 1907b: 452.

Nogagus murrayi Brady, 1883: 136.

Dysgamus ariommus Wilson, 1907a: 713; Wilson, 1950: 201; Rose & Vaissière, 1953: 83; Yamaguti, 1963: 101.

Dysgamus murrayi Leigh-Sharpe, 1934: 28; Yamaguti, 1963: 101.

Dysgamus pacificus Wilson, 1950: 201.

Elytrophora coryphaenae Pearse, 1952: 26; Causey, 1953b: 11; Yamaguti, 1963: 102.

Material examined. — Two females and 2 males (USNM 112874) from gill cavity of dolphinfish, *Coryphaena hippurus* Linnaeus, at 15°N 155°W, collected by United States Fish and Wildlife Service (Honolulu Division). Specimens identified by Dr. A. G. Lewis and on loan from National Museum of Natural History, Smithsonian Institution, Washington, D.C.

Female. — Body as in fig. 67a. Total length (not including setae on caudal ramus) 10.08 mm (9.63-10.52 mm) (n = 2). Cephalothorax subcircu-

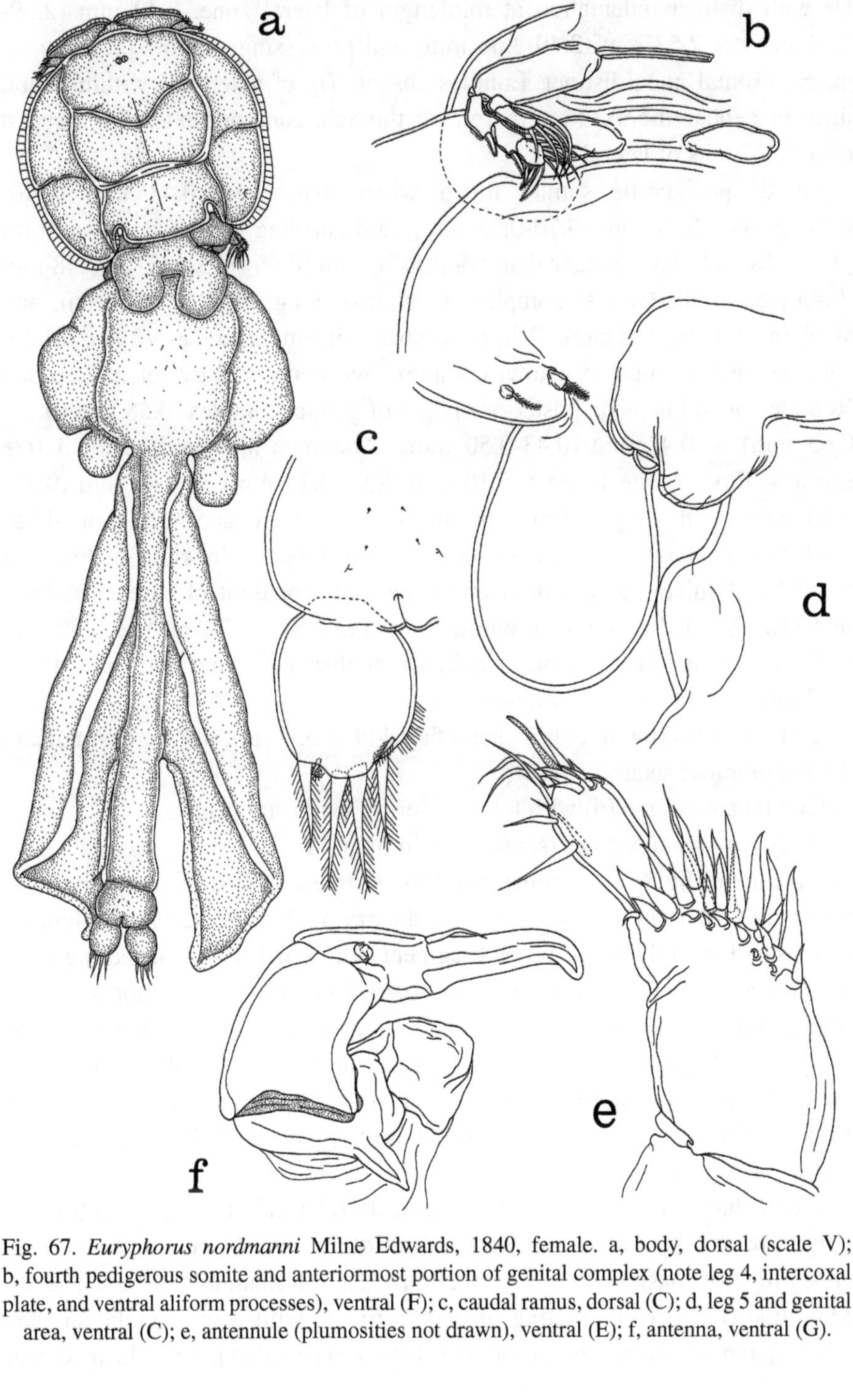

Fig. 67. *Euryphorus nordmanni* Milne Edwards, 1840, female. a, body, dorsal (scale V); b, fourth pedigerous somite and anteriormost portion of genital complex (note leg 4, intercoxal plate, and ventral aliform processes), ventral (F); c, caudal ramus, dorsal (C); d, leg 5 and genital area, ventral (C); e, antennule (plumosities not drawn), ventral (E); f, antenna, ventral (G).

lar with distinct indentation at midlength of lateral zone, 2.42 mm (2.39-2.46 mm) × 2.57 mm (2.49-2.66 mm) and possessing shallow posterior sinuses. Frontal plate distinct. Lunules absent. Tip of antennule within lateral limit of cephalothorax. Free margin of thoracic zone not extending beyond posterior limits of lateral zone.

Fourth pedigerous somite much wider than long, 0.43 mm (0.40-0.47 mm) × 0.71 mm (0.70-0.73 mm), and carrying pair of dorsal aliform plates. Dorsal plates longer than wide, 0.70 mm (0.66-0.73 mm) × 0.48 mm (0.43-0.53 mm). Genital complex wider than long, 2.06 × 2.89 mm, and swollen in medial region. Pair of ventral aliform processes (fig. 67b) located on anterior end of genital complex, and connected by sclerotized bar. Rounded dorsal flaps on posterior margin of genital complex, 0.58 mm (0.53-0.63 mm) × 0.47 mm (0.43-0.50 mm). Abdomen apparently with 3 free somites. First somite longest, 3.10 mm (2.92-3.29 mm) × 0.40 mm (0.37-0.43 mm), with long aliform structures. Abdominal alae 4.96 mm (4.95-4.98 mm) in length, with lateral edge folded dorsally. Junction of first and second abdominal somites demarcated by posterior limit of base of abdominal alae. Second abdominal somite 1.28 mm (1.13-1.43 mm) × 0.50 mm (0.43-0.56 mm). Third somite small, wider than long, and 0.40 mm (0.37-0.43 mm) × 0.63 mm (0.60-0.66 mm).

Caudal ramus (fig. 67c) broad and flat, 449 × 357 μm, bearing 2 small and 4 large plumose setae.

Genital area (fig. 67d) with lobes covering oviducal opening.

Body with small setules (sensilla) as in fig. 67a.

Antennule (fig. 67e) 2-segmented. First segment 173 μm long, with suture on posterior margin near midlength, and carrying 27 setae (20 large plumose and 7 small smooth setae). Second segment cylindrical, 124 μm, bearing 13 + 1 aesthetasc (2 setae sharing common base). Antenna (fig. 67f) apparently 4-segmented. First segment unornamented. Second segment with large posteriorly directed spinelike process. Third segment robust and naked. Claw with short robust seta near base and 1 longer, more slender seta near midlength. Postantennal area (fig. 68a) with 3 usual multibranched setule, but process absent.

Mouth tube (fig. 68b) 270 × 248 μm. Intrabuccal stylet and dentiferous strigil present. Mandible (fig. 68c) with third and fourth sections fused, thus appearing to consist of only 3 sections. Tip of mandible with 12 teeth. Maxillule (fig. 68d) consisting of papilla tipped with 3 setae (1 larger than other 2) and curved process. Heavily sclerotized rounded process located near

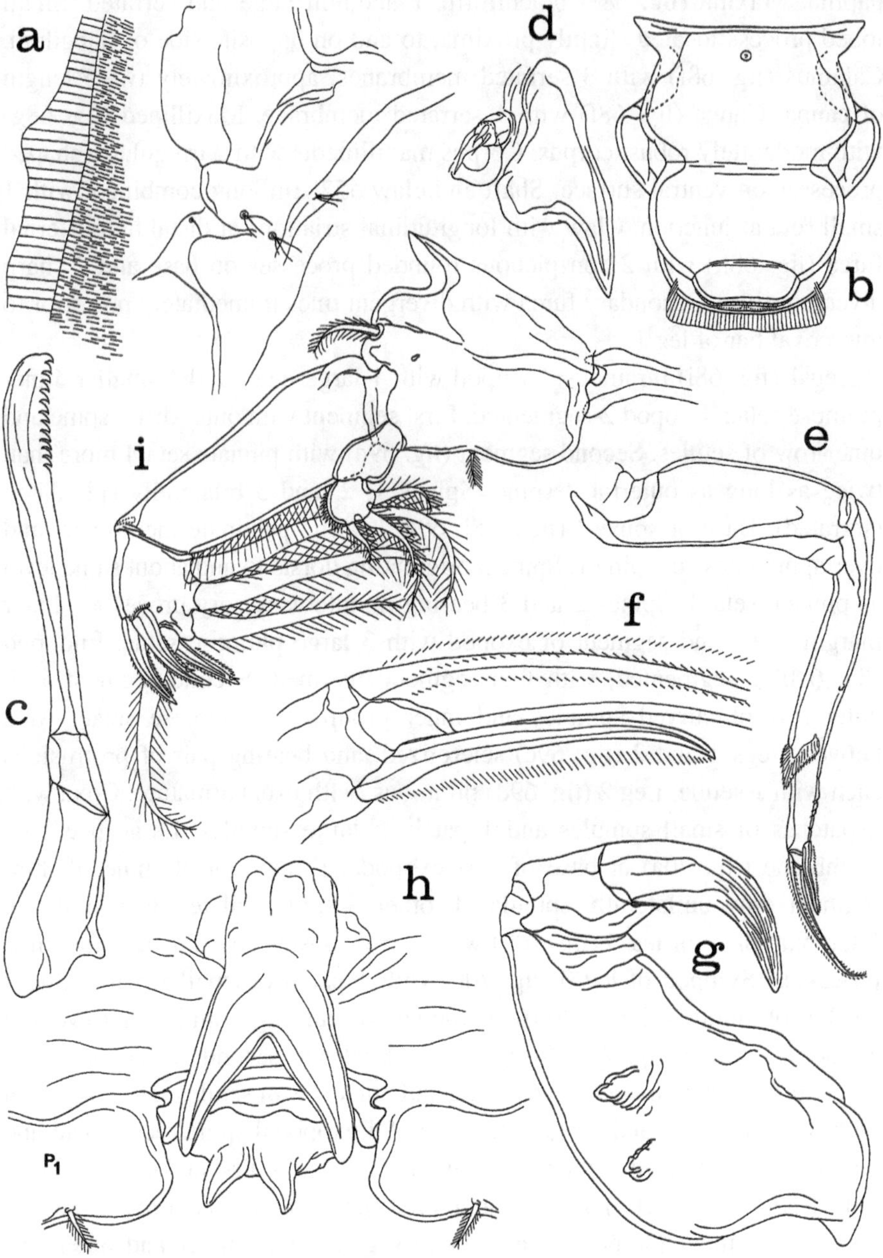

Fig. 68. *Euryphorus nordmanni* Milne Edwards, 1840, female. a, postantennal area, ventral (scale G); b, mouth tube, ventral (J); c, mandible, ventral (U); d, maxillule, ventral (G); e, maxilla, dorsal (B); f, terminal portion of maxilla, dorsal (L); g, maxilliped, ventral (I); h, sternal furca, secondary furca, and sympod of leg 1, ventral (G); i, leg 1, ventral (G).

papilla. Maxilla (fig. 68e) brachiform. Flabellum large and serrated. Small lobed process located slightly proximal to and on opposite side of flabellum. Calamus (fig. 68f) with 3 serrated membranes, approximately twice length of canna. Canna (fig. 68f) with 1 serrated membrane. Maxilliped (fig. 68g) with moderately robust corpus. Corpus maxillipedis with 2 irregularly shaped processes on ventral surface. Shaft and claw 552 μm long combined, with 1 small seta at junction. Claw with longitudinal striations at distal half. Sternal furca (fig. 68h) with 2 conspicuous rounded processes on base and 2 sharp divergent tines. Secondary furca with divergent tines immediately posterior to intercoxal bar of leg 1.

Leg 1 (fig. 68i) biramous. Sympod with 1 large outer and 1 smaller, inner plumose setae. Exopod 2-segmented. First segment with outer distal spine and inner row of setules. Second segment (fig. 69a) with pinnate seta 4 more than twice as long as outermost spine. Spines 1, 2, and 3 bilaterally spinulated (serrated). Base of spine 2 (figs. 68i, 69a) bearing pectinate membrane, and overlapping base of spine 1. Spine 3 originating dorsal to and about same level as pinnate seta 4. Spines 2 and 3 bearing minute accessory processes. Inner margin of second segment of exopod with 3 large plumose setae. Endopod (fig. 69b) 2-segmented, with first segment unarmed. Second segment with outer row of curved spinules and inner group of 3 plumose setae. Area between legs 1 and 2 (fig. 69c) sclerotized, and bearing pair of protrusions each with 1 setule. Leg 2 (fig. 69d) biramous, with usual armature. Coxa with 2 patches of small spinules and 1 patch of large spinules. Large pectinate membrane (fig. 70a) at base of first exopodal spine. Dorsal spine of third segment only unilaterally spinulated; other 3 spines bilaterally spinulated. Intercoxal plate of leg 3 (fig. 70b) with pair of posteriorly directed, furcalike processes. Sympod of leg 3 (fig. 70b) with 1 patch of small spinules and 2 patches of minute spinules. Serrated membrane curving from lateral to ventral surfaces. Exopod (fig. 70c) 3-segmented. First and second segments with 1 outer spine, 1 large inner plumose seta, and inner row of setules. Third segment with 3 outer spines and 5 plumose setae. All exopodal spines acuminate and unilaterally spinulated. Pectinate membrane at base of first and second spines only. Endopod (fig. 70b) 3-segmented. First segment with outer row of setules and 1 large inner plumose seta. Second segment with inner and outer row of setules and 2 inner plumose setae. Third segment with 4 plumose setae. Leg 4 (fig. 70d) with distinct intercoxal plate (see *E. brachypterus*). Sympod large, carrying outer distal seta, and possessing rounded expansion on inner margin; rounded expansion with serrated membrane. Exopod 3-segmented.

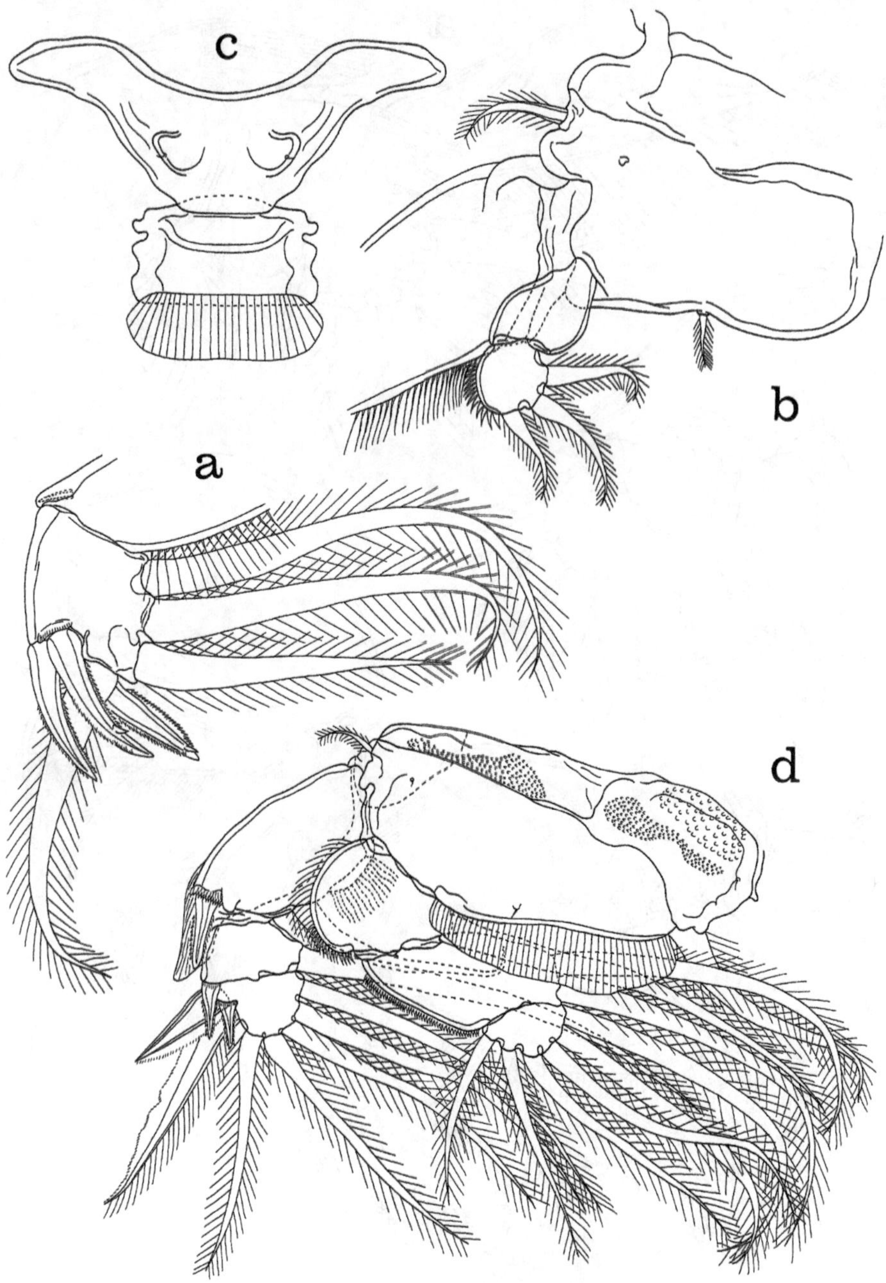

Fig. 69. *Euryphorus nordmanni* Milne Edwards, 1840, female. a, terminal portion of exopod of leg 1, ventral (scale E); b, sympod-exopod joint and endopod of leg 1, ventral (E); c, sclerotized area between legs 1 and 2, and intercoxal plate of leg 2, ventral (B); d, leg 2, ventral (G).

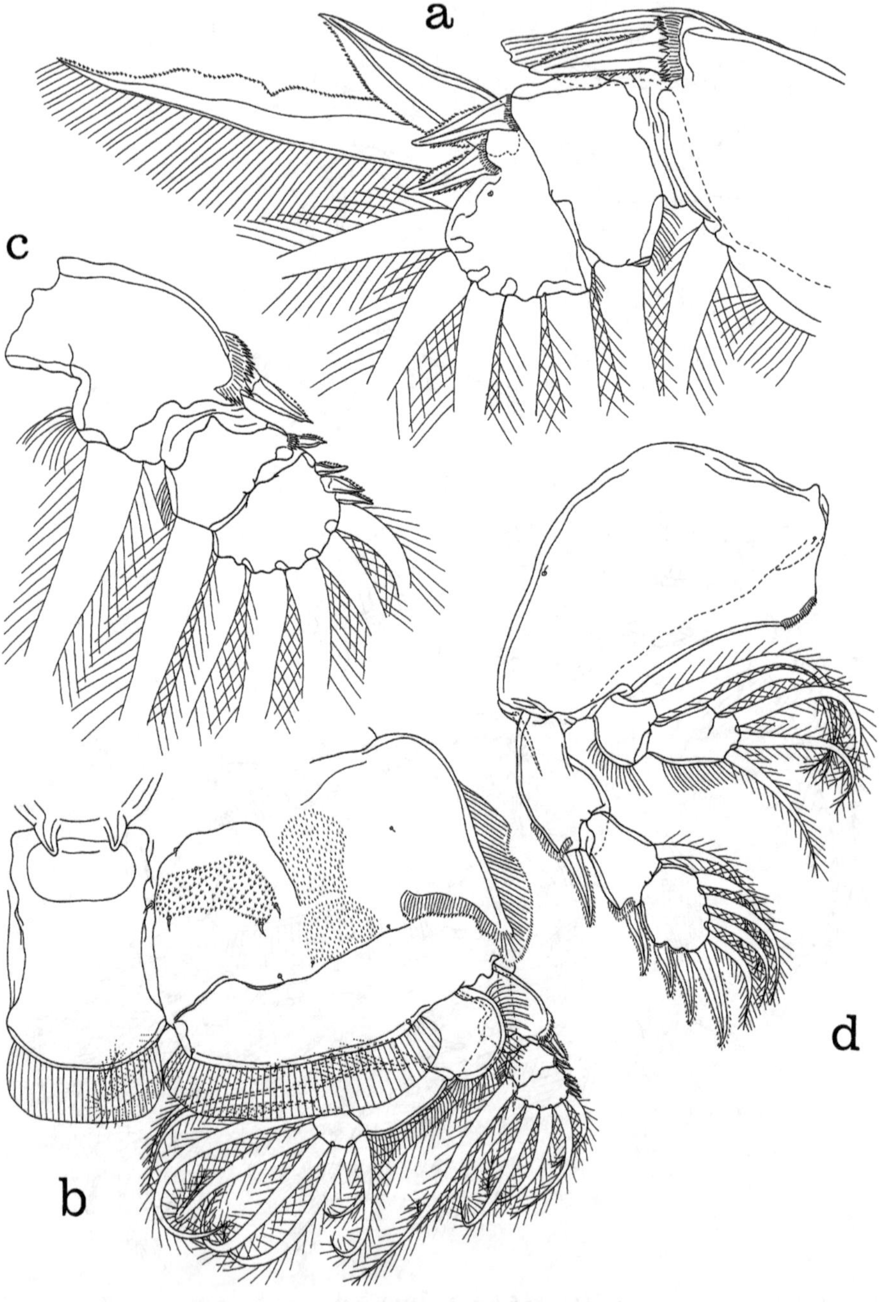

Fig. 70. *Euryphorus nordmanni* Milne Edwards, 1840, female. a, exopod of leg 2, ventral (scale D); b, leg 3, ventral (B); c, exopod of leg 3, ventral (D); d, leg 4, ventral (G).

First segment with outer spine and pectinate membrane. Second segment with outer spine, pectinate membrane, and 1 plumose inner seta. Third segment with 3 spines and 4 plumose setae. Pectinate membranes at bases of outer spines. All spines bilaterally spinulated. Endopod indistinctly 3-segmented. First segment bearing outer row of setules and 1 plumose inner seta. Second segment partially fused to third segment and carrying 1 plumose inner seta. Third segment with 3 plumose setae. Row of setules present on outer margin of second and third segments. Leg 5 (fig. 67d) located on posteroventral area of genital complex, represented by 2 papillae; 1 papilla tipped with 1 plumose seta, and other bearing 3 plumose setae.

Male. — Body as in fig. 71a. Total length 5.38 mm (5.22-5.54 mm) (n = 2). Cephalothorax similar to that in female, but much more slender, 2.16 mm (2.09-2.22 mm) × 2.16 mm (2.12-2.19 mm).

Fourth pedigerous somite much wider than long, 0.43 mm (0.43-0.44 mm) × 0.65 mm (0.63-0.66 mm). Dorsal aliform plates 0.49 mm (0.47-0.52 mm) × 0.31 mm. Genital complex almost as wide as long, 1.13 mm (1.10-1.16 mm) × 1.08 mm (1.06-1.10 mm), and slightly indented on lateral edge near midlength. Abdomen comprising 3 free somites. First somite 0.68 mm (0.66-0.70 mm) × 0.41 mm (0.39-0.43 mm), with aliform expansion, 0.96 × 0.45 mm (0.43-0.46 mm). Second somite wider than long, 0.38 mm (0.33-0.43 mm) × 0.48 mm (0.46-0.50 mm), widest posteriorly. Third somite 0.28 mm (0.27-0.30 mm) × 0.52 mm (0.50-0.54 mm).

Caudal ramus similar to that of female, except smaller, 345 × 288 μm.

Body surface with small setules similar to that in female.

Antennule as in female. Antenna (fig. 71b) similar to that in female, except third segment with corrugated area near distal inner end and curved tine near base of attenuate seta.

Mouth tube, mandible, maxillule, and maxilla as in female. Maxilliped (fig. 71c) similar to that in female, except bilobed protrusion associated with distally directed curved spine and 1 stout curved spinelike process at myxal area. Sternal furca and accessory furca as in female.

Legs 1-4 as in female. Leg 5 (fig. 71d) on posterolateral edge of genital complex. Leg 6 apparently absent, but sixth pedigerous somite may be represented by ventral flap of genital area.

Spermatophore (fig. 71e), attached to posteroventral region of female genital complex, ovoid, 518 × 345 μm.

Remarks. — Since the original description of *Euryphorus nordmanni* by Milne Edwards (1840), this species of parasitic copepod has accumulated

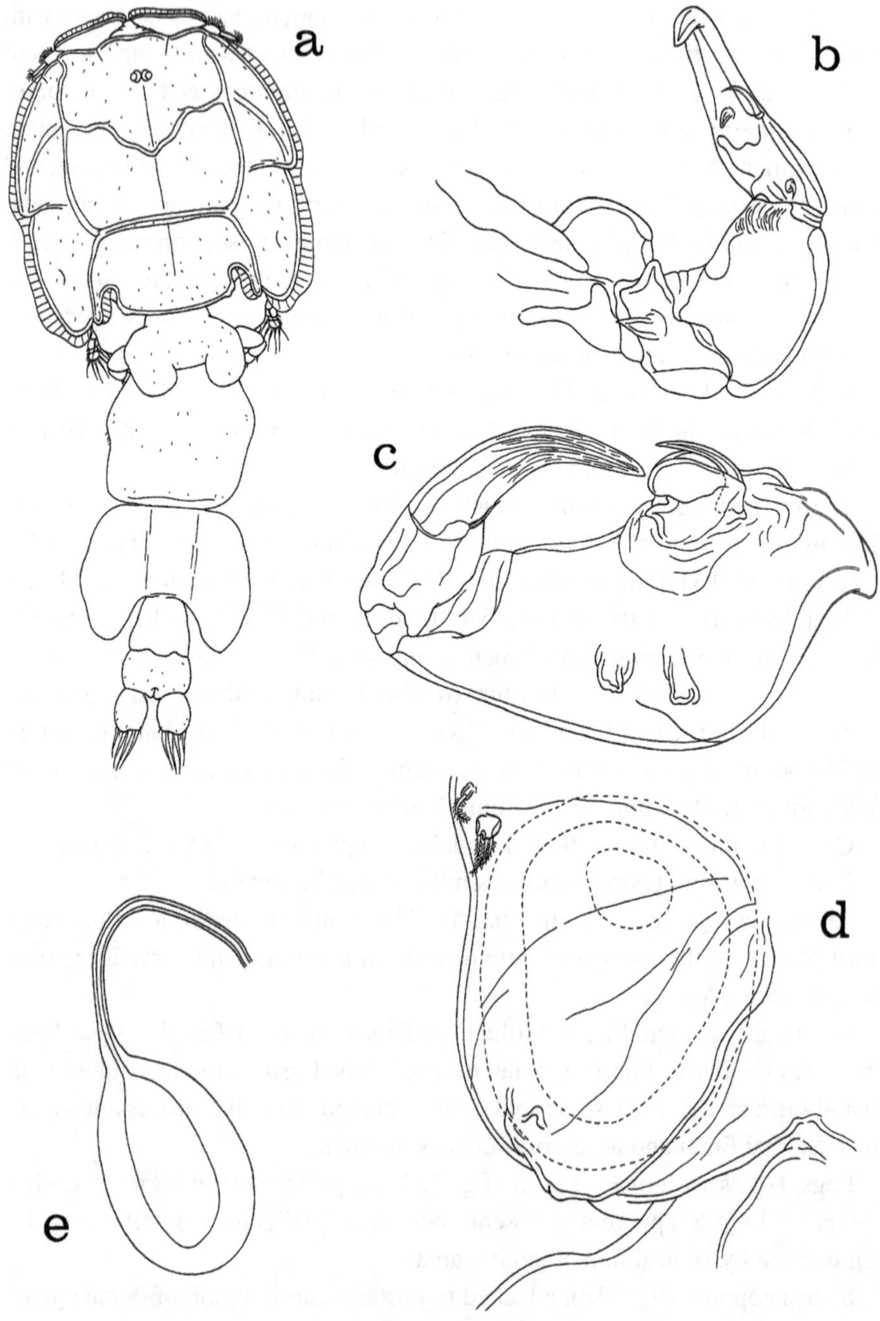

Fig. 71. *Euryphorus nordmanni* Milne Edwards, 1840, male. a, body, dorsal (scale AA); b, antenna, ventral (G); c, maxilliped, ventral (B); d, leg 5 and genital area, ventral (B); e, spermatophore, lateral (C).

numerous synonyms. Much of the taxonomic confusion surrounding this species arose due to the use of the sizes and shapes of the dorsal plates (aliform structures on the fourth pedigerous somite) and outgrowths of the body (alae of the genital complex and of the first abdominal somite) in the female as taxonomic features. However, Shiino (1954b, 1959b) and Kabata & Gusev (1966) showed that these structures appear only at the later stages of development. The majority of the spurious species (e.g., *Dysgamus atlanticus* Steenstrup & Lütken, 1861; *D. ariommus* Wilson, 1907; *D. murrayi* Leigh-Sharpe, 1934; *D. pacificus* Wilson, 1950; *Nogagus errans* Krøyer, 1863; and *N. murrayi* Brady, 1883) were established based on juveniles of *E. nordmanni* (see Heegaard, 1972). Apparently, there is a great deal of intraspecific variation in the shape of the sternal furca within this species. This also added to the taxonomic confusion over *E. nordmanni*.

The type-species can be distinguished from its only congener, *E. brachypterus* (Gerstaecker, 1853), by the shape of the genital complex of the female, segmentation of the abdomen (*E. nordmanni* with 3 free somites and *E. brachypterus* with only 2), abdominal alae of the female much larger in *E. nordmanni*, presence of ventral aliform processes on genital complex of the female of *E. nordmanni*, presence of 2 processes on ventral surface of corpus maxillipedis of *E. nordmanni* (*E. brachypterus* with 2 membranous flaps instead of processes), presence of accessory furca in *E. nordmanni*, and setae of leg 4 not reduced in *E. nordmanni*.

TABLE VIII

Hosts and localities of collections of *Euryphorus nordmanni* Milne Edwards, 1840

Reported binomen of copepod	Host	Locality	Reference
Euryphorus nordmanni Milne Edwards, 1840	unknown	Asia	Milne Edwards, 1840
	Coryphaena hippurus Linnaeus	Ceylon (Sri Lanka)	Kirtisinghe, 1937
	Coryphaena hippurus	03°02′S 09°53′E (Gabon)	Capart, 1959
		11°10′S 13°30′E (Angola)	Capart, 1959
	Coryphaena hippurus	15°N 115°W	Lewis, 1967
		120 miles south of Oahu	Lewis, 1967
		130 miles south of Niihau, Hawaii	Lewis, 1967

TABLE VIII
(Continued)

Reported binomen of copepod	Host	Locality	Reference
	Coryphaena hippurus	19°32′S 65°46′E	Lewis et al., 1969
		00°33′S 80°08′E	Lewis et al., 1969
	Coryphaena equisetis Linnaeus	13°50′N 70°07′E	Lewis et al., 1969
	Coryphaena equisetis	09°46′N 70°06′E	Lewis et al., 1969
		05°48′N 70°03′E	Lewis et al., 1969
	Coryphaena hippurus	Straits of Florida	Burnett-Herkes, 1974
	Coryphaena hippurus	Gilbert Islands	Kazachenko, 1976
		Daito Islands	Kazachenko, 1976
	Coryphaena hippurus	24°04′N 141°56′E	Kazachenko & Avdeev, 1977
	Coryphaena hippurus	Off Trivandrum, India	Pillai, 1985
	Coryphaena equisetis	Off Trivandrum, India	Pillai, 1985
	Neothunnus macropterus (= *Thunnus albacares Bonnaterre*)	Off Trivandrum, India	Pillai, 1985
	Coryphaena hippurus	Brazil	Luque et al., 1999
	Coryphaena hippurus	Taiwan	Lin & Ho, 2002b
Dysgamus atlanticus Steenstrup & Lütken, 1861	free-swimming	15°15′N 25°09′W	T. Scott, 1912
		11°10′N 25°20′W	T. Scott, 1912
		34°02′S 49°07′W	T. Scott, 1912
	free-swimming	29°N 21°W	Heegaard, 1943
		05°30′N 27°W	Heegaard, 1943
Euryphorus nympha Steenstrup & Lütken, 1861	*Lampugus punctulatus* (= *Coryphaena* sp.)	Atlantic Ocean	Steenstrup & Lütken, 1861
	Coryphaena sp.	Atlantic Ocean	Heller, 1865
	Neothynnus macropterus (Temminck & Schlegel)	"Simakatu" (Shimakatsu) Sima, Mie Prefecture	Shiino, 1954b
	Coryphaena hippurus	01°13′S 73°32′E	Shiino, 1958a
	Coryphaena hippurus	Hamajima, Japan	Shiino, 1959b
	Coryphaena hippurus	Hamajima, Japan	Shiino, 1959c
	Thunnus alalunga (Bonnaterre)	06°52′S 74°49′E	Shiino, 1959c
	Coryphaena hippurus	Nanfangao (northeast of Taiwan)	Ho, 1963

TABLE VIII
(Continued)

Reported binomen of copepod	Host	Locality	Reference
	Coryphaena hippurus	"Hikkaduwa"	Kirtisinghe, 1964
	Coryphaena hippurus	Vizhingom and Trivandrum, India	Pillai, 1964a
	Alepisaurus borealis (Gill)	03°18′N 101°54′W	Shiino, 1965
		09°17′N 102°37′W	Shiino, 1965
		00°59′S 101°42′W	Shiino, 1965
	Thunnus obesus Lowe	00°59′S 101°42′W	Shiino, 1965
	Coryphaena hippurus	Ryuku Island, near Tanegashima, Japan	Kabata & Gusev, 1966
Elytrophora coryphaenae Krøyer, 1863	*Coryphaena hippurus*	Atlantic Ocean (near West Indies)	Krøyer, 1863
	Coryphaena hippurus	Hawaii	Bonnet, 1948
	Rachycentron canadum (Linnaeus)	Port Aransas, Texas	Causey, 1953b
	Coryphaena hippurus	Grand Isle, Louisiana	Causey, 1953a
	Coryphaena hippurus	Gulf of Mexico	Causey, 1955
	Coryphaena hippurus	Pacific (Japan)	Yamaguti, 1936b
	Coryphaena hippurus	Texas coast	Causey, 1953b
Electrophora coryphaenae (= *Elytrophora coryphaenae*)	*Coryphaena hippurus*	Texas coast	Pearse, 1952
Nogagus errans Krøyer, 1863	free-swimming	Porto Santo (Atlantic Ocean)	Wilson, 1907b
Nogagus murrayi Brady, 1883	unknown	"off Rio de Janeiro"	Brady, 1883
		North Atlantic	Brady, 1883
Dysgamus murrayi	unknown	unknown	Leigh-Sharpe, 1934
Dysgamus ariommus Wilson, 1907	free-swimming	unknown	Wilson, 1907a
	free-swimming	Romblon Island, Philippines	Wilson, 1950
Dysgamus pacificus Wilson, 1950	free-swimming	north of Marquesas Islands	Wilson, 1950

Although *Euryphorus nordmanni* has been reported from scombrids *Neothynnus macropterus* (Temminck & Schlegel), *Thunnus alalunga* (Bonnaterre), *Thunnus albacares* (Bonnaterre) and *Thunnus obesus* Lowe, a lancetfish *Alepisaurus borealis* (Gill), and *Rachycentron canadum* (Linnaeus) (Shiino, 1954b, 1959c, 1965; Causey, 1953b; Pillai, 1985), this species is primarily a parasite of the branchial cavity of the dolphinfish (also referred to as mahi-mahi) *Coryphaena hippurus* (table VIII). This species of copepod has been found in all the major oceans (North and South Atlantic, Indian, and North and South Pacific Oceans), but has not yet been found outside the area bounded by 40°N and 40°S latitudes.

Euryphorus brachypterus (Gerstaecker, 1853)
(figs. 72-77)

Elytrophora brachyptera Gerstaecker, 1853: 16; Nordmann, 1864: 468; Heller, 1865: 189; Heller, 1866: 754; Stossich, 1880: 257; Valle, 1880: 60; Carus, 1885: 360; Bassett-Smith, 1896a: 158; Bassett-Smith, 1896b: 12; Brian, 1898: 9; Brian, 1899a: 2; Brian, 1899b: 11; Bassett-Smith, 1899: 462; Brian, 1906: 51; Norman & Scott, 1906: 210; Brian, 1908: 3; Brian, 1912: 10; Scott & Scott, 1913: 83; Yamaguti, 1936: 3; Bonnet, 1948: 7; Delamare Deboutteville & Nunes-Ruivo, 1953: 202; Lüling, 1953: 84; Rose & Vaissière, 1953: 85; Shiino, 1954b: 279; Heegaard, 1955: 46; Markewitsch, 1956: 145; Shiino, 1957: 364; Yamaguti, 1963: 102; Lewis, 1967: 42; Hewitt, 1968: 117 (as *E. brachyptera brachyptera*); Lewis et al., 1969: 417; Cressey & Cressey, 1980: 36.

Arnaeus thynni Krøyer, 1863: 231.

Dinematura thynni Krøyer, 1863: 231; Wilson, 1907b: 376.

Elytrophora hemiptera Wilson, 1921: 4; Shiino, 1954b: 283; Shiino, 1958a: 105; Shiino, 1959c: 351; Yamaguti, 1963: 103; Shiino, 1965: 422; Kensley & Grindley, 1973: 91; Pillai, 1985: 494.

Dysgamus longifurcatus Wilson, 1923: 11; Yamaguti, 1963: 101.

Elytrophora atlantica Wilson, 1932: 417; Yamaguti, 1963: 102.

Elytrophora indica Shiino, 1958a: 107; Shiino, 1959c: 351; Yamaguti, 1963: 103; Shiino, 1965: 422; Cressey & Cressey, 1980: 37; Pillai, 1985: 498.

Dysgamus sagamiensis Shiino, 1958b: 161; Kabata & Gusev, 1966: 167; Yamaguti, 1963: 101.

Euryphorus brachypterus Heegaard, 1972: 313; Kabata, 1979: 203; Dojiri, 1983: 177; Jones, 1991: 419; Piasecki, 1993: 167; Boxshall & Montú, 1997; p. 67; Ho et al., 2008: 92.

Material examined. — Three females and 3 males (syntypes, catalog number 929) from Mediterranean Sea in 1853. Specimens borrowed from Zoologisches Museum, Berlin, Germany. Museum label without host name. Gerstaecker designated 22 females and 13 males as syntypes. One female selected as lectotype and 1 male as allolectotype by Dojiri (1983); rest designated as paralectotypes.

Female. — Body as in fig. 72a. Total length (not including setae on caudal ramus) 10.22 mm (9.28-11.20 mm) (n = 3). Cephalothorax subcircular, 4.14 mm (3.78-4.32 mm) × 4.61 mm (4.32-4.99 mm), with shallow posterior sinuses. Lateral margin of cephalothorax with distinct indentation at mid-length. Frontal plate distinct, but narrow. Lunules absent. Tip of antennule well within lateral limit of cephalothorax. Free margin of thoracic zone not extending beyond posterior limit of lateral zone.

Fourth pedigerous somite wider than long, 0.99 × 1.31 mm, and carrying pair of dorsal aliform plates. Dorsal plates longer than wide, 1.38 × 1.12 mm,

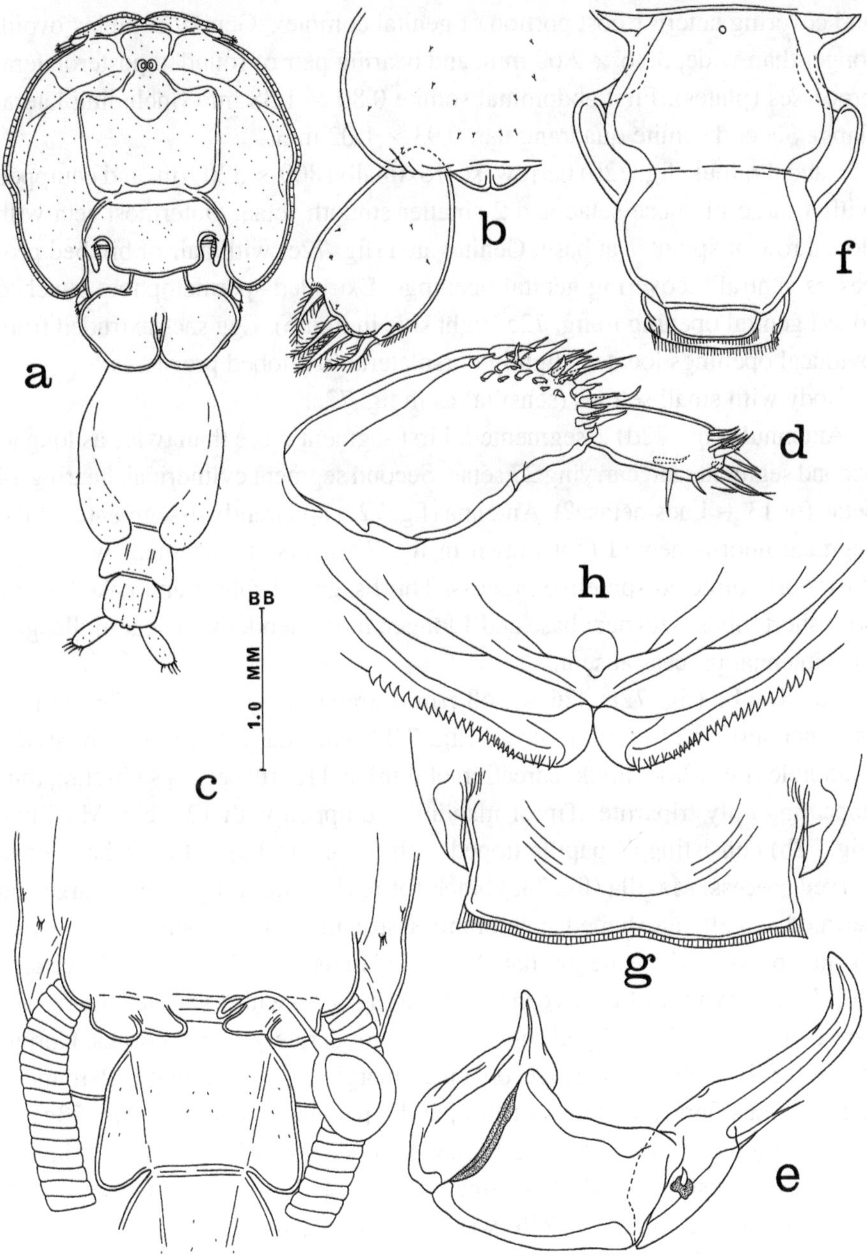

Fig. 72. *Euryphorus brachypterus* (Gerstaecker, 1853), female. a, body, dorsal (scale A); b, caudal ramus, dorsal (F); c, genital area, ventral (BB); d, antennule (plumosities not drawn), dorsal (G); e, antenna, ventral (G); f, mouth tube, ventral (G); g, intrabuccal stylet and frons labri (with labium removed), dorsal (L); h, tip of labium (with labrum removed) and strigil, ventral (Y).

and covering anteriormost portion of genital complex. Genital complex ovoid, longer than wide, 3.36 × 2.62 mm, and bearing pair of rounded posterolateral processes (plates). First abdominal somite 0.80 × 1.38 mm (including lateral flaps). Second somite quadrangular, 0.93 × 1.02 mm.

Caudal ramus (fig. 72b) narrowest proximally, 800 × 520 μm, and equipped with 4 large plumose setae and 2 smaller smooth setae. Outermost seta with dorsal row of spinules at base. Genital area (fig. 72c) with pair of bilobed processes ventrally covering genital openings. Extruded spermatophore attached to left genital opening in fig. 72c (right side in figure). Egg sacs extruded from oviducal openings located slightly dorsolateral to bilobed processes.

Body with small setules (sensilla) as in fig. 72a.

Antennule (fig. 72d) 2-segmented. First segment more than twice as long as second segment, and carrying 27 setae. Second segment cylindrical, bearing 14 setae (or 13 + 1 aesthetasc?). Antenna (fig. 72e) apparently 4-segmented. First segment unornamented (not drawn in fig. 72e). Second segment with large posteriorly directed spinelike process. Third segment robust and naked. Claw with short robust seta near base and 1 longer, more slender seta near midlength. Postantennal process absent.

Mouth tube (fig. 72f) 426 × 368 μm. Intrabuccal stylet (fig. 72g) present on inner surface of labrum. Strigil (fig. 72h) with many dentiform processes. Mandible (fig. 73a) with demarcation of third and fourth sections obscure, thus appearing only tripartite. Tip of mandible equipped with 12 teeth. Maxillule (fig. 73b) consisting of papilla tipped with 3 setae (1 larger than other 2) and curved process. Maxilla (fig. 73c) brachiform. Flabellum (fig. 73c, d) large and serrated. Small fan-shaped membranous structure located slightly proximal to and on opposite side of flabellum. Calamus (fig. 73e) with 3 serrated membranes, twice as long as canna. Canna (fig. 73e) with 1 serrated membrane and 1 or 2 barbs. Maxilliped (fig. 73f) with moderately robust corpus. Corpus maxillipedis with raised ridge on inner margin near midlength, 2 rounded membranous flaps on ventral surface, and 2 patches of denticles (fig. 73g) on dorsal, distal portion. Shaft short; claw recurved with longitudinal striations (grooves) at distal half of claw. Small seta located at junction between shaft and claw. Sternal furca (fig. 73h) with sharp divergent tines.

Leg 1 (figs. 73i, 74a-c) very similar to that in *Euryphorus nordmanni*. Leg 2 (fig. 74d) also similar to that in congener. Proximal portion of coxa of leg 2 with no conspicuous sclerotized ridges or protrusions, but with patch of spinules near intercoxal plate. One patch of rounded denticles near outer distal seta of sympod. Large pectinate membrane (fig. 74e) at base of first

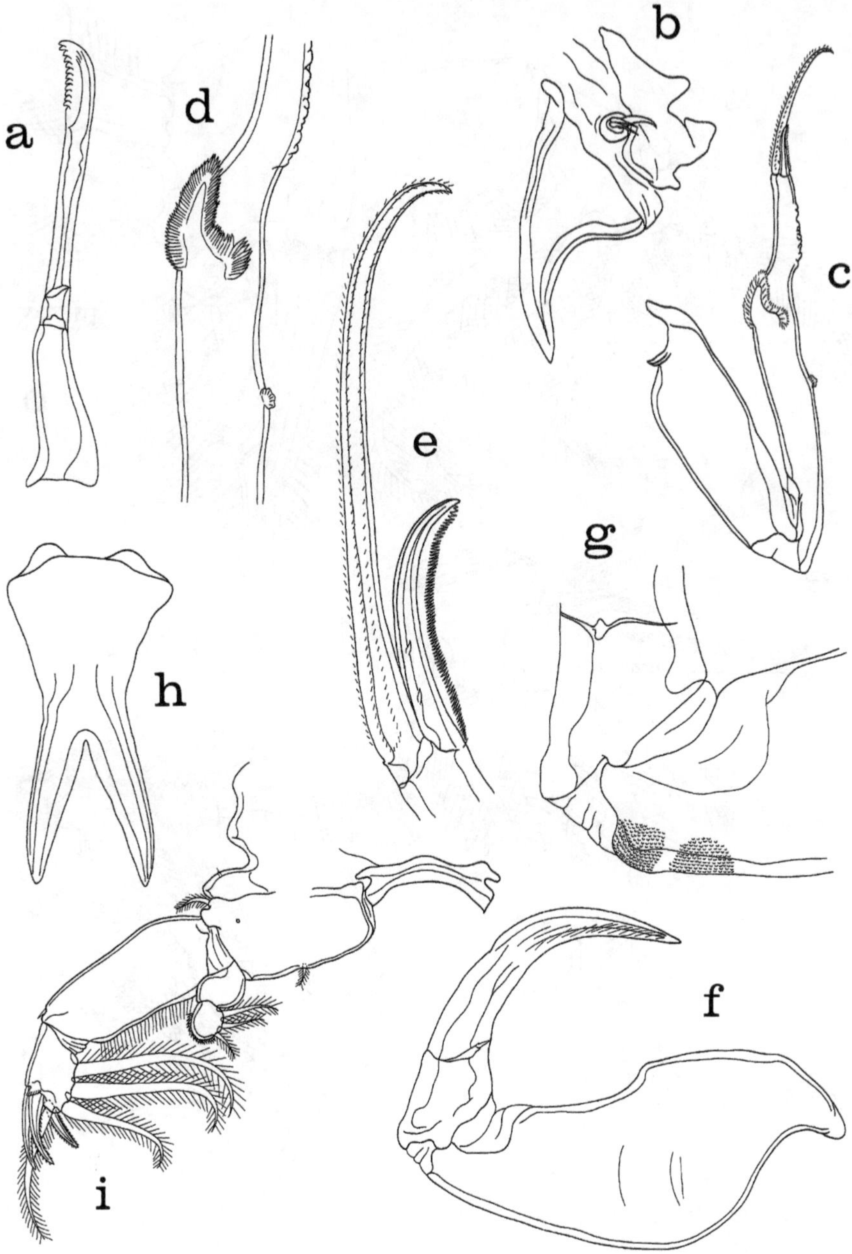

Fig. 73. *Euryphorus brachypterus* (Gerstaecker, 1853), female. a, mandible, ventral (scale E); b, maxillule, ventral (I); c, maxilla, dorsal (F); d, midregion of brachium of maxilla, dorsal (B); e, terminal portion of maxilla, dorsal (E); f, maxilliped, ventral (H); g, joint between corpus maxillipedis and claw, dorsal (I); h, sternal furca, ventral (I); i, leg 1 and intercoxal plate, ventral (F).

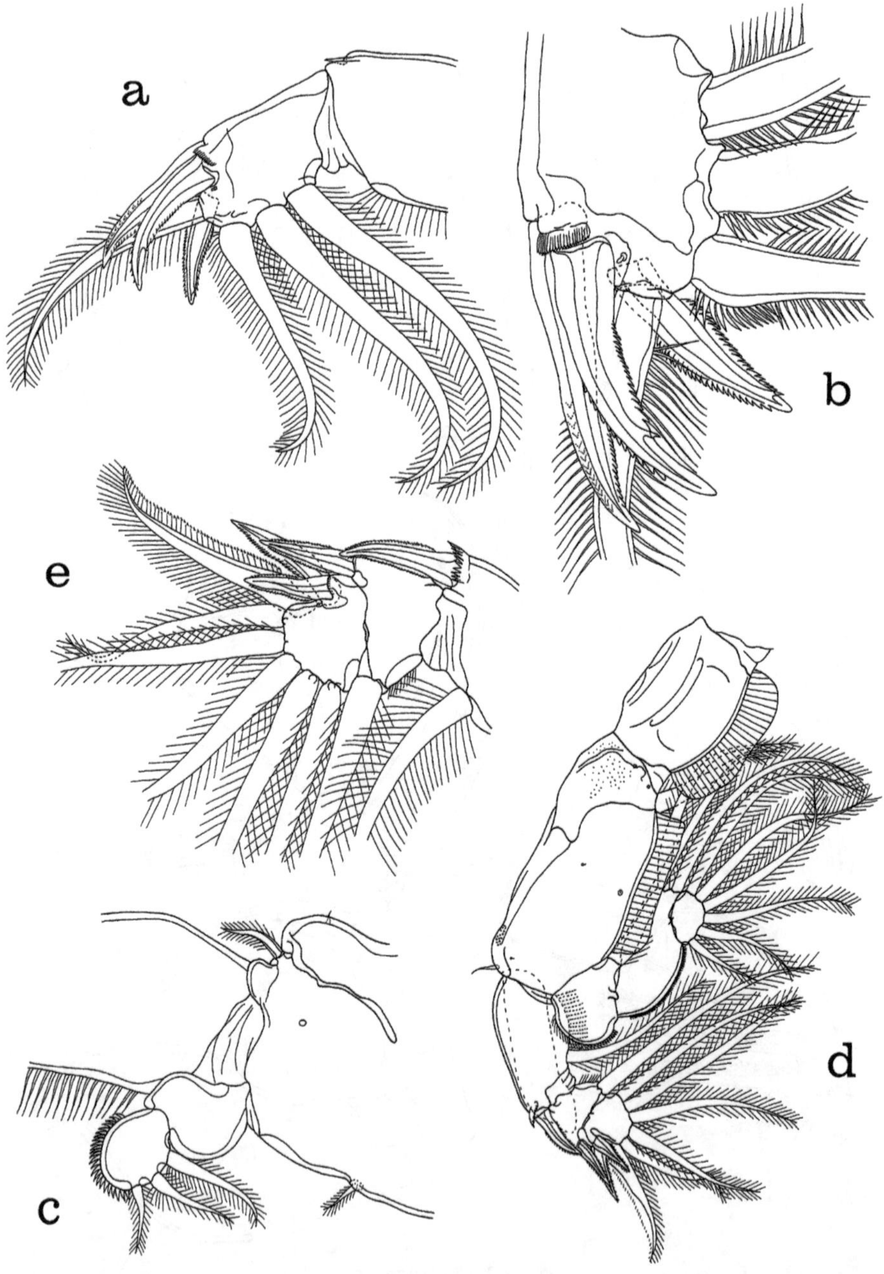

Fig. 74. *Euryphorus brachypterus* (Gerstaecker, 1853), female. a, terminal portion of exopod of leg 1, ventral (scale B); b, tip of exopod of leg 1, ventral (E); c, sympod-exopod joint and endopod of leg 1, ventral (B); d, leg 2 and intercoxal plate, ventral (F); e, exopod of leg 2, ventral (B).

exopodal spine. All 3 spines of second and third exopodal segments extending beyond distal limit of third segment of exopod. Dorsal spine of third segment only unilaterally spinulated. Sympod of leg 3 (fig. 75a) with 1 patch of small spinules and 1 patch of minute spinules. Fringing membrane curving from lateral to ventral surfaces; membrane serrated ventrally. Exopod (fig. 75b) and endopod (fig. 75c) as in *E. nordmanni*. Pectinate membrane at bases of exopodal spines, except distalmost one. Leg 4 (fig. 75d) with distinct intercoxal plate, biramous. Sympod large, carrying outer distal seta, and possessing rounded expansion on inner margin; rounded expansion with patch of small spinules. Exopod (fig. 75e) similar to that in *E. nordmanni*, except setae short. Endopod (fig. 75f) indistinctly 3-segmented. First segment bearing outer row of setules and 1 plumose inner seta. Second segment partially fused with third segment, and carrying 1 plumose inner seta; third segment with 3 plumose setae. Row of setules present on outer margin of second and third segments. Leg 5 (fig. 75g) ventrolaterally located on genital complex near origin of posterolateral processes (plates) and comprised of 2 papillae; anterior papilla tipped with 1 smooth seta, and posterior one bearing 3 smooth setae.

Male. — Body as in fig. 76a. Total length 7.65 mm (6.78-8.51 mm) (n = 2). Cephalothorax as in female, 3.62 mm (3.23-4.00 mm) × 3.31 mm (2.85-3.78 mm), but more slender.

Fourth pedigerous somite much wider than long, 0.62 mm (0.54-0.70 mm) × 1.02 mm (0.93-1.12 mm). Dorsal aliform plates 0.94 mm (0.90-0.99 mm) × 0.82 mm (0.67-0.96 mm). Genital complex almost as wide as long, 1.74 mm (1.54-1.95 mm) × 1.54 mm (1.34-1.73 mm), ovoid, and slightly indented on lateral margin at midlength. Abdomen comprising 2 free somites. First somite 0.56 mm (0.48-0.64 mm) × 0.72 mm (0.58-0.86 mm) (including small posterolateral flaps). Second somite quadrangular, 0.62 mm (0.54-0.70 mm) × 0.72 mm (0.64-0.80 mm). Caudal ramus similar to that of female.

Body surface with small setules similar to that in female.

Antennule as in female. Antenna (fig. 76b) with smaller posteriorly directed spinelike process on second segment. Third segment with corrugated adhesion area near inner distal end. Claw similar to that of female, but with a curved tine on inner margin near base of attenuate seta.

Mouth tube as in female. Mandible (fig. 76c) exhibiting slightly irregular margin at tip. Maxillule as in female. Maxilla (fig. 76d) similar to that in female, except barbs on canna absent. Maxilliped (fig. 76e) with only 1 rounded membranous flap on ventral surface, and bilobed sac-like protrusion associated with distally directed, curved, spinous process at myxal area.

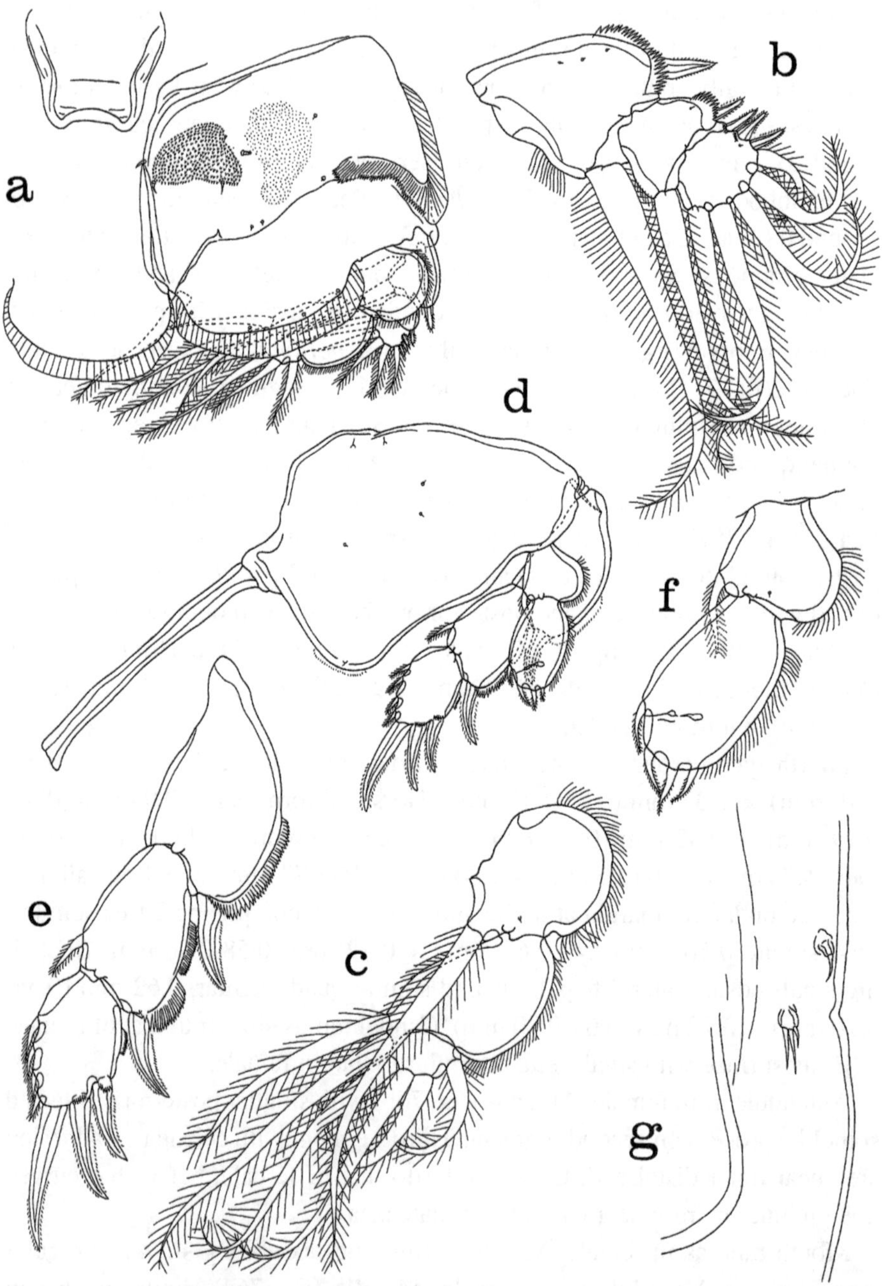

Fig. 75. *Euryphorus brachypterus* (Gerstaecker, 1853), female. a, leg 3, ventral (scale F); b, exopod of leg 3, ventral (G); c, endopod of leg 3, ventral (I); d, leg 4 and intercoxal plate, ventral (C); e, exopod of leg 4, ventral (B); f, endopod of leg 4, ventral (G); g, leg 5, ventral (C).

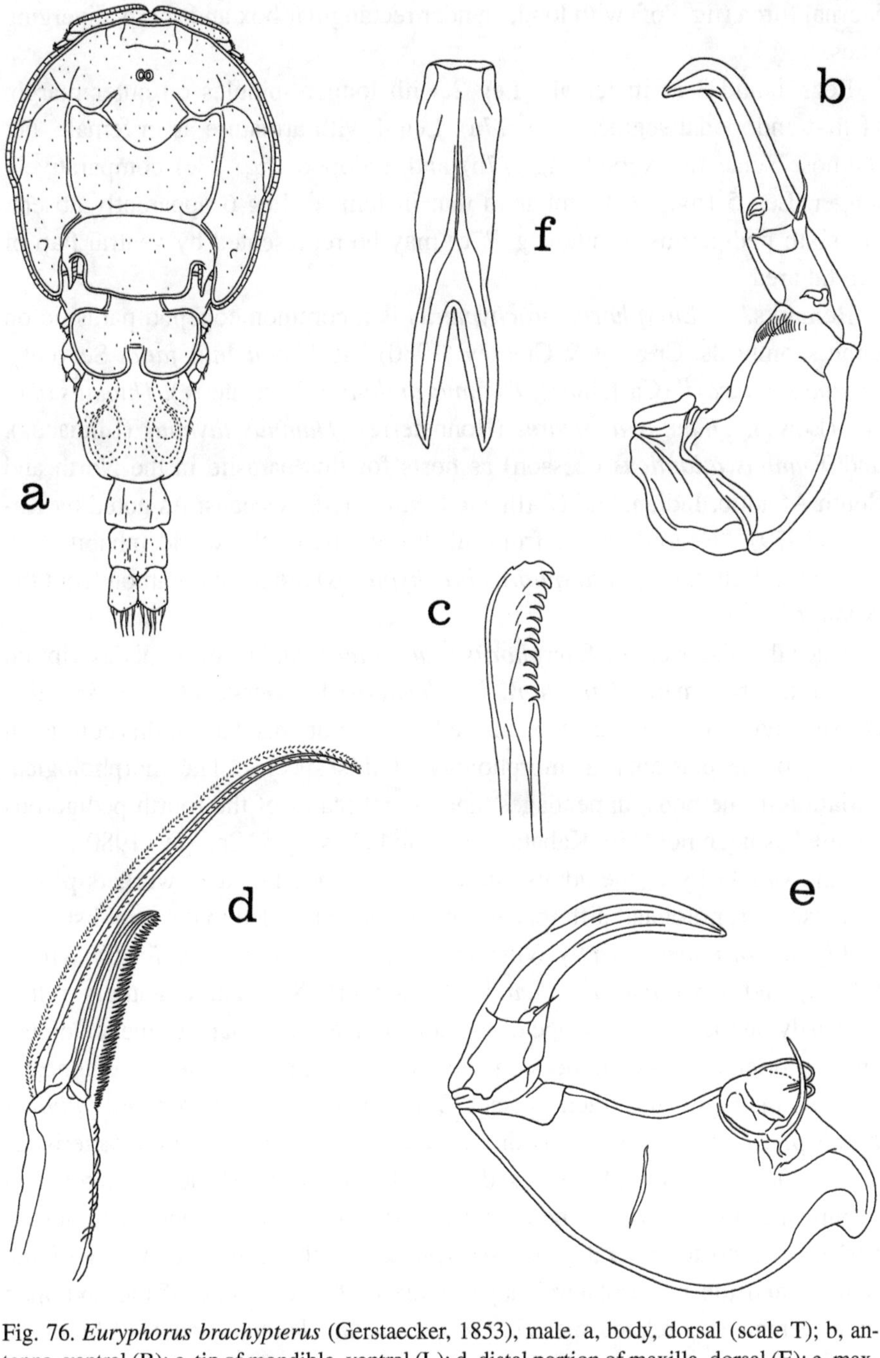

Fig. 76. *Euryphorus brachypterus* (Gerstaecker, 1853), male. a, body, dorsal (scale T); b, antenna, ventral (B); c, tip of mandible, ventral (L); d, distal portion of maxilla, dorsal (E); e, maxilliped, ventral (C); f, sternal furca, ventral (G).

Sternal furca (fig. 76f) with long, slender rectangular box and sharp, diverging tines.

Legs 1 and 3 as in female. Leg 2 with longer spinules on outer margin of first endopodal segment (fig. 77a). Leg 4 with armature as in female, but plumose setae of exopod (fig. 77b) and endopod (fig. 77c) comparatively longer. Leg 5 (fig. 77d) similar to that in female. Leg 6 apparently absent, but sixth pedigerous somite (fig. 77e) may be represented by ventral flap of genital area.

Remarks. — *Euryphorus brachypterus* is a common copepod parasitic on large scombrids. Cressey & Cressey (1980) list *Allothunnus fallai* Serventy, *Thunnus maccoyii* (Castelnau), *Thunnus alalunga* (Bonnaterre), *Thunnus obesus* (Lowe), *Thunnus albacares* (Bonnaterre), *Thunnus thynnus* (Linnaeus), and *Thunnus atlanticus* (Lesson) as hosts for this parasite in the North and South Atlantic, Indian, and North and South Pacific Oceans. As noted by Kabata (1979), "Records come from all the oceans of the world inhabited by the tunas, though they (*Euryphorus brachypterus*) are most common from the Atlantic".

Since the discovery of *Euryphorus brachypterus* and its original description (under the binomen *Elytrophora brachyptera*) by Gerstaecker (1853), this copepod has accumulated an extensive list of synonyms. Part of this confusion is due to the plasticity in morphology of this species. The morphological variation in the body, appendages, and dorsal plates of the fourth pedigerous somite has been noted by Kabata (1979) and Cressey & Cressey (1980).

This variability in the adults and the importance previous workers placed on these morphological differences among specimens led to the establishment of *Elytrophora hemiptera* by Wilson (1921), *Elytrophora atlantica* by Wilson (1932), and *Elytrophora indica* by Shiino (1958a). These authors relied primarily on the size and shape of the dorsal plates, the shape of the abdomen, and the orientation of the dorsal plates on the genital complex as diagnostic features distinguishing their species from each other and from *Euryphorus brachypterus*. However, all the differences described for these characteristics can be attributed to differences due to the maturity of the specimen and to intraspecific variation. The dorsal aliform plates of the fourth pedigerous somite, the posterolateral processes (plates) of the genital complex of the female, and the posterolateral flaps of the first free somite of the abdomen in the female of a related species, *Euryphorus nordmanni* (under the binomen *E. nympha*), continue to develop (increasing in size and changing in shape) with advancing age, eventually resulting in a change in general habitus of the

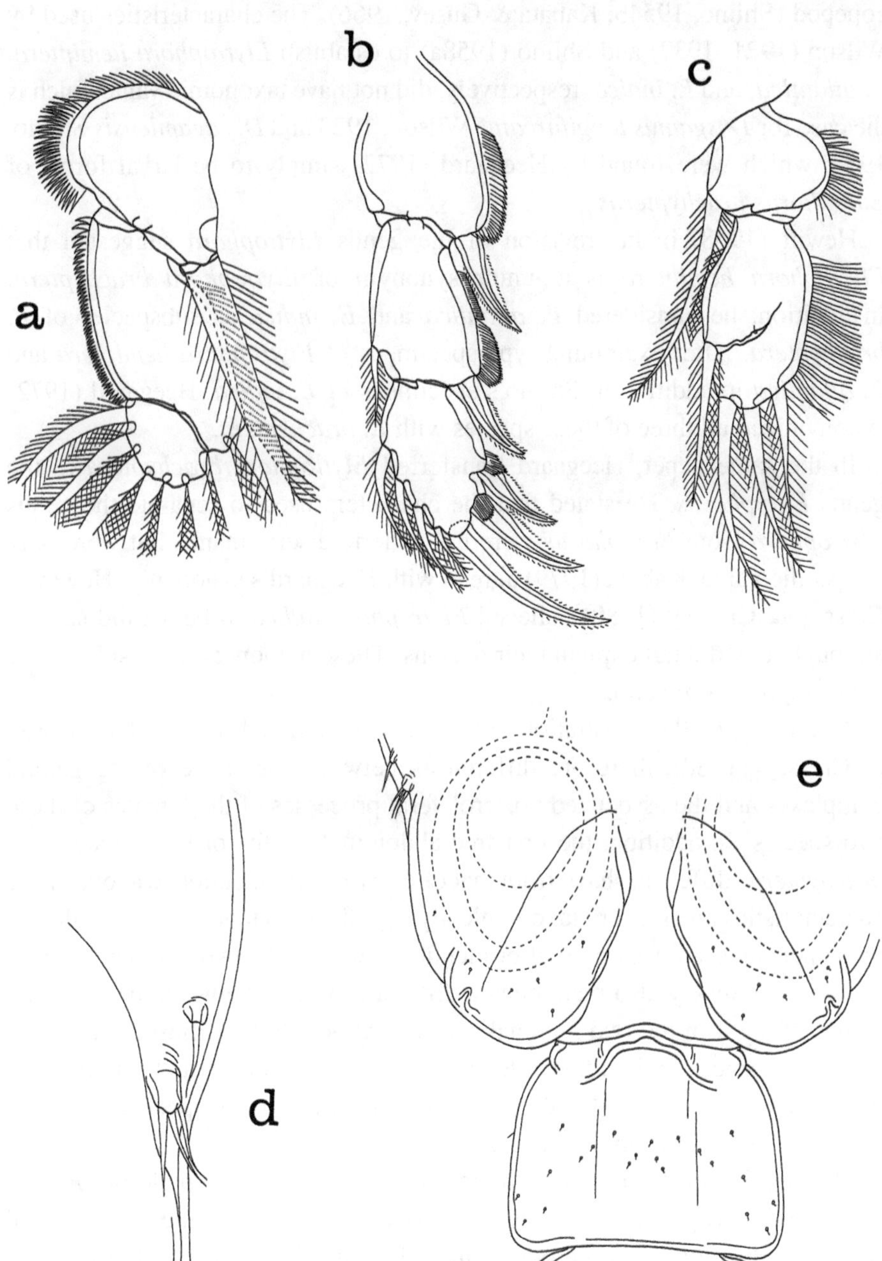

Fig. 77. *Euryphorus brachypterus* (Gerstaecker, 1853), male. a, endopod of leg 2, ventral (scale G); b, exopod of leg 4, ventral (G); c, endopod of leg 4, ventral (J); d, leg 5, ventral (G); e, genital area, ventral (F).

copepod (Shiino, 1954b; Kabata & Gusev, 1966). The characteristics used by Wilson (1921, 1932) and Shiino (1958a) to establish *Elytrophora hemiptera*, *E. atlantica*, and *E. indica*, respectively, did not have taxonomic value. Such is the case for *Dysgamus longifurcatus* Wilson, 1923 and *D. sagamiensis* Shiino, 1958, which were found by Heegaard (1972) simply to be larval forms of *Euryphorus brachypterus*.

Hewitt (1968) in his revision of the genus *Elytrophora* suggested that *Elytrophora hemiptera* is a junior synonym of *Elytrophora brachyptera*. In addition, he considered *E. atlantica* and *E. indica* as subspecies of *E. brachyptera*. After examining type-specimens of *Elytrophora hemiptera* and *E. atlantica* in addition to Shiino's specimens of *E. indica*, Heegaard (1972) synonymized all three of these species with *E. brachyptera*.

In the same paper, Heegaard transferred *Elytrophora brachyptera* to the genus *Euryphorus*. He stated that the characters used to separate the genus *Elytrophora* from *Euryphorus* were not generic discriminants, but were only of specific value. Kabata (1979) agreed with Heegaard's synonymy. However, Cressey & Cressey (1980) believed *Elytrophora indica* to be a valid species although they did not explain their reasons. These authors also considered the genus *Elytrophora* valid.

According to the illustrations of these two nominal species by Cressey & Cressey (1980), there are differences between the shape of the genital complexes and the associated posterolateral processes of the females of these two species. In addition, the first free abdominal somite of the female of *E. brachypterus* differs in shape from that of *E. indica*. As mentioned above, these characteristics are not very taxonomically significant in this genus. In addition, the appendages are very similar or identical between the two nominal species. Cressey & Cressey also mentioned a difference in size between the exopodal spines of leg 4 in the two nominal species. Although their illustrations show slight differences in the relative lengths of the setae, spines, and spinules, the differences are so minor that they could be attributed to intraspecific variation, which is conspicuous in this species.

One significant feature concerning the illustration of the male of "*Elytrophora indica*" (fig. 79a of Cressey & Cressey, 1980) is the absence of the dorsal plates on the fourth pedigerous somite. The absence of these dorsal plates would be taxonomically very important at the specific level. Since Shiino's (1958a) original description of the male distinctly shows the dorsal plates, it can be presumed that Cressey & Cressey (1980) simply overlooked the plates in their male specimens.

In view of the morphological plasticity observed in the adults of *Euryphorus brachypterus* and the morphological differences due to the stage in maturity of the specimens, it appears that Heegaard (1972) and Kabata (1979) are correct in their synonymies. This conclusion is strengthened by the similarity in the details of the appendages illustrated and described by the various authors of these nominal species (Wilson, 1921, 1932; Shiino, 1958a).

Genus **Gloiopotes** Steenstrup & Lütken, 1861

Gloiopotes Steenstrup & Lütken, 1861: 363; Bassett-Smith, 1899: 458; Stebbing, 1900: 670; Wilson, 1907a: 698; Wilson, 1932, p. 415; Yamaguti, 1963: 103; Hewitt, 1964b: 85; Cressey, 1967a: 2; Lewis, 1967: 56; Dojiri, 1983: 192; Boxshall & Montú, 1997: 67; Pillai, 1985: 479; Kazachenko, 2001: 35; Boxshall & Halsey, 2004: 726.

Female. — Cephalothorax subovate, narrower anteriorly, with several rows of setules (only 1 row in posterolateral corner in *G. hygomianus* Steenstrup & Lütken, 1861). Posterior sinuses shallow and slightly constricted posteriorly. Frontal plate without lunules. Fourth pedigerous somite small, with 1 pair of prominent dorsal aliform plates. Genital complex quadrangular or somewhat triangular, and bearing pair of long digitiform posterolateral processes. Long spikelike or spatulate projection (leg 5) originating from each process. Abdomen comprising 2 free somites, each long and cylindrical. Caudal ramus elongate, styliform, bearing setae and spines. Body covered with setules and spines (not conspicuous in *G. hygomianus*).

Antennule 2-segmented. First segment robust, with spinelike process on anteromedial margin near base; bifid process situated on posterior side near insertion of second segment; and anterior margin armed with numerous setae. Second segment elongate, slender, and bearing 14 setae. Antenna 4-segmented, with posteriorly directed spinelike process on second segment. Terminal segment hamate. Postantennal process with at least 1 accessory tine. Mouth tube with intrabuccal stylet and strigil. Mandible with third and fourth sections fused, thus appearing to contain only 3 sections; inner margin of tip with 12 teeth. Maxillule consisting of dentiform process with accessory tine, and adjacent setiferous papilla. Maxilla brachiform. Maxilliped subchelate and prehensile. Sternal furca with accessory tine near base of divergent primary tine; primary tine bifid at tip (except in *G. hygomianus*). Rounded, sclerotized lobe situated on either side of sternal furca occasionally present.

Leg 1 with 2-segmented rami. Terminal exopodal segment with 3 spines; inner 2 spines bifid with middle setiform process. Seta located on inner distal corner, and 3 plumose setae on inner margin. Endopod small, bearing 3 setae at tip. Leg 2 biramous with 3-segmented rami. Leg 3 with large ventral apron, biramous, with 3-segmented rami. Exopod with large clawlike spine fused to basal swelling of first segment; spine with accessory tine. Exopod with formula: I-1; I-1; III, 5. Endopod with formula: 0-0; 0-2; 4. Velum present and second endopodal segment with greatly inflated outer margin. Leg 4 with sympod and 3-segmented exopod with formula: I-0; I-0; III. Leg 5 represented by sclerotized projection on posterolateral process of genital complex; projection tipped with setae and armed with numerous spines. Leg 6 may be represented by setiferous papilla.

Male. — Cephalothorax similar to that in female. Dorsal aliform plates of fourth pedigerous somite comparatively smaller than in female. Genital complex subovate with long spikelike leg 5 on posterolateral corner. Abdomen and caudal ramus similar to that in female. Antenna with corrugated adhesion pad on second segment and accessory claw on inner margin of terminal claw. Maxillule with accessory tine on process. Maxilliped occasionally differing morphologically from that of female. Leg 5 a long projection bearing heavy spinules. Leg 6 represented by 2 spines and 1 plumose seta on posterolateral corner of genital complex.

Type-species. — *Gloiopotes hygomianus* Steenstrup & Lütken, 1861.

Remarks. — *Gloiopotes* Steenstrup & Lütken, 1861 can easily be distinguished from all other caligid genera by the combination of general body morphology, the styliform caudal ramus, the accessory tine on the sternal furca, the structure of the outer spine on the first exopodal segment of leg 3, and the structure of leg 5. The morphology of leg 1 is unique in the family. The bifid spines (with the middle hyaline setiform process) of the terminal exopodal segment have not been reported in any other genus of the Caligidae. The leg 1 endopod of species of *Gloiopotes* is small, 2-segmented, and tipped with 3 setae.

There are five valid species of *Gloiopotes* (Cressey, 1967a): *G. hygomianus* Steenstrup & Lütken, 1861 (type-species); *G. huttoni* (Thomson, 1889); *G. ornatus* Wilson, 1905; *G. watsoni* Kirtisinghe, 1934; and *G. americanus* Cressey, 1967.

Gloiopotes costatus Wilson, 1919, *G. longicaudatus* (Marukawa, 1925), *G. zeugopteri* Rao, 1951 have all been synonymized with *G. huttoni* (Hewitt, 1964b; Cressey, 1967b). *Gloiopotes auriculatus* Barnard, 1957 was listed as a junior synonym of *G. watsoni* by Cressey (1967a). Finally, *G. crassus* described by Bere (1936) was transferred to *Lepeophtheirus* by Shiino (1960a) and Hewitt (1964b). Although Hewitt (1964b) and Lewis (1967) considered

TABLE IX
Hosts and localities of collections of five species of *Gloiopotes* Steenstrup & Lütken, 1861 (tabulated from Cressey, 1967a)

Species	Host	Locality
Gloiopotes hygomianus Steenstrup & Lütken, 1861	*Acanthocybium solandri* (Cuvier)	see table X
Gloiopotes huttoni (Thomson, 1889)	*Tetrapturus audax* (Philippi)	Indo-Pacific
	Istiophorus orientalis [= *Istiophorus platypterus* (Shaw)]	Indo-Pacific
Gloiopotes ornatus Wilson, 1905	*Tetrapturus albidus* Poey	Atlantic Ocean
	Makaira nigricans Lacépède	Atlantic Ocean
Gloiopotes watsoni Kirtisinghe, 1934	*Tetrapturus audax*	Indo-Pacific
	Istiophorus orientalis [= *Istiophorus platyoterys*]	Indo-Pacific
	Makaira mazara (Jordan & Snyder)	Indo-Pacific
	Makaira indica (Jordan & Snyder)	Indo-Pacific
Gloiopotes americanus Cressey, 1967	*Istiophorus americanus* Cuvier & Valenciennes	Atlantic coast of Florida; Gulf of Mexico

G. watsoni a synonym of *G. huttoni*, Cressey (1967b) considered both species valid based primarily on the relative lengths of leg 5 and the posterolateral process of the genital complex in the female.

In his revision of *Gloiopotes*, Cressey (1967b) commented on the host specificity exhibited by members of this genus. The five species of *Gloiopotes* have been collected from hosts belonging to the billfish family Istiophoridae and the scombrid genus *Acanthocybium* (table IX).

Gloiopotes hygomianus Steenstrup & Lütken, 1861 (figs. 78-83)

Gloiopotes hygomianus Steenstrup & Lütken, 1861: 363; Bassett-Smith, 1899: 458; Stebbing, 1900: 670; Wilson, 1907a: 698; Rao, 1951: 254; Shiino, 1960b: 533; Yamaguti, 1963: 103; Hewitt, 1964b: 95; Shiino, 1965: 421; Lewis, 1967: 66; Cressey, 1967a: 8; Lewis et al., 1969: 421; Kazachenko, 1976: 215; Cressey & Cressey, 1980: 38; Dojiri, 1983: 195; Pillai, 1985: 486; Boxshall & Montú, 1997: 67; Ho & Nagasawa, 2001: 3; Morales-Serna, 2012: 51.
Gloiopotes sp., Bonnet, 1948: 7.

Material examined. — Holotype female on loan from Zoologisk Museum, Copenhagen, Denmark, collected from Atlantic Ocean. No host recorded. Loan arranged by Dr. Jean Just. Eight females and 3 males (2 couples in amplexus) (USNM 168212) on loan from National Museum of Natural History, Smithsonian Institution, Washington, D.C. This material collected

"SE of Carolina" 36°15′N 71°49′W (Atlantic Ocean) on 9 October 1957. No host recorded on label, but probably from *Acanthocybium solandri* (Cuvier).

Female. — Body as in fig. 78a. Total length (not including setae on caudal ramus) 14.85 mm (14.11-15.40 mm) (n = 6). Cephalothorax subcircular in outline, narrower anteriorly, longer than wide, 6.84 mm (6.37-7.67 mm) × 5.59 mm (5.31-5.84 mm). Posterior sinuses deep, with conspicuous transparent membrane along outer and inner edges. Frontal plate distinct, with no lunules. Tip of antennule within lateral limit of cephalothorax. Free margin of thoracic zone of cephalothorax almost at same level as posterior limit of lateral zone. Lateral margin of cephalothorax rimmed with spines (fig. 78b) and usual transparent membrane. Posterolateral portion of cephalothorax possessing row of multibranched setule (3 or 4 branches).

Fourth pedigerous somite wider than long, 1.46 × 2.06 mm, and bearing pair of dorsal aliform plates. Dorsal plates very large, longer than wide, 3.34 mm (3.12-3.52 mm) × 1.99 mm (1.79-2.22 mm), and covering almost entire genital complex. Genital complex wider than long, 1.96 × 2.82 mm (excluding posterolateral processes), and possessing small anterior protrusion on each side. Posterolateral process very large, 3.05 × 1.33 mm, extending almost to region of anus, and bearing spatulate leg 5 ventrolaterally. First abdominal somite longer than wide, 1.27 mm (1.20-1.36 mm) × 0.63 mm (0.43-0.76 mm). Second somite much longer than first, 2.02 mm (1.90-2.19 mm) × 0.55 mm (0.46-0.66 mm).

Caudal ramus (fig. 78c, d) very long and slender, 1.26 × 0.13 mm, possessing 2 plumose setae on anterior third and tipped with 1 subterminal plumose seta and terminally with 3 robust, plumose spinelike setae.

Body with small setules (sensilla) as in fig. 78a.

Antennule (fig. 78e) 2-segmented. First segment 380 μm long (measured along nonsetiferous margin), with spinelike process on anteromedial margin near base, 2 small corrugated protrusions (similar to bifid process found in this position in caligids, e.g., *Lepeophtheirus pectoralis*) on posterior side near insertion of second segment, and bearing 25 setae (22 plumose and 3 smooth). Second segment slender, cylindrical, 598 μm long, with 14 setae (2 sharing common base). Antenna (fig. 78f) apparently 4-segmented. First segment unarmed. Second segment with posteriorly directed spinelike process. Third segment robust, with corrugated protrusion on dorsal side (fig. 78g). Fourth segment a claw (fig. 78f-h) bearing robust seta and corrugated ledge proximally, and 1 long, slender seta at midlength. Distal portion of claw brown. Postantennal process (fig. 79a) with small rounded protrusion on outer margin

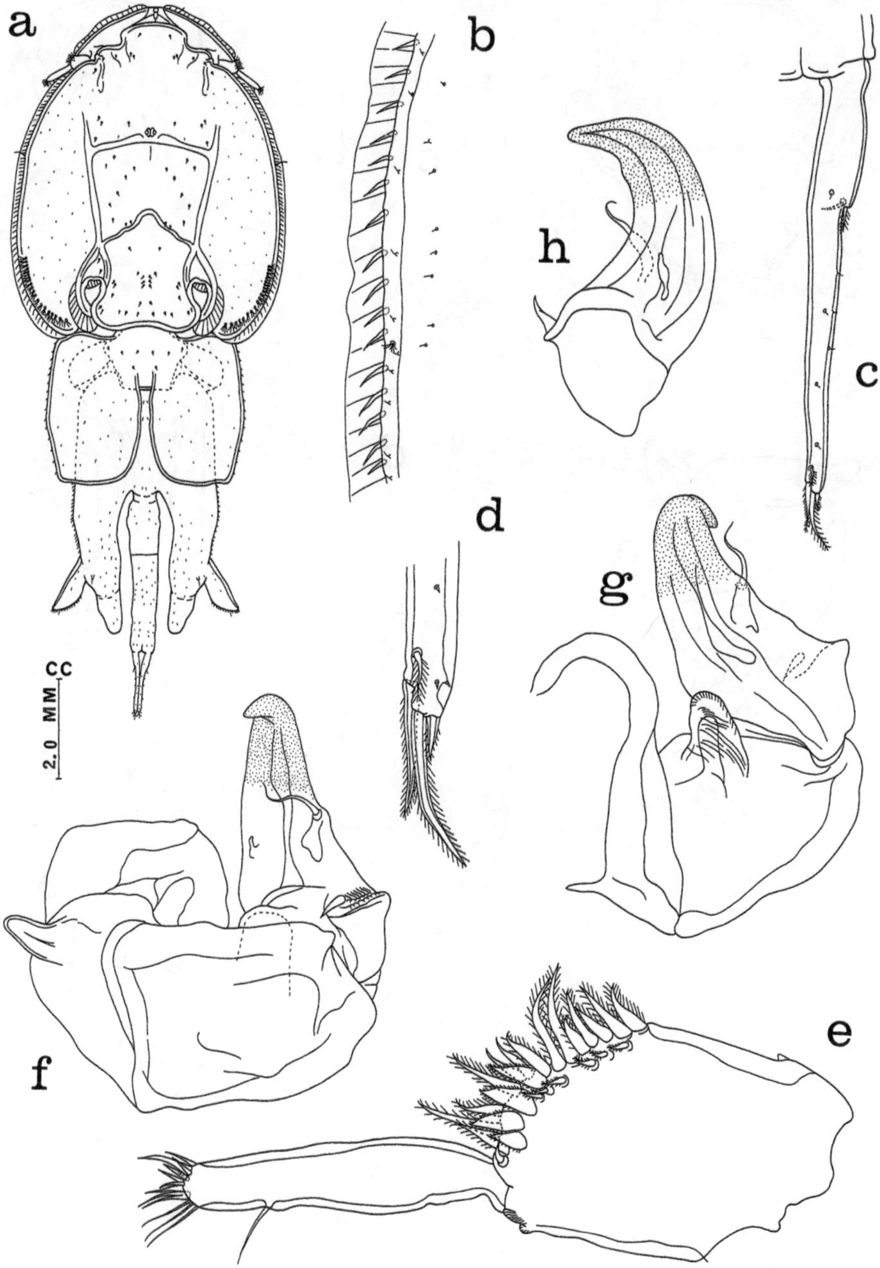

Fig. 78. *Gloiopotes hygomianus* Steenstrup & Lütken, 1861, female. a, body, dorsal (scale CC); b, spinules on lateral margin of cephalothorax, ventral (F); c, caudal ramus, dorsal (F); d, tip of caudal ramus, dorsal (G); e, antennule, ventral (I); f, antenna, ventral (C); g, same, dorsal (C); h, claw of antenna, posterior (C).

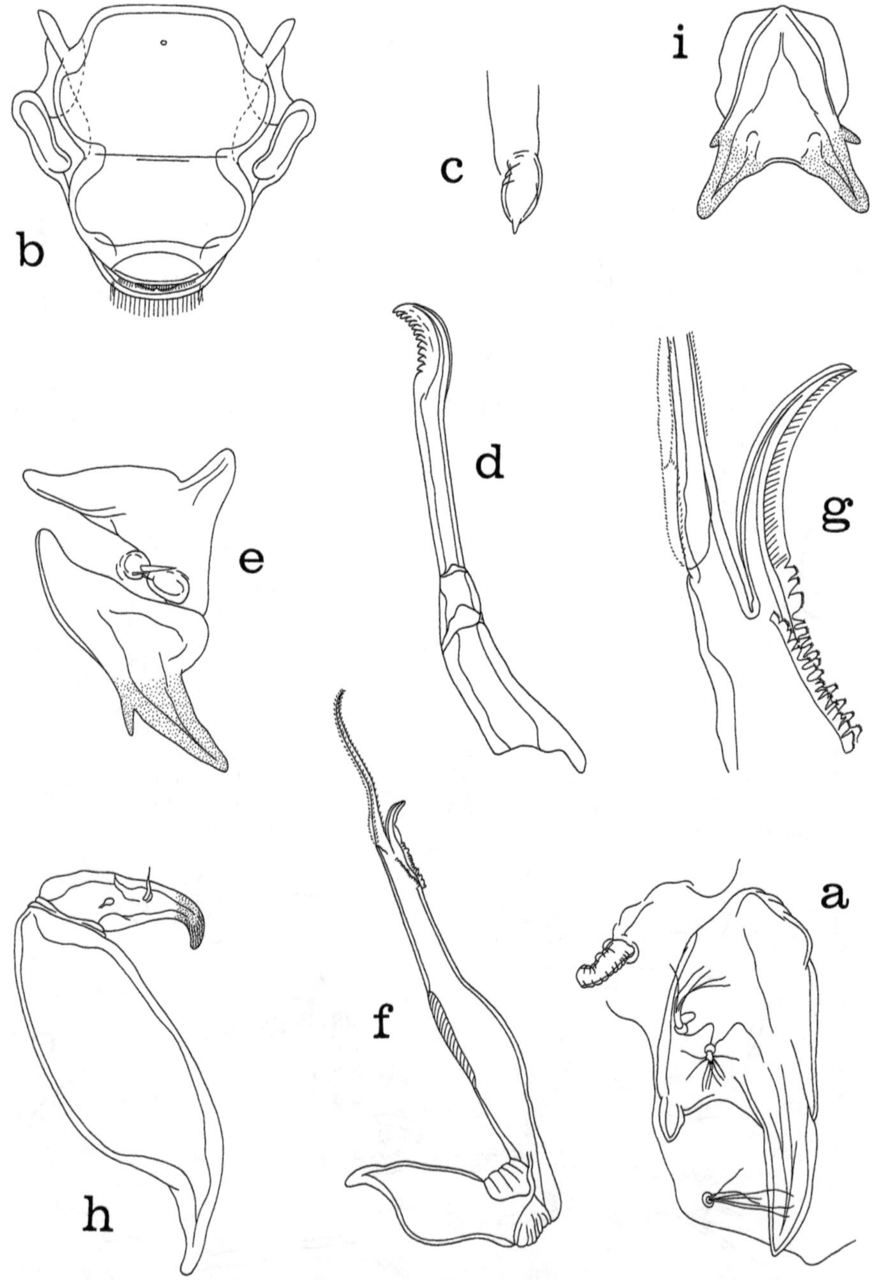

Fig. 79. *Gloiopotes hygomianus* Steenstrup & Lütken, 1861, female. a, postantennal process, ventral (scale G); b, mouth tube, ventral (I); c, intrabuccal stylet, dorsal (Y); d, mandible, ventral (J); e, maxillule, ventral (C); f, maxilla, dorsal (F); g, distal portion of maxilla, dorsal (E); h, maxilliped, ventral (F); i, sternal furca, ventral (H).

near base, 1 large tine (rimmed with flanges) on lateral side, and 1 small tine on medial side. Usual 3 sclerites bearing multibranched setule present. Transparent tube (duct) located medial to base of postantennal process.

Mouth tube (fig. 79b) wider than long, 518 × 564 μm. Intrabuccal stylet (fig. 79c) present on inner surface of labrum. Strigil with many dentiform processes on inner surface of labium. Mandible (fig. 79d) with demarcation of third and fourth sections inconspicuous, thus appearing only tripartite; tip of mandible with 12 teeth and outer hyaline flange. Maxillule (fig. 79e) consisting of papilla tipped with 3 setae (1 larger than other 2), and curved process. Process with accessory tine on medial side; both discolored brown. Maxilla (fig. 79f) brachiform. Flabellum large. Brachium noticeably inflated on side directly opposite of flabellum. Calamus (fig. 79g) with 3 serrated membranes, twice as long as canna. Canna (fig. 79g) flanked by hyaline membranes (1 with minute striations) on each side. Base of canna with irregularly shaped, membranous projections tipped with hairlike structures. Maxilliped (fig. 79h) with moderately robust corpus. Shaft and claw 552 μm long combined. Seta located near junction of shaft and claw. Tip of claw brown and with minute striations. Sternal furca (fig. 79i) with divergent tines and pair of small accessory tines. Tines brown. Pair of posteromedially directed spinelike processes (fig. 80a) located anterior to bases of maxillipeds.

Leg 1 (fig. 80b) biramous. Sympod with 1 outer and 1 inner plumose setae. Exopod 2-segmented. First segment with outer distal spine and inner row of setules. Outer distal spine (fig. 80c) with small spinule at base. Second segment of exopod (fig. 80d) with unilaterally pinnate seta 4 slightly longer than outermost spine. Spine 1 bilaterally spinulated. Spines 2 and 3 bifid with middle setiform process. All 3 spines serrated on outer edges, and with pectinate membranes at their bases. Inner margin of second segment of exopod with 3 large plumose setae. Endopod (fig. 80e) 2-segmented. First segment with setules along outer edge. Second segment with 2 plumose and 1 smooth setae, and patch of setules. Oblique, sclerotized rib (fig. 80f) anterior to intercoxal plate of leg 1 with conspicuous rounded, spatulate protrusion. Area between intercoxal plates of legs 1 and 2 with pair of bumps, each tipped with 1 setule. Leg 2 (fig. 81a) biramous with usual armature. First exopodal spine (fig. 81b) with striated membrane at its base. First 3 exopodal spines serrated and brown at their distal ends. Fourth exopodal spine fringed with hyaline membranes on both sides. Sympod of leg 3 (fig. 81c) with corrugated pad on dorsal surface of anterolateral corner. Exopod (fig. 81d) 3-segmented. First segment with large exopodal spine (fused to basal swelling) on outer

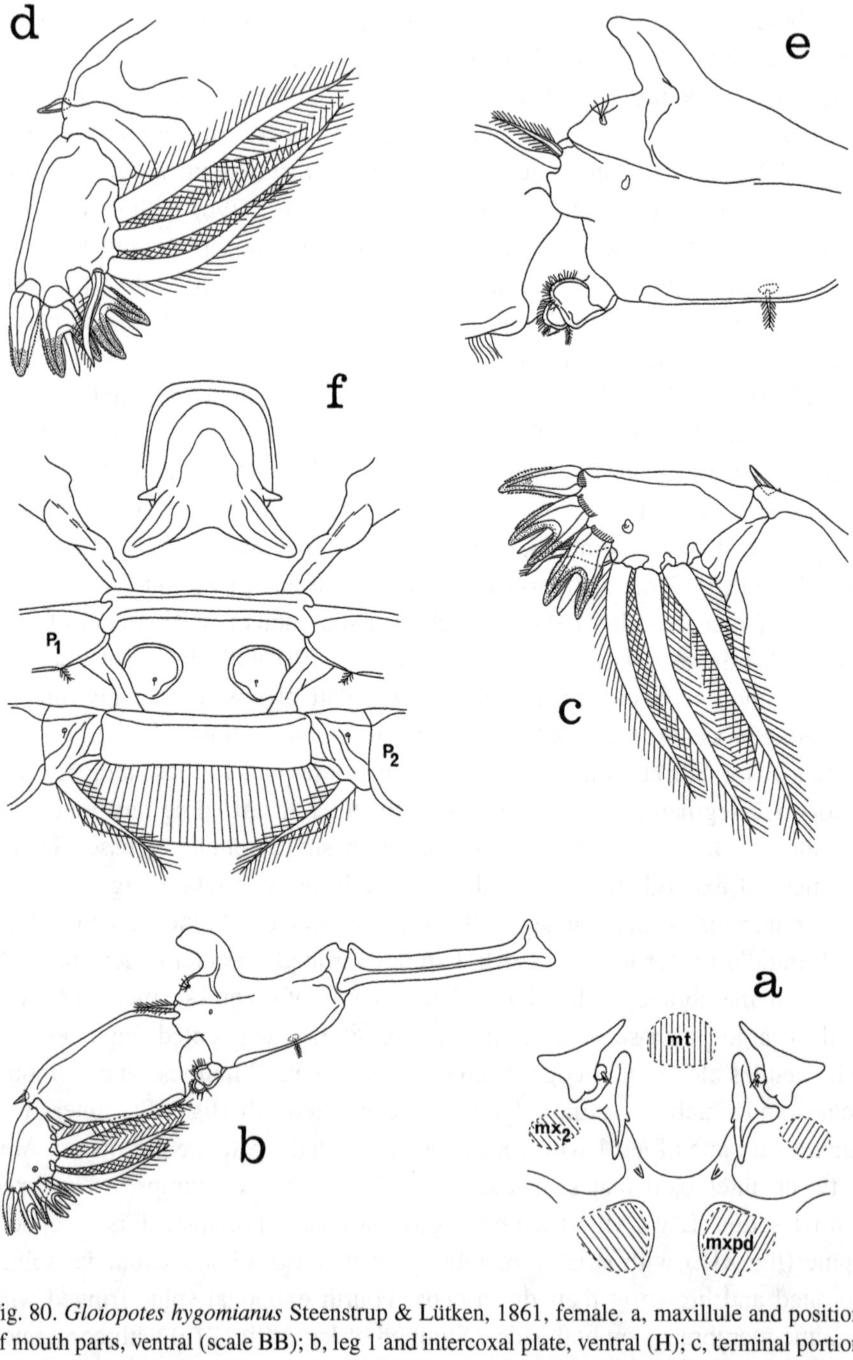

Fig. 80. *Gloiopotes hygomianus* Steenstrup & Lütken, 1861, female. a, maxillule and position of mouth parts, ventral (scale BB); b, leg 1 and intercoxal plate, ventral (H); c, terminal portion of exopod of leg 1, ventral (I); d, same, dorsal (I); e, sympod-exopod joint and endopod of leg 1, ventral (I); f, sternal furca and sclerotized area between legs 1 and 2, ventral (H).

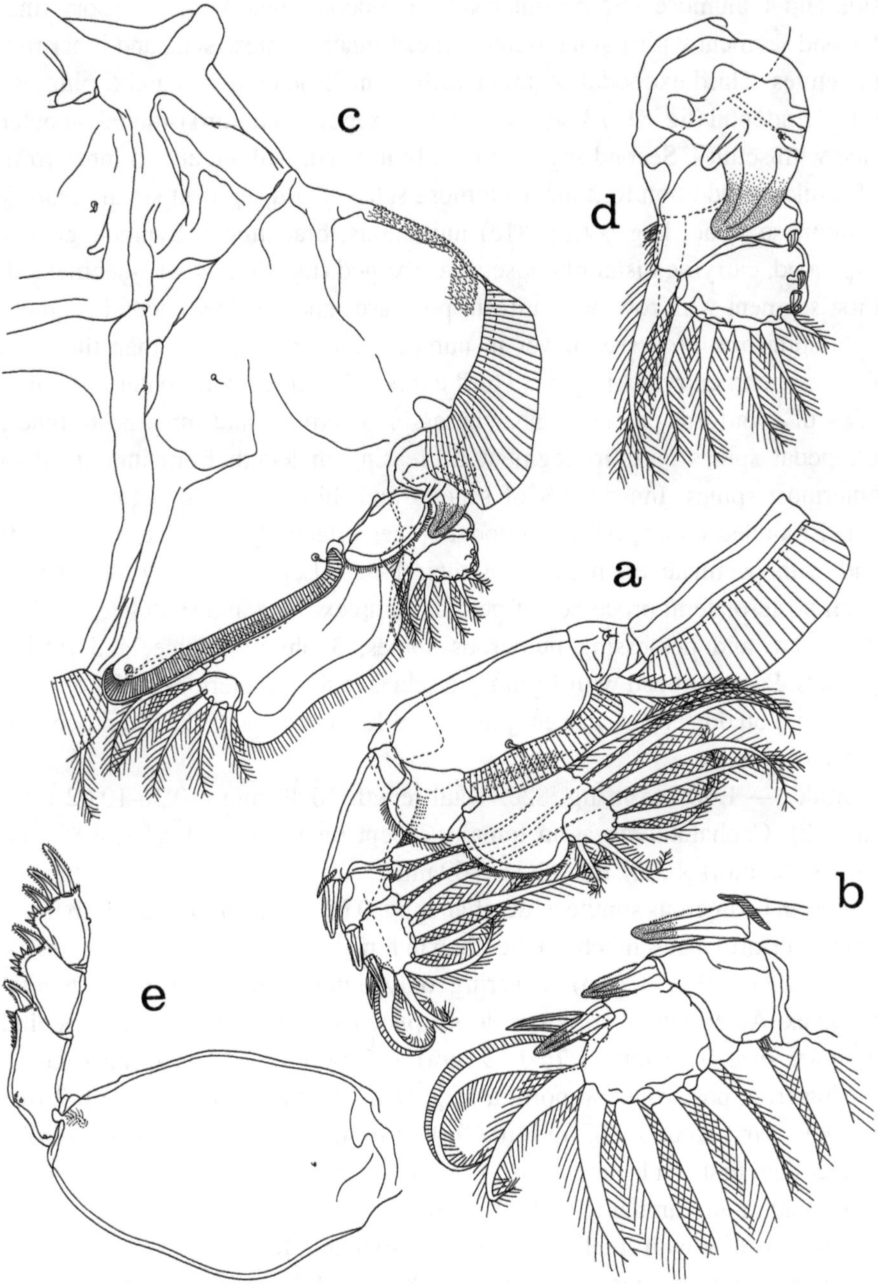

Fig. 81. *Gloiopotes hygomianus* Steenstrup & Lütken, 1861, female. a, leg 2 and intercoxal plate, ventral (scale H); b, exopod of leg 2, ventral (I); c, leg 3, ventral (F); d, exopod of leg 3, ventral (B); e, leg 4, ventral (F).

side and 1 plumose seta on inner side. Exopodal spine with accessory tine. Second segment with 1 small outer spine, 1 inner plumose seta, and inner row of setules. Third exopodal segment with 3 small outer spines and 5 plumose setae. Endopod (fig. 81c) 3-segmented. First segment (velum) rimmed on outer side with setules. Second segment greatly inflated, with outer and inner rows of setules in addition to 2 inner plumose setae. Third segment small, bearing 4 plumose setae. Leg 4 (fig. 81e) uniramous, brachiform. Sympod greatly expanded, carrying distal plumose seta. Exopod (figs. 81e, 82a) 3-segmented. First segment with row of spiniform processes and 1 spine. Second segment with spiniform processes fewer in number, but larger in size than those on first segment, carrying 1 spine. Third exopodal segment with outer and inner rows of spiniform processes near bases of outermost and innermost spines. Exopodal spines of third segment decreasing in length from innermost to outermost spines. Inner sides of second and third exopodal segments with small rounded knobs. All 5 exopodal spines bilaterally spinulated (serrated) and with pectinate membranes at their bases. Leg 5 (fig. 82b, c) arising from posterolateral processes of genital complex, long and spatulate, 1.43 × 0.46 mm, and possessing numerous spines, 3 plumose setae, and ruffled process. Papilla tipped with 1 plumose seta (fig. 82d) (anterior papilla of leg 5 or leg 6?) on dorsal surface near junction of leg 5 and posterolateral process of genital complex.

Male. — Body as in fig. 82e. Total length 10.49 mm (10.36-10.62 mm) (n = 2). Cephalothorax as in female, except much more slender, 4.80 mm (4.75-4.85 mm) × 3.62 mm (3.55-3.69 mm).

Fourth pedigerous somite wider than long, 0.63 × 1.24 mm (1.23-1.26 mm). Dorsal aliform plates much smaller than in female, 1.46 mm (1.43-1.49 mm) × 0.91 mm (0.90-0.93 mm), covering only anterolateral corners of genital complex. Anterolateral corner of dorsal plate (fig. 82f) with 3 spines. Genital complex oval, 1.78 mm (1.76-1.79 mm) × 1.64 mm (1.63-1.66 mm). Leg 5 arising from posterolateral portion of genital complex. Abdomen with 2 free somites. First somite 0.65 mm (0.60-0.70 mm) × 0.53 mm. Second somite longer than first, 1.11 mm (1.10-1.13 mm) × 0.50 mm.

Caudal ramus similar to that of female.

Body surface with small setules similar to that in female.

Antennule as in female. Antenna (fig. 83a, b) different from that of female. First segment with spiniform process. Second segment with 2 raised corrugated adhesion pads and 1 ventrolateral corrugated strip. Claw similar to that in female, but with accessory tine on inner margin near midlength of claw. Base

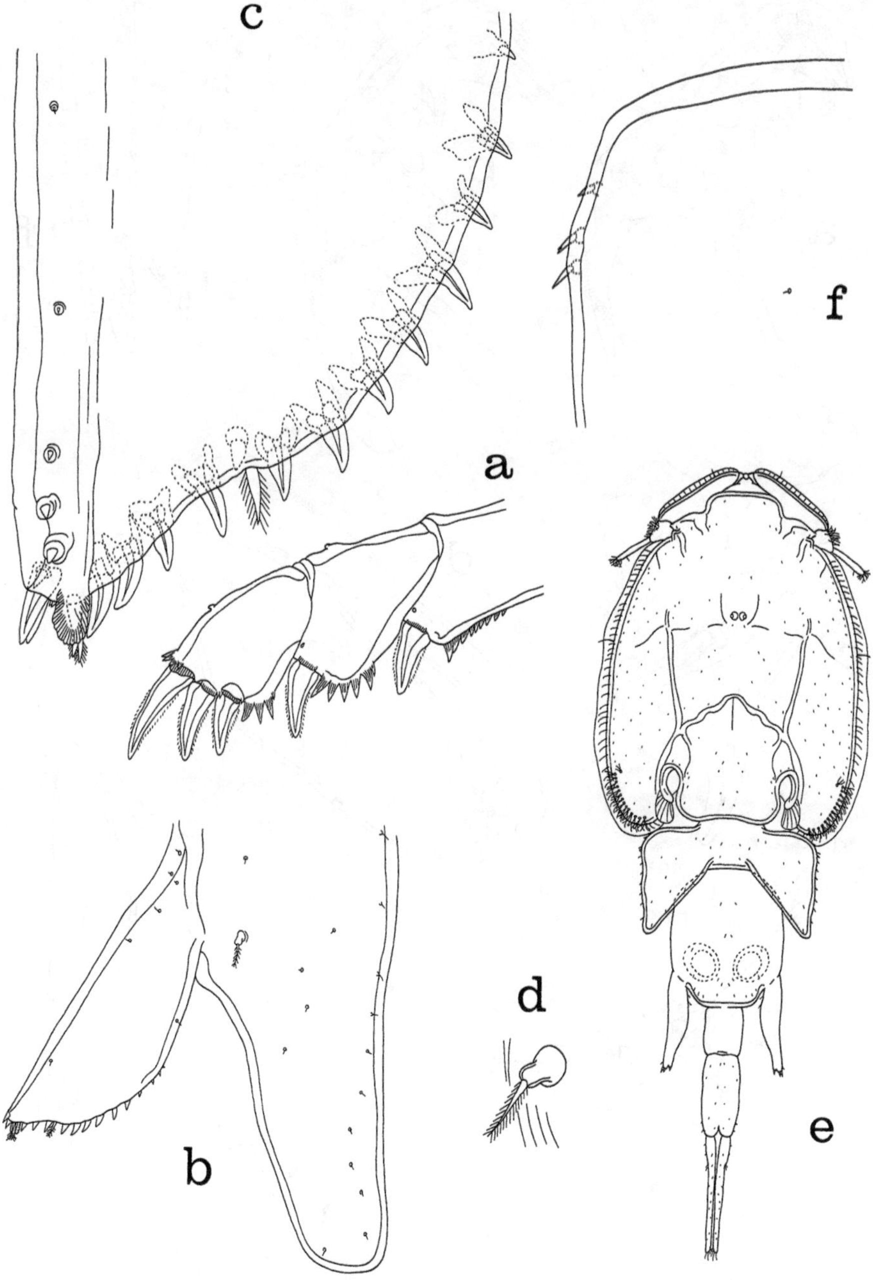

Fig. 82. *Gloiopotes hygomianus* Steenstrup & Lütken, 1861. Female. a, exopod of leg 4, ventral (scale B); b, posterolateral process of genital complex and leg 5 (and leg 6?), dorsal (H); c, tip of leg 5, ventral (J); d, leg 6 (?), dorsal (G). Male. e, body, dorsal (T); f, anterolateral corner of dorsal aliform plate on fourth pedigerous somite, dorsal (F).

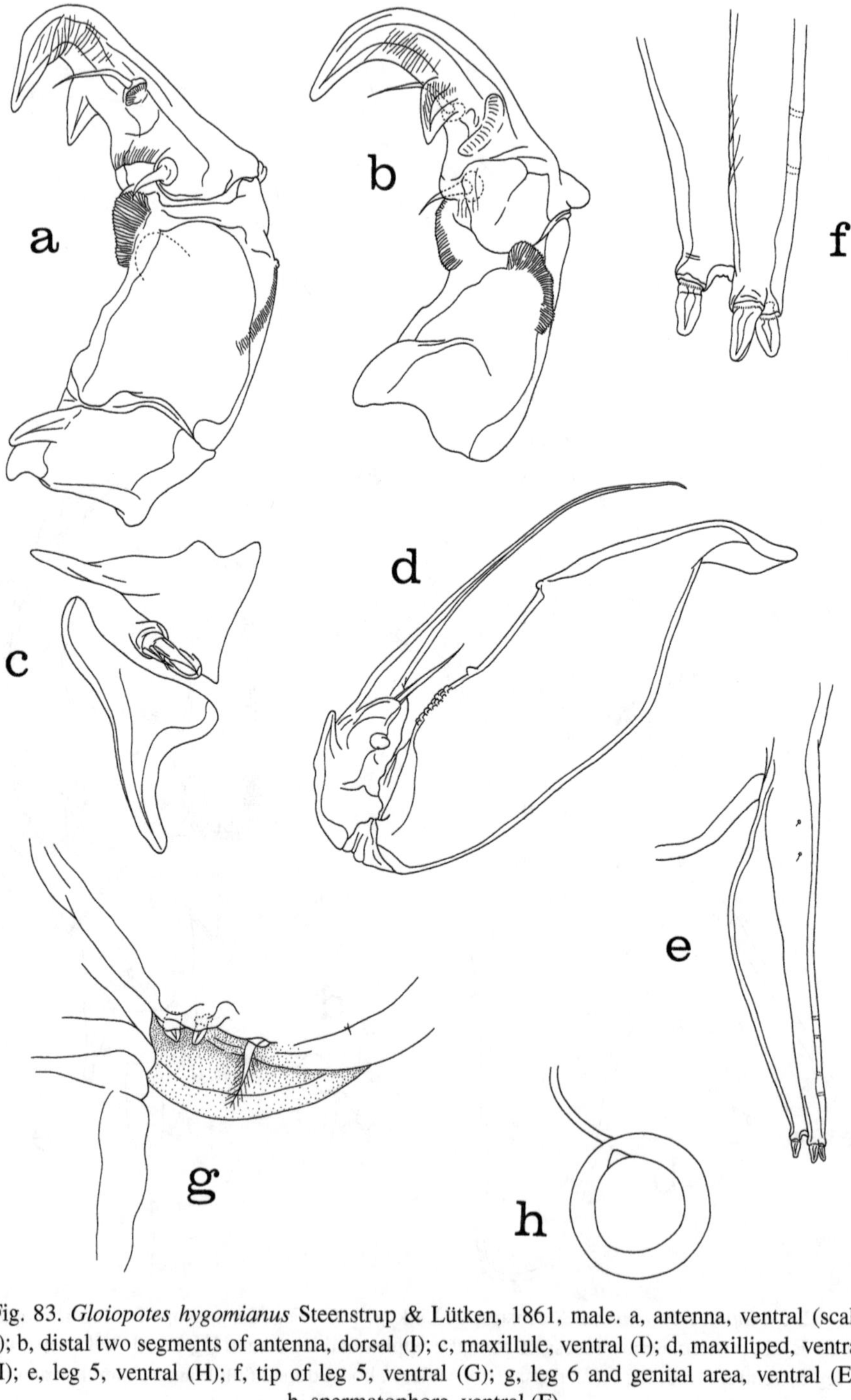

Fig. 83. *Gloiopotes hygomianus* Steenstrup & Lütken, 1861, male. a, antenna, ventral (scale I); b, distal two segments of antenna, dorsal (I); c, maxillule, ventral (I); d, maxilliped, ventral (I); e, leg 5, ventral (H); f, tip of leg 5, ventral (G); g, leg 6 and genital area, ventral (E); h, spermatophore, ventral (F).

of attenuate seta with striated projection. Accessory tine with curved ridge near base in dorsal view. Postantennal process as in female.

Mouth tube and mandible as in female. Maxillule (fig. 83c) without accessory tine. Maxilla as in female. Maxilliped (fig. 83d) different from that of female. Corpus maxillipedis with 2 conical tubercles and patch of denticles on inner margin. Shaft with spiniform projection and hyaline knob on vental surface. Claw highly attenuate. Junction of shaft and claw with 1 slender seta; shaft and claw 930 μm combined. Sternal furca as in female.

Legs 1-4 as in female. Leg 5 (fig. 83e) more than 4 times as long as wide, 1.43 × 0.33 mm, with 3 blunt spines at tip. All 3 spines (fig. 83f) with row of minute spinules at bases. Leg 6 (fig. 83g) consisting of 2 spines and 1 papilla tipped with 1 plumose seta on posteroventral flap of genital complex.

Spermatophore (fig. 83h) spherical, 437 μm in diameter.

Remarks. — Since the discovery of *Gloiopotes hygomianus* by Steenstrup & Lütken (1861), a few keys to the species (five species including *G. hygomianus*) of the genus *Gloiopotes* have been published (Wilson, 1907a; Hewitt, 1964b; Cressey, 1967a). In addition, a few relatively complete (Shiino, 1960b; Lewis, 1967) and brief (Stebbing, 1900; Cressey, 1967a; Cressey & Cressey, 1980) redescriptions have contributed greatly to our knowledge of the external morphology of this species.

The type-species, *Gloiopotes hygomianus*, can be distinguished from its four congeners by its general habitus and by the details of its appendages. Although *G. hygomianus* is the type of its genus, it is the least "typical" of *Gloiopotes*. Based upon host specificity and external morphology, the four congeners [*G. ornatus* Wilson, 1905; *G. americanus* Cressey, 1967; *G. huttoni* (Thompson, 1889); and *G. watsoni* Kirtisinghe, 1934] are more similar to each other than they are to the type-species.

The type-species differs from its congeners in several morphological features. For example, the female of *G. hygomianus* possesses dorsal aliform plates on the fourth pedigerous somite that covers most of the genital complex. In contrast, the dorsal plates of the females of the remaining four species only cover the anteriormost portions of their genital complexes. In the type-species, the posterolateral processes of the genital complex extend beyond the distal limits of leg 5 in the female, unlike those of its congeners. There also exist differences in the appendages, such as the postantennal process (2 tines in *G. hygomianus*, and 3 tines with 1 of the tines often bifid in congeners). The sternal furca of the four species of *Gloiopotes* possess the accessory tines near its base; however, in *G. hygomianus* the primary tine is single. The four congeners possess primary tines that are bifid at their tips. The spines of the second and third exopodal segments of leg 3 are much shorter and stouter in *G. hygomianus* than in the other four species. Leg 5 is spiniform in the females of the four congeners, not spatulate as in the female of *G. hygomianus*.

The major differences between the male of *G. hygomianus* and the males of the four congeners is the claw of the maxilliped. In the type-species, this structure is long and whiplike, almost extending to the proximalmost portion of the corpus maxillipedis. However, in other species of this genus this segment is a normal clawlike structure.

Although Steenstrup & Lütken (1861) never recorded the host for *Gloiopotes hygomianus* collected from the Atlantic Ocean, the host was probably *Acanthocybium solandri* (Cuvier) since this is the only reported host for this species of parasitic copepod. It appears from the literature that *Gloiopotes hygomianus* is very specific for this host species (Shiino, 1960b, 1965; Cressey, 1967a; Lewis, 1967; Lewis et al., 1969; Kazachenko, 1976; Cressey & Cressey, 1980; Ho & Nagasawa, 2001). Bonnet's (1948) collection of an unidentified species of *Gloiopotes* from *Acanthocybium solandri* from the Hawaiian Islands was most probably *G. hygomianus*. The other four species are apparently specific for members of the billfish family Istiophoridae (Cressey, 1967a).

The known zoogeographic distribution of *G. hygomianus* is summarized in table X. Cressey & Cressey (1980) believed that *G. hygomianus* may be found "throughout the range of the host", which has a circumtropical and subtropical distribution.

Genus **Hermilius** Heller, 1865

Hermilius Heller, 1865: 186; Bassett-Smith, 1899: 445; Barnard, 1955b: 100; Pillai, 1963b: 179; Yamaguti, 1963: 82; Kabata, 1964b: 609; Kabata, 1979: 158; Dojiri, 1983: 205; Prabha, 1983: 55; Pillai, 1985: 410; Kazachenko, 2001: 31; Boxshall & Halsey, 2004: 726; Ho & Lin, 2004: 271.

Female. — Cephalothorax large, with lateral zones folded ventrally, comprising about 1/3 to 1/2 total body length, and possessing prominent dorsal ribs. Frontal plates present, without lunules. Posterior sinuses present, but often hidden in dorsal view by anterior portion of genital complex. Fourth pedigerous somite small, dorsally hidden by genital complex. Genital complex large, providing majority of organism's length. Abdomen distinctly or indistinctly comprising 2 free somites. Caudal ramus with setae.

Antennule 2-segmented. First segment robust, with numerous large plumose setae. Second segment cylindrical with 13 + 1 aesthetasc. Antenna without

TABLE X
Localities of collections of *Gloiopotes hygomianus* Steenstrup & Lütken, 1861

Locality	Reference
Atlantic Ocean	Steenstrup & Lütken, 1861
Roviana, New Georgia (Solomon Islands, western south Pacific)	Stebbing, 1900
Cocos Island (eastern Pacific Ocean off Panama)	Shiino, 1960b
11°14.7′N 113°26.2′W (eastern Pacific Ocean)	Shiino, 1965
Hawaii	Lewis, 1967
Line Islands (near Hawaii)	Lewis, 1967
Washington Island (Line Islands)	Lewis, 1967
Fanning Island (Line Islands)	Lewis, 1967
Socorro Island (eastern Pacific)	Cressey, 1967a
Indian Ocean	Cressey, 1967a
Hawaii	Cressey, 1967a
Puerto Rico	Cressey, 1967a
17°18′S 70°05′E (Indian Ocean)	Lewis et al., 1969
20°02′S 79°50′E (Indian Ocean)	Lewis et al., 1969
Pacific Ocean	Kazachenko, 1976
Seychelles	Cressey & Cressey, 1980
Madagascar	Cressey & Cressey, 1980
Ile Amsterdam	Cressey & Cressey, 1980
Christmas Islands (Pacific Ocean)	Cressey & Cressey, 1980
Washington Island	Cressey & Cressey, 1980
Hawaii	Cressey & Cressey, 1980
Socorro Island, Revillagigedo	Cressey & Cressey, 1980
Puerto Rico	Cressey & Cressey, 1980
Nicaragua	Cressey & Cressey, 1980
North Atlantic (30°N 41°W)	Cressey & Cressey, 1980
North Carolina	Cressey & Cressey, 1980
Brazil	Cressey & Cressey, 1980
Canary Islands	Cressey & Cressey, 1980
Cape Verde Islands	Cressey & Cressey, 1980
Azores Island	Cressey & Cressey, 1980
India	Pillai, 1985
Off Wakayama, Japan	Ho & Nagasawa, 2001
36°15′N 71°49′W (Atlantic Ocean)	Present paper (specimens from collection of National Museum of Natural History, Smithsonian Institution)

posteriorly directed spinelike process, and terminating in large claw; accessory claw also present. Postantennal process absent. Mouth tube as in other caligids. Mandible comprising 4 sections, with 12 teeth on inner margin of terminal section. Maxillule comprised of long, slender process and adjacent setiferous papilla (papilla not reported in *H. youngi* by Kabata, 1965). Maxilla brachiform. Maxilliped similar in shape to maxilla, with slender corpus, shaft, and claw. Sternal furca present.

Leg 1 with 2-segmented exopod and vestigial endopod. Sympod with large, plumose seta on outer distal corner near sympod-exopod junction. Terminal exopodal segment with 3 extremely long terminal spines, 1 small seta (seta 4) on inner distal corner, and 3 plumose setae on inner margin. Endopod a spinulose lobe. Leg 2 biramous, with 2-segmented (as in *H. youngi*) or 3-segmented rami. Leg 3 with large ventral apron; rami 2- or 3-segmented. Large spine present on basal swelling of first exopodal segment. Leg 4 exhibiting trend toward reduction from sympod and 2-segmented exopod to 1-segmented, setiferous lobe. Leg 5 represented by setae.

Male. — Body similar to general habitus of males of *Caligus* and *Lepeophtheirus*. Cephalothorax suborbicular with prominent dorsal ribs; lateral zones folded as in female. Fourth pedigerous somite free, and small. Genital complex subovate. Abdomen consisting of 1 free somite. Caudal ramus with 6 plumose setae. Antenna with corrugated adhesion pads; large accessory claw absent. Legs 5 and 6 represented by setae. Other appendages similar to those in female.

Type-species. — *Hermilius pyriventris* Heller, 1865.

Remarks. — In 1865 Heller published a description of a new genus and species of caligid copepod attached to the gills of the catfish "*Arius acutus*" (= *Plicofollis argyropleuron* Valenciennes). The type-species, *H. pyriventris*, was collected at Java. Since then, nine additional species have been reported: *H. longicornis* Bassett-Smith, 1898; *H. armatus* Capart, 1959; *H. helleri* Pillai, 1963; *H. youngi* Kabata, 1964; *H. tachysuri* Pillai & Natarajan, 1977; *H. alatus* Hameed, 1981; *H. pseudari* Hameed, 1981; *H. ariodi* Prabha & Pillai, 1986; and *H. longicaudus* Ho & Kim, 2000. It was suggested by Kabata (1964b) and verified by Cressey (1974) that *H. armatus* be relegated to synonymy with the type-species, *H. pyriventris*. Another species *H. helleri* was relegated to synonymy with *H. pyriventris* by Ho & Kim (2000).

The females of *Hermilius* exhibit an interesting morphological adaptation for their predilection for the gills as the attachment site. The entire cephalothorax is bent ventrally at the mid-dorsal line; thus, giving the head a "bivalve" appearance. As mentioned by Pillai (1963b) and Cressey (1974), this modification is not unusual for copepods that are parasitic on gill filaments, e.g., *Abasia*, *Lernanthropus*, *Metataeniacanthus*, *Protochondria*, and *Pseudocycnus*, among others.

The species of *Hermilius* show a trend toward a gradual reduction of the fourth leg (Kabata, 1964b). This trend is from a 3-segmented fourth leg (sympod with two exopodal segments) to a 1-segmented one. Since this trend is exhibited by species within a single genus, the reliability of the fourth leg when evaluated alone may not be taxonomically significant at the generic level (see discussion on *Pseudocaligus*).

The eight species of this genus have all been found attached to the gills of marine catfish (Ariidae) (Cressey, 1974; Ho & Kim, 2000).

Hermilius pyriventris Heller, 1865
(figs. 84-87)

Hermilius pyriventris Heller, 1865: 187; Bassett-Smith, 1899: 445; Brian, 1924: 32; Barnard, 1955b: 100; Capart, 1959: 91; Pillai, 1961: 123 (*H. longicornis*); Pillai, 1963b: 181; Yamaguti, 1963: 83; Kabata, 1964b: 617; Cressey, 1974: 236; Dojiri, 1983: 209; Leong, 1985: 50; Pillai, 1985: 412; Ho & Sey, 1996: 63; Lin & Ho, 2000: 202; Ho & Kim, 2000: 81; Purivirojkul & Areechon, 2008: 42.
Hermilius armatus Capart, 1959: 90.
Hermilius helleri Pillai, 1963b: 185; Prabha & Pillai, 1986: 39.

Material examined. — Sixteen females (USNM 160551-3, 160570-1) from gill filaments of an unidentified species of sea catfish, *Tachysurus* sp. (family Bagridae), collected at Philippine Islands. Specimens on loan from National Museum of Natural History, Smithsonian Institution, Washington, D.C.

Female. — Body as in fig. 84a, b. Total length (not including setae of caudal ramus) 3.38 mm (2.89-3.62 mm) (n = 8). Cephalothorax almost 1/2 of total body length, 1.51 mm (1.46-1.59 mm) × 0.99 mm (0.93-1.13 mm) (measured from dorsal midline to lateral edge). Lateral portion of cephalothorax folded ventrally to envelope gill filament of host (fig. 84b). Heavily sclerotized dorsal rods providing rigidity to cephalothorax. Posterior sinuses (fig. 84c) relatively shallow and with striated membrane. Fourth pedigerous somite small and hidden in dorsal view by anteriormost portion of genital complex, and in ventral view by ventral apron of leg 3. Genital complex (fig. 84a) widest anteriorly and approximately 1/2 of total body length, 1.70 mm (1.43-1.86 mm) × 1.15 mm (1.00-1.29 mm). Abdomen (fig. 84d) with 2 free somites. First somite as long as wide, 0.22 × 0.22 mm. Second somite wider than long, 0.16 × 0.22 mm.

Caudal ramus (fig. 84e) incompletely fused to anal somite, longer than wide, 138 × 76 μm, and bearing 3 acuminate spines and 3 plumose setae.

Antennule (fig. 84f) 2-segmented. First segment 57 μm long, possessing 27 plumose setae. Second segment 65 μm long, bearing 13 + 1 aesthetasc (2 setae sharing common base). Antenna (fig. 85a) with 5 sections. First

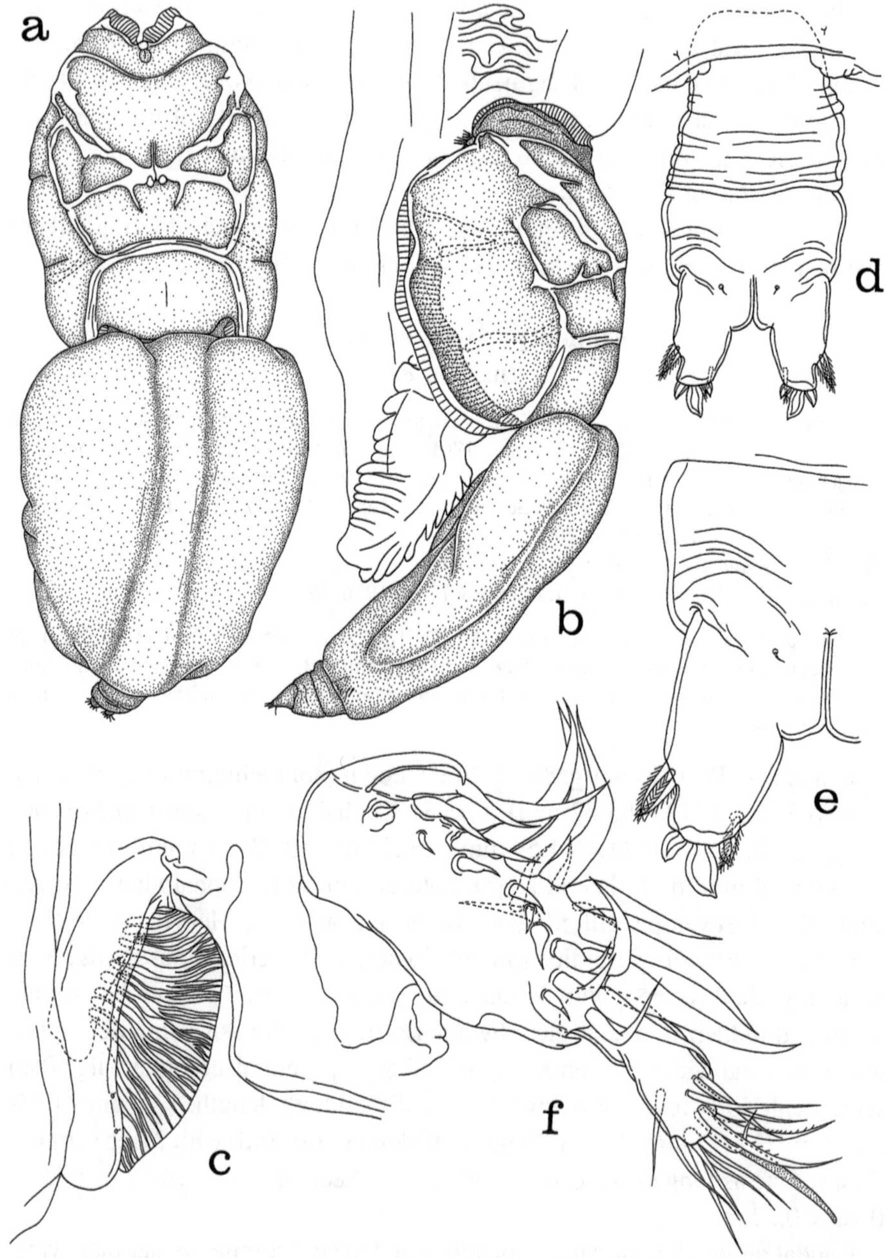

Fig. 84. *Hermilius pyriventris* Heller, 1865, female. a, body, dorsal (scale K); b, body (attached to gill filament of host), lateral (K); c, posterior sinus of cephalothorax, dorsal (D); d, postgenital somites, ventral (G); caudal ramus, ventral (D); f, antennule (plumosities not drawn), ventral (U).

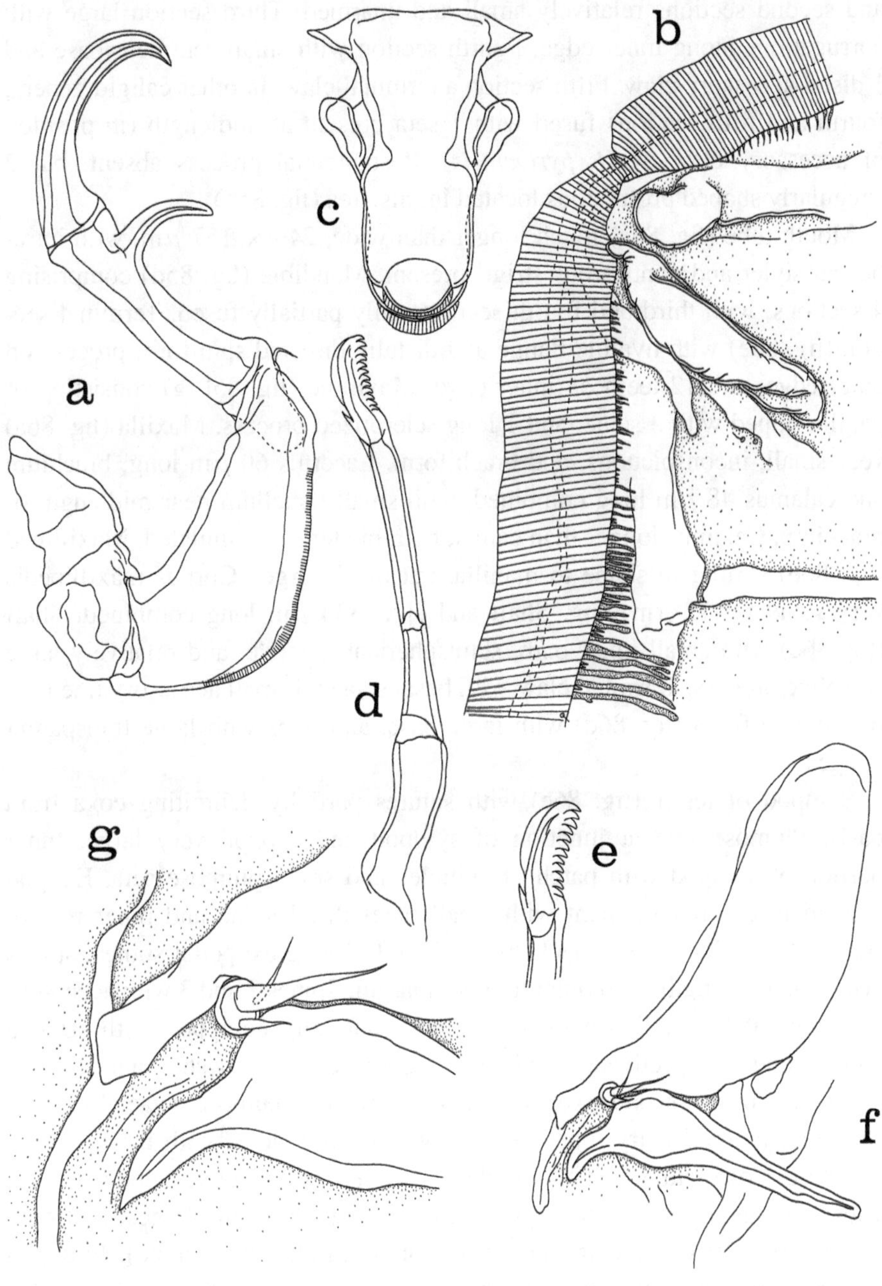

Fig. 85. *Hermilius pyriventris* Heller, 1865, female. a, antenna, ventral (scale G); b, postantennal area, ventral (E); c, mouth tube, ventral (E); d, mandible, ventral (L); e, tip of mandible, ventral (Y); f, maxillule, ventral (D); g, setiferous papilla and basal portion of dentiform process of maxillule, ventral (Y).

and second sections relatively small and unarmed. Third section large with corrugations along inner edge. Fourth section with small seta near base and 1 distal accessory claw. Fifth section a terminal claw. In other caligid genera, fourth and fifth sections fused with 1 seta present at midlength (in position of accessory claw of *H. pyriventris*). Postantennal process absent, but 2 irregularly shaped protrusions located in this area (fig. 85b).

Mouth tube (fig. 85c) much longer than wide, 243 × 157 μm, with intrabuccal stylet and dentiferous strigil present. Mandible (fig. 85d) comprising 4 sections, with third and fourth sections only partially fused. Terminal section (fig. 85e) with hyaline flange and distally directed spiniform process on outer edge, and 12 teeth on inner edge. Maxillule (fig. 85f, g) consisting of papilla tipped with 3 setae, and 1 long sclerotized process. Maxilla (fig. 86a) very small, inconspicuous, and brachiform. Lacertus 60 μm long; brachium and calamus 48 μm long combined, with small flabellum near midlength of brachium. Calamus longer than canna; both bilaterally spinulated. Maxilliped (fig. 86b) similar in shape to maxilla, but much larger. Corpus maxillipedis very slender, 472 μm long. Shaft and claw 334 μm long combined. Shaft (fig. 86c) with small inner barb, semispherical tubercle, and relatively large spinelike process. Terminal claw with basal seta and small accessory tine near tip. Sternal furca (fig. 86d) with large base, and tines with large transparent flanges.

Sympod of leg 1 (fig. 86e) with sutures partially delimiting coxa from basis. Plumose seta at junction of sympod and exopod very large. Inner portion of sympod with patch of spinules and small plumose seta. Exopod 2-segmented. First segment with small outer distal spine and inner row of short setules. Second segment (fig. 86f) with 3 long exopodal spines; spines decreasing in length from outer to inner margin. Spines 2 and 3 with accessory processes at tips. All 3 exopodal spines bilaterally equipped with striated flanges, and with pectinate membranes at bases. Seta 4 very short and smooth. Inner margin of second exopodal segment with 3 plumose setae. Endopod (fig. 86e) a small hairy lobe. Leg 2 (fig. 86g) biramous, with usual caligid armature. Coxa with patch of spinules near intercoxal plate. Exopod (fig. 87a) with 4 bilaterally spinulated spines; fourth exopodal spine largest and only slightly sclerotized (giving appearance of seta rather than spine). Endopod (fig. 86g) with spinules on outer margin of all 3 segments. Sympod of leg 3 (fig. 87b, c) with dorsal corrugated pad on anterolateral corner, and patch of spinules near base of exopod. Exopod incompletely 3-segmented. First segment with large clawlike spine at tip of basal swelling; flanges situated

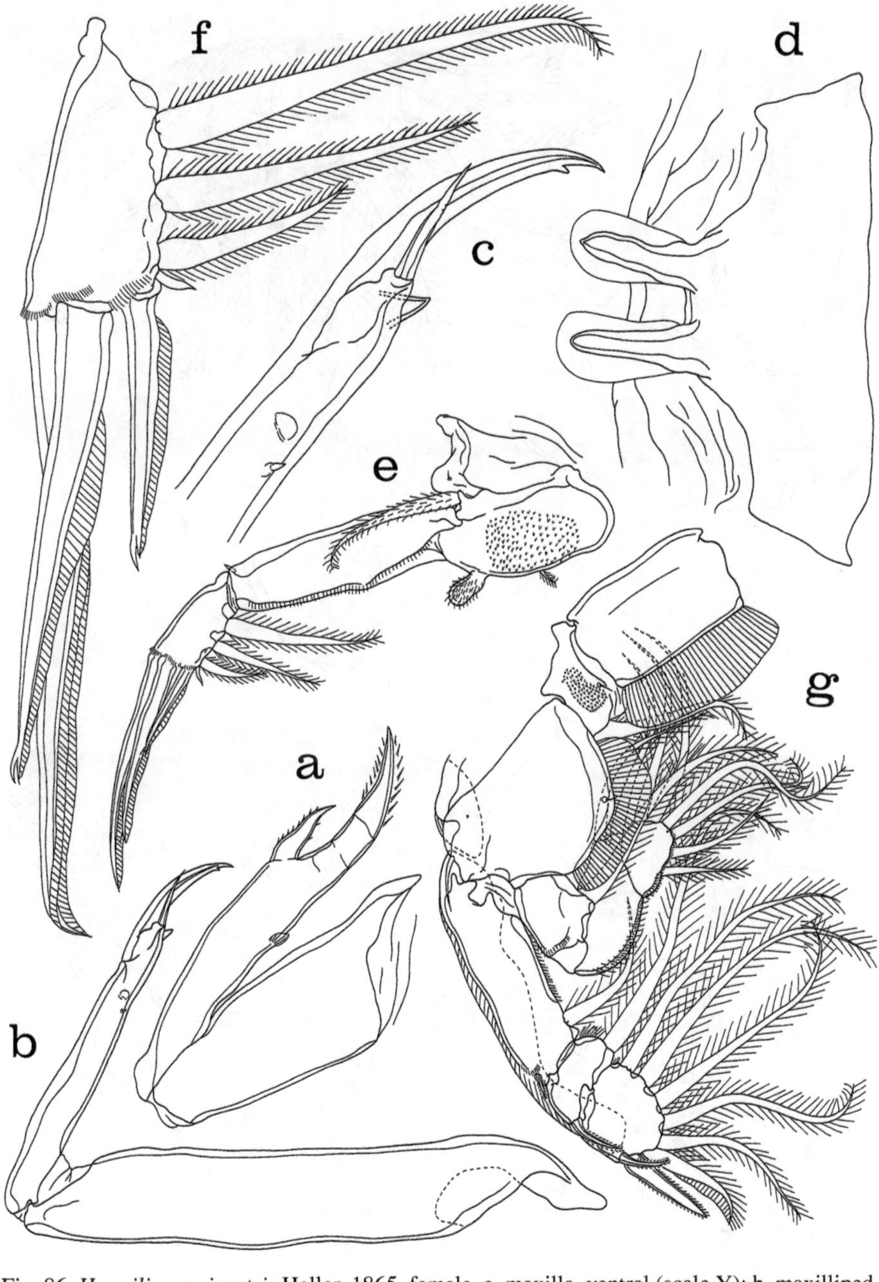

Fig. 86. *Hermilius pyriventris* Heller, 1865, female. a, maxilla, ventral (scale Y); b, maxilliped, ventral (E); c, tip of claw of maxilliped, ventral (L); d, sternal furca, ventral (L); e, leg 1, anterior (J); f, terminal exopodal segment of leg 1, ventral (U); g, leg 2 and intercoxal plate, ventral (J).

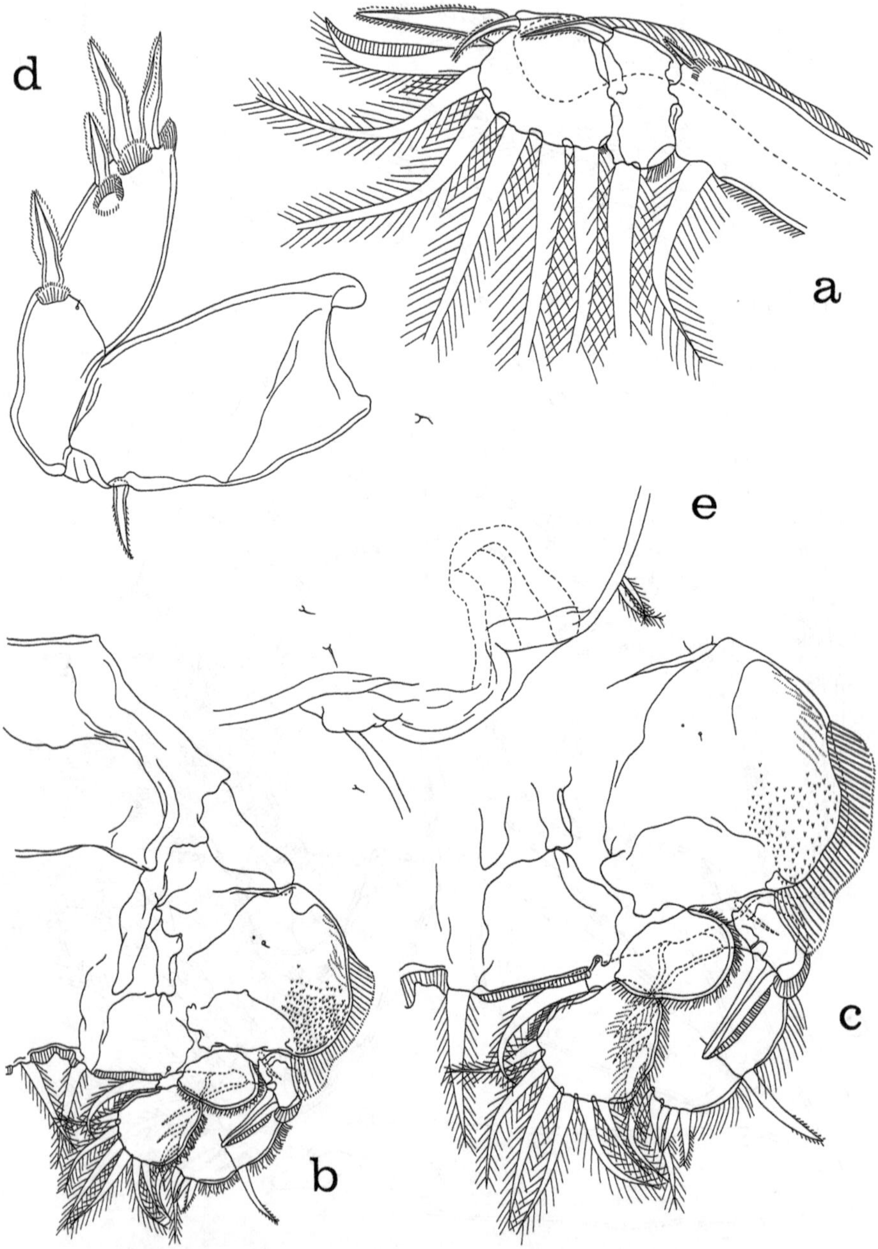

Fig. 87. *Hermilius pyriventris* Heller, 1865, female. a, exopod of leg 2, ventral (scale D); b, leg 3, ventral (J); c, same (D); d, leg 4, ventral (L); e, leg 5 and genital area, ventral (D).

on outer margin of basal swelling and exopodal spine. Second and third segments almost completely fused; second segment with distally spinulated outer spine and 1 inner plumose seta. Third segment with 7 setae; second and third setae smooth, others plumose. Endopod 2-segmented. Outer margin of first segment expanded to form velum, inner margin with plumose seta. Second segment with 6 plumose setae. Leg 4 (fig. 87d) uniramous, brachiform. Sympod robust with distal bilaterally spinulated spine. Exopod 2-segmented; all 4 exopodal spines bilaterally spinulated, and bearing pectinate membranes at bases. Terminal 3 exopodal spines decreasing in length from inner to outer margin (although middle spine longest of 3 in fig. 87d). Leg 5 (fig. 87e) represented by 2 plumose setae on posterolateral corner of genital complex. Leg 6 apparently absent.

Male. — As described by Cressey (1974) and Lin & Ho (2000).

Remarks. — Heller (1865) described *Hermilius pyriventris* as a new genus and species from the gill filaments of the sea catfish "*Arius acutus*" (= *Plicofollis argyropleuron*) from Java. Since its discovery, there have been several records of this species parasitizing various species of the sea catfish families Ariidae and Tachysuridae (= Bagridae) (table XI). All the collections of this species are from the Atlantic and the Indo-Pacific Oceans. The present collection from the Philippine Islands is the same as that reported by Cressey (1974).

Pillai's (1961) redescription of the specimens of *H. pyriventris* later proved to be a congener, *H. longicornis* (Kirtisinghe, 1964). Pillai (1963b) subsequently revised the genus, which included a redescription of *H. pyriventris* (type-species), *H. longicornis*, and the description of a new species, *H. helleri*, which has been relegated to synonymy with *H. pyriventris* by Ho & Kim (2000). A detailed redescription of *H. pyriventris* was presented by Ho & Lin (2004) based on specimens collected from the gill filaments of *Arius maculatus* Thunberg and *Trachinotus blochii* (Lacépède) from Taiwan.

Cressey (1974), apparently unaware of Pillai's (1963b) redescription of *H. pyriventris*, provided an excellent redescription of this type-species along with the first description of a male of this genus. The two main differences between Cressey's (1974) redescription and our description are the spiniform projection on the outer margin of the tip of the mandible (see fig. 85e) and the accessory process on spines 2 and 3 of the terminal segment of the exopod of leg 1 (see fig. 86f). Our description of the fourth leg is consistent with that of Cressey (1974) and Lin & Ho (2000), but differs from Pillai's (1963b) account. Pillai has figured this leg with a sympod and a 1-segmented exopod equipped with five spines. The description of this appendage in the above account is similar to that of *Hermilius helleri* Pillai, 1963.

TABLE XI
Hosts and localities of collections of *Hermilius pyriventris* Heller, 1865

Host	Locality	Reference
"*Arius acutus*" [= *Plicofollis argyropleuron* (Valenciennes)]	Java	Heller, 1865
Arius heudelotii Valenciennes	Mauritania	Brian, 1924
Galeichthys (*Tachysurus*) *feliceps* Valenciennes	Gordon's Bay, South Africa	Barnard, 1955b
	Table Bay, South Africa	Barnard, 1955b
Arius heudelotii	05°56′S 12°03′E	Capart, 1959
	12°30′N 17°09′W	Capart, 1959
Pseudarius platystomus [= *Plicofollis platystomus* (Day)]	Trivandrum, India	Pillai, 1963b
Netuma thalassinus (Rüppell)	Vipingno, Kenya	Cressey, 1974
	Diani, Kenya	Cressey, 1974
(reported as *Arius thalassinus*)	Penang, Malaysia	Leong, 1985
	Trivandrum, India	Pillai, 1985
	Kuwait	Ho & Sey, 1996; Ho & Kim, 2000
"*Netuma macrocephalus*" (= *Arius macrocephalus* Bleeker?)	Vipingno, Kenya	Cressey, 1974
Tachysurus (= *Plicofollis*) sp.	Philippines	Cressey, 1974
Arius sp.	Vizakhapatnam, India	Cressey, 1974
	Gulf of Thailand	Purivirojkul & Areechon, 2008
Arius maculatus (Thunberg)	Taiwan	Lin & Ho, 2000

Hermilius pyriventris can be distinguished from its congeners, except *H. helleri*, by the presence of a distinct segmental suture between the terminal claw and the accessory claw of the antenna. This suture is also possessed by *H. helleri*. In fact, these two species are morphologically similar (Pillai, 1963b), but may be distinguished by the structure of the sternal furca and the armature of the caudal ramus. However, Prabha & Pillai's (1986) redescription of *H. helleri* shows these two features to be identical to those of *H. pyriventris*. Thus, Ho & Kim (2000) proposed to relegate *H. helleri* to synonymy with *H. pyriventris*.

Genus **Kabataella** Prabha & Pillai, 1984

Kabataella Prabha & Pillai, 1984: 1; Boxshall & Montú, 1997: 48; Boxshall & Halsey, 2004: 727.

Female. — Cephalotholax semilunar, with large wing-like posterolateral lobes. Frontal plates and lunules absent. Cephalothoracic shield with shallow and widely open posterior sinuses. Fourth pedigerous somite small, without aliform plates. Genital complex enlarged, subrectangular. Abdomen reduced, comprising 1 free somite. Caudal ramus bearing 3 plumose and 3 pinnate setae. Egg sac long, uniseriate.

Antennule 2-segmented. Antenna 3-segmented; proximal segment without medial spiniform process. Postantennal process absent. Maxillule with stout lobe and small dentiform process; one of three setae also stout. Maxilla 2-segmented and brachiform. Maxilliped with robust coxa (corpus) bearing bifid process in myxal area. Sternal furca absent.

Leg 1 without setae on posterior margin of distal expodal segment; endopod vestigial. Leg 2 biramous, with 2-segmented rami. Leg 3 exopod 2-segmented and endopod reduced to 1 seta; apron reduced in size. Leg 4 uniramous, with 3-segmented exopod. Legs 5 and 6 each represented by 1 seta.

Male. — Unknown.

Type-species. — *Kabataella indica* Prabha & Pillai, 1984.

Remarks. — Because *Kabataella indica* Prabha & Pillai, 1984 is a parasite of the nasal fossae of the saddle grunt, *Pomadasys maculatus* (Bloch), some morphological modifications for life in this unusual habitat are expected. These changes involve development of the cephalothorax into a structure without frontal plates, lunules, postantennal processes, and sternal furca. Furthermore, the apron of leg 3 is reduced and the posterior sinuses of the cephalothoracic shield have become shallow and wide open, due to the formation of a pair of aliform structures that project outwards and ventrally. These morphological changes have adapted this species for life in a pit (i.e., nasal fossae) and no longer allow it to attach and live on a flat surface, such as the body surface of its fish host. Other significant morphological changes for life in the nasal fossae are the reduction of legs 2 and 3. Both rami of leg 2 are 2-segmented and the endopod of leg 3 is reduced to 1 seta. This distinctive combination of characters clearly distinguishes *Kabataella* from all other caligid genera and confirm the validity of this genus.

Kabataella indica Prabha & Pillai, 1984
(figs. 88-90)

Kabataella indica Prabha & Pillai, 1984: 1.

Material examined. — Twelve females (USNM 268272) from nasal cavity of 6 specimens of *Pomadasys maculatus* originally purchased at Royapuran Beach Fish Market, Royapuran Beach, Madras, India, 13°07′N 80°20″E, on 13 September 1966; copepods collected from these fish that were housed in the fish collection of the National Museum of Natural History, Smithsonian Institution, Washington, D.C. by first author (M. Dojiri).

Female. — Body as in fig. 88a. Total length (excluding setae on caudal ramus) 2.27 mm (1.92-2.61 mm) × 1.45 mm (1.24-1.72 mm). Cephalothorax wider than long, 1.01 mm (0.82-1.16 mm) × 1.48 mm (1.32-1.64 mm), with tips of antennules well within lateral limits of cephalothorax; posterior sinuses very shallow; lateral zone expanded into aliform structures that are moderately folded ventrad. Fourth pedigerous somite wider than long, 0.21 mm (0.14-0.26) × 0.49 mm (0.41-0.54 mm), and demarcation with genital complex difficult to distinguish. Genital complex transversely oblong, 0.89 mm (0.81-0.94 mm) × 1.58 mm (1.44-1.85 mm). Abdomen (fig. 88b) indistinguishably fused to posteromedial margin of genital complex, about 0.29 mm (0.28-0.31 mm) × 0.27 mm (0.24-0.29 mm). Caudal ramus (fig. 88b, c) longer than wide, 0.06 mm (0.05-0.06 mm) × 0.03 mm (0.02-0.03 mm), and bearing 3 plumose and 3 naked setae.

Frontal plate absent; small, medial rostral-like protrusion evident between bases of antennules. Antennule (fig. 88d) 2-segmened; first segment bearing approximately 20 naked setae (size and overlapping positions make setae difficult to count); second segment cylindrical and carrying 12 + 1 aesthetasc (2 posterior terminal setae sharing common base). Antenna (fig. 88e) 3-segmented; first segment unarmed, small, and without posteriorly directed spiniform process; second segment stout, longest of 3 segments, and armed with a hyaline, ballonlike structure tipped with a setiform process; terminal segment a curved claw with small hyaline seta on inner margin at about midlength. Postantennal process absent.

Mouth tube apparently without intrabuccal stylet and dentiferous strigil. Mandible (fig. 88f) comprising 4 sections; third and fourth sections almost completely fused; terminal portion with 12 medial teeth. Maxillule (fig. 89a) consisting of small dentiform process and large lobe that bears 3 naked setae. Maxilla (fig. 89b) brachiform; first segment (lacertus) naked; second segment (brachium) with flabellum that consists of striated membrane with serrated margin; both calamus and canna bilaterally serrated. Maxilliped (fig. 89c) 2-segmented; stout first segment with bifid process on inner margin; shaft and claw with small seta at incomplete suture. Sternal furca absent.

Leg 1 (fig. 89d) not highly sclerotized and biramous; sympod with outer and inner plumose setae. Exopod indistinctly 2-segmented; first segment with spine at outer distal end; second segment with outer spine, 3 terminal spines (outermost terminal spine stout and serrated on inner margin as is middle

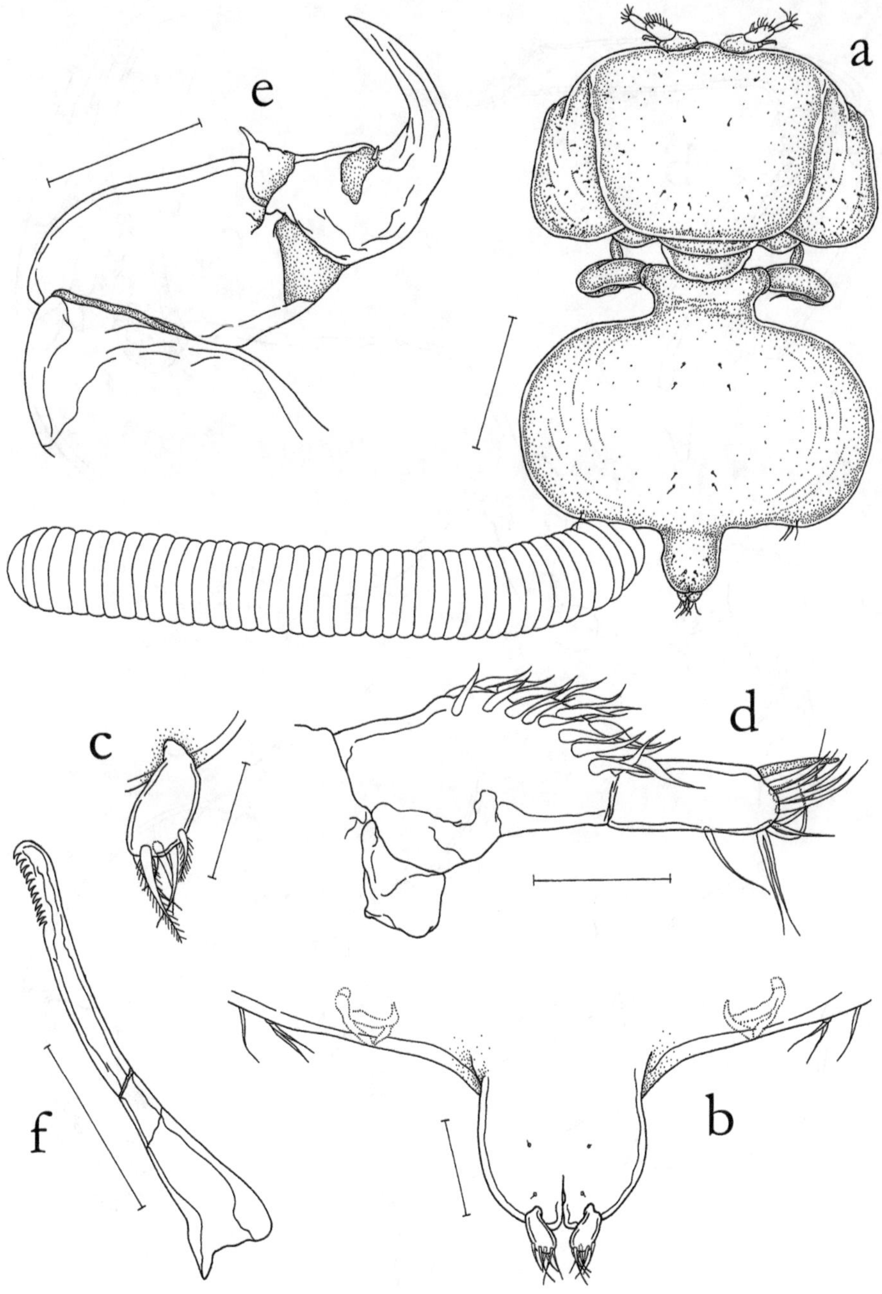

Fig. 88. *Kabataella indica* Prabha & Pillai, 1983, female. a, body, dorsal; b, anal somite and caudal rami, ventral; c, caudal ramus, ventral; d, antennule, ventral; e, antenna, ventral; f, mandible, ventral. Scale bars: 0.5 mm in a; 0.1 mm in b; 0.05 mm in c-f.

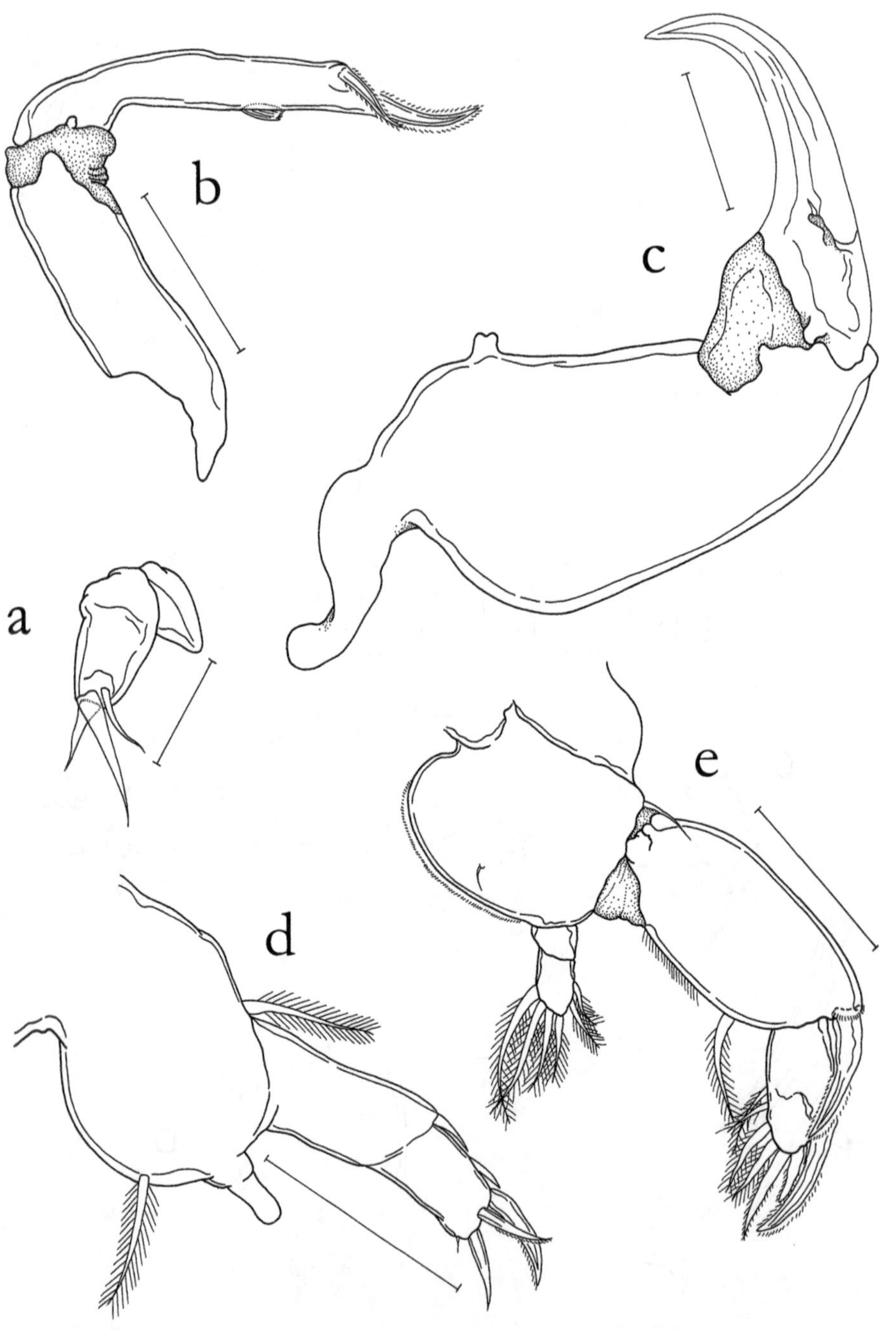

Fig. 89. *Kabataella indica* Prabha & Pillai, 1984, female. a, maxillule, ventral; b, maxilla, ventral; c, maxilliped, ventral; d, leg 1, ventral; e, leg 2, ventral. Scale bars: 0.03 mm in a; 0.1 mm in b, c, e; 0.05 mm in d.

spine; innermost terminal spine apparently naked), and 1 very small hyaline seta (possibly representing seta 4) at inner distal corner. Leg 2 (fig. 89e) biramous with sclerotized intercoxal plate; sympod with outer distal naked seta, 1 inner setule, and inner marginal row of spinules. Exopod 2-segmented; first segment with row of setules and 1 large plumose seta along inner margin and 1 outer distal, bilaterally serrated, stout spine bearing pectinate membrane at its base; terminal segment with partial segmental suture, armed with 1 stout bilaterally serrated spine, 1 small naked seta, 2 spinulated setae (spines?), and 4 plumose setae. Endopod (fig. 89e) also 2-segmented, distinctly smaller than exopod; proximal segment unarmed, and distal segment bearing 6 plumose setae. Leg 3 (fig. 90a, b) with ventral apron reduced in size, not highly sclerotized, fleshy in appearance; intercoxal plate with row of spinules; marginal membranous flange absent; sympod with relatively large plumose seta on outer margin near insertion of exopod, row of spinules on inner margin, and 1 small naked seta on inner proximal corner near junction with intercoxal plate. Exopod (fig. 90b) 2-segmented; first segment with serrated stout spine bearing pectinate membrane at its base; second segment with 4 spines and 2 plumose setae. Endopod represented by 1 plumose seta. Leg 4 (fig. 90c, d) brachiform; sympod slightly inflated, with 1 naked seta on outer distal corner. Exopod 3-segmented (fig. 90d); first and second segments each with 1 serrated spine; third segment with 3 serrated spines; all spines with pectinate membranes at bases. Leg 5 (fig. 88b) comprising 1 naked seta and 1 papilla; latter tipped with 2 naked setae.

Male. — Unknown.

Remarks. — Prabha & Pillai (1984) originally described this species from the nasal fossae of *Pomadays maculatus* from Trivandrum, India, and established a new genus to accommodate it. Their description is detailed, fairly accurate, and complete. Our redescription differs only in a few minor details: (1) caudal ramus actually bears the typical six setae, not the five that was reported by the original authors; (2) Prabha & Pillai (1984) did not mention the structure of the mandible; (3) a very small hyaline seta that most likely represents seta 4 of leg 1 exopod of other caligids is present; (4) 1 small naked seta on the inner proximal corner of the leg 3 sympod at junction with intercoxal plate is present; (5) leg 3 endopod is represented by 1 seta, not 2 as described by the original authors; and (6) leg 5 is represented by 1 seta and a papilla bearing 2 setae. For distinctive features of this unusual species, refer to Remarks of *Kabataella*.

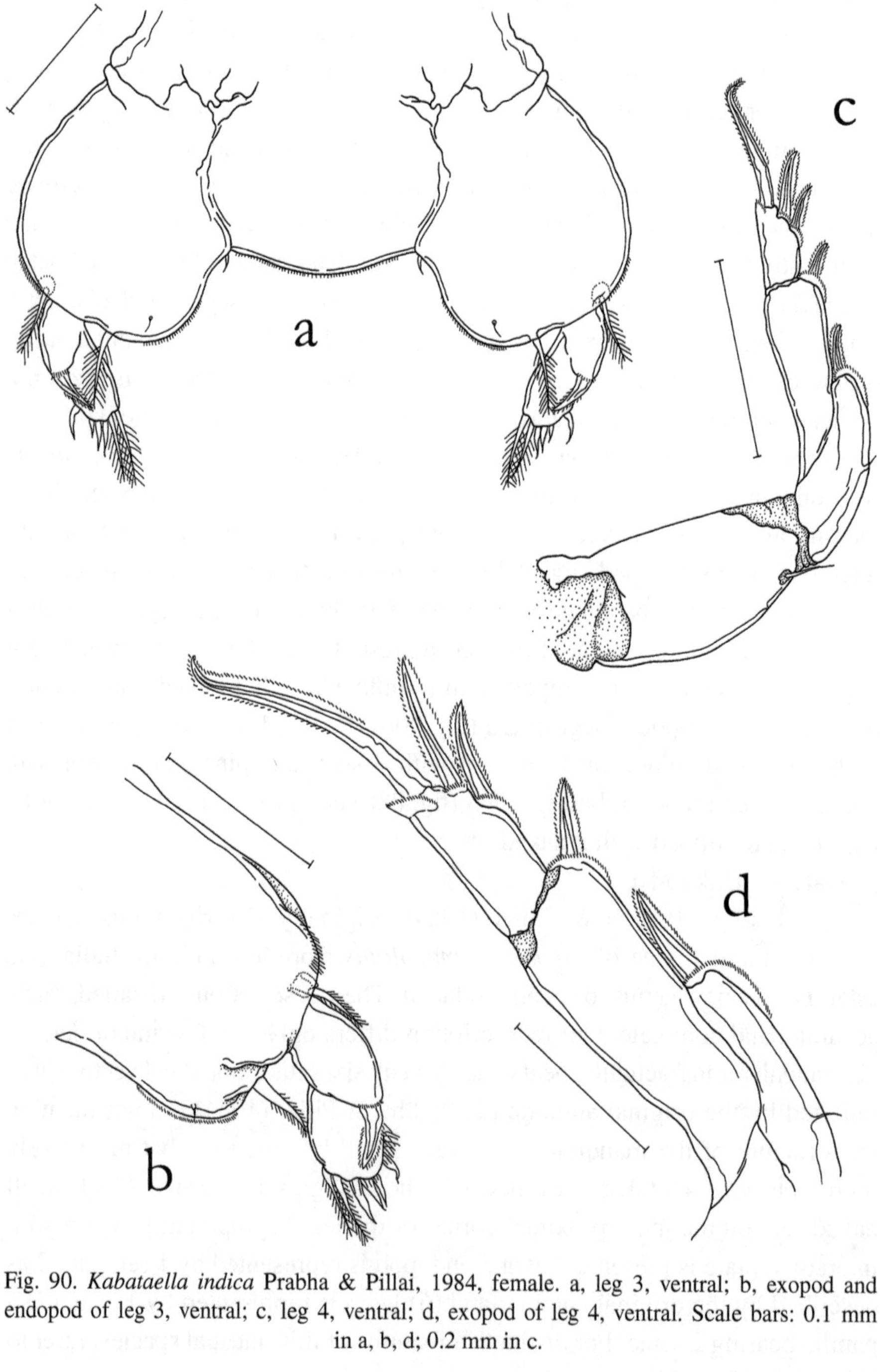

Fig. 90. *Kabataella indica* Prabha & Pillai, 1984, female. a, leg 3, ventral; b, exopod and endopod of leg 3, ventral; c, leg 4, ventral; d, exopod of leg 4, ventral. Scale bars: 0.1 mm in a, b, d; 0.2 mm in c.

Genus **Lepeophtheirus** Nordmann, 1832

Lepeophtheirus Nordmann, 1832: 30; Baird, 1850: 273; Heller, 1865: 180; Brady, 1883: 132; Carus, 1885: 359; Bassett-Smith, 1896a: 157; Bassett-Smith, 1899: 453; Wilson, 1905a: 615; Scott & Scott, 1913: 64; Wilson, 1932: 409; Barnard, 1955a: 250; Markewitsch, 1956: 130; Yamaguti, 1963: 71; Hewitt, 1963: 83; Lewis, 1964a: 178; Lewis, 1967: 75; Kabata, 1973a: 729; Kabata, 1979: 189; Dojiri, 1983: 214; Prabha, 1983: 53; Pillai, 1985: 431; Boxshall & Montú, 1997: 56; Kim, 1998: 692; Ho, 1998; 253; Kazachenko, 2001: 29; Boxshall & Halsey, 2004: 726; Ho & Lin, 2004: 276.

Homoiotes Wilson, 1905a: 661; Yamaguti, 1963: 83; Kabata, 1973b: 1892; Kabata, 1979: 168.

Dentigryps Wilson, 1913, p. 221; Yamaguti, 1963: 80; Lewis, 1964a: 203; Lewis, 1964b: 348; Hewitt, 1971b: 333; Ho & Dojiri, 1977: 96; Kabata, 1979: 164, 170.

Indocaligus Pillai, 1961: 125.

Female. — Cephalothorax subcircular, usually narrower anteriorly. Posterior sinuses present. Frontal plate distinct, and without lunules. Fourth pedigerous somite small, without dorsal plates. Genital complex of various shapes: triangular, quadrangular, or globular, occasionally elongate. Abdomen consisting of 1 or 2 free somites, predominantly shorter than 1/2 length of genital complex, occasionally longer. Caudal ramus with 6 plumose setae.

Antennule and antenna as in *Caligus*. Postantennal process present. Mouth tube and mandible as in *Caligus*. Dentiform process of maxillule usually bifid, occasionally single; adjacent setiferous papilla present. Maxilla brachiform. Maxilliped subchelate and prehensile. Sternal furca with tines occasionally bifid.

Leg 1 with 2-segmented exopod and vestigial endopod. Terminal exopodal segment with 3 apical spines (only 1 extremely large terminal spine in *L. monacanthus* Heller, 1865; *L. quadratus* Krøyer, 1863; and *L. unispinosus* Pearse, 1952), 1 seta on inner distal corner and 3 plumose inner setae. Leg 2 biramous, with 3-segmented rami. Leg 3 with ventral apron and 3-segmented exopod and 2-segmented endopod. Second and third exopodal segments may be fused; clawlike exopodal spine terminally or subterminally situated on basal swelling of first exopodal segment. Leg 4 with sympod and most commonly 3-segmented exopod; occasionally 2-segmented exopod. Spinal formula usually I-0; I-0; III, but can be I-0; IV or I-0; III. Leg 5 represented by setiferous process.

Male. — Cephalothorax similar to that in female. Genital complex comparatively smaller and somewhat oval in outline. Antenna with corrugated adhesion pads and frequently with accessory tines on terminal claw. Dentiform process of maxillule occasionally with corrugated area and additional hyaline digitiform process. Maxilliped often different from that of female with corpus bearing patches of denticles. Legs 5 and 6 represented by setiferous processes.

Type-species. — *Lepeophtheirus pectoralis* Müller, 1776.

Remarks. — Nordmann (1832) was the first to establish the genus *Lepeophtheirus*. The main distinction between species of this genus and *Caligus* is the absence of the lunules in *Lepeophtheirus*; otherwise, the members of this genus are morphologically very similar to those in the genus *Caligus*. This resemblance is reflected in the general habitus and morphology of their appendages.

The overall zoogeographic distribution of the genera differ: *Lepeophtheirus* prefers temperate waters and *Caligus* the warmer waters of tropical and subtropical regions (Kabata, 1970, 1979). The approximately 100 species of *Lepeophtheirus* are parasitic on marine fishes, predominantly teleosts; however, there are a few host records from elasmobranchs (Kabata, 1979).

Kabata (1973a) conducted a study on ten species of *Lepeophtheirus* from fishes of British Columbia that resulted in a better understanding of the taxonomically important characters of the members of this genus, and set the trend for other investigators to follow (Ho & Dojiri, 1977; Dojiri, 1979; Dojiri, 1983; and the present work). Kabata recognized two groups within *Lepeophtheirus* (though not considered taxonomic subdivisions within the genus): the species with short abdomens (shorter than 1/2 the length of the genital complex) and species with long abdomens (equal to or greater than the length of the genital complex). The majority of the species fall into the former group.

This evolutionary trend toward a reduction in the size of the abdomen has a few exceptions [e.g., *L. salmonis* (Krøyer, 1838); *L. sturionis* (Krøyer, 1838); *L. thompsoni* Baird, 1850; *L. pollachius* Bassett-Smith, 1896; *L. parvicruris* Fraser, 1920; *L. longiventris* Yü & Wu, 1932; *L. marceps* Wilson, 1944; and *L. hidekoi* Ho, 1961]. The culmination to this evolutionary trend can be seen in species of the closely related genus *Anuretes*. It has been argued (e.g., Pillai, 1967; Ho & Dojiri, 1977) that since a distinct generic boundary could not be drawn between these two genera based on the reduction of the abdomen, then perhaps *Anuretes* should be considered a synonym of *Lepeophtheirus* (Ho & Dojiri, 1977). However, a combination of features and the structure of leg 3 provide diagnostic features for *Anuretes* and affords some validity to this genus (see discussion on *Anuretes*).

There are three genera once considered valid, but no longer recognized as distinct from *Lepeophtheirus*. The first of these genera, *Homoiotes* Wilson, 1905, was found to be an artificial taxon comprised of three different species of *Lepeophtheirus* (Kabata, 1973b; also refer to discussion on *Homoiotes*). The second genus *Indocaligus* was described by Pillai (1961) to accommodate one species *Indocaligus echinus* from *Pseudarius jatius* [= *Cephalocassis jatia* (Hamilton)], but was soon relegated to synonymy with *Lepeophtheirus longipalpus* Bassett-Smith, 1898 (Pillai, 1963c; also see discussion on *Indocaligus*).

Finally, the genus *Dentigryps* Wilson, 1913 with its type-species *D. curtus* was established. Since then, several species have been attributed to this genus. The diagnostic feature distinguishing *Dentigryps* from *Lepeophtheirus* was the presence of long spikelike fifth legs in the females of *Dentigryps*. The taxonomic value of this feature has been questioned (Pillai, 1966; Hewitt, 1971b; Kabata, 1979). Lewis (1964b) supported the retention of this genus, as did Ho & Dojiri (1977). For reasons discussed later (refer to discussion on *Dentigryps*), this genus can no longer be considered distinct from *Lepeophtheirus*.

Lepeophtheirus pectoralis (Müller, 1776)
(figs. 91-95)

Lernaea pectoralis Müller, 1776: 227; Müller, 1788: 41.

Caligus pectoralis Blainville, 1823: 129 (see Boxshall, 1974a); Cuvier, 1817: 36 (see Boxshall, 1974a); Krøyer, 1837: 628; Krøyer, 1838: 8; Milne Edwards, 1840: 454; Thompson, 1847: 247; Claus, 1864: 365.

Lepeophtheirus pectoralis Nordmann, 1832: 30; Burmeister, 1835: 330; Krøyer, 1837: 200; Baird, 1850: 275; Krøyer, 1863: 215; Olsson, 1869: 10; Valle, 1880: 59; Carus, 1885: 359; Bassett-Smith, 1896a: 158; T. Scott, 1897: 155; Bassett-Smith, 1899: 454; T. Scott, 1900: 150; A. Scott, 1901a: 192; Wilson, 1905a: 616; Brian, 1906: 47; Norman & Scott, 1906: 208; Pearson, 1906: 165; Scott & Scott, 1913: 64; Hansen, 1923: 30; Stephensen, 1929: 8; Pesta, 1934: 22; Schuurmans Stekhoven, 1934, p. 336; Schuurmans Stekhoven, 1935: 4; Oorde-de Lint & Schuurmans Stekhoven, 1936: 136; Schuurmans Stekhoven, 1936: 7; Stephensen, 1940: 3; Heegaard, 1947a, p. 103; Markewitsch, 1956: 135; Bresciani & Lützen, 1962: 405; Bruce et al., 1963: 131; Yamaguti, 1963: 71; O'riordan, 1966: 372; Stillger, 1968: 276; Mann, 1970: 178; Boxshall, 1974a: 445; Boxshall, 1974b: 681; Boxshall, 1974c: 361; Boxshall, 1974d: 359; Boxshall, 1974e: 373; Boxshall, 1977: 411; Kabata, 1979: 190; Dojiri, 1983: 220; Boxshall, 1986: 160; Zeddam et al., 1988: 143; Boxshall, 1990: 169; Anstensrud, 1990a: 247; Anstensrud, 1990b: 627; Anstensrud, 1990c: 141; Anstensrud, 1990d: 269; Anstensrud, 1992: 31; Kabata, 2003: 70; Cavaleiro & Santos, 2007: 1218.

Material examined. — Eight females from flounder, *Platichthys flesus* (Linnaeus), and 9 males from plaice, *Pleuronectes platessa* Linnaeus, collected at Robin Hood's Bay, Whitby, Yorkshire, England, on August 1973 by Dr. G. A. Boxshall, Natural History Museum, London, London, England.

Female. — Body as in fig. 91a-c. Total length (not including setae on caudal ramus) 5.44 mm (5.28-5.91 mm) (n = 8). Cephalothorax subcircular (compare fig. 91a and b) measuring 2.54 mm (2.46-2.62 mm) × 2.25 mm (2.03-2.42 mm), with shallow posterior sinuses. Lunules absent. Antennule not extending beyond lateral limit of cephalothorax. Free margin of thoracic zone extending only slightly beyond tip of lateral zone.

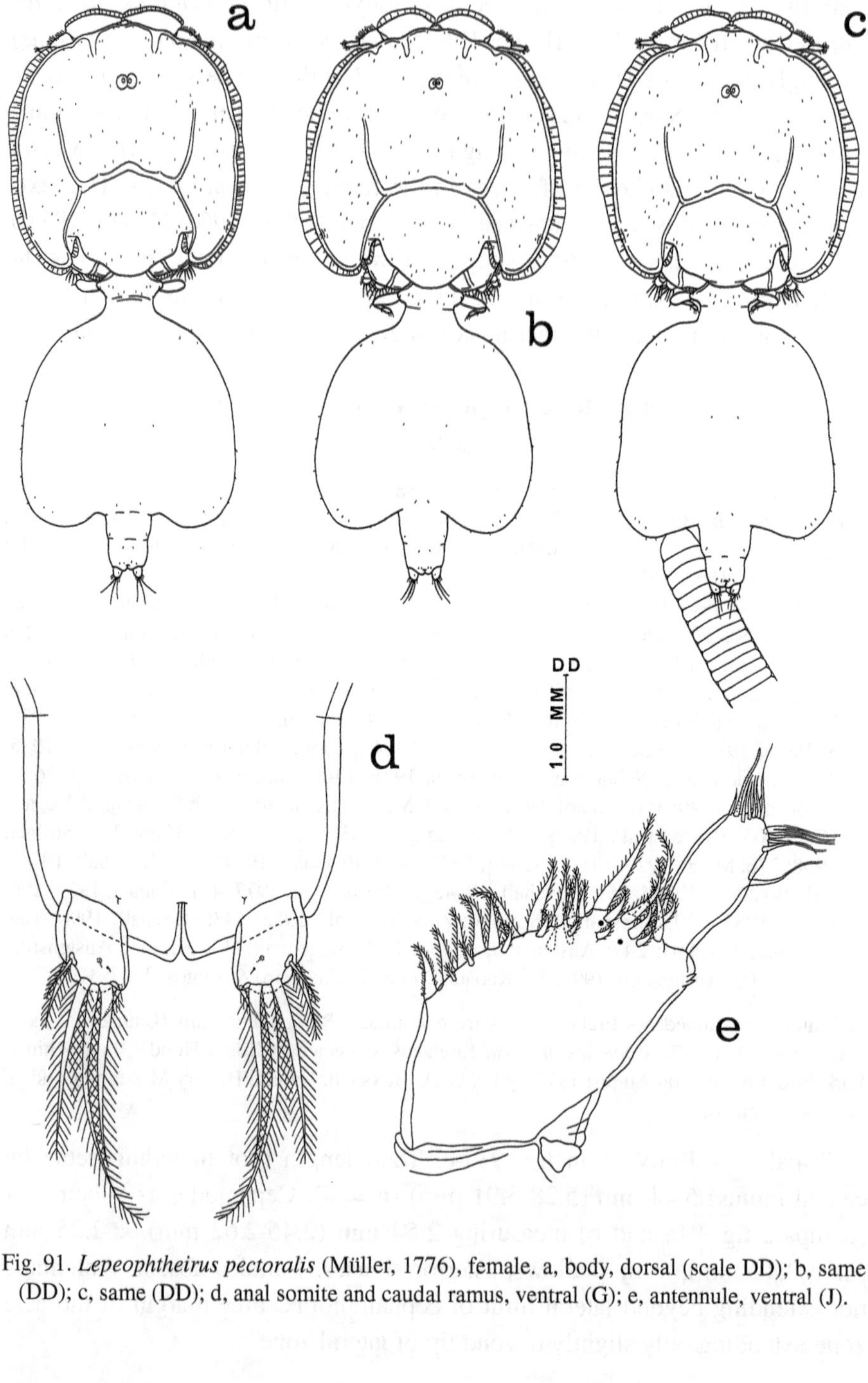

Fig. 91. *Lepeophtheirus pectoralis* (Müller, 1776), female. a, body, dorsal (scale DD); b, same (DD); c, same (DD); d, anal somite and caudal ramus, ventral (G); e, antennule, ventral (J).

Fourth pedigerous somite wider than long, not clearly delimited from genital complex. Genital complex subrectangular, narrower anteriorly (compare fig. 91a and c). Abdomen indistinctly comprising 2 free somites, demarcation of 2 somites indicated by indentation. First somite 0.27 mm (0.23-0.30 mm) × 0.46 mm (0.43-0.50 mm) and second somite 0.29 mm (0.27-0.30 mm) × 0.38 mm (0.37-0.43 mm). Two somites combined shorter than 1/2 length of genital complex.

Caudal ramus (fig. 91d) longer than wide, 119 × 92 μm, bearing 6 plumose setae.

Body with small setules (sensilla) as in fig. 91a-c.

Egg sacs 5.20 mm (4.78-5.61 mm) × 0.48 mm (0.46-0.50 mm), containing approximately 41 eggs (35-47 eggs) (n = 2).

Antennule (fig. 91e) 2-segmented. First segment 334 μm long (measured along setiferous margin), bearing 27 plumose setae along anterior margin and bifid process at posterodistal corner. Second segment 150 μm long, with 1 seta on posterior margin at about midlength, and 12 naked setae and 1 aesthetasc at tip. Antenna (fig. 92a) apparently 4-segmented. First segment small and unarmed. Second segment carrying spinelike process directed posteriorly. Third segment with dorsal corrugated pad and a small inner-distal pad. Claw uncinate with proximal seta and another seta at about midlength. Postantennal process (fig. 92b) curved with rather blunt tip. Two papillae located on basal portion and another situated posteromedially to process; all 3 bearing various numbers of setules.

Mouth tube (fig. 92c) 322 × 299 μm, possessing intrabuccal stylet (fig. 92d) on inner surface of labrum. Strigil present (indicated by dotted line in fig. 92c). Mandible (fig. 92e) with 4 sections and armed with 12 teeth. Sections 86, 16, 89, and 62 μm long from proximal to distal end (measured along outer margin). Maxillule (fig. 92f) comprised of papilla and 2 tines; former tipped with 3 setae. Medial tine longer than lateral. Base with small laterally directed conical process. Medial tine with lateral and ventral flange, and lateral tine with flanges on lateral margins only. Maxilla (fig. 92g) brachiform. Flabellum, situated near midpoint of brachium, often doubling up on itself at its distal end giving appearance of spine (see Boxshall, 1974a, fig. 8D). Calamus with 3 serrated membranes and canna with only 1 serrated membrane. Maxilliped (fig. 92h, i) with relatively robust corpus bearing 2 patches of denticles on dorsal surface (fig. 92i). Shaft and claw measuring 518 μm along axis, with 1 seta near midlength. Sternal furca (fig. 92j) with parallel or very slightly diverging tines. Tines sharp and edged with marginal flanges.

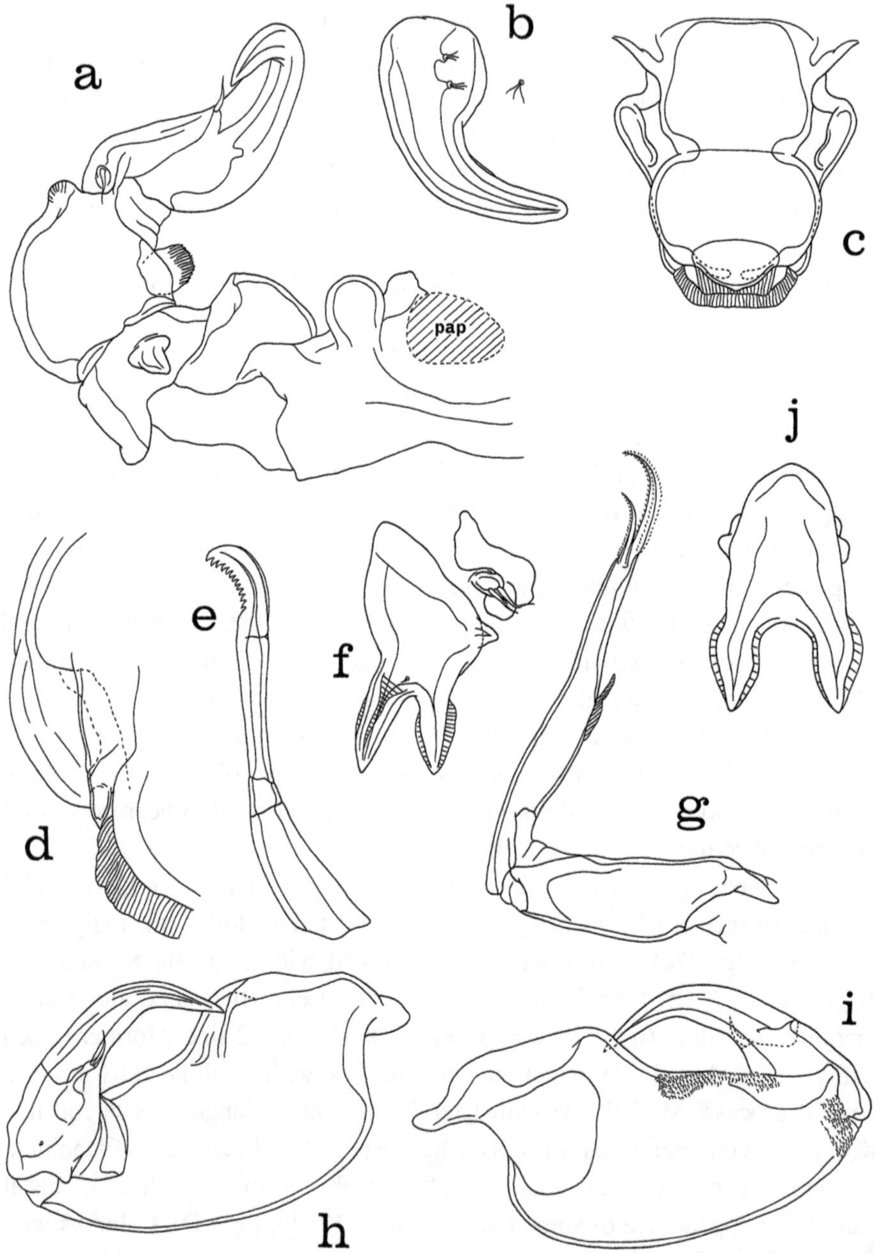

Fig. 92. *Lepeophtheirus pectoralis* (Müller, 1776), female. a, antenna and position of postantennal process, posteroventral (scale G); b, postantennal process, ventral (J); c, mouth tube, ventral (G); d, intrabuccal stylet and part of labrum (with labium removed), dorsal (U); e, mandible, ventral (D); f, maxillule, ventral (G); g, maxilla, dorsal (B); h, maxilliped, posteroventral (I); i, same, dorsal (I); j, sternal furca, ventral (B).

Base of plumose seta at junction of sympod and exopod of leg 1 (fig. 93a) not covered by ventral protrusion. Pinnate seta 4 of exopod of leg 1 (fig. 93b) shorter than outermost spine. Spines 1, 2, and 3 with 2 rows of spinules. Spines 2 and 3 bearing accessory processes at distal ends. Endopod slender, tipped with 2 minute processes. Area between legs 1 and 2 (fig. 93c) with pair of rounded swellings, each tipped with 1 setule. Proximal portion of coxa of leg 2 (fig. 93d) with conical process equipped with 1 setule. No spines of exopod of leg 2 (fig. 93e) extending beyond distal limits of terminal segment. Spines of second and third exopodal segments not serrated. All other features of leg 2 typical for caligids. Ventral apron of leg 3 bearing dorsal corrugated pad (dotted lines in figure) on outer margin as in fig. 94a. Leg 3 (fig. 94b) as in *Caligus curtus*, except exopodal spine situated subterminally on basal swelling and presence of inner seta on first exopodal segment. One specimen examined with 3 spines and 5 setae on third segment of exopod instead of usual 3 spines and 4 setae. Leg 4 exopod (fig. 94c) 2-segmented, with innermost spine of terminal segment more than twice as long as other 2 and middle spine slightly longer than outermost spine. Innermost spine slightly curved, but straight (fig. 94d) on 1 leg of 1 specimen. Leg 5 (fig. 94e), located at posteroventral surface of genital complex, a rounded process tipped with 3 plumose setae, in addition to 1 small papilla equipped with single plumose seta.

Male. — Body as in fig. 94f. Total length 2.61 mm (2.39-2.69 mm) (n = 9). Cephalothorax similar in shape to that of female, measuring 1.70 mm (1.56-1.83 mm) × 1.46 mm (1.29-1.59 mm).

Fourth pedigerous somite wider than long, 0.20 mm (0.17-0.23 mm) × 0.38 mm (0.33-0.40 mm). Genital complex oval, longer than wide, and measuring 0.48 mm (0.40-0.53 mm) × 0.49 mm (0.43-0.53 mm) excluding legs 5 and 6. Spermatophores indicated by dotted lines in fig. 94f. Abdomen (fig. 95a) consisting of 1 free somite, 0.25 mm (0.20-0.27 mm) × 0.25 mm (0.23-0.27 mm).

Caudal ramus (fig. 95a) similar to that of female, but shorter, 113 × 92 μm. Body surface with small setules similar to that of female.

Antennule like that of female, except 2 additional small setae (at positions indicated by dots in fig. 91e). Antenna (fig. 95b, c) 3-segmented. First and second segments with numerous adhesion pads as figured. Claw (fig. 95d, e) 146 μm along it axis, recurved, and equipped with 1 seta and 2 flanges on ventral surface, 1 seta on inner margin at midlength, and conical flange on dorsal surface. Postantennal process (fig. 95f) similar to that of female, but slightly sharper.

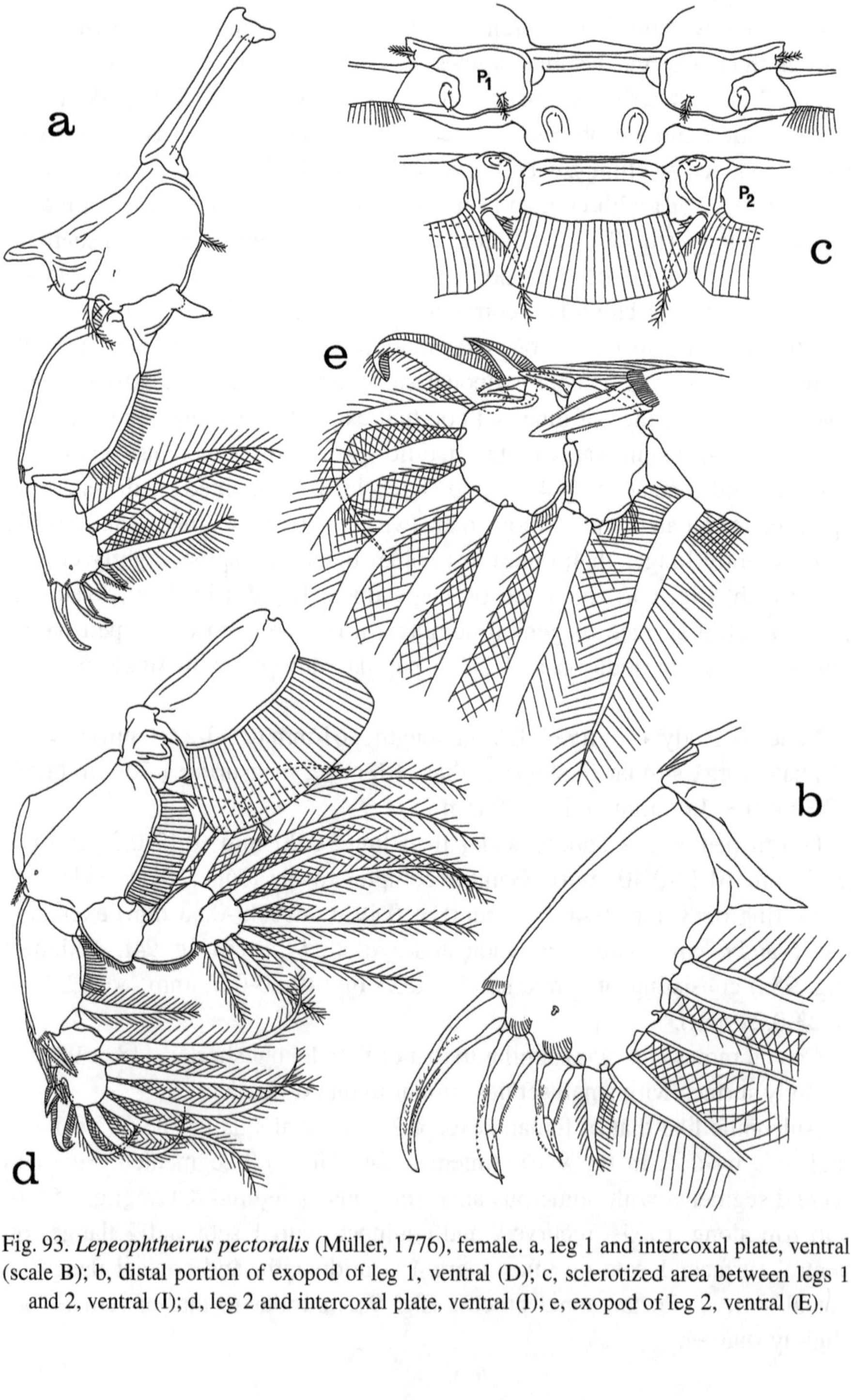

Fig. 93. *Lepeophtheirus pectoralis* (Müller, 1776), female. a, leg 1 and intercoxal plate, ventral (scale B); b, distal portion of exopod of leg 1, ventral (D); c, sclerotized area between legs 1 and 2, ventral (I); d, leg 2 and intercoxal plate, ventral (I); e, exopod of leg 2, ventral (E).

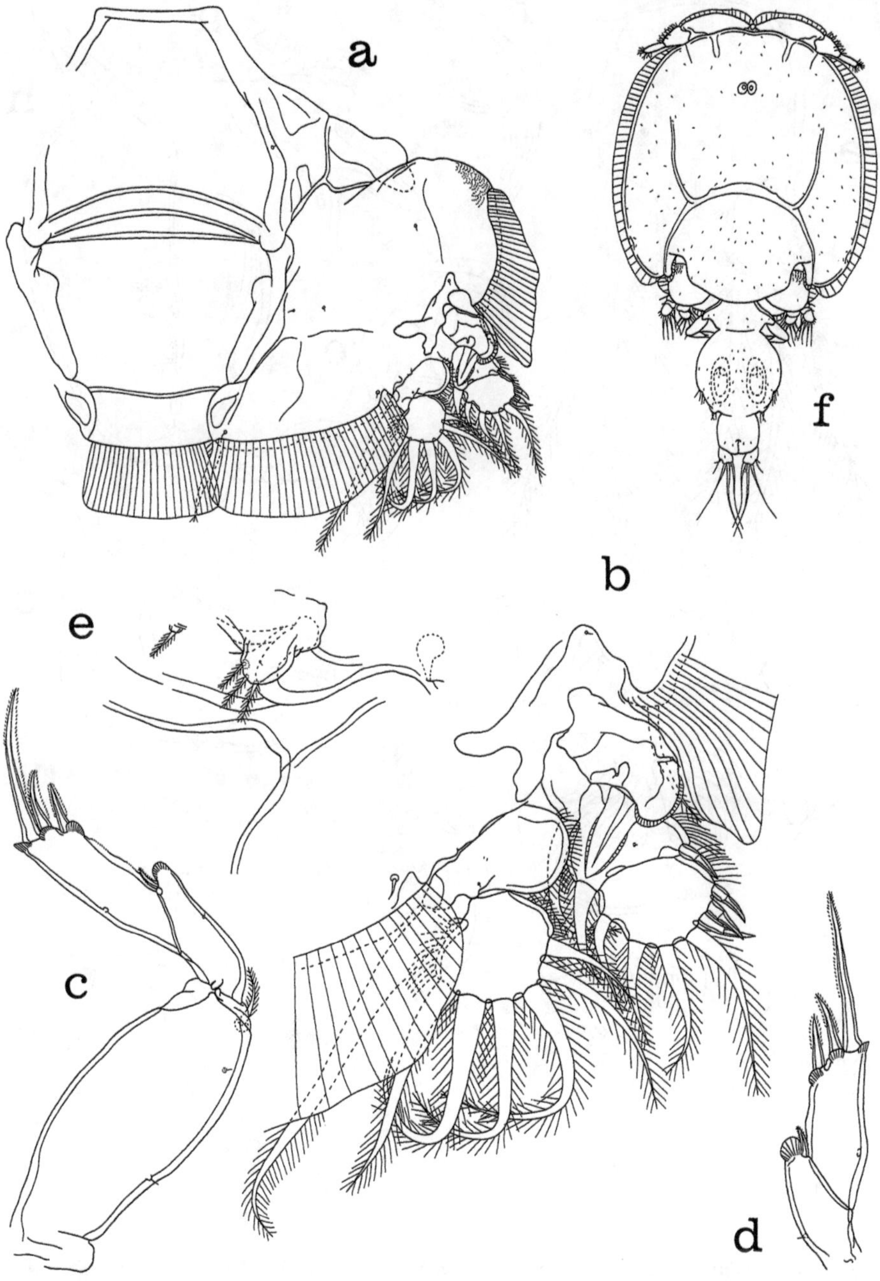

Fig. 94. *Lepeophtheirus pectoralis* (Müller, 1776). Female. a, leg 3, ventral (scale I); b, exopod and endopod of leg 3, ventral (E); c, leg 4, ventral (J); d, exopod of leg 4, ventral (J); e, leg 5 and genital area (I). Male. f, body, dorsal (BB).

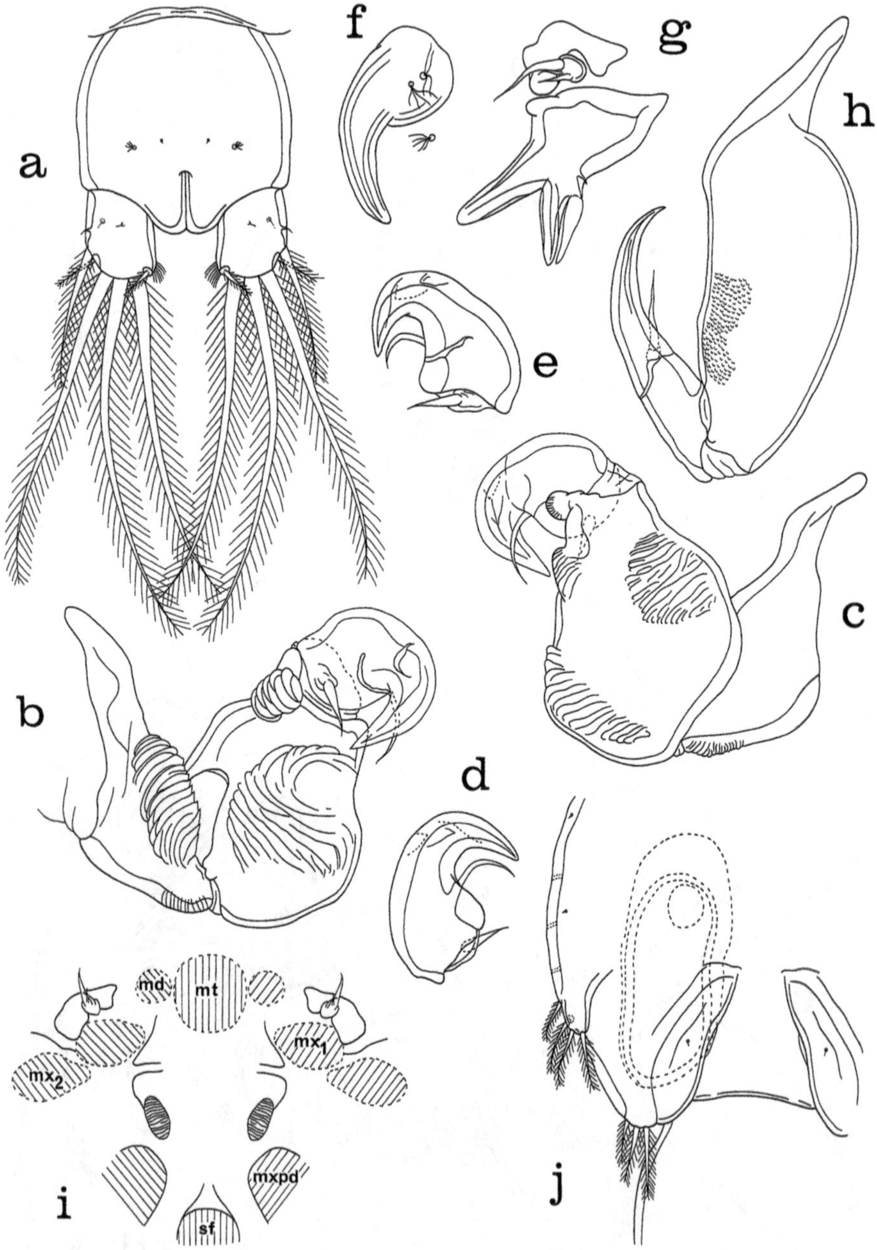

Fig. 95. *Lepeophtheirus pectoralis* (Müller, 1776), male. a, anal somite and caudal rami, dorsal (scale G); b, antenna, ventro-inner (E); c, same, dorso-outer (E); d, claw of antenna, dorsal (E); e, same, ventral (E); f, postantennal process, ventral (J); g, maxillule, ventral (J); h, maxilliped, dorsal (J); i, corrugated pad and position of mouth parts, ventral (B); j, legs 5 and 6, and genital area, ventral (G).

Mouth tube and mandible as in female. Maxillule (fig. 95g) with lateral tine longer than medial. Posteriorly directed process at base rounded. Hyaline process and minute knob located near base of medial tine. Maxilla like that of female. Maxilliped (fig. 95h) possessing more slender corpus with only 1 patch of denticles, but shorter claw, 270 μm, with relatively longer seta. Position of mouth parts indicated in fig. 95i, with pair of oval corrugated pads situated posteromedially to bases of maxilla. Sternal furca as in female, but often having curved tines (see Kabata, 1979, fig. 637).

Legs 1-4 as in female. Leg 5 (fig. 95j) positioned ventrolaterally on genital complex, somewhat triangular, with 1 lateral plumose seta at its base, and 3 plumose setae at its apex. Leg 6 a swelling with 3 plumose setae.

Remarks. — *Lepeophtheirus pectoralis* has been known now for over two centuries. Like so many organisms that have been known for so long, this species has a long list of citations in the literature. Unfortunately, as Boxshall (1974a) mentioned, many of these records may be misidentifications of *Lepeophtheirus thompsoni* Baird, 1850, a parasite of members of the flatfish family Bothidae. Although a relatively complete synonymy is given for this species, many of the earlier citations may be unreliable.

Müller (1776) was the first to describe this species. At that time, however, it was placed in the genus *Lernaea*, a group of highly modified copepods parasitic on freshwater fishes. It was then transferred to the genus *Caligus* presumably by Cuvier (1817), and finally was established as the type of the genus *Lepeophtheirus* (Nordmann, 1832).

The most recent redescriptions of *L. pectoralis* have been provided by Boxshall (1974a) and Kabata (1979). We have included a thorough description for the sake of completeness of the revision, to confirm or amend the work of previous investigators, and to clarify details of some appendages.

The developmental stages (Boxshall, 1974b), population dynamics dealing with seasonal variation and dispersion pattern (Boxshall, 1974c, e), histopathological effects on the host (Boxshall, 1977), and reproductive behavior (Anstensrud, 1990a, 1990b, 1990c, 1990d, 1992) have been studied for this species.

Lepeophtheirus pectoralis is primarily found in the Atlantic waters off Europe, in the White Sea, Iceland, and Faroes (Boxshall, 1974a; Kabata, 1979). This species is predominantly parasitic on species of the flatfish family Pleuronectidae, although *L. pectoralis* did occur fortuitously on *Raja clavata* and *Gadus morhua* (Boxshall, 1974d). Kabata (1979) listed four pleuronectid species as favored hosts for this parasite. They are the flounder *Platichthys flesus* (Linnaeus); the plaice *Pleuronectes platessa* Linnaeus; the dab *Limanda limanda* (Linnaeus); and the Arctic flounder *Liopsetta glacialis* Pallas.

The ovigerous females of this species are primarily found on the pectoral and pelvic fins of their hosts. However, according to Boxshall (1974d) the nonovigerous females, males, and juveniles were found on the external body surface and fins, and the wall of the branchial cavity.

Lepeophtheirus curtus (Wilson, 1913)
(figs. 96-98)

Dentigryps curtus Wilson, 1913: 222; Wilson, 1935a: 331; Heegaard, 1945b: 15; Yeatman, 1957: 346; Yamaguti, 1963: 81; Lewis, 1964a: 204; Lewis, 1964b: 358.
Homoiotes bermudensis Heegaard, 1943: 21; Heegaard, 1945b: 15.
Dentigryps bermudensis Yamaguti, 1963: 81.
Lepeophtheirus curtus Hewitt, 1971b: 333; Dojiri, 1983: 232; Luque et al., 1999: 12.

Material examined. — One female holotype (USNM 43595) and 2 paratypes (both females) (USNM 42328) from the mouth and branchial cavity of yellowfin grouper, *Mycteroperca venenosa apua* (Bloch), at Bermuda Island on 27 July 1903. Twelve females (USNM 69789) from "skin" (external body surface) of *Mycteroperca venenosa* (Linnaeus) at Dry Tortugas, Florida, collected by O. L. Williams, identified by C. B. Wilson, and confirmed by A. G. Lewis. All material on loan from National Museum of Natural History, Smithsonian Institution, Washington, D.C.

Female. — Body as in fig. 96a. Total length (not including setae on caudal ramus) 3.14 mm (2.06-3.49 mm) (n = 7). Cephalothorax subcircular, narrower anteriorly, slightly longer than wide, 2.39 mm (2.26-2.59 mm) × 2.28 mm (2.09-2.39 mm). Posterior sinuses relatively shallow. Anterolateral portion of cephalothorax, near base of antennule, with conspicuous notch. Free margin of thoracic zone of cephalothorax extending beyond posterior limit of lateral zone and covering most of fourth pedigerous somite. Frontal plate distinct, with no lunules. Tip of antennule within lateral limit of cephalothorax.

Fourth pedigerous somite partially hidden in dorsal view by posterior extension of thoracic zone of cephalothorax, wider than long, 0.17 mm (0.13-0.21 mm) × 0.65 mm (0.60-0.66 mm). Genital complex much wider than long, 0.60 mm (0.53-0.63 mm) × 1.23 mm (1.13-1.33 mm), with conical posterolateral processes, and with 1 pair of small digitiform processes on posterior margin dorsal to insertion of egg sacs. Abdomen (fig. 96b) consisting of 1 free somite, 0.22 mm (0.17-0.24 mm) × 0.20 mm (0.18-0.22 mm) (excluding posterolateral processes). Posterolateral process of abdomen rounded, approximately 84 μm (69-92 μm) × 34 μm.

Caudal ramus (fig. 96b) 3 times as long as wide, 167 × 54 μm, wider posteriorly, and bearing 6 plumose setae.

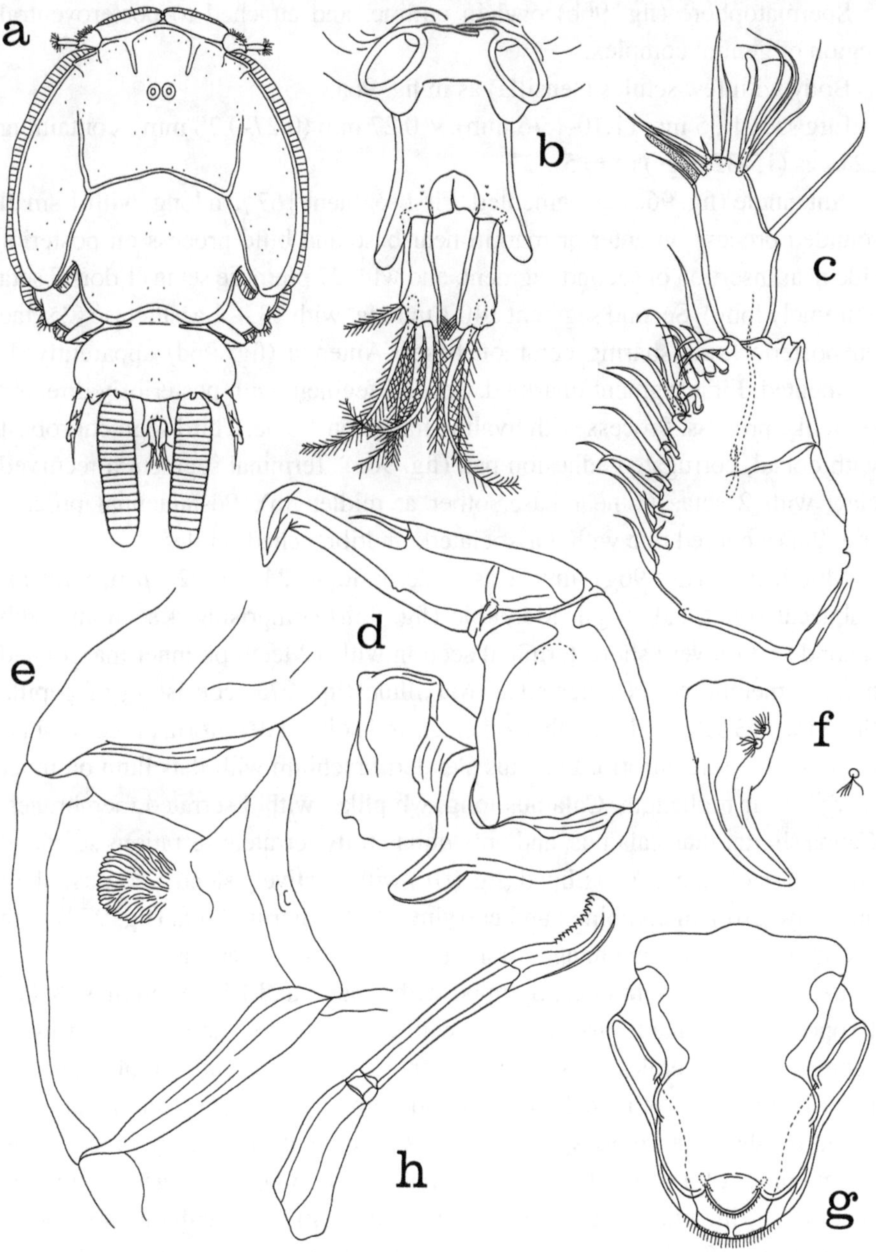

Fig. 96. *Lepeophtheirus curtus* (Wilson, 1913), female. a, body, dorsal (scale AA); b, anal somite and caudal rami (left side missing 4 setae), ventral (B); c, antennule (plumosities not drawn), ventral (E); d, antenna, ventral (E); e, third segment (with corrugated pad) of antenna, dorsal (P); f, postantennal process, ventral (E); g, mouth tube, ventral (O); h, mandible, ventral (P).

Spermatophore (fig. 96b) oval in outline, and attached to posteroventral region of genital complex.

Body with few setules (sensilla) as in fig. 96a.

Egg sacs 1.35 mm (1.10-1.76 mm) × 0.27 mm (0.27-0.28 mm), containing 22 eggs (15-32 eggs) (n = 5).

Antennule (fig. 96c) 2-segmented. First segment 167 μm long, with 1 small rounded process on anterior margin near base and bifid process on posterior side near insertion of second segment, and with 27 plumose setae (1 dorsal seta extremely long). Second segment 140 μm long, with 13 + 1 aesthetasc (2 setae on posterior side sharing common base). Antenna (fig. 96d) apparently 4-segmented. First segment unarmed. Second segment with posteriorly directed spinelike process. Process with hyaline flange on 1 side. Third segment robust with dorsal, corrugated adhesion pad (fig. 96e). Terminal segment a recurved claw with 2 setae (1 near base, other at midlength). Postantennal process (fig. 96f) a curved tine with 3 associated, multibranched setules.

Mouth tube (fig. 96g) almost as wide as long, 243 × 221 μm, with intrabuccal stylet and strigil. Mandible (fig. 96h) comprising 4 sections, with second section very short; terminal section with 12 teeth on inner margin and hyaline membrane on outer edge. Maxillule (fig. 97a) consisting of papilla tipped with 3 setae (1 larger than other 2), and pointed dentiform process. Maxilla (fig. 97b) brachiform. Lacertus slender; brachium with flabellum on inner margin near midlength. Calamus long, whiplike, with 3 serrated membranes. Canna shorter than calamus, and only unilaterally serrated; serrations surround entire base of canna. Maxilliped (fig. 97c) with relatively slender corpus. Shaft and claw 216 μm combined and carrying 1 seta. Sternal furca (fig. 97d) with divergent tines; hyaline flanges along both margins of each tine.

Leg 1 (fig. 97e) biramous. Sympod with 1 outer and 1 inner plumose setae. Exopod 2-segmented. First segment with outer distal spine and inner row of setules; row of setules only on distal end of segment, and not along entire inner margin as in other caligids. Second segment (fig. 97f) with pinnate seta 4 shorter than outermost spine. Spines 2 and 3 with accessory processes at distal ends. All 3 spines bilaterally serrated. Inner margin of second segment of exopod with 3 plumose setae. Endopod (fig. 97g) small with minute process at tip. Leg 2 (fig. 98a) biramous, with usual armature. All 3 exopodal spines relatively large and bilaterally serrated. First exopodal spine with pectinate membrane at base. Two large setae next to exopodal spine of terminal segment; 1 bilaterally possessing transparent striated membrane, other with transparent membrane on outer margin and plumosities on inner margin. Sympod of leg 3

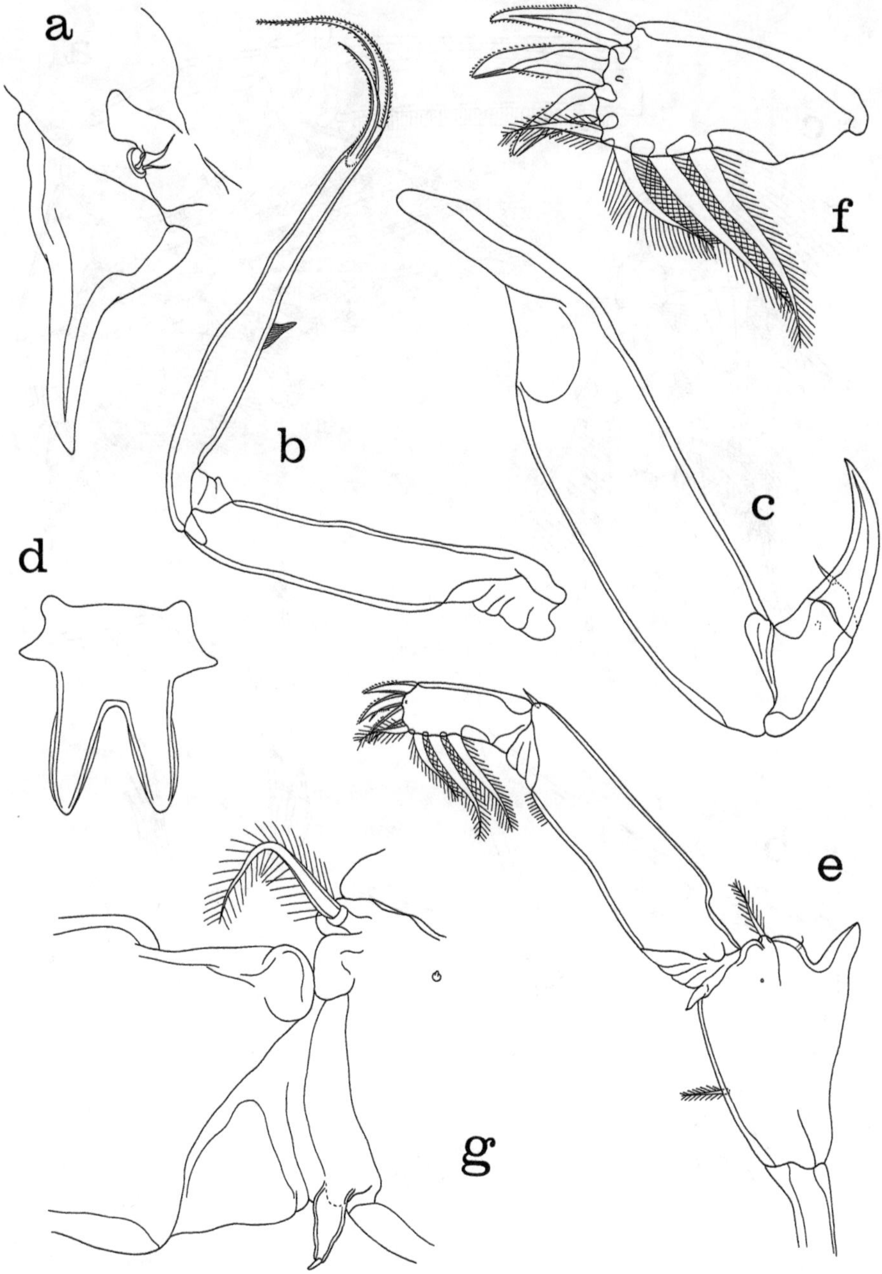

Fig. 97. *Lepeophtheirus curtus* (Wilson, 1913), female. a, maxillule, ventral (scale E); b, maxilla, dorsal (N); c, maxilliped, dorsal (O); d, sternal furca, ventral (O); e, leg 1, ventral (N); f, terminal exopodal segment of leg 1, ventral (R); g, sympod-exopod joint of leg 1, ventral (L).

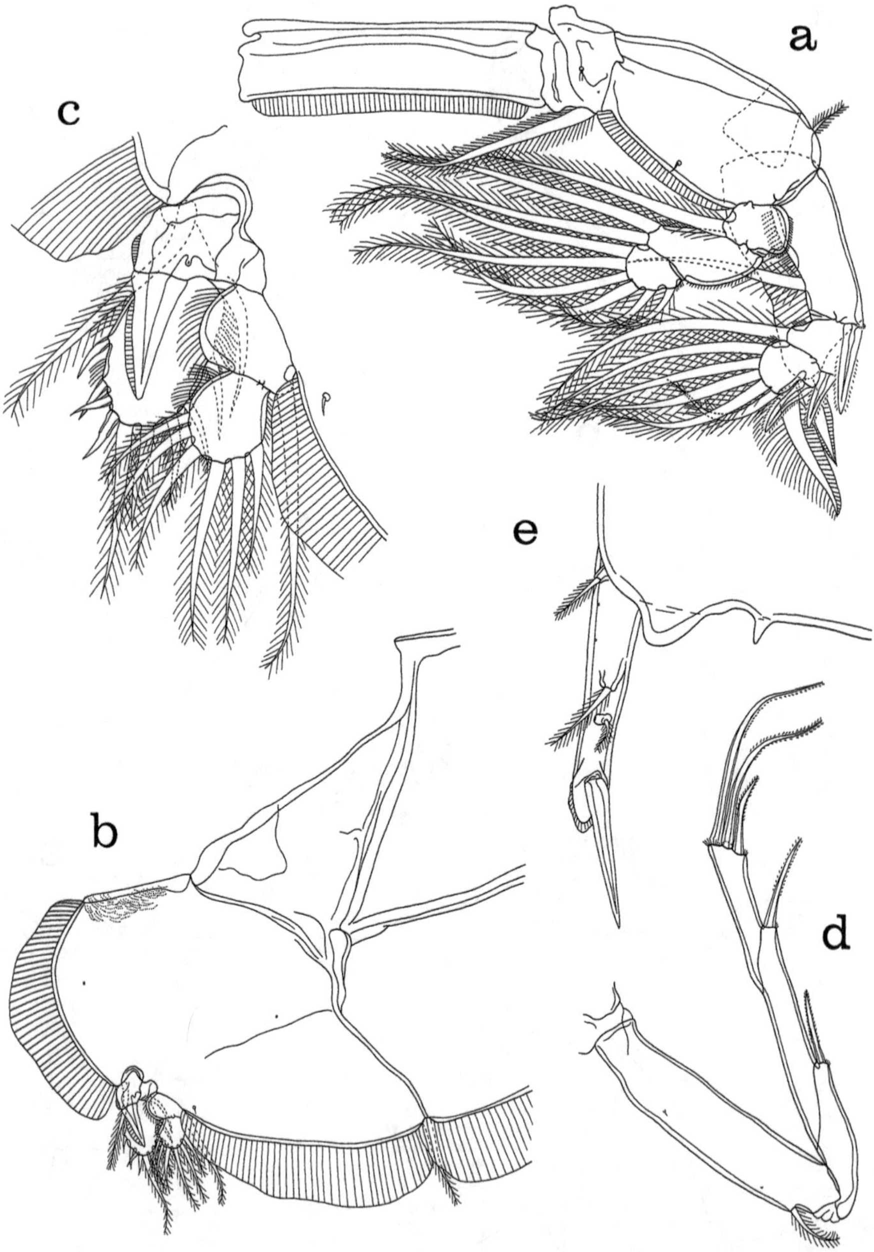

Fig. 98. *Lepeophtheirus curtus* (Wilson, 1913), female. a, leg 2 and intercoxal plate, ventral (scale B); b, leg 3, ventral (I); c, exopod and endopod of leg 3, ventral (D); d, leg 4, dorsal (B); e, leg 5, dorsal (B).

(fig. 98b) with corrugated pad on dorsal surface of anterolateral corner. One long plumose seta (usual seta on outer margin of basis) near insertion of exopod. Exopod and endopod (fig. 98c) 2-segmented. First segment of exopod with large straight spine on outer side and 1 plumose seta on inner side; basal swelling and exopodal spine with hyaline flanges on outer margins. Second segment of exopod with outer row of setules, 4 short spines, 5 plumose setae and inner row of setules. First segment of endopod with outer row of setules and 1 inner plumose seta. Second segment with outer and inner row of setules and 6 plumose setae. Leg 4 (fig. 98d) uniramous, brachiform. Sympod slender, carrying distal plumose seta. Exopod 3-segmented. Three spines of terminal segment decreasing in length from inner to outer margin. Innermost spine only unilaterally serrated; remaining 4 spines bilaterally serrated. All 5 spines with pectinate membrane at base. Leg 5 (fig. 98e) on posterolateral corner of genital complex, long and spikelike, extending beyond posterior limit of abdomen, and carrying 3 plumose setae and 1 large terminal spine. Leg 6 (fig. 98e) may be represented by small conical process medial to base of leg 5.

Male. — As described by Lewis (1964b).

Remarks. — *Lepeophtheirus curtus* was first described under the binomen *Dentigryps curtus* by Wilson (1913) from the grouper *Mycteroperca venenosa apua* (Bloch) collected in Bermuda. Wilson (1935a) obtained 17 additional females from *M. venenosa apua* from Dry Tortugas, Florida. Yeatman (1957) provided a redescription of this species, which he collected from two species, *Mycteroperca tigris* (Valenciennes) and *M. venenosa apua*, at Bermuda. Although Yeatman's redescription is much more accurate than Wilson's original account, it lacks the detail found in the redescription of Lewis (1964b). Lewis examined several specimens from an unidentified species of grouper from Tennessee Reef and Ft. Lauderdale, Florida along with specimens collected by O. L. Williams (USNM 69789) and ones collected by C. B. Wilson (USNM 64040). In addition, Lewis provided a complete description of the male.

Heegaard (1943) described a new species, *Homoiotes bermudensis* from "*Thynnus pelamys*" [= *Euthynnus pelamis* (Linnaeus) or *Katsuwonus pelamis* (Linnaeus)] in Bermuda. In a subsequent paper, Heegaard (1945b) synonymized *H. bermudensis* with *Dentigryps curtus*. The male was first described by Heegaard (1943) as *Homoiotes bermudensis*.

Hewitt (1971b) synonymized *Dentigryps* with *Lepeophtheirus* (see discussion on *Dentigryps*). A list of hosts and localities of collection of the seven species formerly assigned to *Dentigryps* is provided (table XII).

TABLE XII

Hosts and localities of collections of species formerly assigned to *Dentigryps* Wilson, 1913 (now considered members of *Lepeophtheirus* Nordmann, 1832)

Species	Host	Locality	Reference
L. curtus (Wilson, 1913)	*Mycteroperca venenosa apua* (Bloch)	Bermuda	Wilson, 1913
	Mycteroperca venenosa apua	Dry Tortugas, Florida	Wilson, 1935a
	Thynnus pelamys [= *Katsuwonus pelamis* (Linnaeus)]	St. Davis Island, Bermuda	Heegaard, 1943 (reported under the binomen *Homoiotes bermudensis*)
	Mycteroperca venenosa apua	Bermuda	Yeatman, 1957
	Mycteroperca tigris (Valenciennes)	Bermuda	Yeatman, 1957
	unidentified grouper	Florida	Lewis, 1964b
L. spinifer Kirtisinghe, 1937	*Chorinemus* (= *Scomberoides*) sp.	Ceylon (Sri Lanka)	Kirtisinghe, 1937
	Rachycentron canadum (Linnaeus)	India	Rangnekar, 1959
	Rachycentron canadum	India	Pillai, 1961
	Chorinemus tala [= *Scomberoides tala* (Cuvier)]	India	Pillai, 1961
	Chorinemus (= *Scomberoides*) sp.	Ceylon (Sri Lanka)	Kirtisinghe, 1964
	Chorinemus lysan [= *Scomberoides lysan* (Forsskål)]	India	Pillai, 1966
L. lichiae Barnard, 1948	*Lichia amia* (Linnaeus)	Natal, South Africa	Barnard, 1948
L. lewisi (Lewis, 1964) (name proposed for *Dentigryps bifurcatus* Lewis, 1964 by Hewitt, 1971)	*Acanthurus olivaceus* Bloch & Schneider	Honolulu Aquarium, Hawaii	Lewis, 1964a
	Naso hexacanthus (Bleeker)	Oahu, Hawaii	Lewis, 1964a
	Acanthurus triostegus sandvicensis Streets	Oahu, Hawaii	Lewis, 1964a
	Fistularia petimba Lacépède	Oahu, Hawaii	Lewis, 1964b
	Myripristis sp.	Oahu, Hawaii	Lewis, 1964b
	Bodianus bilunulatus (Lacépède)	Honolulu Aquarium, Hawaii	Lewis, 1967

TABLE XII
(Continued)

Species	Host	Locality	Reference
	Chaetodon fremblii Bennett	Oahu, Hawaii	Lewis, 1967
	Aulostomus chinensis (Linnaeus)	Oahu, Hawaii	Lewis, 1967
L. litus (Lewis, 1964)	*Plectropomus leopardus* (Lacépède)	Enewetak Atoll	Lewis, 1964b
	Epinephelus fuscoguttatus (Forsskål)	Enewetak Atoll	Lewis, 1964b
	Aulostomus chinensis	Enewetak Atoll	Lewis, 1964b
	Balistoides viridescens (Bloch)	Enewetak Atoll	Lewis, 1964b
	Cromileptes altivelis (Cuvier & Valenciennes)	Heron Island, Australia	Kabata, 1965a
	Plectropomus leopardus	Wistari Reef, Australia	Ho & Dojiri, 1977
L. ulua (Lewis, 1964)	*Caranx melampygus* (?) Cuvier & Valenciennes	Oahu, Hawaii	Lewis, 1964b, 1967
	Caranx ignobilis (Forsskål)	Heron Island, Australia	Ho & Dojiri, 1977
L. longicauda (Cressey, 1966)	*Caranx hippos* (Linnaeus)	Nosy Bé, Madagascar	Cressey, 1966

Lepeophtheirus parvicruris Fraser, 1920
(figs. 99-101)

Lepeophtheirus parvicruris Fraser, 1920: 56; Yamaguti, 1963: 76; Kabata, 1973a: 743; Dojiri, 1983: 237; Kabata, 1988: 18; Poly & Mah, 2002: 1123.
Pseudolepeophtheirus longicauda (?) Markewitsch, 1940: 14; Markewitsch, 1956: 142; Yamaguti, 1963: 84.

Material examined. — Eleven females from branchial cavity of starry flounder, *Platichthys stellatus* (Pallas) (USNM 54088), from Prince William Sound, Alaska. This material obtained from fish housed in ichthyological collection at National Museum of Natural History, Smithsonian Institution, Washington, D.C.

Female. — Body as in fig. 99a. Total length (not including setae on caudal ramus) 7.43 mm (7.01-8.27 mm) (n = 6). Cephalothorax subcircular, 2.58 mm (2.36-2.82 mm) × 2.25 mm (2.12-2.46 mm) with shallow posterior sinuses.

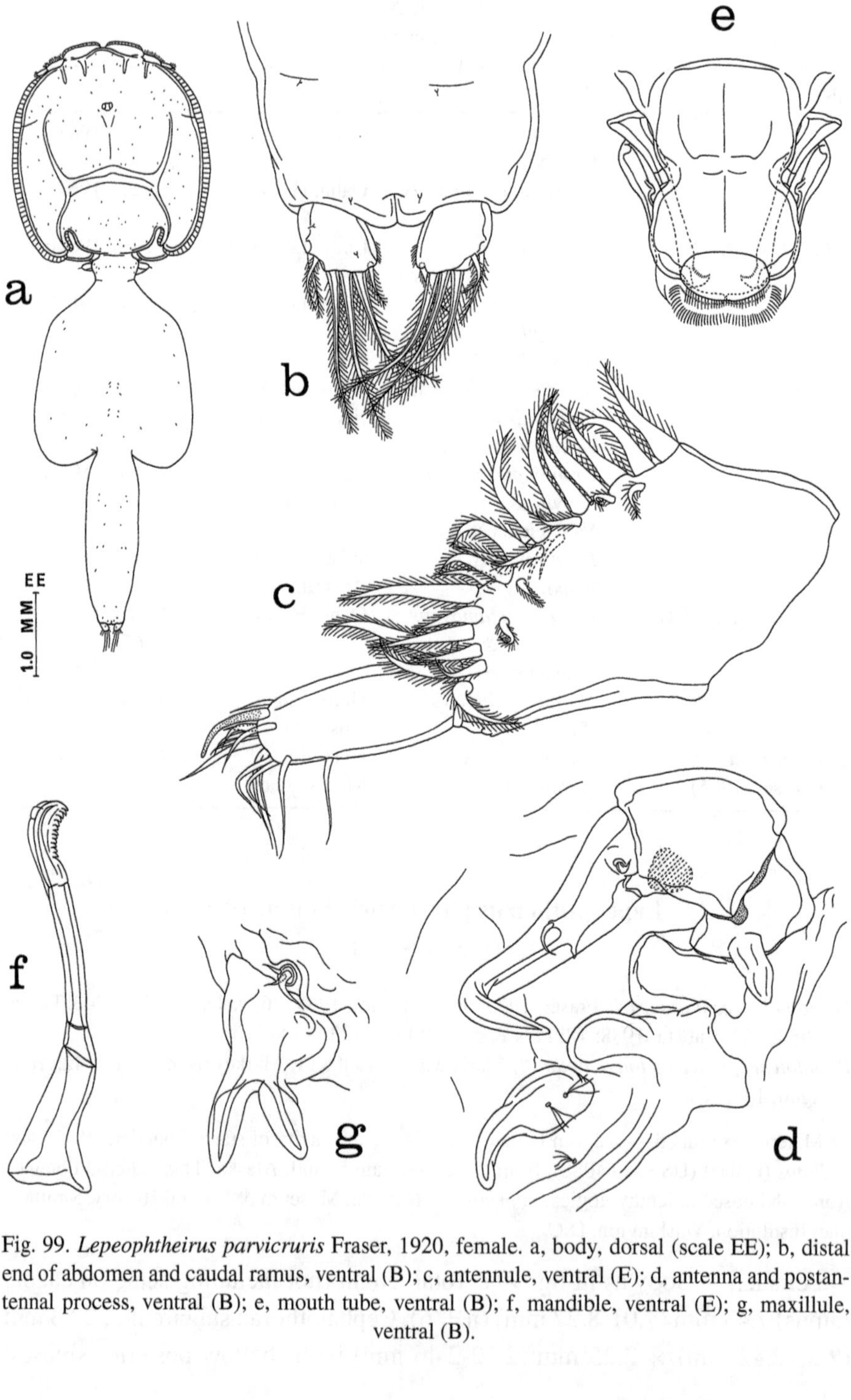

Fig. 99. *Lepeophtheirus parvicruris* Fraser, 1920, female. a, body, dorsal (scale EE); b, distal end of abdomen and caudal ramus, ventral (B); c, antennule, ventral (E); d, antenna and postantennal process, ventral (B); e, mouth tube, ventral (B); f, mandible, ventral (E); g, maxillule, ventral (B).

Free margin of thoracic zone at same level as posterior tip of lateral zone. Antennule not extending beyond lateral limit of cephalothorax.

Fourth pedigerous somite small, wider than long, and not clearly delimited from genital complex. Genital complex about as large as cephalothorax, 2.67 mm (2.46-2.89 mm) × 2.01 mm (1.86-2.22 mm), narrower anteriorly, and with rounded protrusions on posterolateral corners. Abdomen consisting of 1 free somite, much longer than wide, 2.33 mm (2.09-2.49 mm) × 0.68 mm (0.63-0.73 mm).

Caudal ramus (fig. 99b) little longer than wide, 119 × 86 μm, carrying 6 plumose setae.

Antennule (fig. 99c) 2-segmented, with first segment 189 μm long and possessing 27 plumose setae. Second segment cylindrical, 135 μm long, and with usual armature of 13 + 1 aesthetasc. Antenna (fig. 99d) apparently 4-segmented. Second segment with stout, posteriorly directed, spinelike process. Third segment with dorsal adhesion pad. Claw recurved and with usual 2 setae. Postantennal process (fig. 99d) curved and with 3 groups of associated setules.

Mouth tube (fig. 99e) longer than wide, 357 × 299 μm, with intrabuccal stylet present on inner surface of labrum. Strigil located on distal end of labium. Mandible (fig. 99f) composed of 4 sections, with 12 teeth on inner margin and folded (or ripped) transparent membrane on outer margin. Maxillule (fig. 99g) consisting of papilla and bifid process; papilla tipped with 3 setae. Medial tine longer than lateral, both possessing flanges. Small knob present near base. Maxilla (fig. 100a) brachiform, usual shape, except relatively large flabellum on brachium. Calamus longer than canna, both with usual serrated membranes. Maxilliped (fig. 100b, c) with indentation on corpus (may receive tip of claw when closed) and 2 patches of denticles. Shaft and claw curved, 541 μm long combined. Sternal furca (fig. 100d) with narrow base and broad, blunt, slightly diverging tines.

Pinnate seta at junction of sympod and exopod of leg 1 (fig. 100e) not covered by ventral protrusion. First segment of exopod with usual short seta on outer distal corner, and long setules on inner margin. Pinnate seta 4 shorter than outermost spine. Spines 1, 2, and 3 bilaterally spinulated, with pectinate membranes at their bases. Spines 2 and 3 with accessory processes at their distal ends. Endopod with 2 minute apical processes. Area between legs 1 and 2 (fig. 100f) with rounded protrusions bearing setules. Leg 2 (fig. 100g) of usual form. First spine of third segment of exopod not extending beyond distal limits of terminal segment. Apron of leg 3 (fig. 101a) with dorsal adhesion pad on lateral corner. Exopod spine with marginal flange, terminally located

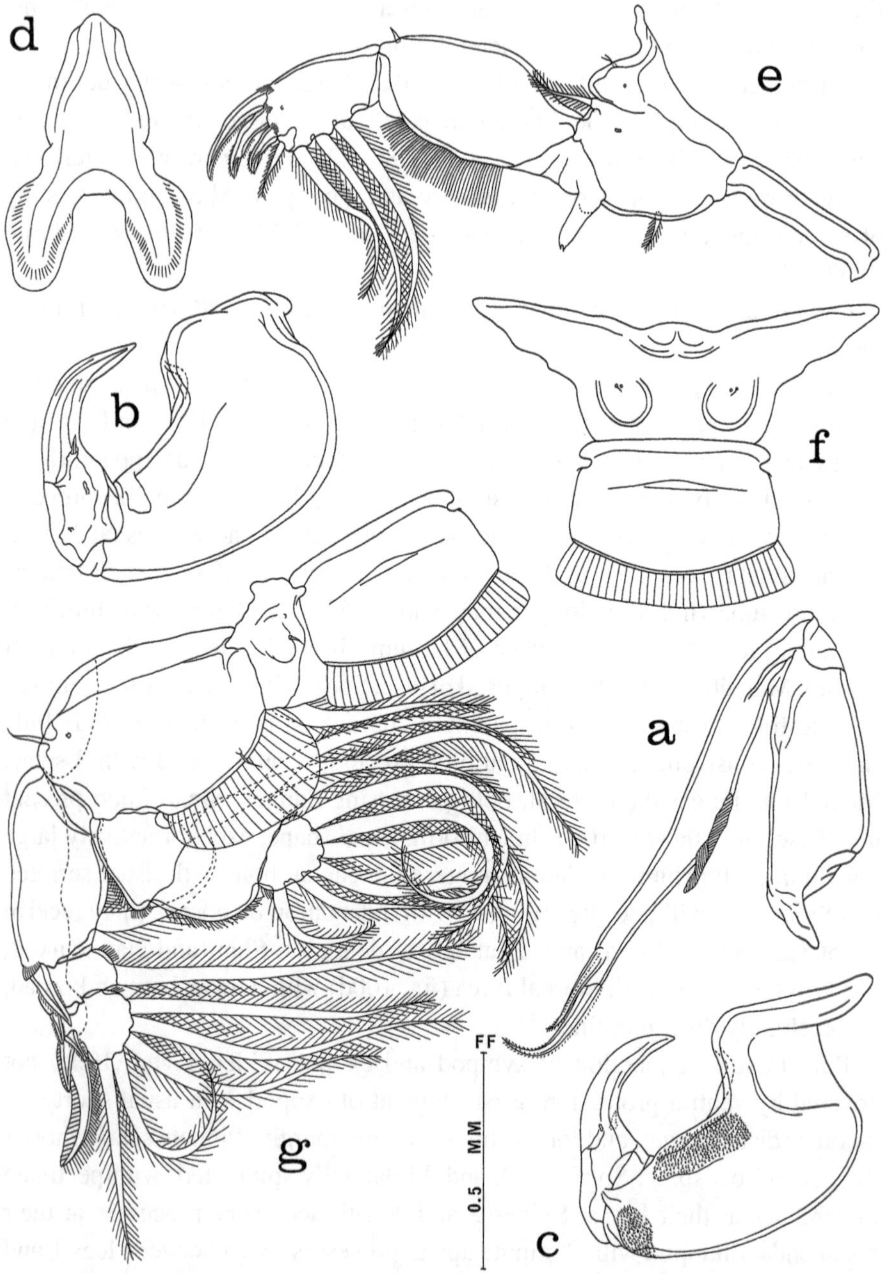

Fig. 100. *Lepeophtheirus parvicruris* Fraser, 1920, female. a, maxilla, dorsal (scale B); b, maxilliped, ventral (FF); c, same, dorsal (FF); d, sternal furca, ventral (B); e, leg 1 and intercoxal plate, ventral (B); f, sclerotized area between legs 1 and 2, and intercoxal plate of leg 2, ventral (B); g, leg 2 and intercoxal plate, ventral (B).

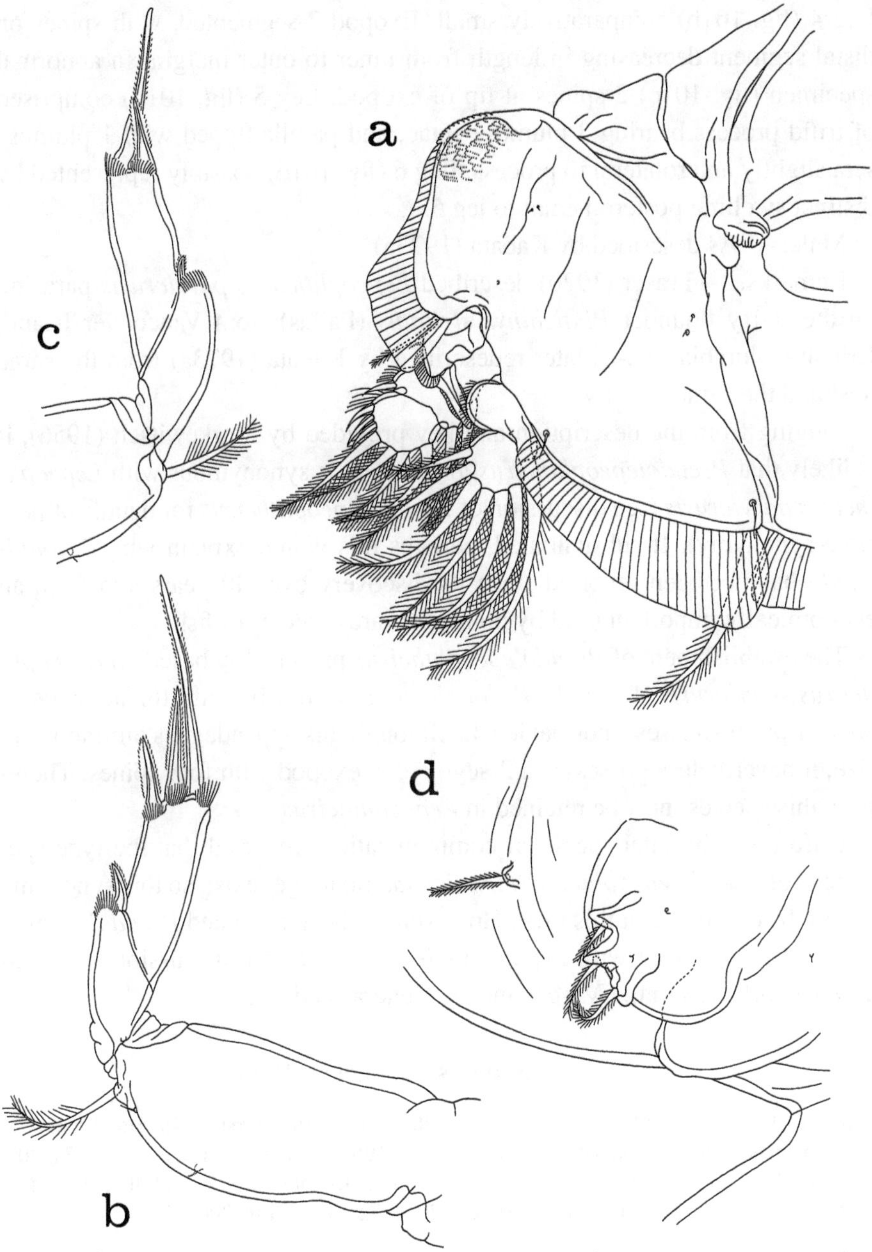

Fig. 101. *Lepeophtheirus parvicruris* Fraser, 1920, female. a, leg 3, ventral (scale B); b, leg 4, ventral (E); c, leg 4 with abnormal terminal armature, ventral (E); d, leg 5 and genital area, ventral (B).

on basal swelling. Hyaline membrane on outer distal corner of basal swelling. Leg 4 (fig. 101b) comparatively small. Exopod 2-segmented, with spines on distal segment decreasing in length from inner to outer margin. In abnormal specimen (fig. 101c) 2 spines at tip of exopod. Leg 5 (fig. 101d) comprised of trifid process bearing 3 plumose setae, and papilla tipped with 1 plumose seta slightly anterolateral to process. Leg 6 (fig. 101d) possibly represented by distinct swelling posteromedial to leg 5.

Male. — As described by Kabata (1973a).

Remarks. — Fraser (1920) described *Lepeophtheirus parvicruris* parasitic on the starry flounder *Platichthys stellatus* (Pallas) from Vancouver Island, British Columbia. It was later redescribed by Kabata (1973a) from the same host and the same locality.

Judging from the description and key provided by Markewitsch (1956), it is likely that *Pseudolepeophtheirus longicauda* is synonymous with *Lepeophtheirus parvicruris* (see discussion of *Pseudolepeophtheirus* for details of possible synonymy). If substantiated, perhaps this would explain why *P. longicauda* has not been reported since its discovery over 40 years ago from an economically important (and by no means rare) species of fish.

The establishment of *Pseudolepeophtheirus* presumably based on *Lepeophtheirus parvicruris* (formerly *P. longicauda*) would be unfortunate since *L. parvicruris* possesses a normal leg 4. Although this appendage is diminutive in size, it nevertheless possesses a 2-segmented exopod with four spines. Therefore, this species must be retained in *Lepeophtheirus*.

Unfortunately, Titar (personal communication) reported that the type-specimens of *Pseudolepeophtheirus longicauda* no longer exist, so this synonymy cannot be confirmed at this time. However, we have relegated *Pseudolepeophtheirus* to synonymy with *Lepeophtheirus* (refer to the discussion on *Pseudolepeophtheirus* in the Miscellaneous Genera section).

Genus **Mappates** Rangnekar, 1958

Mappates Rangnekar, 1958: 303; Yamaguti, 1963: 86; Kabata, 1964a: 648; Ho, 1966: 759; Kirtisinghe, 1964: 77; Kabata, 1965a: 30; Pillai, 1967: 363; Hameed & Pillai, 1973a: 407; Pillai, 1977: 58; Kabata, 1979: 158; Dojiri, 1983: 244; Prabha, 1983: 57; Pillai, 1985: 466; Kazachenko, 2001: 32; Boxshall & Halsey, 2004: 726; Ho & Lin, 2004: 292.

Female. — Cephalothorax suboval, comprising majority of total body length. Dorsal transverse rib (normally located at about midlength in other caligid genera) situated near level of posterior limit of lateral zone of cephalothorax. Posteromedian lobe of cephalothorax with rounded free margin extended posteriorly, completely covering fourth pedigerous somite and part or all of genital complex in dorsal view. Posterior sinuses hidden dorsally by posteromedian lobe of cephalothorax. Frontal plate distinct. Lunules absent. Fourth pedigerous somite hidden in ventral view by ventral apron of third leg. Abdomen small, but distinct.

Antennule, antenna, mouth tube, mandible, maxilla, and maxilliped as in *Lepeophtheirus* and *Caligus*. Postantennal process and sternal furca absent. Maxillule consisting of setiferous papilla and weakly sclerotized lobelike process.

Legs 1 and 2 as in *Lepeophtheirus* and *Caligus*. Leg 3 biramous, with 2-segmented rami. First segment of exopod represented by large spine; second exopodal segment with 6 or 7 setae. First segment of endopod a lobe with outer setules; second segment bearing 3 setae. Exopod of leg 4 without sutures indicating segmentation and possessing total of 5 long setiform spines. Leg 5 represented by papilla tipped with 3 setae and 1 seta somewhat removed from papilla.

Male. — Unknown.

Type-species. — *Mappates plataxus* Rangnekar, 1958.

Remarks. — Rangnekar (1958) established *Mappates* based on *M. plataxus*, parasitic on the gill filaments of *Platax teira* (Forsskål) in Bombay. The main distinctions among females of *Mappates*, *Pseudanuretes*, and *Eirgos* (= *Anuretes*) cited by Rangnekar in her remarks section are the extension of the median lobe of the cephalothorax (most highly developed in *Mappates*), segmentation of the abdomen (reported to be 2-segmented in *Mappates* in the original description, but subsequently found to be incorrect by Hameed & Pillai, 1973a), and the presence or absence of the dentiform processes (i.e., postantennal process, process of maxillule, and sternal furca).

Yamaguti (1963) established a new subfamily Mappatinae for *Mappates*. This subfamily cannot now be accepted (Ho, 1966; also refer to discussion on *Anuretes*).

Mappates remained a monotypic genus for only a few years, until the discovery of *M. alter* from *Platax novemaculeatus* [= *Zabidius novemaculeatus* (McCulloch)], a congener of the type-host, from Queensland, Australia (Kabata, 1964a). Although morphologically very similar to the type-species, several diagnostic characters separate *M. alter* from *M. plataxus* (e.g., the shape of the genital complex, the number of free abdominal somites, the size of the posteromedian area of the cephalothorax, the size of the process of the maxillule, the armature of the second exopodal segment of leg 3, and the shape of the posterior margin of the ventral apron).

Several authors have questioned the validity of *Mappates* (Kirtisinghe, 1964; Kabata, 1965a; Pillai, 1967; Hameed & Pillai, 1973a). The discussions centered around the vagueness of the generic definition of this and related genera, and the presumed absence of a distinct generic boundary separating *Mappates* from these genera [for a detailed discussion of this problem see Pillai (1967, 1977); also refer to discussion on *Anuretes*].

The unique feature of this genus, distinguishing it from all other genera of the Caligidae, is the unusual pattern displayed by the dorsal ribs (supporting rods) of the cephalothorax. The pattern is almost identical in the two known species. Other characteristics in combination help to identify *Mappates* from the remaining *Anuretes* group. These features include the weakly sclerotized, lobelike process of the maxillule, the segmentation and armature of legs 1-5, and the absence of the postantennal process and sternal furca (see Ho & Lin, 2004: 292). In addition, the hosts for the two known species of this genus are confamilial.

Mappates plataxus Rangnekar, 1958
(figs. 102-105)

Mappates plataxus Rangnekar, 1957: 12; Rangnekar, 1958: 303; Yamaguti, 1963: 86; Kabata, 1964a: 641; Ho, 1966: 759; Pillai, 1967: 362; Hameed & Pillai, 1973a: 403; Pillai, 1977: 53; Kabata, 1979: 162; Dojiri, 1983: 247; Prabha & Pillai, 1983: 36; Pillai, 1985: 467; Ho & Lin, 2004: 292; Samotylova, 2010: 99.
Eirgos plataxus Kirtisinghe, 1964: 77.

Material examined. — One paratype female (C 3516/1) from gill filaments of *Platax teira* (Forsskål) collected at Bombay Harbor. Material on loan from Zoological Survey of India, Calcutta, India.

Female. — Body as in fig. 102a, b. Total length (not including setae of caudal ramus) 3.29 mm (n = 1). Cephalothorax large in comparison to body, 3.19 × 1.93 mm, with no posterior sinuses, and greatly elongated thoracic zone covering anterior half of genital complex. Dorsal transverse rib almost at same level as posterior tip of lateral zone. Antennule not extending to lateral limit of cephalothorax.

Fourth pedigerous somite wider than long, approximately 0.13 × 0.53 mm, hidden in dorsal view by elongation of thoracic zone, and completely covered ventrally by apron of third leg. Genital complex ovoid, much wider than long, 0.66 × 1.16 mm. Abdomen (partially mutilated when received from museum) apparently fused to genital complex, and highly reduced in size.

Caudal ramus (fig. 102c) much longer than wide, 78 × 46 μm, and bearing 5 plumose setae.

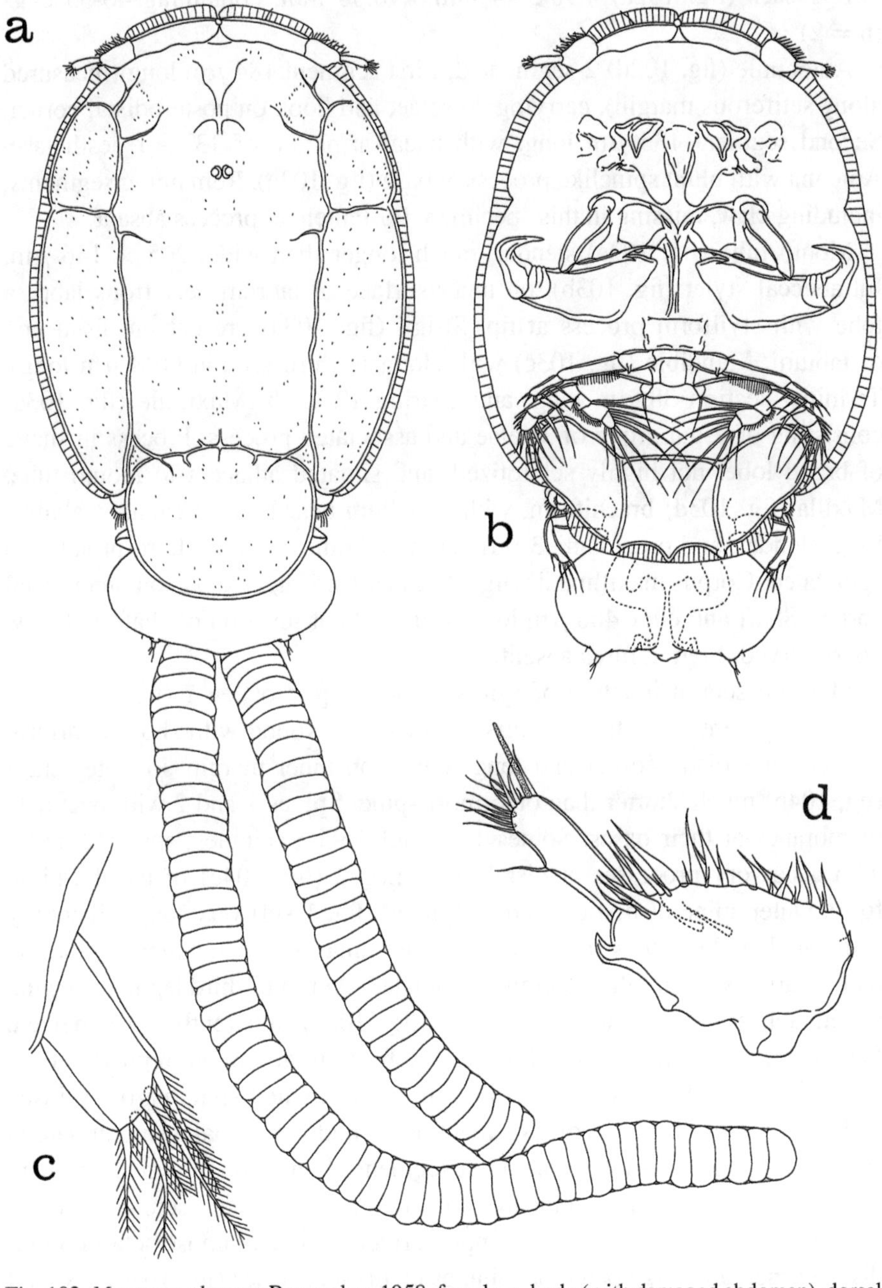

Fig. 102. *Mappates plataxus* Rangnekar, 1958, female. a, body (with damaged abdomen), dorsal (scale M); b, same (with antenna missing), ventral (M); c, caudal ramus, dorsal (P); d, antennule (plumosities not drawn), ventral (O).

Egg sacs (fig. 102a) 4.98-5.44 mm × 0.30 mm, containing 46-55 eggs (n = 2).

Antennule (fig. 102d) 2-segmented. First segment 184 μm long (measured along setiferous margin), carrying 26 setae, and hook on posterodistal corner. Second segment 135 μm long, with usual armature of 13 + 1 aesthetasc. Antenna with blunt spinelike process at base (fig. 103a). Remaining segments, including claw, missing in this specimen. Postantennal process absent.

Mouth tube (fig. 103a) slender, much longer than wide, 265 × 146 μm. Intrabuccal stylet (fig. 103b), on inner surface of labrum near frons labri, a lobe with styliform process at tip. Strigil (fig. 103a) present on distal end of labium. Mandible (fig. 103c) with elongate third section (108 μm long). Terminal section 38 μm long, and bearing 12 teeth. Maxillule (fig. 103a) consisting of usual group of 3 setae and associated process. Process in shape of broad lobe, not highly sclerotized, and situated adjacent to mouth tube. Maxilla (fig. 103d) brachiform, with flabellum near base of canna. Calamus long, slender, and possessing 3 serrated membranes. Canna short, broad, and spinulated. Corpus maxillipedis (fig. 103e) with corrugated pad on outer distal corner. Shaft and claw 460 μm long, usual seta at junction of shaft and claw not observed. Sternal furca absent.

Plumose seta at junction of sympod and exopod of leg 1 (fig. 104a) not covered by ventral protrusion. First segment of exopod with short spiniform seta at outer distal corner and long setules on inner margin. Pinnate seta 4 (fig. 104b) much shorter than outermost spine. Spines 1 and 2 with pectinate membranes at their bases. Spines 1, 2, and 3 with spinules; spines 2 and 3 with accessory processes at distal ends. Leg 2 (fig. 104c) of usual caligid form. Outer spine of first exopodal segment (fig. 104d) very long, bilaterally spinulated, and reaching to distal limit of terminal segment. Other 3 spines on outer margin shorter; all bilaterally spinulated. Apron of third leg not carrying dorsal adhesion pads. Rami of leg 3 (fig. 105a) apparently 2-segmented. First segment of exopod indistinguishably fused to apron, but represented by exopodal spine. Basal swelling and terminally situated spine partially fused and highly sclerotized. Second segment of exopod with usual 3 smooth lateral setae and 4 inner plumose setae. First segment of endopod unarmed, but with usual outer hairs. Distal segment with 3 plumose setae and setules on outer margin. Leg 4 (fig. 105b) with sympod armed with 1 plumose seta on outer distal corner, and large patch of spinules on dorsal surface (fig. 105c). Exopod bearing 5 setiform spines, with all 3 segments fused. Leg 5 (fig. 105d) a lobe armed with 3 plumose setae, situated on posterolateral protrusion of genital complex. Additional plumose seta situated near base of protrusion.

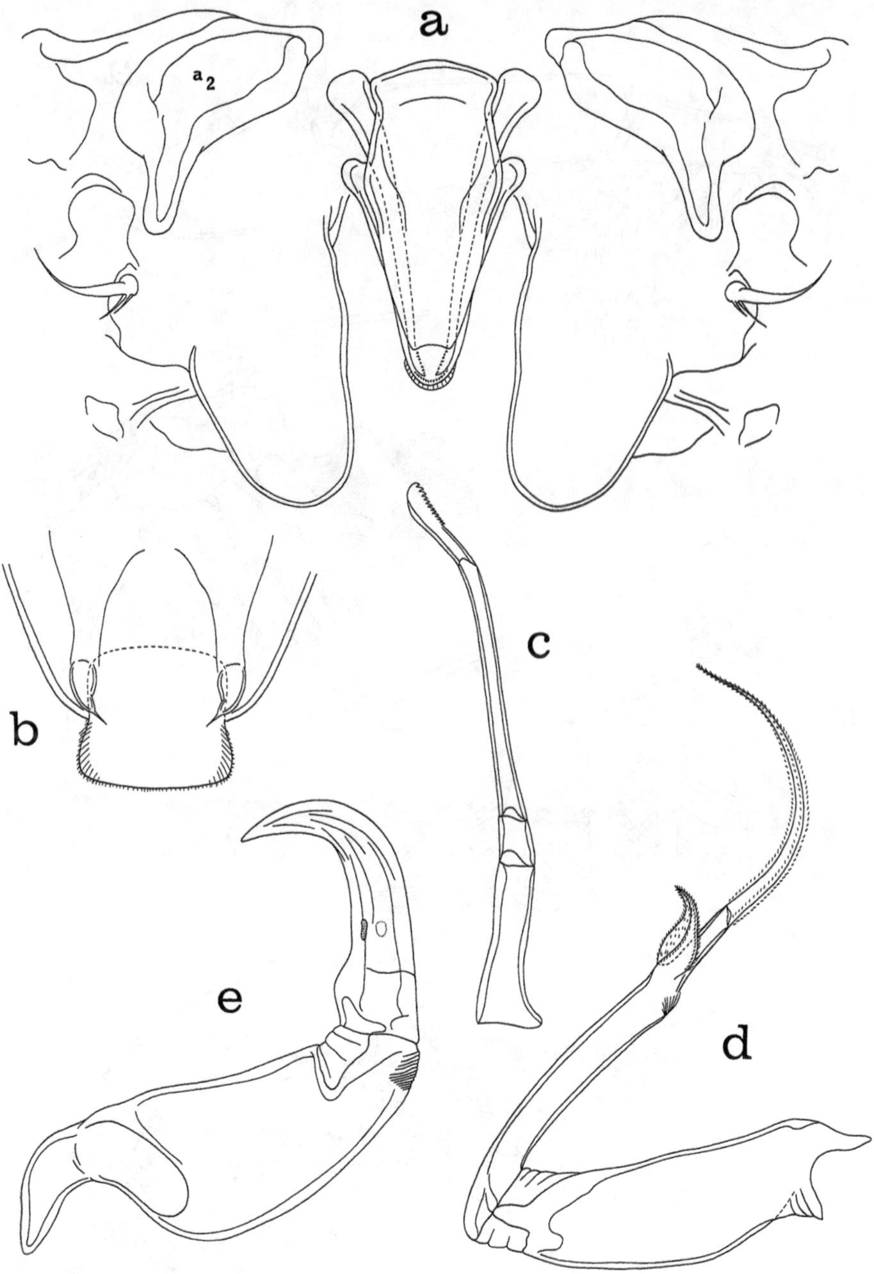

Fig. 103. *Mappates plataxus* Rangnekar, 1958, female. a, base of antenna (with posteriorly directed spinelike process), mouth tube, and maxillule, ventral (scale O); b, distal portion of labrum (with labium removed) and intrabuccal stylet, dorsal (W); c, mandible, ventral (P); d, maxilla, dorsal (N); e, maxilliped, dorsal (X).

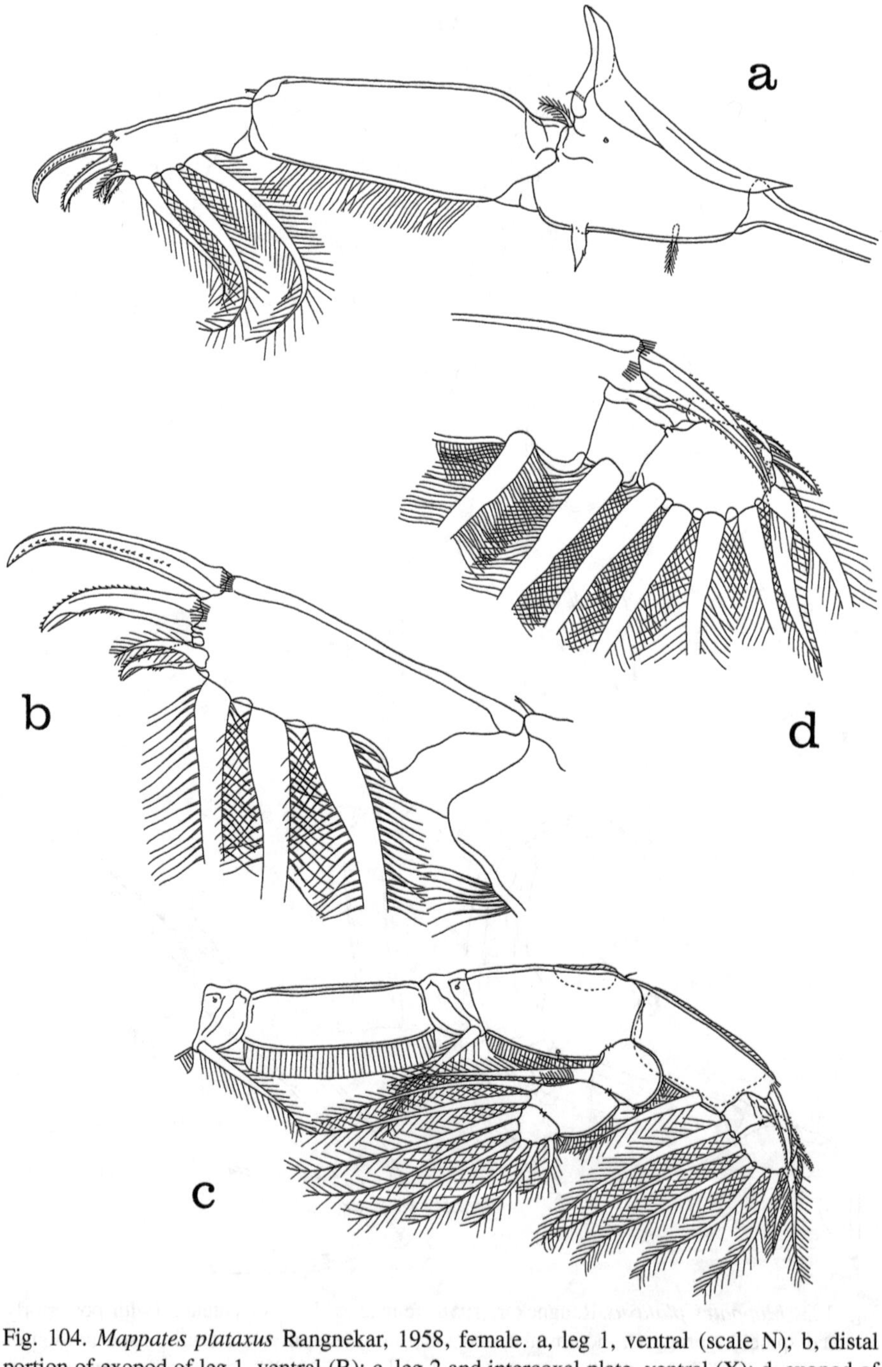

Fig. 104. *Mappates plataxus* Rangnekar, 1958, female. a, leg 1, ventral (scale N); b, distal portion of exopod of leg 1, ventral (R); c, leg 2 and intercoxal plate, ventral (X); d, exopod of leg 2, ventral (O).

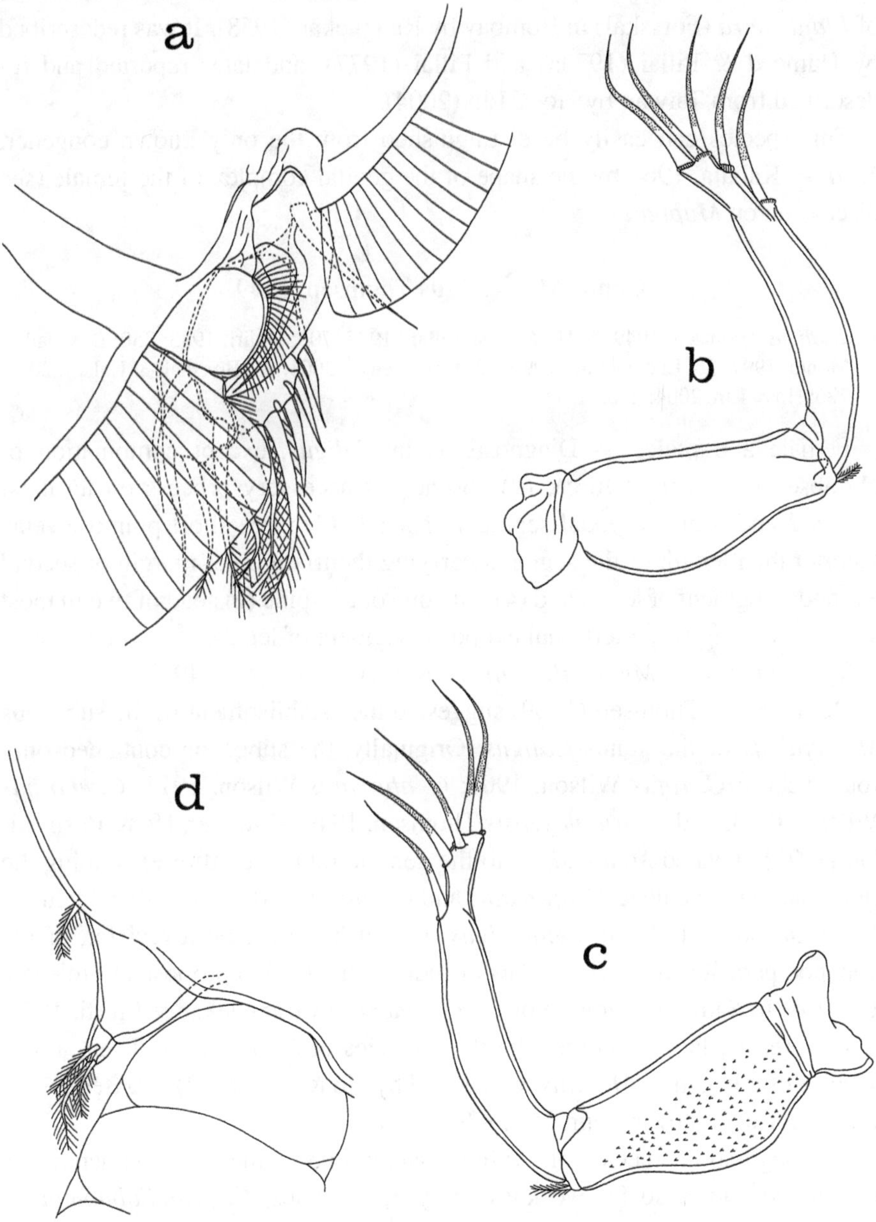

Fig. 105. *Mappates plataxus* Rangnekar, 1958, female. a, exopod and endopod of leg 3, ventral (scale R); b, leg 4, ventral (N); c, same, dorsal (N); d, leg 5 and genital area, dorsal (R).

Male. — Unknown.

Remarks. — *Mappates plataxus* was first described from the gill filaments of *Platax teira* (Forsskål) in Bombay by Rangnekar (1958). It was redescribed by Hameed & Pillai (1973a) and Pillai (1977), and later reported and re-described from Taiwan by Ho & Lin (2004).

This species can easily be distinguished from the only known congener, *M. alter* Kabata, 1964 by the shape of the genital complex of the female (see discussion on *Mappates*).

Genus **Metacaligus** Thomsen, 1949

Metacaligus Thomsen, 1949: 3; Ho & Bashirullah, 1977: 703; Dojiri, 1983: 250; Boxshall & Montú, 1997: 56; Luque et al., 1999: 12; Kazachenko, 2001: 25; Boxshall & Halsey, 2004: 726; Ho & Lin, 2004: 296.

Female and male. — Diagnosis as in *Caligus*, except combination of (1) absence of sternal furca; (2) absence of accessory process on terminal spines 2 and 3 on exopodal segment of leg 1; (3) three short plumose setae (shorter than length of the segment carrying them) on inner margin of second exopodal segment of leg 1; and (4) only one outer spine (instead of two in most species of *Caligus*) on terminal exopodal segment of leg 2.

Type-species. — *Metacaligus uruguayensis* Thomsen, 1949.

Remarks. — Thomsen (1949) suggested the establishment of the subgenus *Metacaligus* of the genus *Caligus*. Originally, the subgenus contained only four species: *C. rufus* Wilson, 1908; *C. afurcatus* Wilson, 1913; *C. enormis* Wilson, 1913; and *C. uruguayensis* Thomsen, 1949. However, Ho & Bashirullah (1977) elevated *Metacaligus* to the generic rank and, after examining the type-material, excluded *C. afurcatus* and *C. enormis*. Moreover, they included *C. hilsae* Shen, 1947 in *Metacaligus*. But in his work on a revision of the copepod parasites of marine fishes of India, Pillai (1985) proposed to relegate *C. hilsae* to a junior synonym of *C. unguidentatus* Rangnekar & Murti, 1950. Therefore, the latter becomes the third species of *Metacaligus*. Then, a new species of this genus, *M. latus*, was added by Ho & Lin (2002). Consequently, four species are now recognized in this genus.

Although some species of *Caligus* display one of the four characteristics mentioned above, so far we know only one species, *C. grandiabdominalis* Yamaguti, 1954, that exhibits two of the four features (i.e., characters 1 and 4) mentioned above and no species possesses more than two. Thus, we support Ho & Bashirullah's (1977) rationale for the elevation of *Metacaligus* to the generic rank.

Among the four species of *Metacaligus*, *M. uruguayensis* has the widest distribution, being reported from India (Gnanamuthu, 1950), Sri Lanka (Kirtisinghe, 1964), Venezuela (Ho & Bashirullah, 1977), Brazil (Boxshall & Montú, 1997), and Taiwan (Lin & Ho, 2000). As to the remaining three species, while *M. unguidentatus* is known from India (Rangnekar & Murti, 1950) and China (Shen, 1957), the other two species are known only from their originally reported locality, i.e., *M. rufus* from Woods Hole, Massachusetts (Wilson, 1908) and *M. latus* from Chiayi, Taiwan (Ho & Lin, 2002).

Metacaligus uruguayensis Thomsen, 1949
(figs. 106-108)

Caligus (*Metacaligus*) *uruguayensis* Thomsen, 1949: 3.
Metacaligus uruguayensis Ho & Bashirullah, 1977: 707; Dojiri, 1983: 250; Boxshall & Montú, 1997: 56; Luque et al., 1999: 12; Lin & Ho, 2000: 206; Ho & Lin, 2004: 302.
Caligus uruguayensis Yamaguti, 1963: 62; Pillai, 1985: 365.
Caligus longicervicis Gnanamuthu, 1950a: 115; Yamaguti, 1963: 55; Kirtisinghe, 1964: 65; Pillai, 1966: 125.
Caligus cunicephalus (not Gnanamuthu, 1950) Pillai, 1963: 77.

Material examined. — None. Description given below is modified from Ho & Lin (2004).

Female. — Body as in fig. 106a. Total length (not including setae on caudal ramus) 4.16 mm (3.46-5.22 mm) (n = 10). Cephalothoracic shield longer than wide, measuring 1.67 mm (1.52-2.06 mm) × 1.52 mm (1.40-1.72 mm), with relatively deep posterior sinuses. Sensory pit located near posterolateral corner. Antennule not extending beyond lateral limit of cephalothorax. Free margin of thoracic zone extending beyond posterior tip of lateral zone. Lunules present, located ventrally.

Fourth pedigerous somite wider than long, 0.38 mm (0.26-0.44 mm) × 0.47 mm (0.42-0.52 mm). Genital complex (fig. 106a) ovoid, longer than wide, 1.49 mm (1.16-1.78 mm) × 1.25 mm (1.06-1.52 mm). Abdomen consisting of 1 free somite, long, not clearly delimited from genital complex, 0.58 mm (0.40-0.86 mm) × 0.37 mm (0.30-0.44 mm).

Caudal ramus (fig. 106f) longer than wide, 160 × 76 μm, and armed with 3 short and 3 long plumose setae.

Antennule (fig. 106b) 2-segmented; first segment with 27 plumose setae on anterodistal surface; distal segment with subterminal seta on posterior margin and 11 setae plus 2 aesthetascs on distal margin. Antenna (fig. 106c) indistinctly 4-segmented, with basal 2 segments appearing like one segment and unarmed. Third segment also unarmed. Terminal claw stocky, bearing 1 small seta in basal region and another small one close to middle region. Postantennal process (fig. 106c) small, but sharply pointed, carrying 2 basal

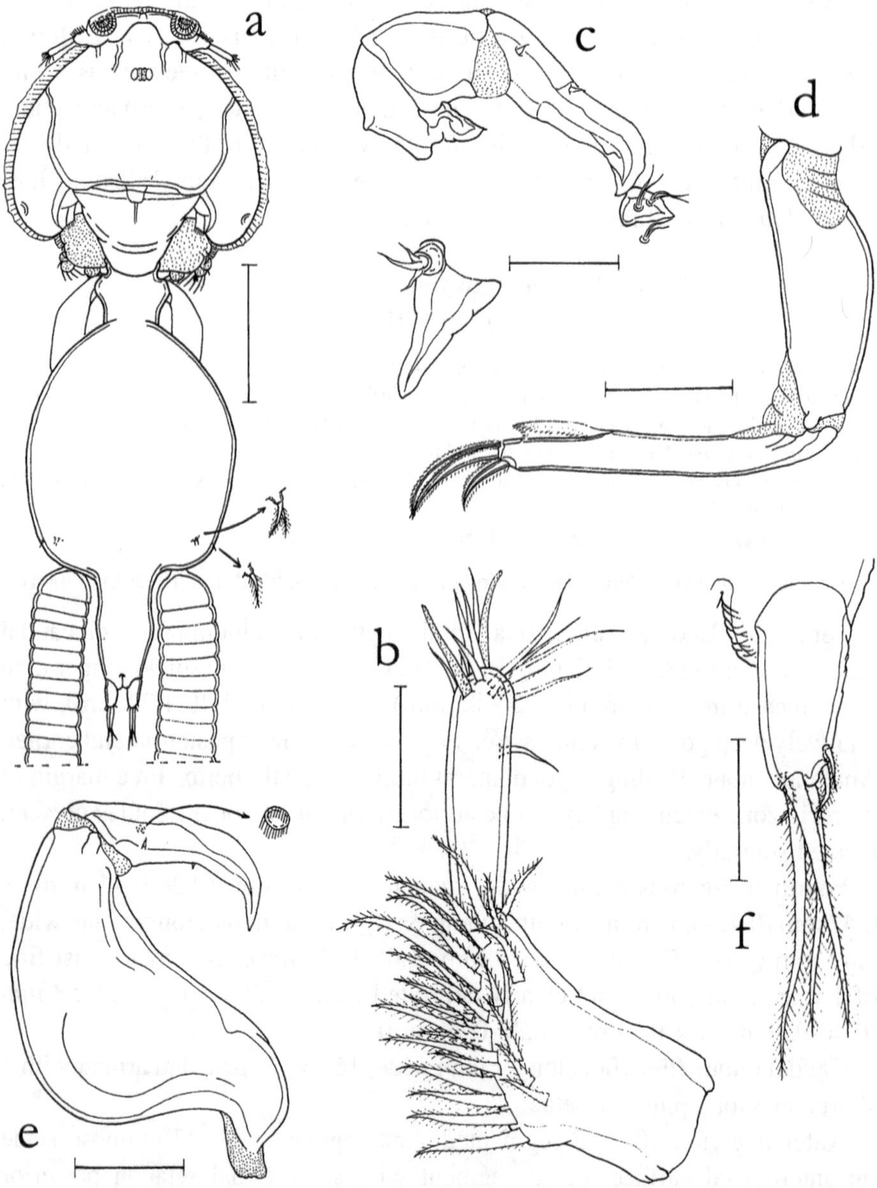

Fig. 106. *Metacaligus uruguayensis* Thomsen, 1949, female. a, body, dorsal; b, antennule, ventral; c, antenna, postantennal process, and maxillule, ventral; d, maxilla, ventral; e, maxilliped, ventral; f, caudal ramus, dorsal. Scale bars: 1 mm in a, 0.1 mm in b and c; 0.15 mm in d and f; 0.2 mm in e.

papillae with each bearing 2 setules; another similar papilla located nearby on sternum.

Mouth tube longer than wide. Intrabuccal stylet and strigil present. Mandible (not figured) with 4 sections, bearing 12 teeth on medial margin of distal blade. Maxillule (fig. 106c) comprising stout dentiform process and a papilla tipped with 3 setae. Maxilla (fig. 106d) 2-segmented; proximal segment (lacertus) large, but short and unarmed; distal segment (brachium) slender, with large flabellum located close to tip; subterminal canna bilaterally spinulate, but terminal calamus with additional row of spinules on side. Maxilliped (fig. 106e) 2-segmented; proximal segment (corpus) robust, with roundly protruded outer-basal margin; distal claw sharply pointed, bearing small seta and micropit in basal region and another small, medial seta in middle region. Sternal furca absent.

Leg 1 (fig. 107a) sympod with long, plumose outer seta and another small, plumose inner seta, in addition to a papilla bearing 2 long setules near outer margin of coxa. First segment of exopod with very short spinules on inner margin and short spiniform seta at outer distal corner. On distal segment of exopod, spine 1 twice as long as spine 3; spines 2 and 3 bilaterally serrate and without accessory process; simple seta 4 shorter than spine 1, but longer than spines 2 and 3. Three medial plumose setae shorter than exopodal segment carrying them. Vestigial endopod a small conical protrusion tipped with tiny setule. Leg 2 (fig. 107b) typical for caligid, except bearing only one small outer spine on third exopodal segment. Leg 3 (fig. 107c) also typical for caligid. Leg 4 (fig. 107d) sympod longer than 2-segmented exopod; proximal exopodal segment longer than distal segment, with 1 spine on former and 4 spines on latter; all spines fringed with spinules on both sides except for innermost spine, which is bilaterally spinulate. Leg 5 (fig. 106a) represented by 2 papillae; 1 (outer) papilla tipped with 1 plumose seta, and other (inner) with 2 plumose setae. Leg 6 not seen.

Male. — Body as in fig. 108a. Total length 2.78 mm (2.56-3.00 mm) (n = 3). Cephalothorax slightly longer than wide, 1.52 mm (1.44-1.60 mm) × 1.23 mm (1.12-1.34 mm).

Fourth pedigerous somite much wider than long, 0.11 mm (0.10-0.12 mm) × 0.34 mm (0.28-0.40 mm). Genital complex distinctly longer than wide, 0.68 mm (0.64-0.72 mm) × 0.33 mm (0.30-0.36 mm). Abdomen 2-segmented; first free somite 0.22 mm (0.20-0.24 mm) × 0.20 mm (0.18-0.22 mm) and anal somite 0.23 mm (0.22-0.26 mm) × 0.19 mm (0.18-0.20 mm).

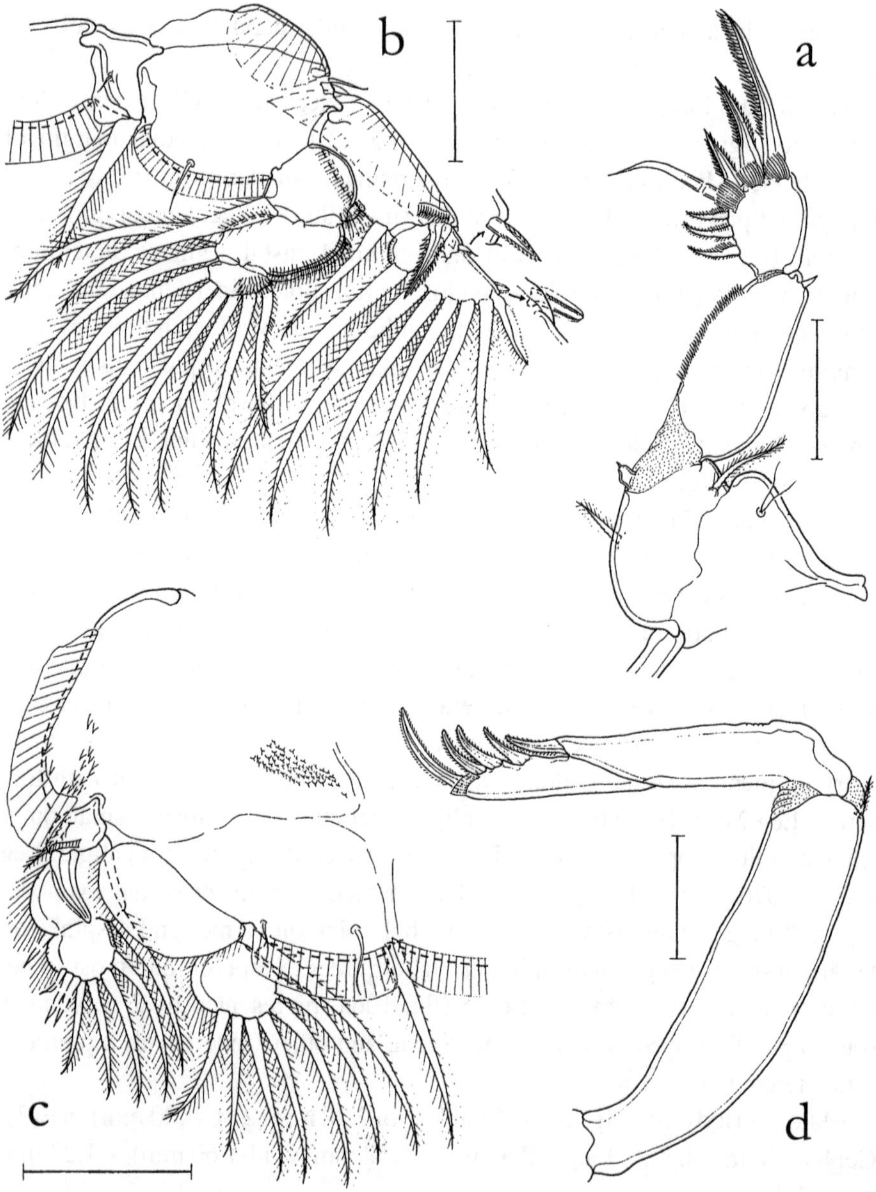

Fig. 107. *Metacaligus uruguayensis* Thomsen, 1949, female. a, leg 1, ventral; b, leg 2, ventral; c, leg 3, ventral; d, leg 4, ventral. Scale bars:0.15 mm in a; 0.2 mm in b and c; 0.3 mm in d.

Caudal ramus similar to that of female except slightly longer, 170 × 65 μm. Antennule like that of female. Antenna (fig. 108b, c) 3-segmented; proximal segment unarmed; middle segment largest, with 2 corrugated pads on me-

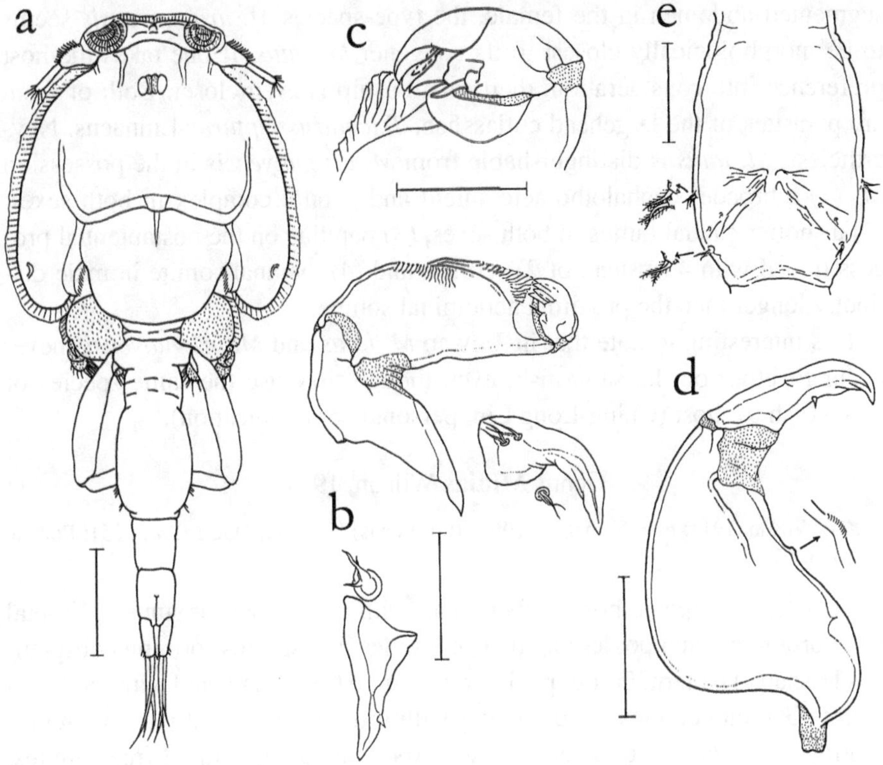

Fig. 108. *Metacaligus uruguayensis* Thomsen, 1949, male. a, body, dorsal; b, antenna, postantennal process, and maxillule, ventral; c, distal two segments of antenna, posterior; d, maxilliped, ventral; e, genital complex, ventral. Scale bars: 0.5 mm in a; 0.1 mm in b; 40 μm in c; 0.2 mm in d and e.

dial surface; terminal segment a small sclerotized structure tipped with 4 overlapping cuticular lamellae and carrying 2 spiniform setae (one on each side) (fig. 108c). Postantennal process (fig. 108b) better developed than in female.

Mouth tube and mandible as in female. Dentiform process of maxillule (fig. 108b) bearing a small subterminal hyaline process. Maxilla as in female. Corpus of maxilliped (fig. 108d) bearing a small myxal protrusion with serrated edge.

Legs 1-5 as in female. Leg 6 (fig. 108e) represented by ventral flap near genital area. Ventral flap with 2 pinnate setae at posterolateral corner and 2 setule-bearing papillae on medial margin.

Remarks. — This species was recently redescribed from specimens collected from *Lobotes surinamensis* (Bloch) and *Trichiurus lepturus* Linnaeus from Taiwan by Ho & Lin (2004). With the combination of a minute outer spine on the middle segment of the exopod of leg 2 in both sexes and a 1-

segmented abdomen in the female, the type-species *M. uruguayensis* seems to be morphologically closest to its congener *M. latus*. If one takes the host preference into consideration, their relationship is even closer. Both of them are parasites of the largehead cutlassfish, *Trichiurus lepturus* Linnaeus. Nevertheless, *M. latus* is distinguishable from *M. uruguayensis* in the possession of: (1) a broader cephalothoracic shield and genital complex in both sexes, (2) a shorter caudal ramus in both sexes, (3) papillae on the postantennal process tipped with 4 (instead of 2) setules, and (4) the anal somite in male distinctly longer than the proximal abdominal somite.

It is interesting to note that in Taiwan *M. latus* and *M. uruguayensis* never occur together on the same fish, even though they use the same species of cutlassfish for host (Ching-Long Lin, personal communication).

Genus **Midias** Wilson, 1911

Midias Wilson, 1911: 625; Yamaguti, 1963: 106; Lewis, 1967: 92; Dojiri, 1983: 251; Prabha, 1983: 53.

Female. — Cephalothorax subcircular, with deep posterior sinuses. Frontal plate prominent and possessing pair of lunules. Dorsal ribs forming conspicuous H-shaped pattern. Fourth pedigerous somite free, small, and without dorsal plates. Genital complex quadrangular with prominent conical process (representing leg 5) on posterolateral corner. Abdomen comprising 2 free somites. First somite with bulbous protrusion on lateral side, extending to lateral limit of genital complex. Second somite with spinulose posterolateral processes. Caudal ramus with 6 setae.

Antennule 2-segmented as in *Caligus*. Antenna 4-segmented. Second segment with posteriorly directed spinelike process. Terminal segment hamate. Postantennal process present. Mouth tube with intrabuccal stylet and strigil. Mandible comprising 4 sections, with 12 teeth on inner margin of terminal section. Maxillule consisting of dentiform process and adjacent setiferous papilla. Maxilla brachiform. Maxilliped subchelate and prehensile. Sternal furca present.

Legs 1-4 as in *Caligus*. Basal swelling of first exopodal segment of leg 3 with 2 large clawlike spines. Leg 4 with inflated sympod and 3-segmented exopod. Leg 5 represented by 2 conical projections; smaller one bearing 1 seta, larger one with 3 setae. Leg 6 absent.

Male. — Cephalothorax similar to that in female. Genital complex suborbicular. Antenna with corrugated adhesion pads. Terminal segment hamate, without accessory tines. Dentiform process of maxillule with corrugated pad. Corpus maxillipedis more robust than in female. Sympod of leg 4 more slender. Other appendages as in female. Leg 5 a triangular process with 3 setae. Leg 6 smaller than leg 5, located near ventral flap of genital area, and bearing 3 setae.

Type-species. — *Midias lobodes* Wilson, 1911.

Remarks. — The genus *Midias* was placed in the subfamily Euryphorinae (subsequently elevated to the familial rank) by Wilson (1911) due to the possession of dorsal plates on the fourth pedigerous somite, the inflated nature of the genital complex and anterior portion of the abdomen, and the caudal process of the posterior region of the abdomen near the insertion of the caudal ramus in the female. Lewis (1967) and Kabata (1979), however, objected to this familial placement of *Midias*, and included this genus in the Caligidae.

The inflated nature of the genital complex and the abdomen in the female is not unique. In fact, females in a number of genera currently recognized in the Caligidae possess a similar degree of modification in these two tagmata. They are *Abasia*, *Parapetalus*, *Parechetus*, and *Sinocaligus*. Moreover, the various appendages possessed by *Midias* are nearly identical to those of *Caligus* and *Lepeophtheirus* with the presence of the lunules suggesting a closer affinity to the genus *Caligus*.

Although Kabata (1979) considered *Midias* as a junior synonym of *Caligus* and *M. lobodes* simply as an unusual species of the latter, *Midias* is here upheld based on: (1) the bulbous lateral expansion on the first abdominal somite of the female, (2) the spinulose posterolateral processes of the second free abdominal somite, (3) the two large clawlike spines on the basal swelling of the first exopodal segment of leg 3, and (4) the setiferous conical projection representing leg 5 in the female. Although females of other species of *Caligus* have broad abdomens (e.g., *C. constrictus* Heller, 1865; *C. cybii* Bassett-Smith, 1898; *C. eventilis* Leigh-Sharpe, 1934; *C. berychis* Wilson, 1936; *C. fortis* Kabata, 1965; among others), no species possess the broad expansion displayed by the female of *M. lobodes*, except *C. grandiabdominalis* Yamaguti, 1954 and *C. kurochkini* Kazachenko, 1975, both of which may have to be assigned to other caligid genera. The posterolateral processes on the second abdominal somite and the two large clawlike spines on the first exopodal segment of leg 3 are characteristics that are not found in members of *Caligus*. The conical projection of leg 5 is found in only two known species of *Caligus*, *C. cordiventris* Shiino, 1952 and *C. tylosuri* (Rangnekar, 1956).

Rangnekar (1956) described what was once thought to be the second species, *Midias carangis*, in this genus from the longnose trevally *Caranx nigrescens* Day [= *Carangoides chrysophrys* (Cuvier)] from Bombay waters. Although similar in general habitus and similar in many of the appendages, these two species differ in four major features. *Midias carangis* has no posterolateral processes on the second free somite of the abdomen. The exopodal spine on the basal swelling of the first segment of leg 3 is a single clawlike structure and not double as in *M. lobodes*. The sympod of leg 4 is not inflated, but slender in *M. carangis*. Leg 5 in *M. lobodes* is triangular, but is figured as a rounded lobe in the female of *M. carangis*. Pillai (1961) synonymized *M. carangis* with *Caligus constrictus* Heller, 1865, which he collected from *Carangoides malabaricus* Bloch from India.

Since *Midias* can be clearly delimited from *Caligus* by the four enumerated characters mentioned above, the genus should be recognized as valid.

Midias lobodes Wilson, 1911
(figs. 109-113)

Midias lobodes Wilson, 1911: 626; Wilson, 1913: 225; Causey, 1953b: 11; Rangnekar, 1956: 52; Shiino, 1958a: 98; Shiino, 1963: 343; Yamaguti, 1963: 107; Kirtisinghe, 1964: 71; Pillai, 1966: 130; Lewis, 1967: 94; Lewis et al., 1969: 421; Dojiri, 1983: 256; Prabha & Pillai, 1983: 20.

Caligus lobodes Kabata, 1979: 170.

Material examined. — One female lectotype (USNM 39613), 1 allolectotype (USNM 112846), and 8 paralectotypes (1 mature, 1 immature, 4 preadult females, and 2 males) (USNM 112847) from outside surface of head of great barracuda, *Sphyraena barracuda* (Walbaum), caught at Tortugas, Florida, on 12 June 1908 by Dr. Edwin Linton. Lectotype, allolectotype, and paralectotypes designated by Dr. Alan G. Lewis. Twenty-one mature females, 1 immature female, 6 males, and 5 cephalothoraces (USNM 53503) labeled as cotypes from outside surface of *Sphyraena barracuda* collected at Montego Bay, Jamaica, on 15 July 1910 by Dr. C. B. Wilson. All material on loan from National Museum of Natural History, Smithsonian Institution, Washington, D.C.

Female. — Body large and as in fig. 109a. Total length (not including setae on caudal ramus) 9.40 mm (8.90-10.16 mm) (n = 6). Cephalothorax subcircular, longer than wide, 3.91 mm (3.75-3.98 mm) × 3.78 mm (3.52-4.05 mm), with deep posterior sinuses. Free margin of thoracic zone at approximately same level as posterior tip of lateral zone. Frontal plate carrying pair of lunules ventrally. Tip of antennule almost extending to lateral limit of cephalothorax.

Fourth pedigerous somite wider than long, 0.70 mm (0.63-0.76 mm) × 1.39 mm (1.29-1.46 mm). Genital complex quadrangular, wider than long, 2.43 mm (2.19-2.86 mm) × 2.57 mm (2.32-2.89 mm). Abdomen fused to genital complex and distinctly comprising 2 free somites. First somite expanded outward to lateral limits of genital complex, much wider than long, 1.42 mm (1.29-1.59 mm) × 2.38 mm (2.09-2.69 mm). Second somite

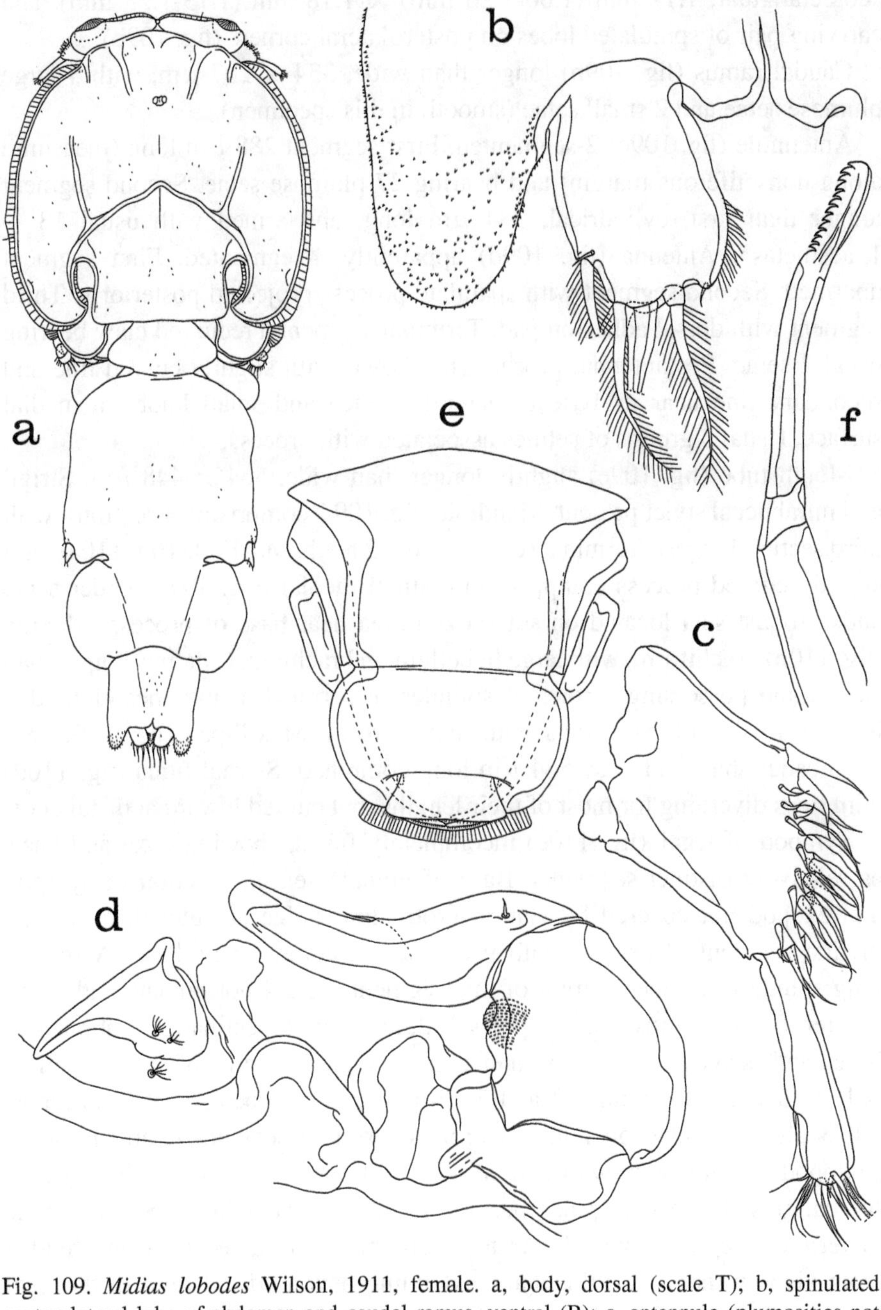

Fig. 109. *Midias lobodes* Wilson, 1911, female. a, body, dorsal (scale T); b, spinulated posterolateral lobe of abdomen and caudal ramus, ventral (B); c, antennule (plumosities not drawn), ventral (B); d, antenna and postantennal process, ventro-inner (B); e, mouth tube, ventral (B); f, mandible, ventral (E).

subrectangular, 1.17 mm (1.06-1.26 mm) × 1.18 mm (1.13-1.23 mm), and carrying pair of spinulated lobes on posterolateral corners (fig. 109b).

Caudal ramus (fig. 109b) longer than wide, 334 × 207 μm, with 4 large plumose setae and 2 small setae (smooth in this specimen).

Antennule (fig. 109c) 2-segmented. First segment 288 μm long (measured along nonsetiferous margin) and bearing 27 plumose setae. Second segment longer than first, cylindrical, 334 μm long, and armed with usual 13 + 1 aesthetasc. Antenna (fig. 109d) apparently 4-segmented. First segment unarmed. Second segment with spatulate process projected posteriorly. Third segment with dorsal adhesion pad. Terminal segment a recurved claw bearing usual 2 setae. Postantennal process (fig. 109d) with slightly curved tine and secondary tine near its base on lateral surface and small knob on medial surface. Usual 3 groups of setules associated with process.

Mouth tube (fig. 109e) slightly longer than wide, 564 × 448 μm. Strigil and intrabuccal stylet present. Mandible (fig. 109f) comprising 4 sections, with third section longest. Terminal section with 12 teeth. Maxillule (fig. 110a) with slightly curved process equipped with small medial tine. Two slender setae and 1 robust seta located on sclerotized area near base of process. Maxilla (fig. 110b) brachiform, with large flabellum on brachium. Calamus longer than canna, and possessing 3 rows of spinules or serrated membranes (1 hidden in view in fig. 110b). Canna with only 2 rows. Maxilliped (fig. 110c) not diagnostic; shaft and claw 564 μm long combined. Sternal furca (fig. 110d) with tines diverging for most of their lengths, but curved inward at distal ends.

Sympod of leg 1 (fig. 110e) incompletely fused, showing coxa and basis as relatively distinct segments. Base of pinnate seta at junction of sympod and exopod not covered by ventral protrusion. Spine at outer distal corner of first segment of exopod with pectinate membrane at its base. A row of long spinules on inner margin of this segment. Seta 4 not pinnate and much shorter than outermost spine. Spines 1, 2, and 3 with pectinate membranes at bases and heavily toothed on inner margin. Outer margins of spines toothed only at distal ends. Spines 2 and 3 with accessory processes at distal ends. All 3 plumose setae on inner margin of terminal segment relatively stout. Endopod 1-segmented, with 2 small knobs at apex. Leg 2 (fig. 110f) typical for caligids. Exopodal spines heavily serrated or spinulated. Second spine of terminal segment with 2 serrated membranes (1 folded over and hidden from view in fig. 111a); this spine extending beyond distal limit of terminal segment. First segment of endopod (fig. 110f) with spinules (instead of setules) along outer margin. Sympod of leg 3 (fig. 111b) with patch of denticles medially, 1 row of denticles at base of exopodal spine, and corrugated adhesion

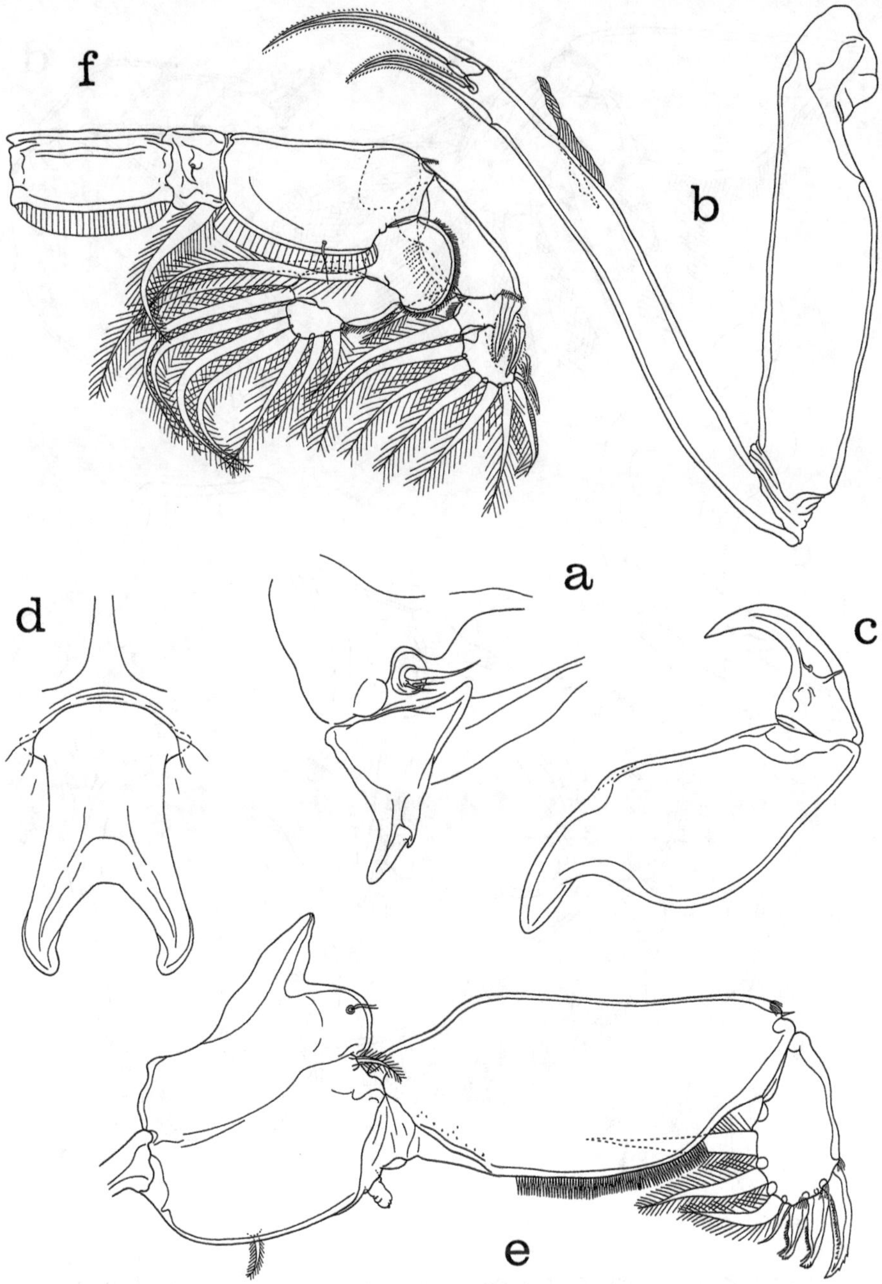

Fig. 110. *Midias lobodes* Wilson, 1911, female. a, maxillule, ventral (scale B); b, maxilla, ventral (B); c, maxilliped, ventral (F); d, sternal furca, ventral (B); e, leg 1, ventral (B); f, leg 2 and intercoxal plate, ventral (F).

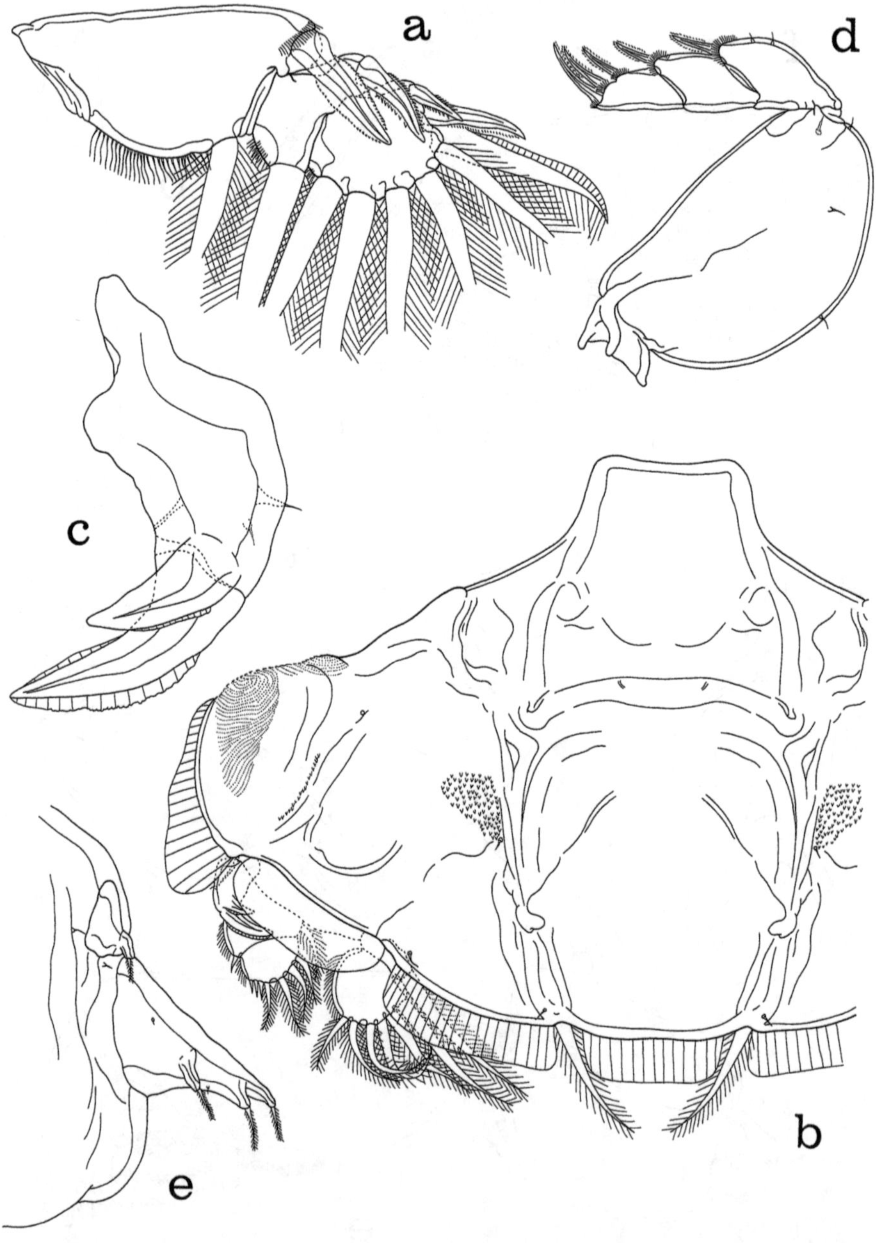

Fig. 111. *Midias lobodes* Wilson, 1911, female. a, exopod of leg 2, ventral (scale B); b, leg 3, ventral (F); c, spines and basal swelling of first exopodal segment of leg 3, ventral (E); d, leg 4, ventral (F); e, leg 5, dorsal (B).

pad dorsally. Exopod 3-segmented, with first segment represented by basal swelling (fig. 111c) with 2 spines. Exopodal spines slightly curved and with transparent flanges. Velum present between exopod and endopod. Endopod apparently 2-segmented (excluding velum), with usual armament. Sympod of leg 4 (fig. 111d) robust, 851 × 552 μm. Exopod shorter than sympod and 3-segmented, with spines of terminal segment decreasing in length from inner to outer margin. All 5 spines with pectinate membranes at bases. Leg 5 (fig. 111e) located on posterolateral corner of genital complex and comprised of 2 processes. Anterior process small and tipped with 1 small plumose seta. Posterior process a conspicuous triangular projection bearing 3 plumose setae.

Male. — Body as in fig. 112a. Total length 6.34 mm (6.21-6.61 mm) (n = 4). Cephalothorax as in female, but smaller, 3.49 (3.42-3.59 mm) × 3.32 mm (3.15-3.42 mm).

Fourth pedigerous somite 0.44 mm (0.43-0.46 mm) × 1.01 mm (0.96-1.06 mm). Genital complex subcircular, 1.24 mm (1.20-1.29 mm) × 1.61 mm (1.49-1.69 mm). Abdomen apparently consisting of 1 free somite, 1.11 mm (1.06-1.20 mm) × 0.97 mm (0.93-1.00 mm), with posterolateral spinulated processes as in female.

Caudal ramus (fig. 112b) 253 × 196 μm, with armature as in female.

Antennule similar to that in female, except 28-29 plumose setae on basal segment. Antenna (fig. 112c, d) with corrugated adhesion pads. Claw similar to that in female. Postantennal process (fig. 112e) with relatively large secondary tine and large medial knob at its base.

Mouth tube and mandible as in female. Maxillule (fig. 112f) different from that of female, possessing oval corrugated pad on ventral surface of dentiform process. Maxilla as in female. Corpus maxillipedis (fig. 113a) plump and with depression presumably housing tip of claw when closed. Shaft and claw combined 494 μm long, with usual seta near junction between them. Sternal furca (fig. 113b) with greatly divergent tines.

Legs 1-4 similar to those in female. Transparent knoblike structure (connected to duct?) near bases of spines 2 and 3 of exopod of leg 1 (fig. 113c) much more prominent than in female. Leg 4 (fig. 113d) similar to that in female, but sympod much more slender, 713 × 264 μm. Leg 5 (fig. 113e, f) a triangular structure bearing 3 setae (2 at apex broken off in specimen figured). Leg 6 (fig. 113e, g) a smaller triangular lobe with 1 small smooth seta, 1 intermediate-sized smooth seta, and 1 terminal seta (largest of 3).

Remarks. — Although Wilson's (1911) original description of *Midias lobodes* was not very detailed, the unusual inflated appearance (by presence

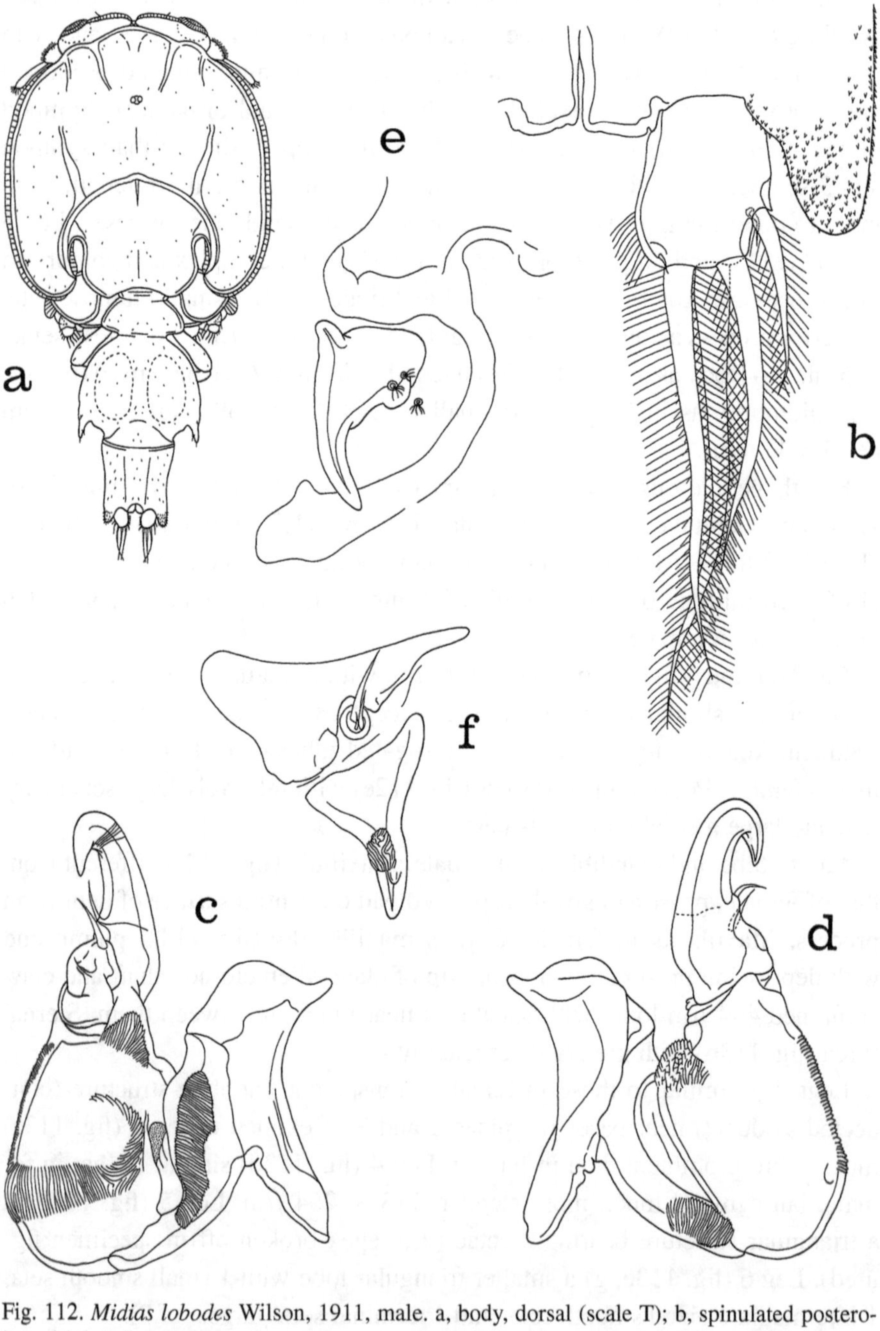

Fig. 112. *Midias lobodes* Wilson, 1911, male. a, body, dorsal (scale T); b, spinulated postero-lateral lobe of abdomen and caudal ramus, dorsal (B); c, antenna, postero-inner (B); d, same, dorso-outer (B); e, postantennal process, ventral (B); f, maxillule, ventral (B).

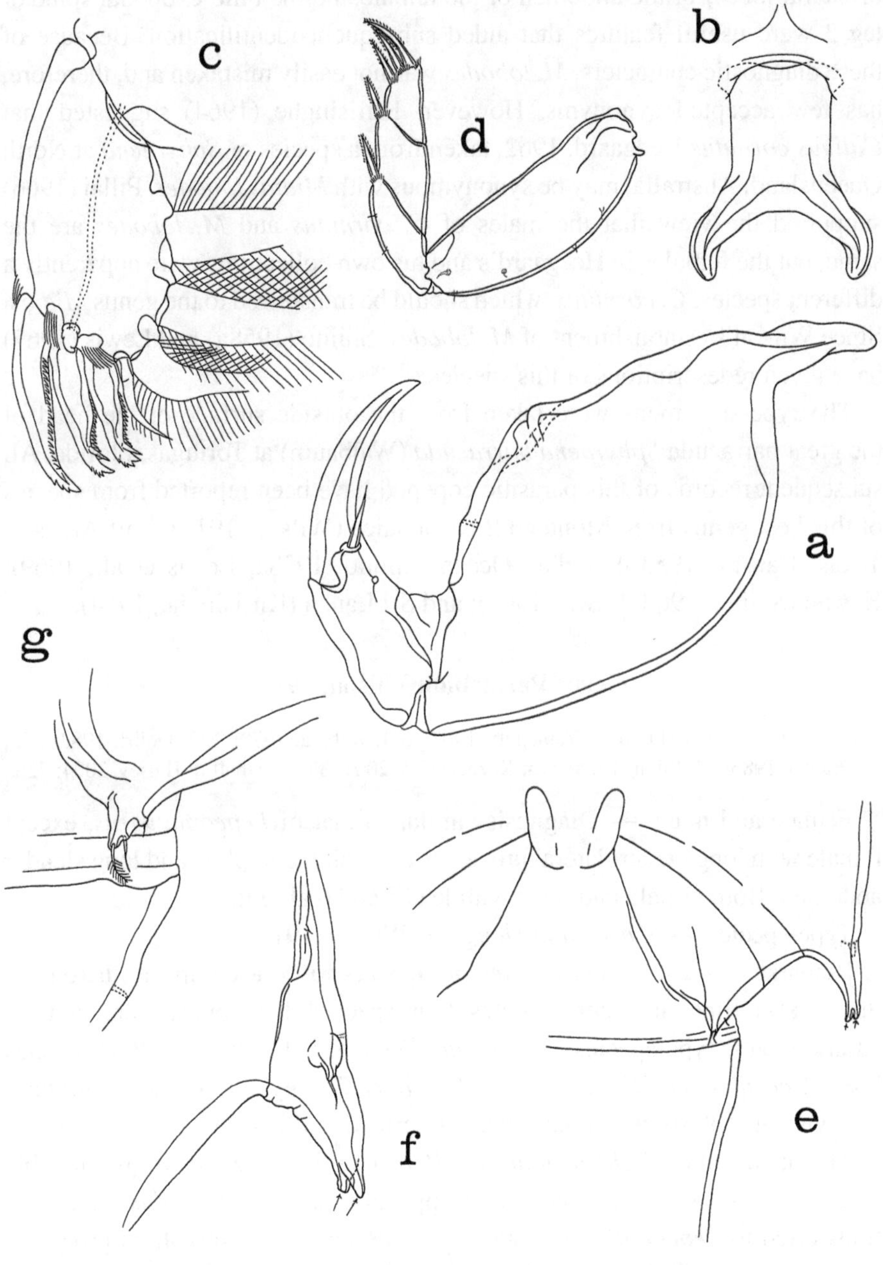

Fig. 113. *Midias lobodes* Wilson, 1911, male. a, maxilliped, ventral (scale B); b, sternal furca, ventral (B); c, terminal exopodal segment of leg 1, ventral (E); d, leg 4, ventral (F); e, legs 5 and 6 (arrows indicating positions of broken setae), and genital area, ventral (F); f, leg 5, dorsal (B); g, leg 6, ventral (B).

of lateral lobes) of the abdomen of the female and the bifid exopodal spine of leg 3 were useful features that aided subsequent identification. Because of these diagnostic characters, *M. lobodes* was not easily mistaken and, therefore, has few accepted synonyms. However, Kirtisinghe (1964) suggested that *Caligus cornutus* Heegaard, 1962, taken from a species of *Sphyraena* at North Queensland, Australia, may be synonymous with *Midias lobodes*. Pillai (1966) supported the view that the males of *C. cornutus* and *M. lobodes* are the same, but the females in Heegaard's and his own collections were apparently a different species, *C. cornutus*, which should be transferred to the genus *Midias*. Since Wilson's establishment of *M. lobodes*, Shiino (1958a) and Lewis (1967) have given redescriptions of this species.

The type-specimens were taken from the outside surface of the head of the great barracuda *Sphyraena barracuda* (Walbaum) at Tortugas, Florida. All subsequent records of this parasitic copepod have been reported from species of this host genus from Montego Bay, Jamaica (Wilson, 1913); Port Aransas, Texas (Causey, 1953b); Indian Ocean (Shiino, 1958a; Lewis et al., 1969); Hawaii (Shiino, 1963; Lewis, 1967); and Sri Lanka (Kirtisinghe, 1964).

Genus **Paralebion** Wilson, 1911

Paralebion Wilson, 1911: 628; Yamaguti, 1963: 104; Kabata, 1979: 200; Dojiri, 1983: 262; Prabha, 1983: 54; Pillai, 1985: 515; Kazachenko, 2001: 35; Boxshall & Halsey, 2004: 726.

Female and male. — Diagnosis similar to that of *Lepeophtheirus*, except female with long posterolateral processes on genital complex, and long slender abdomen. Both female and male with legs 5 and 6 absent.

Type-species. — *Paralebion elongatus* Wilson, 1911.

Remarks. — The genus *Paralebion* appears to be a composite taxon. At the present time, it accommodates four species that appear unrelated. In addition to the type-species, *P. elongatus* Wilson, 1911, there are three species [i.e., *P. curticaudus* Wilson, 1913; *P. pearsei* Causey, 1953; and *P. aliuncus* (Rangnekar, 1955)], which have been described in this genus.

The inclusion of *P. curticaudus* and *P. pearsei* in this genus is questionable (see discussion on *P. elongatus*). It appears that *P. curticaudus* should be transferred to *Lepeophtheirus* (due to the absence of posterolateral processes on the genital complex of the female) and *P. pearsei* possibly to *Tuxophorus* (due to the presence of dorsal aliform plates on the fourth pedigerous somite).

Paralebion aliuncus was originally described under the binomen *Diphyllogaster aliuncus* by Rangnekar (1955), but subsequently transferred to *Par-*

alebion by Pillai & Padmanabhan (1963). No reason was given for this transfer, but presumably the basis for this action was due to the presence of long posterolateral processes on the genital complex and the long, slender abdomen in the female. Both of these features are shared in common with *P. elongatus*. A more detailed description of the appendages of *P. aliuncus* needs to be completed before a definitive conclusion can be reached on this transfer.

Paralebion elongatus possesses some unusual features, such as the accessory process near the base of the maxillule, the accessory tines on either side of the sternal furca, and the short, broad rami of leg 3. *Paralebion elongatus* is morphologically very similar to species of *Lepeophtheirus*. The dorsal ribs of the cephalothorax and the long posterolateral processes of the genital complex of the *P. elongatus* female are unusual for the genus *Lepeophtheirus*, and therefore the genus *Paralebion* is retained.

Although there are a few species of *Lepeophtheirus* (e.g., *L. salmonis* Krøyer, 1863 and *L. pravipes* Wilson, 1912) with small posterolateral lobes on the genital complex in the female, there have been only three species of this genus described in which these processes are long. The first of these species is *L. longiventralis* described by Yü & Wu (1932) from the pleuronectid flatfish *Verasper variegatus* (Temminck & Schlegel) from Chefoo, China and by Ho et al. (2004) from the same host species from Japan. According to the illustrations by the discoverers and subsequent workers, *L. longiventralis* is similar in general habitus to *P. aliuncus*. However, based on the redescription made by Ho et al. (2004), *L. longiventralis* is distinctly different from *P. aliuncus*. The major difference is found in leg 4. While the 2-segmented exopod of this leg has an armature of I-0; III in *L. longiventralis*, it is I-0; IV in *P. aliuncus*. Additionally, the female of *L. longiventralis* has a distinct leg 5 and the male possesses both legs 5 and 6. The second species, *Lepeophtheirus elongatus* collected from a whaler shark, *Galeolamna greyi* Owen, at Rossiter Bay in Western Australia, was described by Heegaard (1962). It appears, based on his illustrations and the host, that this species is synonymous with *Paralebion elongatus*. The female of the third species, *Lepeophtheirus crassus*, described by Bere (1936) possesses distinct posterolateral processes, but also bears a long dactyliform leg 5 in the female and legs 5 and 6 in the male, distinguishing it from *Paralebion elongatus*.

The presence of long, digitiform posterolateral processes on the genital complex in the female and the absence of legs 5 and 6 in both sexes easily distinguish *Paralebion* from species of *Lepeophtheirus*.

Paralebion elongatus Wilson, 1911
(figs. 114-119)

Paralebion elongatus Wilson, 1911: 628; Wilson, 1935a: 331; Bere, 1936: 593; Pearse, 1948: 127; Vaissière, 1959: 551; Yamaguti, 1963: 104; Cressey, 1967b: 5; Cressey, 1970: 2; Dojiri, 1983: 264; Prabha, 1983: 54; Pillai, 1985: 516; Benz et al., 1992: 1027.

Material examined. — Two females (USNM 39553) labeled "holotypes" from Chesapeake Bay, Maryland. Twelve females, 2 female urosomes, 2 males, 1 male urosome, and 3 cephalothoraces labeled "cotypes" (USNM 89803) from "mouth of shark" from Chesapeake Bay, Maryland. This material on loan from National Museum of Natural History, Smithsonian Institution, Washington, D.C.

Female. — Body as in fig. 114a. Total length (not including setae on caudal ramus) 9.83 mm (n = 1) (others with caudal rami broken off). Cephalothorax only about 1/3 total length, subquadrangular or subcircular, 2.86 mm (2.49-3.19 mm) × 3.23 mm (2.92-3.39 mm) based on 5 specimens. Dorsal ribs distinctive. Relatively deep posterior sinuses. Free margin of thoracic zone about at same level as posterior tip of lateral zone. Antennule well within limits of cephalothorax.

Fourth pedigerous somite relatively long, 1.12 mm (0.93-1.26 mm) × 1.01 mm (0.93-1.06 mm), not clearly delimited from genital complex. Genital complex with pair of posterolateral processes, 2.20 mm (1.89-2.42 mm) long (excluding processes) and 3.31 mm (2.92-3.69 mm) long (including processes) × 1.81 mm (1.66-1.96 mm). Tips of posterolateral processes extending to midlength of abdomen. Abdomen very long, 2.56 mm (2.39-2.76 mm) × 0.67 mm (0.56-0.73 mm), with constriction near midlength and segmental boundary near posterior end.

Caudal ramus (fig. 114b) longer than wide, 472 × 184 μm, carrying 6 plumose setae; irregularly shaped in some preserved specimens (fig. 114c).

Antennule (fig. 114d) 2-segmented. First segment robust, 178 μm long (measured along nonsetiferous margin), and bearing 27 plumose setae. Second segment 232 μm long, with usual 13 + 1 aesthetasc (fig. 114e). Antenna (fig. 114f) apparently 4-segmented. First segment unarmed. Second segment with posteriorly directed spinelike process. Third segment with usual dorsal adhesion pad. Claw (fig. 114f, g) strongly recurved with usual 2 setae and additional minute seta. Flange present near tip of claw. Postantennal process (fig. 115a) greatly curved, with blunt process near base, and 3 groups of associated setules.

Mouth tube (fig. 115b) longer than wide, 552 × 146 μm. Intrabuccal stylet and strigil present. Mandible (fig. 115c) with 4 indistinct sections, with 12 teeth on terminal section. Maxillule (fig. 115d) with slightly curved process and palp tipped with 3 setae. Striated cuticular process in close proximity to

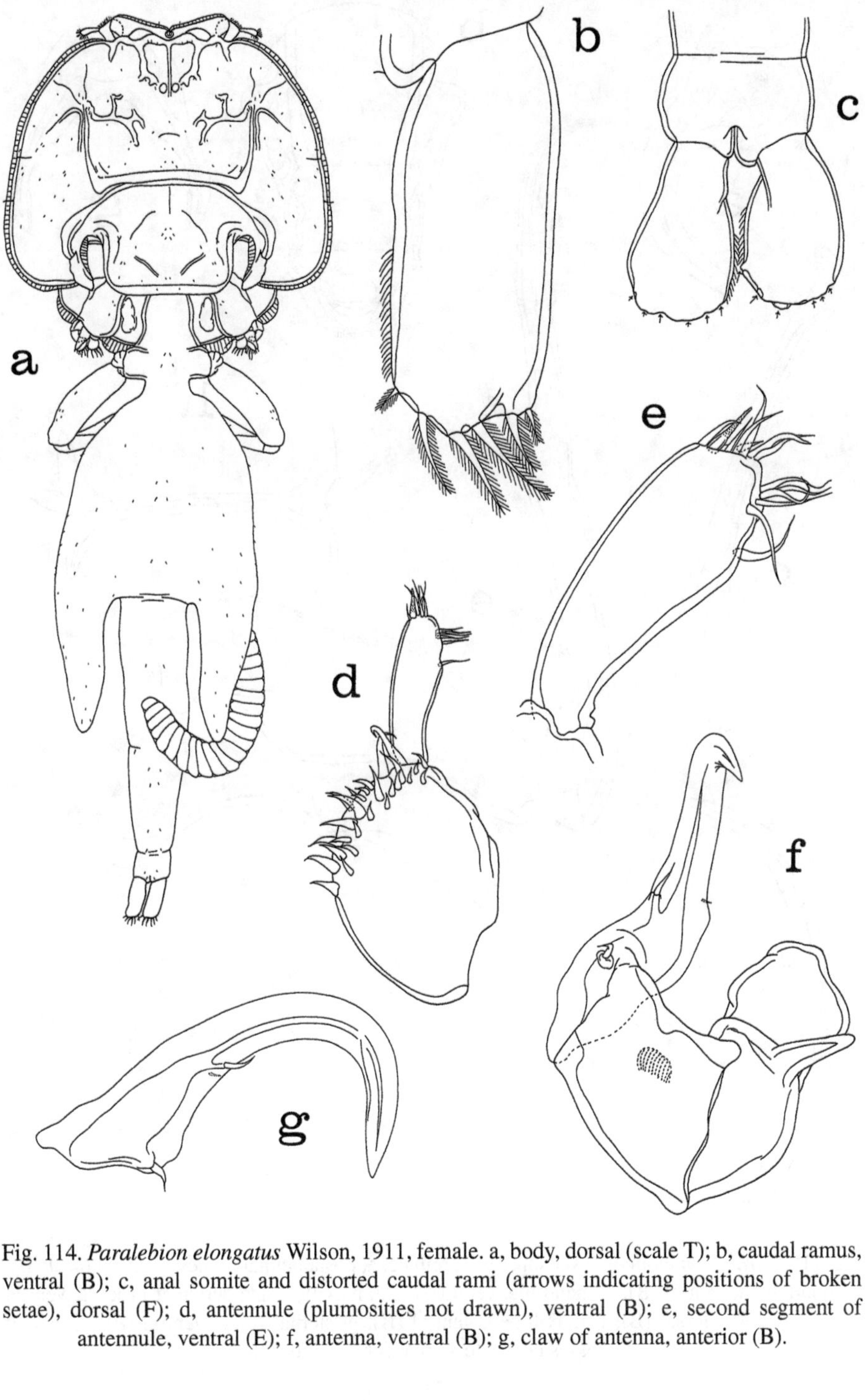

Fig. 114. *Paralebion elongatus* Wilson, 1911, female. a, body, dorsal (scale T); b, caudal ramus, ventral (B); c, anal somite and distorted caudal rami (arrows indicating positions of broken setae), dorsal (F); d, antennule (plumosities not drawn), ventral (B); e, second segment of antennule, ventral (E); f, antenna, ventral (B); g, claw of antenna, anterior (B).

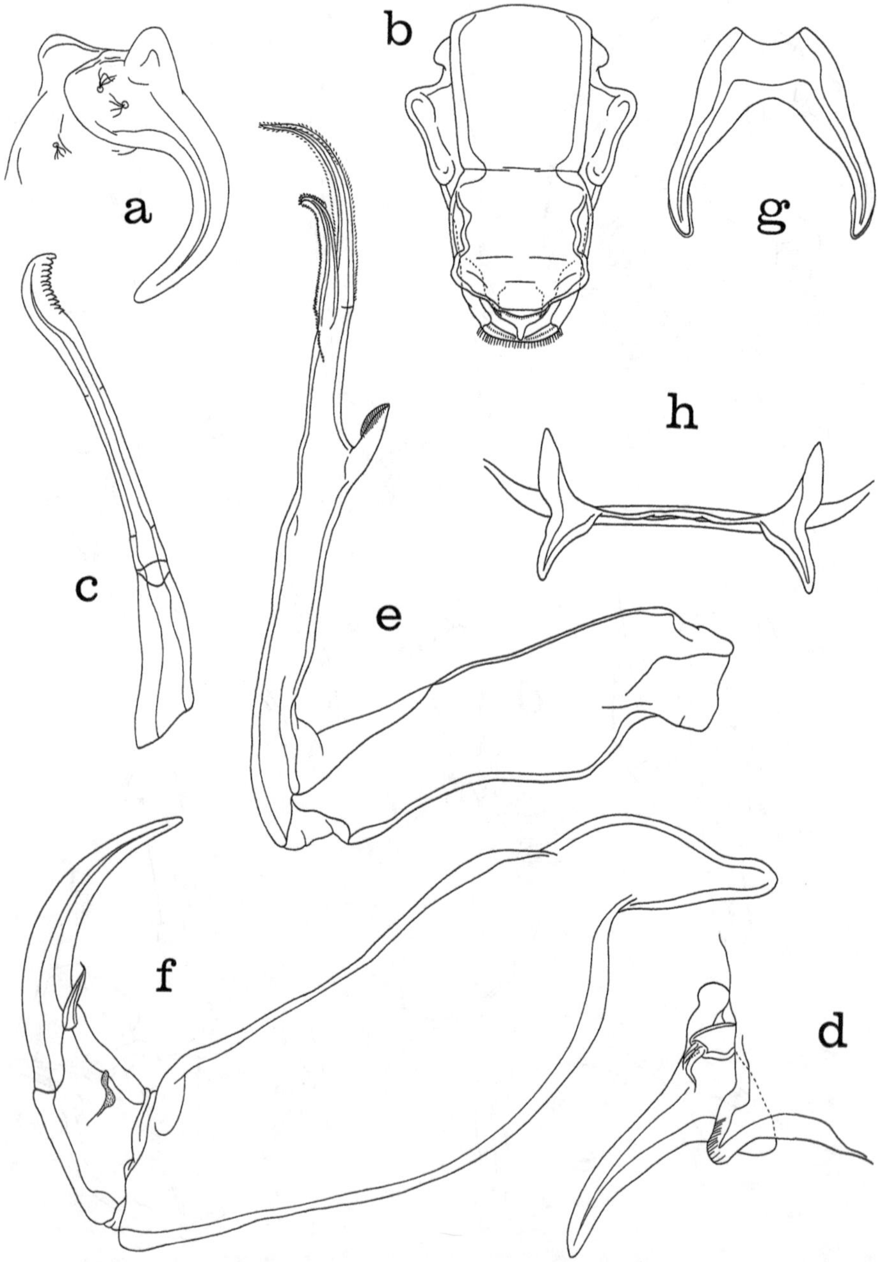

Fig. 115. *Paralebion elongatus* Wilson, 1911, female. a, postantennal process, ventral (scale B); b, mouth tube, ventral (B); c, mandible, ventral (E); d, maxillule and cuticular process, ventral (B); e, maxilla, dorsal (B); f, maxilliped, ventral (B); g, sternal furca, ventral (F); h, pair of processes lateral to sternal furca, ventral (F).

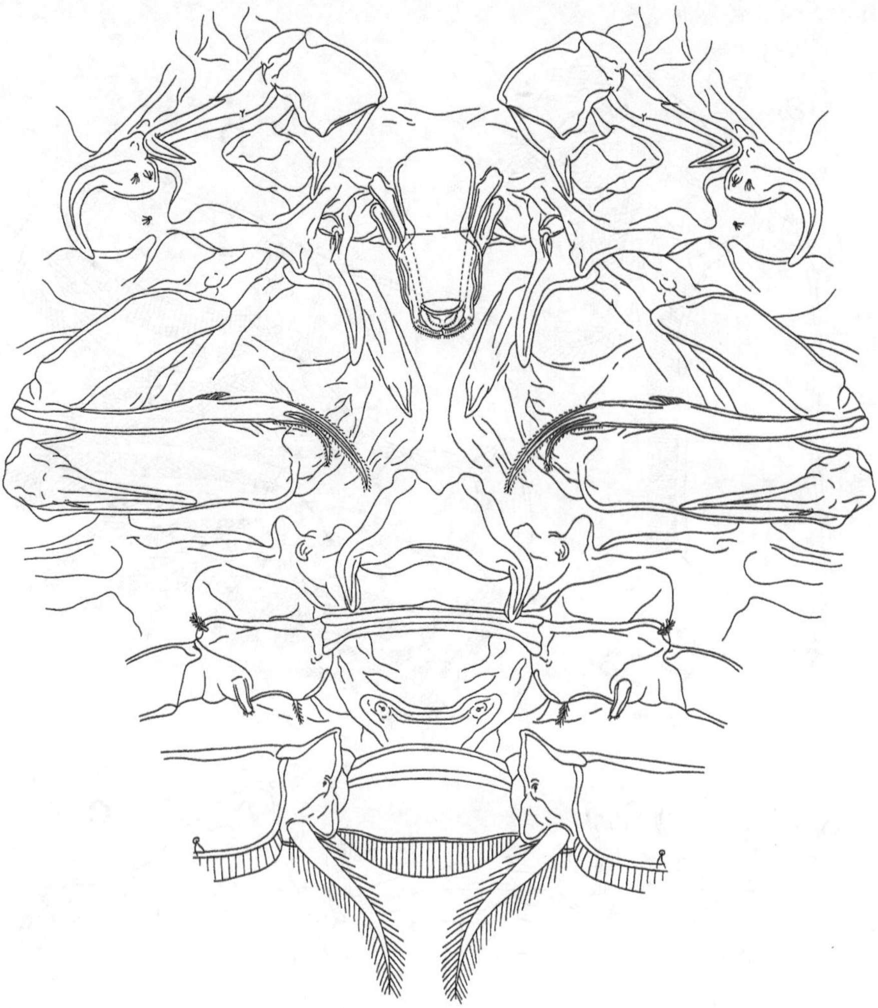

Fig. 116. *Paralebion elongatus* Wilson, 1911, female. Cephalothoracic area, ventral (scale F).

maxillule. Maxilla (fig. 115e) with spinelike process tipped with flabellum near 2/3 length of brachium. Calamus longer than canna. Canna with large serrated membrane. Maxilliped (fig. 115f) of usual caligid form. Shaft and claw 541 μm long combined. Sternal furca (fig. 115g) with narrow, diverging tines. Flanges present at tips of tines. Cuticular process (figs. 115h, 116) present on each side of sternal furca. Pair of posteriorly directed spinelike processes (fig. 116) located posteromedial to bases of maxilla.

Plumose seta near junction of sympod and exopod of leg 1 (fig. 117a) not covered by ventral protrusion. First segment of exopod with spine and small

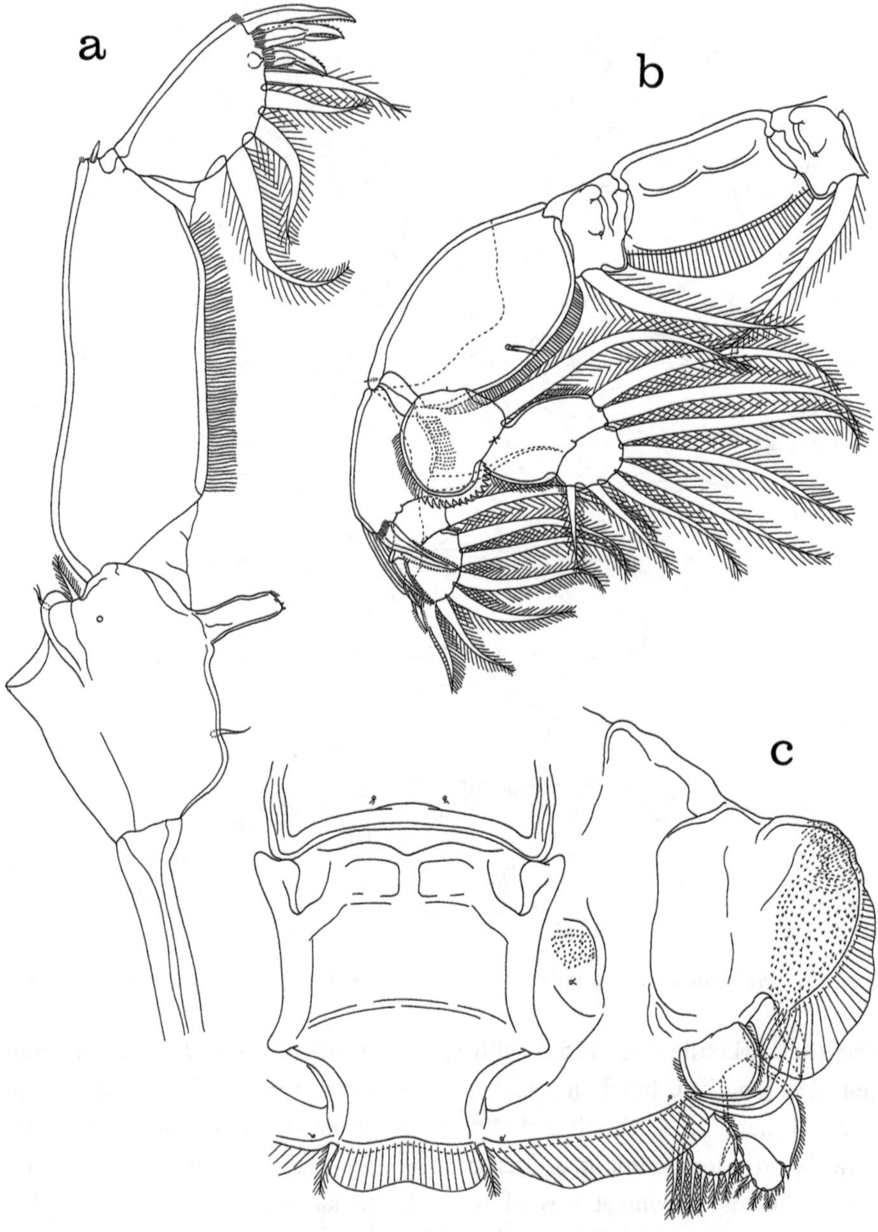

Fig. 117. *Paralebion elongatus* Wilson, 1911, female. a, leg 1, ventral (scale B); b, leg 2 and intercoxal plate, ventral (F); c, leg 3, ventral (F).

patch of spinules on outer distal corner, and usual setules on inner margin. Pinnate seta 4 much longer than outermost spine. Spines 1, 2, and 3 spinulated and with pectinate membranes at their bases. Spines 2 and 3 with accessory processes. Nipplelike hyaline process present near base of innermost spine. Endopod with 2 minute processes and patch of setules at tip. Second leg (fig. 117b) of usual form. Spine of first exopodal segment large and heavily serrated. Spine of second exopodal segment and outermost spine of third segment even with or slightly extending beyond distal limit of third segment. Second outer spine of terminal segment bilaterally spinulated and projecting beyond limit of third segment. First segment of endopod with row of spinelike processes along outer margin. Apron of leg 3 (fig. 117c) with 2 patches of minute spinules and dorsal adhesion pad. Exopodal spine of leg 3 (fig. 118a) with flange on inner margin, terminally located on basal swelling. Next 2 exopodal segments along with 2 segments of endopod with usual armament. Velum situated ventrally between exopod and endopod arising from apron and seemingly unconnected to either ramus. Leg 4 exopod (fig. 118b) 3-segmented with outer margin of second and third segments with serrated membrane. Length of innermost spine of terminal segment twice as long as other 2. Legs 5 and 6 absent; genital area as in fig. 118c.

Male. — Body as in fig. 118d. Total length 5.73 mm (5.68-5.78 mm) (n = 2). Cephalothorax similar to that in female, 2.66 mm (2.62-2.69 mm) × 2.76 mm (2.69-2.82 mm).

Fourth pedigerous somite longer than wide, 0.80 × 0.78 mm (0.76-0.80 mm). Genital complex elongate and rather slender, 1.08 mm (1.06-1.10 mm) × 0.71 mm (0.70-0.73 mm). Abdomen comprising 2 free somites. First somite shorter than anal somite, combined somites 0.80 × 0.50 mm.

Caudal ramus (fig. 118e) 483 × 230 μm, with apparently 6 setae (most setae broken off in specimens; arrows indicate positions of setae).

Antennule as in female. Antenna (figs. 118f, 119a) 3-segmented. First and second segments with large patches of corrugated adhesion pads. Terminal segment (fig. 119b-e) with 4 tines. Main tine (slightly larger than others) possessing flange along main axis; 2 smaller tines flank its base, and large, striated tine arising ventrally. Postantennal process as in female.

Mouth tube and mandible as in female. Maxillule (fig. 119f) different from that of female, possessing corrugated pad near distal end. Maxilla, maxilliped, and sternal furca as in female. Pair of corrugated pads (fig. 119g) posteromedial to bases of maxilla.

Legs 1-4 as in female. Legs 5 and 6 apparently absent. Genital area (fig. 119h) a posteroventral flap on genital complex.

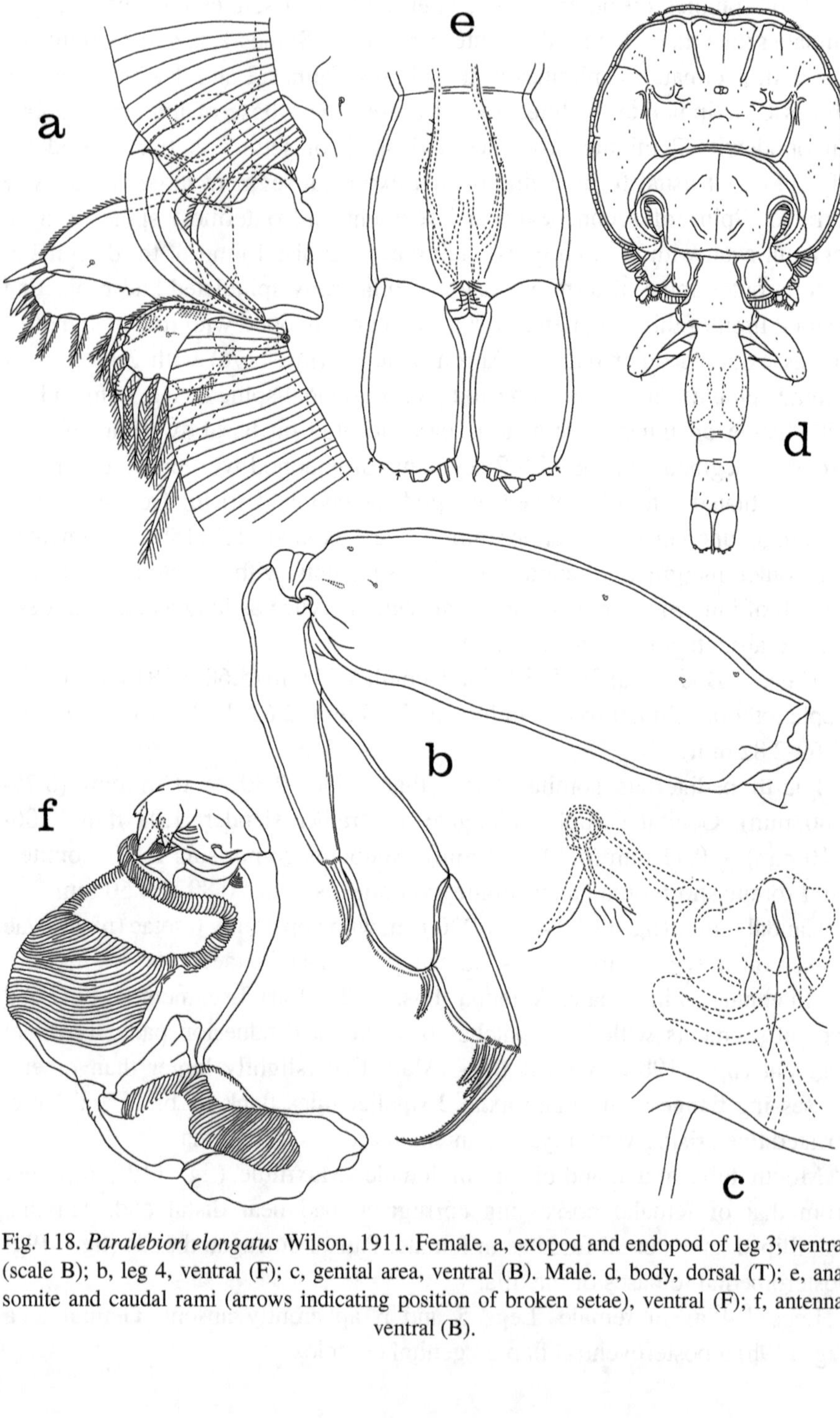

Fig. 118. *Paralebion elongatus* Wilson, 1911. Female. a, exopod and endopod of leg 3, ventral (scale B); b, leg 4, ventral (F); c, genital area, ventral (B). Male. d, body, dorsal (T); e, anal somite and caudal rami (arrows indicating position of broken setae), ventral (F); f, antenna, ventral (B).

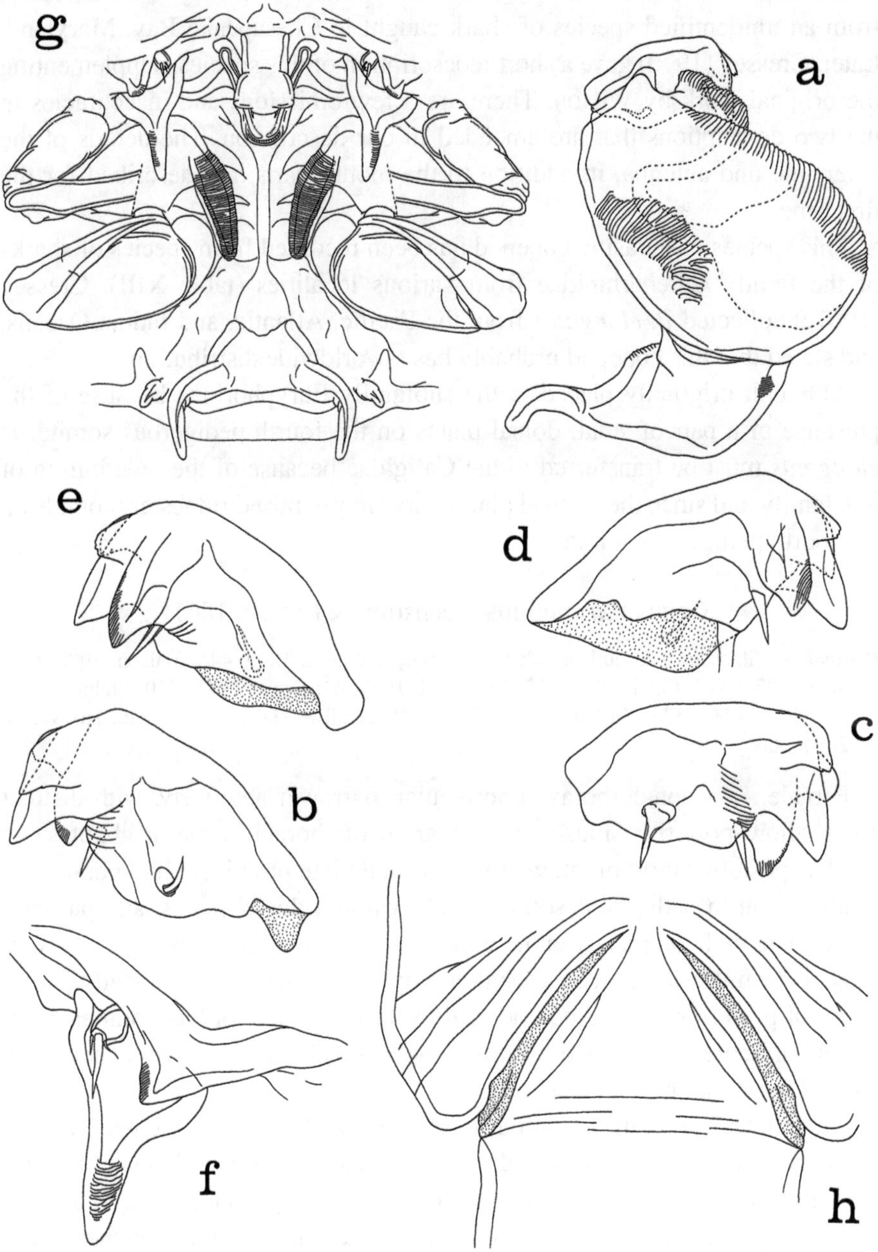

Fig. 119. *Paralebion elongatus* Wilson, 1911, male. a, antenna, dorsal (scale B); b, claw of antenna, ventral (E); c, same, ventral (E); d, same, dorsal (E); e, same, dorsal (E); f, maxillule, ventral (B); g, oral and sternal areas, ventral (F); h, genital area, ventral (B).

Remarks. — *Paralebion elongatus* was first described by Wilson (1911) from an unidentified species of shark caught in Chesapeake Bay, Maryland. Later, Cressey (1970) gave a short redescription of this species, supplementing the original work by Wilson. There are a few omissions and inaccuracies in the two descriptions that are amended in our description. The details of the antennule and antenna, in addition to the mouth parts, are described for the first time.

This species of parasitic copepod has been recorded from species of sharks of the family Carcharhinidae from various localities (table XIII). Cressey (1967c) collected *P. elongatus* from the Pacific, Atlantic, and Indian Oceans, and stated that this copepod probably has a worldwide distribution.

Although originally placed in the subfamily Euryphorinae because of the presence of a pair of small dorsal plates on the fourth pedigerous somite, *P. elongatus* must be transferred to the Caligidae because of the redefinition of this family and since the "dorsal plates" are simply raised ridges or mounds on the fourth pedigerous somite.

Genus **Parapetalus** Steenstrup & Lütken, 1861

Parapetalus Steenstrup & Lütken, 1861: 365; Bassett-Smith, 1899: 445; Wilson, 1908: 606; Shen, 1958: 142; Pillai, 1962: 287; Yamaguti, 1963: 64; Dojiri, 1983: 270; Prabha, 1983: 52; Pillai, 1985: 374; Kazachenko, 2001: 25; Boxshall & Halsey, 2004: 726; Ho & Lin, 2004: 306.

Female. — Cephalothorax suborbicular, narrower anteriorly, with distinct but shallow posterior sinuses. Free margin of thoracic zone at about same level as posterior limit of lateral zone. Frontal plate prominent and possessing lunules. Fourth pedigerous somite small, without dorsal plates, and partially fused to genital complex. Genital complex with lateral aliform expansions, as large as or much larger than cephalothorax, and posteriorly extended to form 1 pair of posterior lobes. Abdomen with varying number of free somites, most commonly consisting of 1 free somite, with broad lateral aliform expansions. Caudal ramus short, with 6 setae.

Antennule 2-segmented, similar to *Caligus*. Antenna 4-segmented. Second segment with posteriorly directed process. Terminal segment hamate. Postantennal process present. Mouth tube with intrabuccal stylet and strigil. Mandible comprising 4 sections, with 12 teeth on inner margin of terminal section. Maxillule comprised of dentiform process and adjacent setiferous papilla. Maxilla brachiform. Maxilliped subchelate and prehensile. Sternal furca present.

Leg 1 with 2-segmented exopod and vestigial endopod. Terminal segment of exopod armed with 3 spines at tip, 1 smooth seta on inner distal corner,

TABLE XIII
Hosts and localities of collections of *Paralebion elongatus* Wilson, 1911

Host	Locality	Reference
shark	Chesapeake Bay, Maryland	Wilson, 1911
Hypoprion brevirostris [= *Negaprion brevirostris* (Poey)]	Dry Tortugas, Florida	Wilson, 1935a
bay shark	Lemon Bay, Gulf of Mexico	Bere, 1936
dusky shark	Lemon Bay	Bere, 1936
spottip ground shark	Lemon Bay	Bere, 1936
Scoliodon terrae-novae [= *Rhizoprionodon terraenovae* (Richardson)]	Beaufort, North Carolina	Pearse, 1948
Negaprion brevirostris	Goree, Senegal	Vaissière, 1959
Carcharhinus leucas (Valenciennes)	Goree, Senegal	Vaissière, 1959
Carcharhinus leucas	Nosy Bé, Madagascar	Cressey, 1967b
	Durban, South Africa	Cressey, 1967b
Carcharhinus milberti [= *Carcharhinus plumbeus* (Nardo)]	west coast of Florida	Cressey, 1970
Carcharhinus leucas	west coast of Florida	Cressey, 1970
Carcharhinus limbatus (Müller & Henle)	west coast of Florida	Cressey, 1970
Negaprion brevirostris	west coast of Florida	Cressey, 1970
Scoliodon terrae-novae [= *Rhizoprionodon terraenovae*]	Trivandrum, India	Pillai, 1985
Negaprion brevirostris	Trivandrum, India	Pillai, 1985
Carcharias commersoni [= *Carcharhinus melanopterus* (Quoy & Gaimard)]	Trivandrum, India	Pillai, 1985
Carcharias obscurus [= *Carcharhinus obscurus* (Lesueur)]	Trivandrum, India	Pillai, 1985
Carcharhinus leucas	Trivandrum, India	Pillai, 1985
Carcharhinus limbatus	Trivandrum, India	Pillai, 1985
Carcharhinus milberti (= *Carcharhinus plumbeus*)	Trivandrum, India	Pillai, 1985
Triaenodon obesus (Rüppell)	Hawaii	Benz et al., 1992

and 3 plumose setae on inner margin. Leg 2 with 3-segmented rami (exopod of *P. occidentalis* Wilson, 1908, figured as 2-segmented by Pillai, 1962). Outer margin of endopodal segment covered with spinules. Leg 3 with large ventral apron. Exopod 3-segmented, and armed with clawlike spine on basal swelling of first segment. Endopod 2-segmented; outer margin of first segment expanded to form velum. Leg 4 with sympod and 3-segmented exopod, and displaying formula of I-0; I-0; III. Leg 5 represented by setae. Leg 6, when present, consisting of setae.

Male. — Cephalothorax and fourth pedigerous somite similar to those of female. Genital complex small and suboval. Abdomen cylindrical and comprising 1 to 3 free somites. Both genital complex and abdomen without aliform expansions. Caudal ramus as in female.

Antenna with corrugated adhesion pad (adhesions pads apparently absent in *P. occidentalis* Wilson, 1908 according to Pillai, 1962). Legs 5 and 6 represented by setae. All other appendages as in female.

Type-species. — *Parapetalus orientalis* Steenstrup & Lütken, 1861.

Remarks. — In 1957, Shen described a new caligid that he named "*Sinocaligus denticulatus* gen. & sp. nov." from a slender rainbow sardine, *Dussumieria elopsides* Bleeker (referred to as *Dussumieria hasseltii* in the original report), taken from Hainan in the South China Sea. Later, when he discovered that *Sinocaligus denticulatus* was attributable to *Parapetalus*, Shen (1958) proposed to change the name from "*Sinocaligus denticulatus* Shen" to "*Parapetalus denticulatus* (Shen)". However, noticing the structural difference in the genital complex and abdomen of *P. denticulatus* with the type species of the genus, Pillai (1962) created a new genus *Pseudopetalus* to accommodate *Parapetalus denticulatus* together with two other congeners *P. caudatus* Gnanamuthu, 1950 and *P. formicoides* (Redkar, Rangnekar & Murti, 1949). Since the genus name *Sinocaligus* Shen, 1957 has priority over *Pseudopetalus* Pillai, 1962, the three species are now known as *S. caudatus*, *S. denticulatus*, and *S. formicoides*.

The distinction that Pillai attempts to make between a winged abdomen versus a flattened abdomen in the female is an argument concerning the process by which the adult structure arrived at its final state. Examination of only the adults will tell us little about the mechanism by which the abdomen became lamelliform.

Pillai (1962) suggested that the rudiments of the fifth and sometimes sixth legs in the female may act as indicators of the boundaries of the genital complex. Consequently, if the fifth leg is located on the margin of the lamelliform structure, then the position of the leg suggests that the genital complex was simply dorsoventrally flattened. However, if the fifth leg is located considerably medial to the edge of the lamelliform structure, then the position of the fifth leg suggests the original boundary of the genital complex, and the lamelliform structure must then represent an aliform expansion. Unfortunately, the abdomen has no such markers; therefore, it is not possible to distinguish a dorsoventrally flattened abdomen as opposed to one with lateral aliform expansions by examining only the adults.

TABLE XIV

A comparison between the females of *Parapetalus* Steenstrup & Lütken, 1861 and *Sinocaligus* Shen, 1957

Parapetalus	*Sinocaligus*
Genital complex with aliform extensions (*P. hirsutus* does not seem to have an expanded genital complex, but possesses posterior processes)	Genital complex greatly elongate and triangular (anteriormost portion forming neck with fourth pedigerous somite)
Sternal furca with relatively thin tines	Sternal furca a large box with tapering, sharp tines
Sympod of leg 1 without spinules	Sympod of leg 1 with spinules (except in *S. dussumieri*)
3 plumose setae on inner margin of terminal exopodal segment of leg 1	3 small spines instead of 3 plumose setae (except in *S. formicoides* var. *denticulatus* and *S. dussumieri*)
Outer margin of second segment of endopod of leg 2 with spinules	Outer margin of second segment of endopod of leg 2 with large conical tooth-like projections
Exopod of leg 4 3-segmented	Exopod of leg 4 2-segmented
Leg 5 represented by setae	Legs 5 and 6 absent

There is, however, a major difference between the aliform expansions on the abdomen of the female of *P. orientalis* from all other females of *Parapetalus* and *Sinocaligus*. In all other species of these two genera, the abdomen appears as a flat, thin lamelliform structure, but in the female of *P. orientalis* only the anterior half possesses aliform, posteriorly directed lateral processes; the posterior half of the abdomen remains as the typical, cylindrical caligid structure.

Based upon the shape of the genital complex and the details of the appendages of the females (table XIV), *Sinocaligus* does seem distinct from *Parapetalus*. Although the characteristics listed in table XIV seem relatively minor and individually not of generic value, collectively they help distinguish the species of these two genera and afford some vailidity to *Sinocaligus*.

There are presently five species recognized in *Parapetalus*. Although Byrnes (1986) reported a new species, *Parapetalus spinosus* Byrnes, 1986, from the Samson fish, *Seriola hippos* Günther, caught off Coffs Harbour, Australia, with the absence of aliform expansions on either the genital complex or abdomen, we conclude that it is not a species of *Parapetalus*. Rather, it is most likely a species of *Caligus* bearing close resemblance to *Caligus parapetalopsis* Hameed & Pillai, 1973. The five recognized species of *Parapetalus* can be easily distinguished by the characters outlined in table XV. It is apparent from a

TABLE XV

Diagnostic features, hosts, and localities of collections of all five described species of *Parapetalus* Steenstrup & Lütken, 1861. PAP = postantennal process; MX_1P – dentiform process of maxilla; and SF = sternal furca

Species	Genital complex	Abdomen	PAP	MX_1P	SF	Host	Locality
P. orientalis Steenstrup & Lütken, 1861	with large lateral aliform expansions	with digitiform processes	single tine	single tine	divergent tines	primarily from carangid *Alectis indicus* (Rüppell)	Philippines; India
P. hirsutus (Bassett-Smith, 1898)	with very narrow aliform expansions	with lateral aliform expansions	single tine	single tine	parallel tines	*Eleutheronema* and *Polydactylus* (reported as *Polynemus*)	India; Sri Lanka; Java; Taiwan
P. occidentalis Wilson, 1908	with large lateral aliform expansions	with broad lateral aliform expansions	single tine	single tine	divergent tines	*Rachycentron canadum* (Linnaeus)	India; Sri Lanka; North Carolina; Louisiana; Taiwan
P. gunteri Pearse, 1952	with large lateral aliform expansions	with broad lateral aliform expansions	single tine	single tine	divergent tines	*Rachycentron canadum*	India; Texas coast; Australia
P. longipennatus Rangnekar, 1956	with lateral aliform expansions	with broad lateral aliform expansions	with accessory tine	with accessory tine	parallel tines	*Caranx nigrescens* Day and *Seriolina nigrofasciata* (Rüppell)	Bombay; Taiwan

detailed comparison of the species that *P. occidentalis* and *P. gunteri* are morphologically similar. The characteristics used by Pearse (1952) (e.g., indentation on lateral margin of cephalothorax, length of egg sacs, etc.) are not taxonomically very important. Kabata (1966a) suggested that *P. gunteri* may have to be relegated to synonymy with *P. occidentalis*. In addition to being morphologically identical, both nominal species infest the same species of host, *Rachycentron canadum* (Linnaeus), and have similar geographic distributions (Indian Ocean and Gulf of Mexico). Pillai (1962), however, recognized the validity of *P. gunteri* primarily due to differences in the antenna of the male, dentiform process of the maxillule (maxilla of Pillai, 1962), and maxilla (first maxilliped of Pillai, 1962). Nevertheless, subsequently in 1985, Pillai adopted Kabata's (1966) suggestion and relegated *P. gunteri* to synonymy with *P. occidentalis*. This treatment was also followed by Ho & Lin (2004). More detailed examinations of *P. gunteri* and *P. occidentalis* may show that the differences listed by Pillai (1962) are simply the result of intraspecific variation. It is likely that these two species are synonymous.

Hameed & Pillai (1973b) described *Caligus parapetalopsis* from the branchial cavity of *Zonichthys nigrofasciata* [= *Seriolina nigrofasciata* (Rüppell)] collected off Kerala. Both the widely expanded genital complex and abdomen suggest a closer affinity to *Parapetalus* than to *Caligus*. In fact Hameed & Pillai's (1973b) *Caligus parapetalopsis* has been treated as a valid species of *Parapetalus* by Ho & Lin (2010b).

Kabata (1979) in his "text fig. 48G" figures *Parapetalus*, but labeled and described it as "*Pseudopetalus*" (now known as *Sinocaligus*).

Parapetalus orientalis Steenstrup & Lütken, 1861
(figs. 120-122)

Parapetalus orientalis Steenstrup & Lütken, 1861: 365; Heller, 1865: 179; Bassett-Smith, 1899: 445; Wilson, 1908: 610; Rangnekar & Murti, 1950: 48; Rangnekar, 1956: 48; Shen, 1958: 142; Pillai, 1962: 290; Yamaguti, 1963: 65; Dojiri, 1983: 277; Prabha & Pillai, 1983: 16; Pillai, 1985: 382; Samotylova, 2010: 99.

Material examined. — Eight females from inner surface of operculum of 4 specimens of carangid, *Alectis indica* (Rüppell), from Manila Market, Philippines. Specimens collected from fish ("Albatross" collection: USNM 191810-12) housed in the ichthyological collection at National Museum of Natural History, Smithsonian Institution, Washington, D.C.

Female. — Body modified as in fig. 120a. Total length (not including setae on caudal ramus) 2.92 mm (2.90-2.94 mm) (n = 2); remaining 6 specimens either too distorted or damaged. Cephalothorax subcircular, about as long as wide, 1.06 mm (1.03-1.10 mm) × 1.11 mm (1.06-1.16 mm), with shallow posterior sinuses. Free margin of thoracic zone extending slightly beyond posterior tip of lateral zone. Lunules located ventrally on distinct frontal plate. Tip of antennule almost extending to lateral limit of cephalothorax.

Fourth pedigerous somite wider than long, 0.17 mm (0.12-0.23 mm) × 0.33 mm (0.31-0.36 mm). Genital complex with large aliform expansions, length (excluding alae) 0.80 mm and (including alae) 1.34 mm (1.33-1.36 mm) and width (including alae) 1.66 mm (1.59-1.73 mm). In some specimens, lateral expansions flat, appearing as wings; in others, expansions fat, giving genital complex a globose appearance (may be artifact of clearing agent). Abdomen divided into 3 sections (somites?). First part approximately 402 μm long, with aliform, posteriorly directed lateral processes. Abdominal aliform processes 1.10 mm (1.03-1.16 mm) long, overlapping posteromedial portion of genital complex, and extending beyond distal tips of caudal ramus. Second and third sections separated by slight constriction (presumably a segmental boundary). Second section 0.24 mm × 0.17 mm (0.16-0.18 mm) and terminal portion (anal somite) 0.16 mm (0.15-0.16 mm) × 0.15 mm.

Caudal ramus (fig. 120b) narrow proximally, widest distally, 89 × 68 μm, bearing an inner row of setules and usual 6 plumose setae.

Egg sac 2.89 × 0.34 mm, containing 24 eggs (n = 1).

Antennule (fig. 120c) 2-segmented. First segment 65 μm long (measured along nonsetiferous margin), with small bifid process on posterodistal corner, and carrying 28 plumose setae. Second segment much longer than first, 178 μm long, wider distally than proximally, and bearing usual 13 + 1 aesthetasc. Antenna (fig. 120d) apparently 4-segmented. First segment irregular in shape and unornamented. Second segment with rosettelike, corrugated, posteriorly directed process. Third segment robust, possessing usual dorsal corrugated adhesion pad. Claw with minute hyaline knob on inner margin and usual 2 setae. Postantennal process (fig. 120e) with large base and rounded knob near medial corner, and bearing straight robust tine tipped with conical protuberance. Two sclerites with setules and 2 sclerites with minute stubs (broken setules?) located on or near base of process.

Mouth tube (fig. 120f) 200 × 130 μm, possessing both intrabuccal stylet and strigil. Mandible (fig. 120g) comprising 4 sections, with 12 teeth on inner margin and transparent membrane on outer margin of terminal section. Maxillule (fig. 121a) with slightly curved process, and 1 large and 2 smaller setae originating from circular sclerotized base. Maxilla (fig. 121b) brachiform with flabellum located slightly distal to midlength of brachium. Calamus with 3 serrated membranes and longer than canna. Canna with 2 serrated

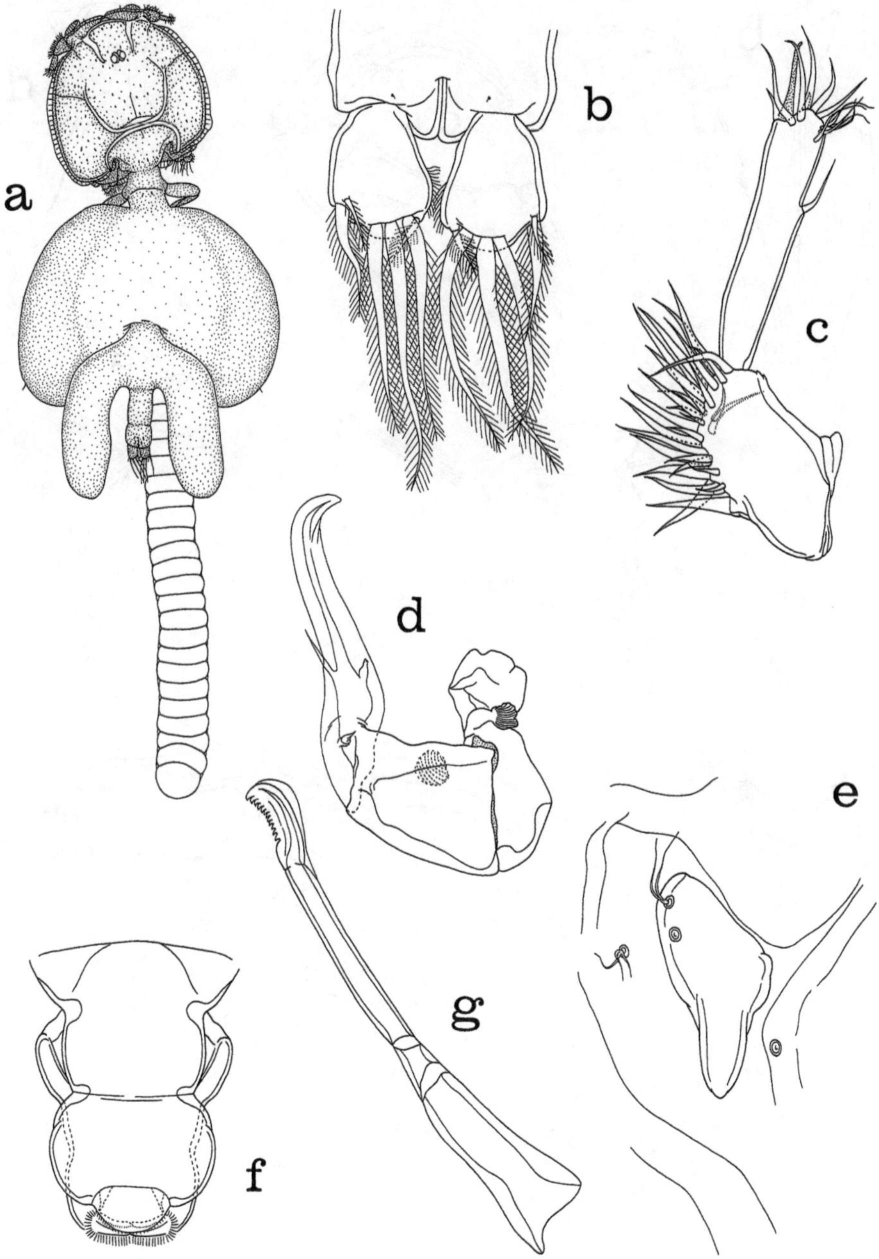

Fig. 120. *Parapetalus orientalis* Steenstrup & Lütken, 1861, female. a, body, dorsal (scale AA); b, caudal rami, ventral (E); c, antennule (plumosities not drawn), ventral (E); d, antenna, ventral (E); e, postantennal process, ventral (L); f, mouth tube, ventral (E); g, mandible, ventral (L).

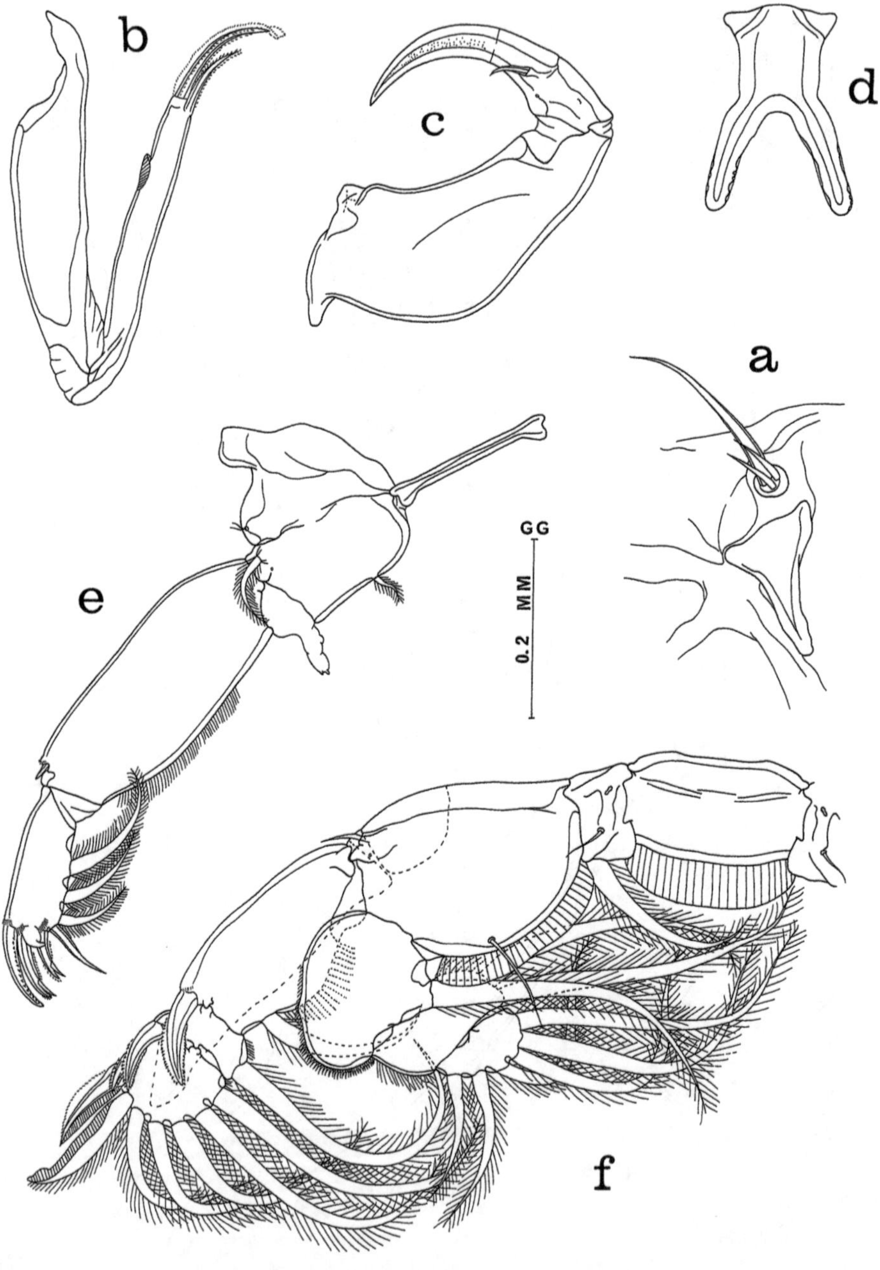

Fig. 121. *Parapetalus orientalis* Steenstrup & Lütken, 1861, female. a, maxillule, ventral (scale Z); b, maxilla, dorsal (E); c, maxilliped, ventral (GG); d, sternal furca, ventral (E); e, leg 1 and intercoxal plate, ventral (E); f, leg 2 and intercoxal plate, ventral (E).

membranes. Corpus maxillipedis (fig. 121c) relatively slender. Shaft and claw of maxilliped comparatively large, 275 μm long combined, strongly curved, and bearing usual seta. Suture present on claw distal to seta. Sternal furca (fig. 121d) with slender, widely divergent tines.

Sympod of leg 1 (fig. 121e) with suture lines incompletely dividing it into 2 segments. Plumose seta near sympod-exopod junction not covered by ventral protrusion. First segment of exopod with short spine and single row of spinules on outer distal corner, and 1 row of setules on inner margin. Seta 4 of terminal exopodal segment not pinnate and shorter than outermost spine. Spines 1, 2, and 3 spinulated and with pectinate membranes at bases. Spines 2 and 3 with accessory processes at distal ends. Balloonlike, hyaline structure tipped with digitiform projection located between bases of spines 2 and 3. Usual 3 plumose setae on inner margin. Endopod with 2 minute knobs at tip. Leg 2 (fig. 121f) of usual caligid form. Spines of first and second exopodal segments bilaterally serrated and curved. First spine of terminal segment curved, bilaterally serrated, and reaching distal limit of terminal segment of exopod. Outer margin of first segment of endopod with row of spinules. Sympod of leg 3 (fig. 122a) with 5 large tubercles near intercoxal plate, 1 row of small spinules placed longitudinally from basal swelling toward anterodistal corner of sympod, and dorsal corrugated adhesion pad. Exopod of leg 3 3-segmented. First segment a basal swelling with terminally situated spine. Second and third segments with usual armament. Endopod 2-segmented, with usual armament. Velum located between exopod and endopod. Leg 4 (fig. 122b) with 3-segmented exopod. Spines of terminal segment decreasing in length from inner to outer margin. All 5 spines of exopod bilaterally spinulated. Spine of first segment with transparent membrane at base; other spines with pectinate membranes at base. Leg 5 (fig. 120a) represented by 1 seta near posterodistal corner of genital alae. Leg 6 (fig. 122c) represented by 2 setae near egg-laying apparatus.

Male. — Unknown.

Remarks. — Steenstrup & Lütken (1861) described *Parapetalus orientalis* as a new genus and species related to, but different from, the genus *Caligus* by the possession of "membranous wings" on the genital complex and abdomen in the female. Heller (1865) and Bassett-Smith (1899) subsequently reported this species. Pillai (1962) revised the genus and provided a description of this species.

So far, *P. orientalis* has been primarily reported from the Indian Ocean, although the present specimens are from the Philippine Islands and was

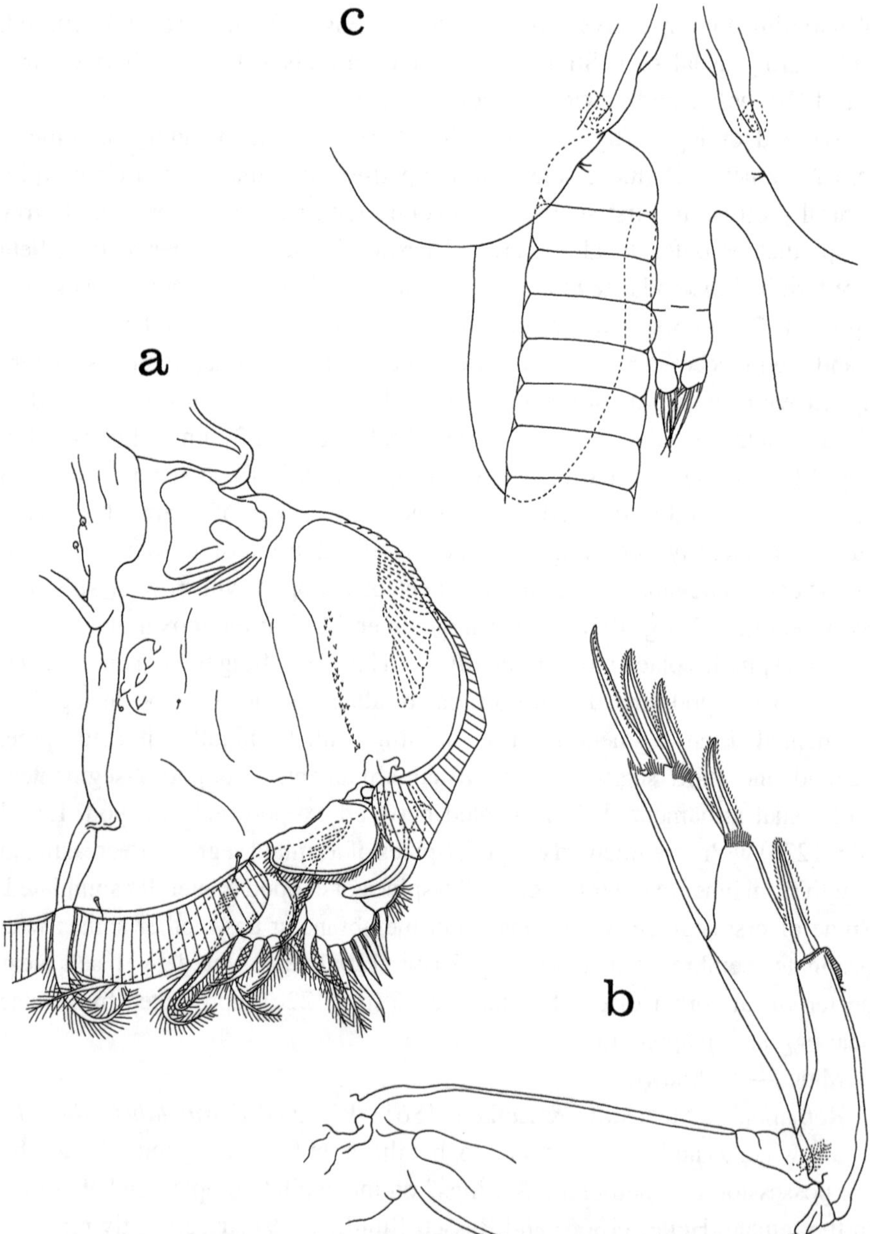

Fig. 122. *Parapetalus orientalis* Steenstrup & Lütken, 1861, female. a, leg 3, ventral (scale E); b, leg 4, ventral (E); c, legs 5 and 6, and genital area, ventral (F).

collected from the carangid *Alectis indica* (Rüppell), the same host as that of Pillai (1962).

Genus **Parechetus** Pillai, 1962

Parechetus Pillai, 1962: 287; also refer to Pillai, 1961: 118; Dojiri, 1983: 281; Prabha, 1983: 52; Pillai, 1985: 396; Kazachenko, 2001: 26; Boxshall & Halsey, 2004: 726.

Female. — Cephalothorax orbicular; posterior sinuses present. Frontal plate with lunules. Fourth pedigerous somite without plates, and forming long slender neck in conjunction with narrow anterior portion of genital complex. Posterior portion of genital complex conical, and bearing posterolateral alae (winglike extensions). Abdomen about as long as cephalothorax, and possessing pair of alae extending beyond tip of caudal ramus. Caudal ramus small and setiferous.

Antennule, antenna, mouth tube, and mandible presumably similar to other caligid genera (e.g., *Caligus* and *Lepeophtheirus*). Postantennal process present. Dentiform process of maxillule absent; setiferous papilla present. Maxilla brachiform. Maxilliped subchelate and prehensile. Sternal furca present.

Leg 1 with 2-segmented exopod and vestigial endopod; terminal exopodal segment with 3 spines, 1 plumose seta on distal inner corner, and 1 small spinule (instead of usual 3 large plumose setae). Leg 2 biramous, with 3-segmented rami. Leg 3 with large ventral apron, biramous with 3-segmented exopod and 2-segmented endopod; exopod with large clawlike spine on basal swelling of first segment. Leg 4 with 3-segmented exopod. Legs 5 and 6 apparently absent.

Male. — Unknown.

Type-species. — *Parechetus carangis* (Bassett-Smith, 1898). Unfortunately, neither the type-specimen of this species (under the binomen *Caligodes carangis*) nor Pillai's specimen could be obtained. The generic diagnosis and subsequent comparisons with other species are based on the literature.

Remarks. — In his revision of the genera of *Parapetalus* and *Pseudopetalus* (now known as *Sinocaligus*), Pillai (1962) established a new genus to include *Caligodes carangis* Bassett-Smith, 1898, collected near Aden. Pillai's main reason for establishing the new genus was that it exhibited the combined characteristics of three other genera (*Caligodes*, *Echetus*, and *Parapetalus*).

Quidor (1913) described a new species, *Caligodes lamarcki*, that he distinguished from *C. carangis* (now considered a junior synonym of *Parechetus carangis*) by the differences in the relative sizes of the tagmata between the two species. Since size is a feature of dubious taxonomic merit, particularly when no other differences seem to exist, Heegaard (1945b) suggested that these two species were synonymous. The general morphology and the structure of leg 1 suggest that the synonymy appears to be well founded. Moreover, both species were collected in the Gulf of Aden.

Another species, *Parechetus constrictus* Kirtisinghe, 1964, was described from the floor of the buccal cavity of *Caranx melampygus* Cuvier from Colombo, Sri Lanka, by Kirtisinghe (1964). The main diagnostic features of *P. constrictus* were the shorter neck and the presence of a transverse suture distinguishing the fourth pedigerous somite from the genital complex in the female. This suture, however, may simply be a fold in the cuticle at the bend in the neck (see Kirtisinghe, 1964: 75, fig. 81). The locality of collection, the general habitus, the illustrations of appendages provided by Kirtisinghe, and the similarity in hosts suggest it is likely that *P. constrictus* is a junior synonym of *P. carangis*.

The long neck, small conical genital complex, posterolateral alae of the genital complex, the abdominal alae, and the structural details of the maxilliped and leg 1 distinguish *Parechetus* from all other caligid genera. This genus has the three plumose setae missing on the exopod of leg 1 (Pillai, 1961). Apparently one of the three has been reduced to a small spinule (other two are absent). In addition, the sympod of leg 1 possesses numerous digitiform processes. Based on these two characteristics, *Parechetus* is similar to *Sinocaligus*.

Genus **Pseudanuretes** Yamaguti, 1936

Pseudanuretes Yamaguti, 1936a: 17; Yamaguti, 1963: 68; Kabata, 1965a: 29; Pillai, 1967: 362; Hameed & Pillai, 1973a: 408; Pillai, 1977: 58; Kabata, 1979: 158; Dojiri, 1983: 283; Prabha, 1983: 57; Pillai, 1985: 464; Kazachenko, 2001: 32; Boxshall & Halsey, 2004: 726.

Female. — Cephalothorax suboval, and much longer and wider than genital complex. Free margin of thoracic zone of cephalothorax extending beyond posterior limits of lateral zone and usually covering anterodorsal portion of genital complex. Cephalic and thoracic zones of cephalothorax almost completely fused, with only vestiges of dorsal transverse rib present. Frontal plate distinct. Lunules and posterior sinuses absent. Fourth pedigerous somite and anterior portion of genital complex usually covered dorsally by posteromedian area of cephalothorax, but occasionally exposed. Abdomen either absent or greatly reduced. Caudal ramus small, bearing 4 or 5 setae. Antennule as in *Caligus* or *Lepeophtheirus*. Claw of antenna with accessory tine. Postantennal process, process of maxillule, and sternal furca absent. Maxillary whip present; rarely absent.

Leg 1 with pinnate seta 4 located between spines 2 and 3 of second exopodal segment. Leg 2 biramous, with 3-segmented rami. Leg 3 reduced, biramous. Exopod 2-segmented with first segment represented by sclerotized basal swelling and large spine; second segment with 7-8 setae. Endopod 2-segmented, with first segment bearing outer setules; second segment with 5-6 setae. Leg 4 exhibiting varying degrees of reduction from brachiform structure (2-segmented exopod in *P. pomacanthi* Lewis, 1968, 1-segmented exopod in *P. fortipedis* Kabata, 1965) to setiform stub (in *P. chaetodontis* Yamaguti, 1936). Leg 5 represented by 2 groups of setae on posterolateral corner of genital complex.

Male. — Cephalothorax and appendages similar to those of female. Genital complex suboval. Antenna with very short claw.

Type-species. — *Pseudanuretes chaetodontis* Yamaguti, 1936.

Remarks. — *Pseudanuretes* was first established by Yamaguti (1936a) to accommodate the type-species, *P. chaetodontis*. As the specific name of this species implies, the specimens were collected from the gills of a species of butterflyfish, *Chaetodon*, found in Japan. Six species have been added since the discovery of this genus: *P. schmitti* Rangnekar, 1957 from *Platax teira* (Forsskål), presumably from India; *P. fortipedis* Kabata, 1965 from *Chaetodontoplus personifer* from Australia; *P. pomacanthi* Lewis, 1968 from *Pomacanthus imperator* (Bloch) from Eniwetok (Enewetak) Atoll; *P. pomacanthodi* Prabha & Pillai, 1983 from "*Pomacanthodes imperator*" [= *Pomacanthus imperator* (Bloch)] from India; *P. indicus* Prabha & Pillai, 1986 from "*Pomacanthodes imperator*" from India; and *P. papernai* Kabata & Deets, 1988 from *Pomacanthus imperator* from Gulf of Aqaba. All but one species of this genus have been obtained from the angelfishes (subfamily Chaetodontinae) and butterflyfishes (subfamily Pomacanthinae) of the Chaetodontidae (refer to Shiino, 1976); others place these two fish groups in separate families (Robins et al., 1980). *Pseudanuretes schmitti* was collected from the batfish *Platax* (subfamily Platacinae: family Ephippidae) closely related to the chaetodontid fishes.

The validity of *Pseudanuretes* has been questioned (Price, 1966; Pillai, 1967, 1977; Hameed & Pillai, 1973a). This discussion will not be repeated here (see comments on *Anuretes*). However, Kabata (1965a) concluded that *Pseudanuretes* is a valid genus. We are in agreement with Kabata that the species of this genus constitute a natural grouping. Certainly, the choice of hosts reflects affinity as do the similarities in the morphology of the appendages.

The species of *Pseudanuretes* are almost identical with the most conspicuous difference among the species displayed in the segmentation and armature of leg 4. If, as noted by Kabata (1965a), these species are considered congeners, then the progressive reduction of the fourth leg to a mere vestige must be considered a taxonomic character of specific rather than of generic rank. Therefore, we have relegated *Pseudocaligus* and *Pseudolepeophtheirus* to synonymy with their sister genera (i.e., *Caligus* and *Lepeophtheirus*, respectively) (see discussion on *Pseudocaligus* and *Pseudolepeophtheirus*).

Admittedly, the generic diagnosis that we have provided is very restrictive; perhaps in the future it will prove to be too exclusive. In the past, however, the generic diagnoses proposed by other researchers have been too broad and inclusive and the generic boundaries among caligid genera were vague and difficult to delineate.

Unfortunately, there are three species that are problematic. The first of these is *Anuretes parvulus* Wilson, 1913, collected from *Pomacanthus arcuatus* (Linnaeus) at Dry Tortugas, Florida. Like the described species of *Pseudanuretes*, it does not possess a postantennal process, dentiform process of the maxillule, and the sternal furca. In addition, as in *Pseudanuretes*, it possesses an accessory tine on the claw of the antenna of the female and was collected from a species of *Pomacanthus*. Unlike other species of *Pseudanuretes*, however, the fourth pedigerous somite of the female is fully exposed in dorsal view; the armature of the terminal exopodal segment of leg 2 and endopodal segment of leg 3 is different; and the maxillary whip is absent. Although Kabata (1965a) considered the maxillary whip taxonomically important in species of *Pseudanuretes*, both *Anuretes anomalus* Pillai, 1967 and *Anuretes rotundigenitalis* Hameed, 1976 possess this structure. Consequently, it is not a structure unique to *Pseudanuretes*. The inclusion of *A. parvulus* with *Pseudanuretes* would slightly disrupt the closely allied species of this genus. This species, however, is here considered an unusual species of *Pseudanuretes*, which exhibits close ties with *Anuretes*. Yamaguti (1936a) was the first to recognize the similarity between this species and other members of *Pseudanuretes*.

The second problematic species is *P. schmitti* described from the gills of *Platax teira*. Although Kabata (1965a) expressed some doubt as to the generic affinity of this species, *P. schmitti* appears to belong to *Pseudanuretes*. The general habitus is like that of other members. In addition, the dentiform process of the maxillule and the sternal furca are absent. Kabata stated "...Rangnekar's species differed from both *P. chaetodontis* and *P. fortipedis* by having post-antennary process and by the absence of the maxillary whip". However, if the structure reported as the postantennal process (maxillule of Rangnekar) is interpreted as the maxillary whip (the position, size, and shape support this claim), the reasons for its exclusion from *Pseudanuretes* are reduced considerably. The generic boundaries may have to be broadened for its inclusion. However, an examination of the type-specimens of *P. schmittii* is necessary before any conclusions are reached, particularly since Rangnekar's (1957) original description lacks sufficient detail for comparisons with other species of *Pseudanuretes*.

The third problematic species is *Anuretes chelatus* Prabha & Pillai, 1986 reported from the gills of *Pomacanthodes imperator* [= *Pomacanthus imperator* (Bloch)] from Trivandrum, India. As typically found in the species of *Pseudanuretes*, *A. chelatus* lacks a postantennal process, dentiform process of the maxillule, and abdomen. And, unlike typical species of *Anuretes*, it possesses a well-developed accessory tine on the terminal claw of the antenna, and carries a much reduced leg 4 with a 1-segmented exopod armed with 3 setae. These are two more character states that are found in species of *Pseudanuretes*. Although Ho & Lin (2000) have suggested transferring *A. chelatus* to *Pseudanuretes*, the possession of a well-developed sternal furca in this species would make this transfer questionable. In that paper, Ho & Lin have also suggested transferring *A. parvulus* and *A. fedderini* Price, 1966 to *Pseudanuretes*. While we can accept the transfer of the former species, the latter species can only be treated as a species inquirenda. Price's (1966) original description of *A. fedderini* is not detailed enough to confirm the validity or the true identity of this species.

Pseudanuretes chaetodontis Yamaguti, 1936
(figs. 123-125)

Pseudanuretes chaetodontis Yamaguti, 1936a: 16; Rangnekar, 1957: 12; Yamaguti, 1963: 69; Kabata, 1965a: 25; Kabata, 1979: 170; Dojiri, 1983: 288.

Material examined. — Four females from gills of spotted-faced anglefish, *Chaetodontoplus personifer* McCulloch, collected at Heron Island, Queensland, Australia, on 12 January 1964 by Mr. P. C. Young and identified by Dr. Z. Kabata. This material on loan from Natural History Museum, London, England.

Female. — Body as in fig. 123a, b and consisting primarily of cephalothorax and genital complex. Total length (not including setae on caudal ramus)

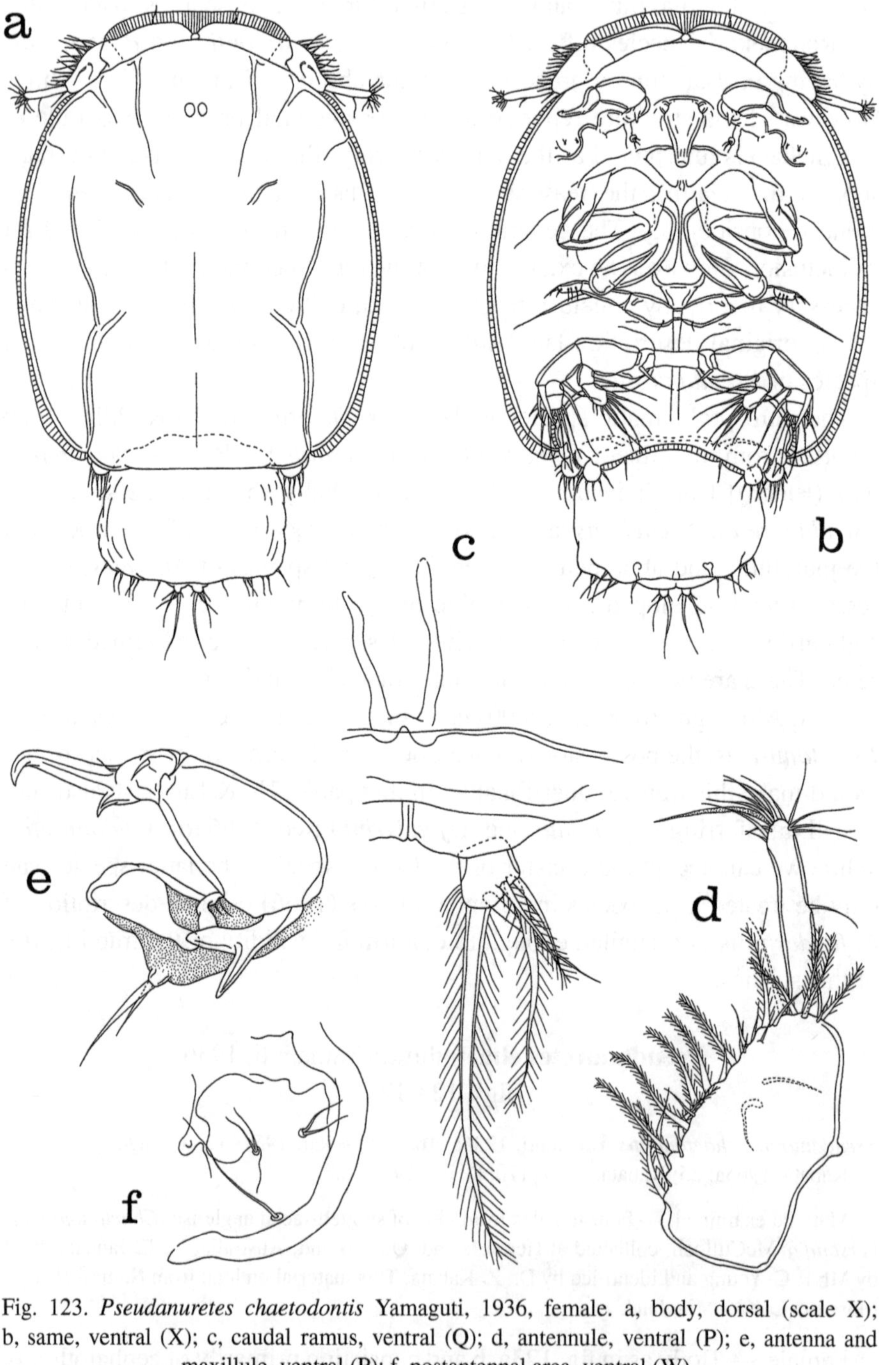

Fig. 123. *Pseudanuretes chaetodontis* Yamaguti, 1936, female. a, body, dorsal (scale X); b, same, ventral (X); c, caudal ramus, ventral (Q); d, antennule, ventral (P); e, antenna and maxillule, ventral (P); f, postantennal area, ventral (W).

only 0.96 mm (0.93-1.00 mm) (n = 4). Cephalothorax very large, oval, 0.78 mm (0.76-0.82 mm) × 0.62 mm (0.60-0.66 mm), and with no discernible posterior sinuses. Transparent membrane on margin of cephalothorax present, although inconspicuous or hidden in dorsal view. Free margin of thoracic zone extending beyond posterior tip of lateral zone, and covering anterior portion of genital complex. Frontal plate relatively large and rimmed with rather wide transparent membrane. Antennule extending almost to lateral limit of cephalothorax.

Fourth pedigerous somite fused to genital complex (fig. 123b). Genital complex somewhat rectangular (slightly distorted in specimen figured), wider than long, 0.28 mm (0.24-0.32 mm) × 0.35 mm (0.31-0.39 mm). Abdomen (fig. 123c) very small, almost completely fused to genital complex.

Caudal ramus (fig. 123c) small, longer than wide, 18 × 16 μm, and bearing 5 setae (3 plumose and 2 smooth).

Antennule (fig. 123d) 2-segmented. First segment 105 μm long (measured along setiferous margin), bearing 27 setae (1 seta broken off and represented by dotted line in figure). Plumose setae much larger than smooth setae. Second segment slender, 84 μm long, with usual 13 + 1 aesthetasc. Antenna (fig. 123e) apparently 4-segmented. First segment unarmed; second with large posteriorly directed, sharp, spinelike process. Third segment robust, with dorsal adhesion pad (not drawn in figure). Claw recurved and possessing 2 setae and accessory tine near base. Postantennal area (fig. 123f) without process, but carrying 3 groups of setules contained within sclerotized semicircle.

Mouth tube (fig. 124a) slender, 138 × 73 μm, possessing intrabuccal stylet on labrum and strigil on labium. Mandible (fig. 124b) with extremely long, slender third section. Fourth section equipped with 12 small teeth. Maxillule (fig. 123e) represented by papilla tipped with 2 setae posterior to base of antenna. Maxilla (fig. 124c) with large bilaterally spinulated calamus (spinulation on other side hidden from view in figure) and very tiny, smooth canna. Maxillary whip (fig. 124d) lateral to bases of maxilla (see fig. 123b), lightly sclerotized, and 92 μm long. Maxilliped (fig. 124e) with robust corpus carrying 2 tubercles on inner margin. Shaft and claw 135 μm long and bearing 1 ventral seta and 1 dorso-inner tubercle. Sternal furca absent.

Leg 1 (fig. 124f) with relatively distinct coxa and basis. Plumose seta at junction of sympod and exopod not covered by ventral protrusion. Basis with usual plumose seta and with patch of denticles on posteroventral surface. Seta on outer distal corner of first exopodal segment relatively long and slender. Seta 4 smooth, shorter than outermost spine, and located between spines 2 and

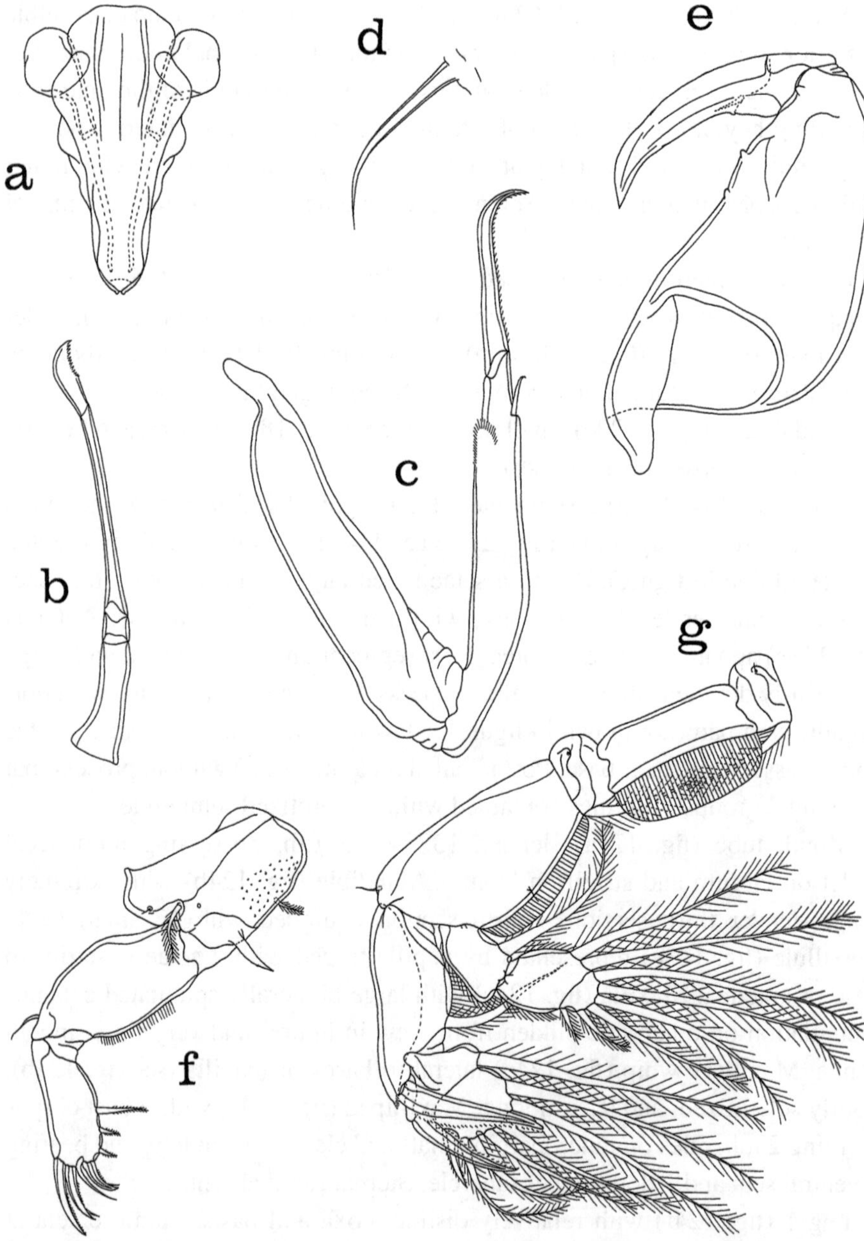

Fig. 124. *Pseudanuretes chaetodontis* Yamaguti, 1936, female. a, mouth tube, ventral (scale R); b, mandible, ventral (Q); c, maxilla, dorsal (Q); d, maxillary whip, ventral (R); e, maxilliped, dorsal (R); f, leg 1, ventral (R); g, leg 2 and intercoxal plate, ventral (R).

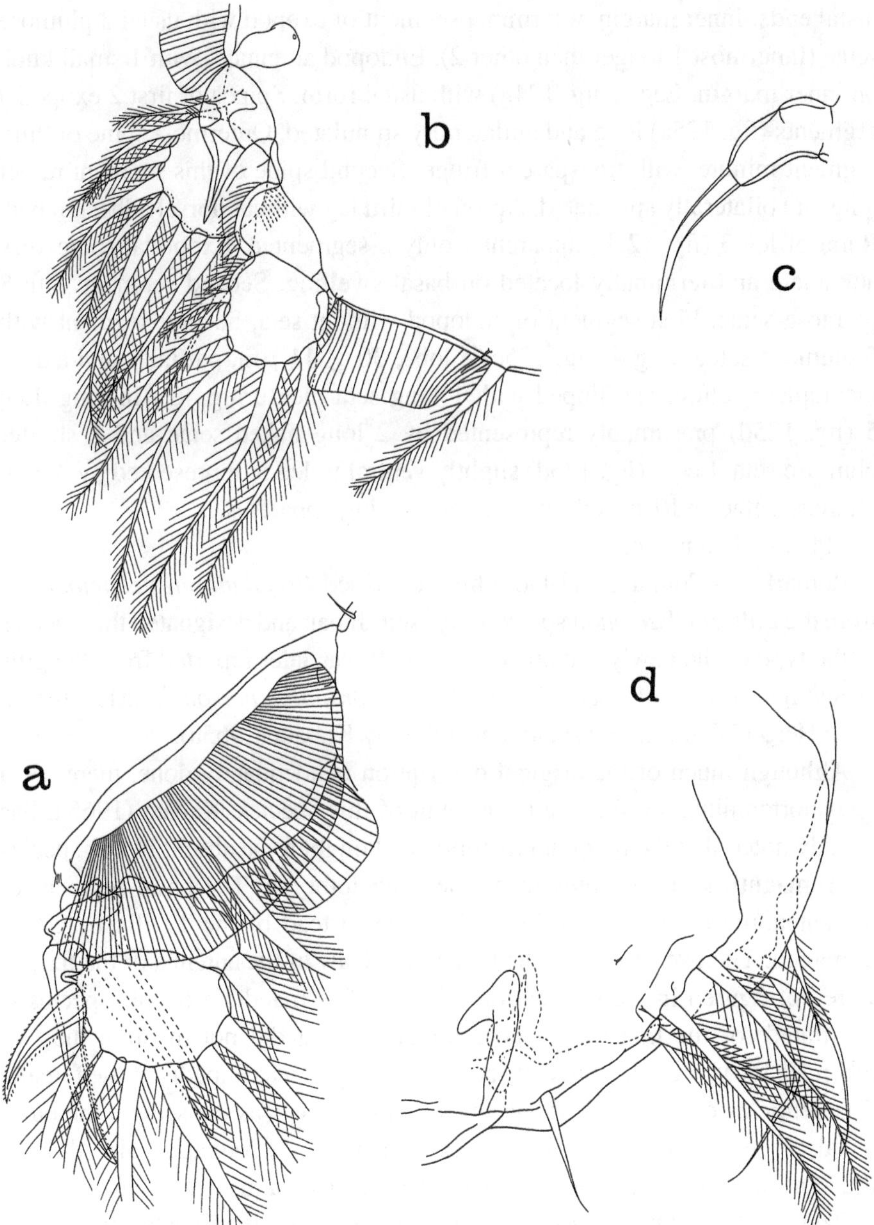

Fig. 125. *Pseudanuretes chaetodontis* Yamaguti, 1936, female. a, exopod of leg 2, dorsal (scale Q); b, exopod and endopod of leg 3, ventral (Q); c, leg 4, ventral (Q); d, legs 5 and 6, and genital area, ventral (Q).

3. Spines 1, 2, and 3 spinulated, and spines 2 and 3 with secondary processes at distal ends. Inner margin of terminal segment of exopod with usual 3 plumose setae (innermost 1 longer than other 2). Endopod attenuate, with 1 small knob on inner margin. Leg 2 (fig. 124g) with usual form. Spines of first 2 exopodal segments (fig. 125a) long and unilaterally spinulated. Outermost spine of third segment minute, with transparent fringe. Second spine of this segment rather long and bilaterally spinulated. Apron of third leg without dorsal adhesion pad. Rami of leg 3 (fig. 125b) apparently only 2-segmented. Exopodal spine very attentuate and terminally located on basal swelling. Second segment with 8 plumose setae. First segment of endopod without seta. Second segment with 5 plumose setae. Leg 4 (fig. 125c) a lobe, 30 × 11 μm, partially divided by incomplete suture, and tipped with 1 long seta measuring 46 μm long. Leg 5 (fig. 125d) presumably represented by 2 long naked setae and 1 shorter plumose seta. Leg 6 (fig. 125d) slightly ventral to leg 5, represented by 3 long plumose setae and 1 naked seta near egg-laying apparatus.

Male. — Unknown.

Remarks. — Yamaguti (1936a) first described *Pseudanuretes chaetodontis* from the gills of *Chaetodon* sp. from Kyushu, Japan and designated this species as the type of the newly established genus. It was later reported from the gills of another species of chaetodontid, *Chaetodontoplus personifer* McCulloch, from Heron Island, Queensland, Australia, by Kabata (1965a).

Although much of the original description is adequately done, there are a few shortcomings to Yamaguti's account of this species. Kabata (1965a) has already noted that the marginal membrane, thought to be absent in this species by Yamaguti, is indeed present on the cephalothorax. The basal segment of the antennule of the female bears 27 setae instead of the 17 illustrated by Yamaguti. The two setae near the base of the claw of the antenna of the female were not figured in the original description. The mandible of this species is illustrated for the first time. The original account did not mention the two tubercles located on the inner surface of the corpus maxillipedis of the female. The denticles on the basis of leg 1 were omitted, as was the seta on the outer distal corner of the first exopodal segment of leg 1. Yamaguti also did not illustrate an outer, plumose seta (usual seta originating from basis) of leg 3. The caudal ramus possesses 5 setae as described by Kabata (1965a) for *Pseudanuretes fortipedis*, and not four setae as figured by Yamaguti.

The most important observational error by Yamaguti in his original description of this species is the omission of the minute outer spine of the terminal segment of the exopod of leg 2. As mentioned earlier, leg 2 appears to be a relatively conservative feature in members of the Caligidae, so the armature of the exopod of leg 2 may be a useful generic discriminant in the case of this genus. Unfortunately, the descriptions of leg 2 of *P. schmitti* by Rangnekar (1957) and *P. fortipedis* by Kabata (1965a) are not sufficient for comparative purposes. However, comparisons with the descriptions of leg 2 of *P. pomacanthodi* by Prabha & Pillai (1983), *P. indicus* by Prabha & Pillai (1986), and *P. papernai* by Kabata & Deets (1988) revealed that there is a significant difference. While the armature formula of the terminal exopodal segment of leg 2 is I, 6 in *P. chaetodontis*, *P. indicus*, and *P. papernai*, it is I, 7 in *P. pomacanthodi*.

It is difficult to distinguish leg 5 from leg 6 in the females of *P. chaetodontis* and *P. fortipedis* (Kabata, 1965a). However, a group of 3 setae is slightly anterodorsal to a second group of setae (leg 6), which includes the smooth seta next to the egg-laying apparatus. Most likely this anterior group represents leg 5 and the more posterior group leg 6.

Pseudanuretes chaetodontis may be distinguished from *P. fortipedis*, *P. indicus*, *P. papernai*, *P. pomacanthi*, and *P. pomacanthodi* by the morphology of leg 4. All of the five named species possess a brachiform fourth leg, while leg 4 of *P. chaetodontis* consists of a lobe tipped with a single seta. In this respect, the type-species closely resembles *P. schmitti*. In fact, there are a number of features that these two species apparently share. But, according to Rangnekar (1957), in *P. schmitti* the postantennal process (Rangnekar's maxillule), although minute, is present. This process, however, may simply be a misidentification of the maxillary whip. In addition, the fourth leg possesses an additional seta at the tip in *P. schmitti*.

Although not much is known about the host preference of this species, it has so far been recorded only from species of the family Chaetodontidae occurring off Kyushu, Japan and Queensland, Australia.

Genus **Pseudechetus** Prabha & Pillai, 1979

Pseudechetus Prabha & Pillai, 1979: 428; Dojiri, 1983: 293; Prabha, 1983: 52; Pillai, 1985: 871; Kazachenko, 2001: 26; Boxshall & Halsey, 2004: 726.

Female. — For diagnosis refer to Prabha & Pillai (1979).

Male. — Unknown.

Type-species. — *Pseudechetus fimbriatus* Prabha & Pillai, 1979.

Remarks. — The type and only described species of *Pseudechetus*, *P. fimbriatus*, was described by Prabha & Pillai (1979) from the "throat" of an unidentified species of carangid. Like the morphologically similar genus

Parechetus, whose members also infest carangids, *Pseudechetus* possesses a *Caligus*-like cephalothorax (including lunules) and an elongate, slender neck formed by the fourth pedigerous somite and an anterior extension of the genital complex in the female. However, the lamelliform posterolateral processes of the genital complex of the female are single in *Parechetus*, but double and very slender in *Pseudechetus*. *Parechetus* bears a pair of aliform lamellae on the abdomen of the female; these processes are slender and digitiform in *Pseudechetus*. *Pseudechetus* possesses a maxillule with a bifid dentiform process (absent in *Parechetus*), a pair of anterior accessory tines on the sternal furca, and 3 plumose setae on the inner margin of the terminal exopodal segment of leg 1 (as opposed to only 1 spinule in this position in *Parechetus*).

Genus **Pupulina** Beneden, 1892

Pupulina Beneden, 1892: 254; C. B. Wilson, 1935b: 593; M. S. Wilson, 1952: 246; Yamaguti, 1963: 84; Dojiri, 1983: 314; Prabha, 1983: 56; Pillai, 1985: 422; Kazachenko, 2001: 33; Boxshall & Halsey, 2004: 726.

Diphyllogaster (?) Brian, 1899c; see Pillai & Padmanabhan, 1963: 91; Prabha, 1983: 55; Kazachenko, 2001: 31; also refer to discussion on *Diphyllogaster*.

Papulina [sic] Pillai & Padmanabhan, 1963: 93; Pillai, 1964b: 237.

Female. — Cephalothorax suborbicular, narrower anteriorly, with distinct posterior sinuses. Frontal plate distinct, without lunules. Dorsal ribs in prominent H-shaped pattern. Fourth pedigerous somite small, free, and without dorsal plates. Genital complex smaller than cephalothorax, not greatly expanded, and possessing pair of posterolateral processes. Abdomen indistinctly comprising 3 free somites. Caudal ramus elongate, slender, bearing setae. Body covered with small spinules.

Antennule 2-segmented. First segment with numerous setae. Second segment cylindrical with several smaller setae. Antenna 4-segmented. Second segment with posteriorly directed spinelike process. Terminal segment hamate. Postantennal process present. Mouth tube with intrabuccal stylet and strigil. Mandible with 12 teeth on inner margin of terminal section. Maxillule consisting of dentiform projection and adjacent setiferous papilla. Small conical process posteromedial to dentiform projection of maxillule present. Maxilla brachiform. Maxilliped subchelate and prehensile. Dentiform or membranous process situated immediately posterior to maxilliped. Sternal furca absent.

Leg 1 biramous, with 2-segmented rami. Terminal exopodal segment with 3 spines, 1 seta on inner distal corner, and 3 large, inner plumose setae. Second endopodal segment with 3 plumose setae. Leg 2 biramous with 3-segmented rami. First and second endopodal segments with inflated outer margins. Leg

3 with large ventral apron, biramous, and bearing 3-segmented rami. Exopod with formula I-1; I-1; III, 4. First exopodal segment largest of 3; basal swelling absent, and exopodal spine of first segment not enlarged to clawlike structure. Endopod with formula 0-0; 0-2; 4. Outer margin of first endopodal segment greatly expanded. Leg 4 with 3-segmented exopod displaying formula I-0; I-0; III (IV, if including small innermost spine). Leg 5 represented by setae on posterolateral process. Leg 6 absent.

Male. — Cephalothorax similar in shape to that of female. Genital complex oval, small. Abdomen distinctly comprising 2 free somites. Antenna with spinelike process and adhesion pads on second segment. Terminal segment a claw with proximal accessory tine. Process of maxillule with accessory process. Corpus maxillipedis with prominent tubercle in myxal area. Second and third spines of exopod of leg 2 may be different from that of female. Legs 5 and 6 represented by 2 groups of setae. All other appendages as in female.

Type-species. — *Pupulina flores* Beneden, 1892.

Remarks. — *Pupulina* was first established by Beneden (1892). This genus presently contains two species (*P. brevicauda* M. S. Wilson, 1952 and *P. minor* M. S. Wilson, 1952) in addition to the first described species, *P. flores*. The members of this genus appear to be specific for species of the ray family Mobulidae. They have been collected from widely scattered localities (table XVI).

The members of *Pupulina* have already been diagnosed and clearly distinguished from other caligid genera (M. S. Wilson, 1952). The salient features of this genus are: (1) the presence of posterolateral processes on the genital complex of the female, (2) the presence of a small conical process posteromedial to the dentiform projection of the maxillule, (3) the presence of a dentiform or membranous process immediately posterior to the maxilliped, (4) a well-developed endopod of leg 1, (5) inflated outer margin of the first and second endopodal segments of leg 2 and first endopodal segment of leg 3, (6) distinctly 3-segmented rami of leg 3, and (7) the armature of the exopod of leg 3.

Pupulina flores Beneden, 1892
(figs. 126-133)

Pupulina flores Beneden, 1892: 254; C. B. Wilson, 1935a: 331; C. B. Wilson, 1935b: 594; Bere, 1936: 590; M. S. Wilson, 1952: 248; Pillai & Padmanabhan, 1963: 91; Dojiri, 1983: 318.
Lepeophtheirus floresi [sic] Bassett-Smith, 1899: 455.
Lepeophtheirus flores Wilson, 1905a: 617.

Material examined. — Five females, 1 immature female, and 2 males (USNM 60439) collected from *Manta birostris* (Walbaum) during Gifford Pinchot Expedition near Galápagos

Islands on 16 June 1929. Specimens identified by C. B. Wilson and used in redescriptions by both C. B. Wilson and M. S. Wilson. Material on loan from National Museum of Natural History, Smithsonian Institution, Washington, D.C.

Female. — Body as in fig. 126a. Total length (not including setae on caudal ramus) 13.96 mm (13.58-14.34 mm) (n = 2). Cephalothorax subcircular, narrower anteriorly, almost as wide as long, 6.09 mm (5.88-6.34 mm) × 5.93 mm (5.64-6.37 mm) (n = 4). Posterior sinuses shallow, with transparent, striated membranes along outer and inner edge. Frontal plate relatively small, with no lunules. Tip of antennule well within lateral limit of cephalothorax. Free margin of thoracic zone of cephalothorax extending beyond posterior limit of lateral zone. Lateral margin of cephalothorax rimmed with long setules and usual transparent membrane. Posterolateral corner of cephalothorax bearing few rows of long setules.

Fourth pedigerous somite wider than long, 0.66 mm (0.60-0.70 mm) × 1.89 mm (1.86-1.93 mm). Genital complex wider than long, 2.14 mm (2.03-2.26 mm) × 2.97 mm (2.79-3.19 mm) (excluding posterolateral processes). Posterolateral processes large, 1.07 mm (1.03-1.16 mm) × 0.90 mm (0.80-1.03 mm), and extending approximately to midlength of abdomen. Abdomen indistinctly comprising 3 free somites, slender, longer than genital complex

TABLE XVI

Hosts and localities of collections of the three species of *Pupulina* Beneden, 1892

Species	Host	Locality	Reference
P. brevicauda M. S. Wilson, 1952	*Mobula lucasana* Beebe and Tee-Van [= *Mobula thurstoni* (Lloyd)]	Santa Catalina, California	M. S. Wilson, 1952
	Mobula diabolus (Shaw) [= *Mobula mobular* (Bonnaterre)]	Trivandrum, India	Pillai & Padmanabhan, 1963
P. minor M. S. Wilson, 1952	*Mobula lucasana* [= *Mobula thurstoni*]	Santa Catalina, California	M. S. Wilson, 1952
	Mobula diabolus [= *Mobula mobular*]	Trivandrum, India	Pillai, 1964b
P. flores Beneden, 1892	*Manta* (?)	Azores	Beneden, 1892
	Manta birostris (Walbaum)	Galápagos	C. B. Wilson, 1935b
	Manta birostris	Galápagos	M. S. Wilson, 1952 (same specimens as C. B. Wilson, 1935)
	"probably *Mobula hypostoma*"	Lemon Bay, Florida	Bere, 1936

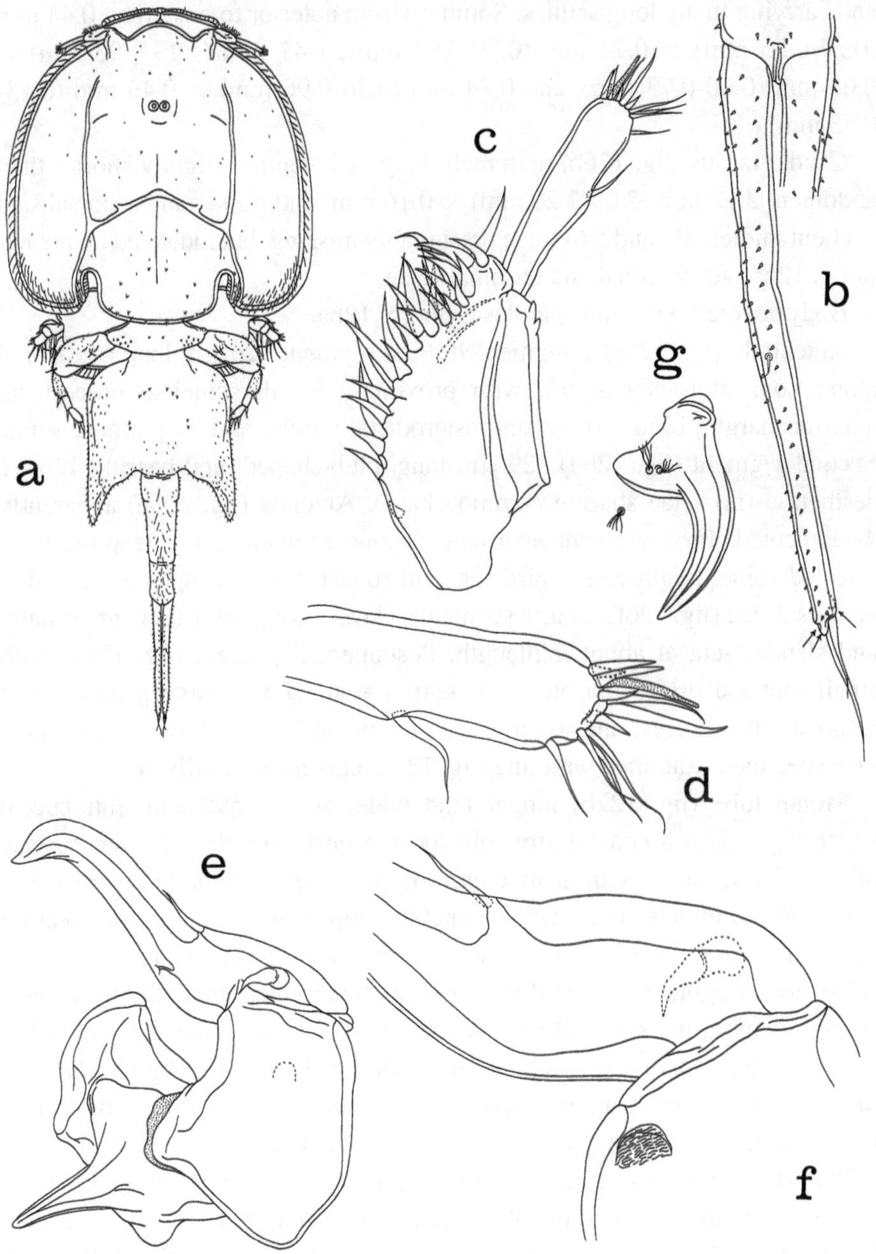

Fig. 126. *Pupulina flores* Beneden, 1892, female. a, body, dorsal (scale CC); b, caudal ramus (arrows indicating positions of broken setae), ventral (H); c, antennule, (plumosities not drawn) ventral (C); d, distal portion of antennule, ventral (G); e, antenna, ventral (C); f, junction of third segment (showing dorsal corrugated pad) and claw of antenna, dorsal (G); g, postantennal process, ventral (F).

and carrying many long setules. Somites (from anterior to posterior) 0.44 mm (0.27-0.56 mm) × 0.74 mm (0.70-0.83 mm), 1.41 mm (1.13-1.93 mm) × 0.66 mm (0.60-0.73 mm), and 0.84 mm (0.76-0.96 mm) × 0.46 mm (0.43-0.53 mm).

Caudal ramus (fig. 126b) extremely long and slender, slightly shorter than abdomen, 2.17 mm (2.09-2.26 mm) × 0.16 mm, and possessing 1 dorsal seta at about midlength, and 5 (3 setae presumably missing, but indicated by arrows in fig. 126b) subterminal and terminal setae.

Body covered with tiny spinules as in fig. 126a.

Antennule (fig. 126c) 2-segmented. First segment 426 μm long (measured along nonsetiferous margin), with proximal, dorsal, spinelike process on anterior margin, bifid process on posterodistal corner, and 29 plumose setae. Second segment (fig. 126d) 529 μm long, club-shaped, and bearing 12 + 1 aesthetasc (no setae sharing common base). Antenna (fig. 126e) apparently 4-segmented. First segment unarmed. Second segment carrying posteriorly directed spinelike process. Third segment robust, with corrugated protrusion on dorsal side (fig. 126f). Fourth segment a claw bearing robust seta proximally and slender seta at about midlength. Postantennal process (fig. 126g) with small rounded protrusion on outer margin near base, 1 large recurved tine rimmed with flanges, and usual 3 sclerites possessing multibranched setule. One specimen examined with tine (fig. 127a) curved outwardly at tip.

Mouth tube (fig. 127b) longer than wide, 805 × 598 μm. Intrabuccal stylet (fig. 127c) a conical structure located on both sides of inner surface of frons labri. Strigil with many dentiform processes present on inner surface of labium. Mandible (fig. 127d) distinctly comprising 4 sections and bearing 12 teeth. Maxillule (fig. 127e) consisting of papilla tipped with 3 setae, and relatively straight, large dentiform process. Sclerotized area slightly antero-outer to papilla transformed into slender spinelike tine [compare with other caligids, e.g., *Lepeophtheirus pectoralis* (Müller, 1776) or *Caligodes laciniatus* (Krøyer, 1863)]. Cuticular spinelike process located slightly posteromedial to dentiform process of maxillule. Maxilla (fig. 127f) brachiform. Flabellum (fig. 127g) a serrated membrane, greatly curled proximally. Calamus (fig. 127h) with 3 serrated membranes, approximately twice length of canna. Canna (fig. 127h) with 2 serrated membranes. Maxilliped (fig. 128a) with moderately slender corpus. Corpus maxillipedis with small spinulated tubercle (fig. 128b) on inner side. Shaft and claw 930 μm long combined. Seta located near junction of shaft and claw. Sternal furca absent; however, 1 pair of large cuticular spines (fig. 128c, d) located posterior to bases of maxillipeds present.

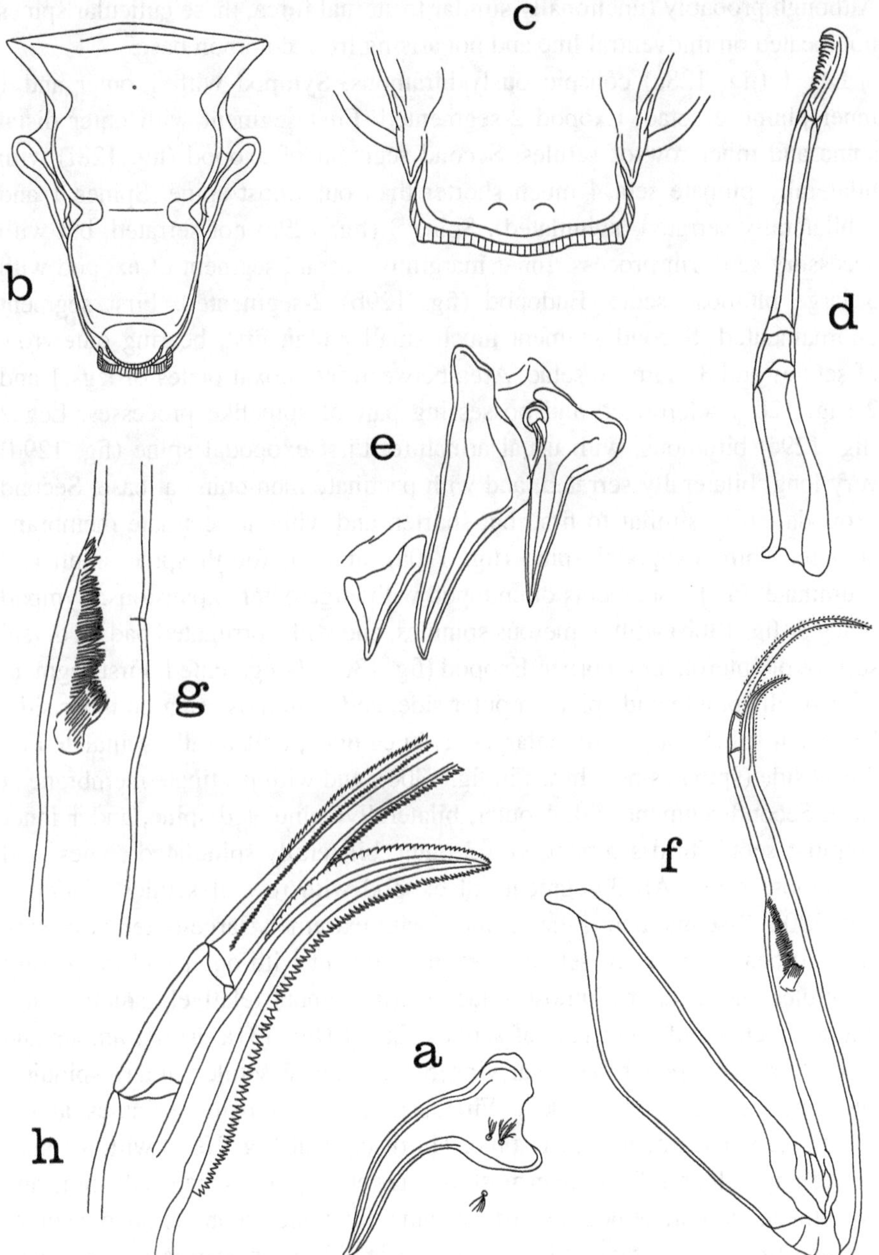

Fig. 127. *Pupulina flores* Beneden, 1892, female. a, postantennal process, ventral (scale F); b, mouth tube, ventral (C); c, intrabuccal stylet and frons labri (with labium removed), dorsal (D); d, mandible, ventral (G); e, maxillule and cuticular processes, ventral (F); f, maxilla, dorsal (F); g, flabellum of maxilla, dorsal (B); h, distal portion of maxilla, dorsal (E).

Although probably functionally similar to sternal furca, these cuticular spines not located on midventral line and not arising from common base.

Leg 1 (fig. 128e) conspicuously biramous. Sympod with 1 outer and 1 inner plumose setae. Exopod 2-segmented. First segment with outer distal spine and inner row of setules. Second segment of exopod (fig. 128f) with bilaterally pinnate seta 4 much shorter than outermost spine. Spines 1 and 2 bilaterally serrated (spinulated). Spine 3 (fig. 129a) not serrated, but with accessory setiform process. Inner margin of second segment of exopod with 3 large plumose setae. Endopod (fig. 129b) 2-segmented. First segment unornamented. Second segment much smaller than first, bearing outer row of setules and 3 plumose setae. Area between intercoxal plates of legs 1 and 2 (fig. 128c) sclerotized and possessing pair of spinelike processes. Leg 2 (fig. 129c) biramous, with usual armature. First exopodal spine (fig. 129d) very long, bilaterally serrated, and with pectinate membrane at base. Second exopodal spine similar to first, but shorter, and without pectinate membrane at base. Third exopodal spine (fig. 130a) smooth; fourth spine small and acuminate. First 2 segments of endopod with large outer expansions. Sympod of leg 3 (fig. 130b) with numerous spinules, and with corrugated pad on dorsal surface of anterolateral corner. Exopod (fig. 130c) 3-segmented. First segment with small spinule and spine on outer side, and 1 plumose seta on inner side. Exopodal spine not greatly enlarged as in caligids, unilaterally spinulated on dorsal side (spinules not shown in fig. 130c), and with pectinate membrane at base. Second segment with 1 outer, bilaterally spinulated spine, and 1 inner plumose seta. Third segment with 3 outer, bilaterally spinulated spines, and 4 plumose setae. All 3 segments of exopod with row of setules. Endopod (fig. 130d) 3-segmented. First segment with extremely large outer expansion (velum) bearing rows of setules. Second segment with outer and inner rows of setules and 2 inner plumose setae. Third segment relatively small with 4 plumose setae and inner row of setules. Leg 4 (fig. 131a, b) uniramous and brachiform. Sympod robust, carrying 5 long dorsal setules, many spinules, and small distal plumose seta. First segment with many spinules and 1 spine. Second segment with 1 spine. Third segment (fig. 131c) with 3 spines decreasing in length from innermost to outermost spines. One small spine and 1 pectinate membrane near innermost spine. First and second exopodal spines with pectinate membrane at bases. Leg 5 (fig. 131d) consisting of 1 plumose seta and group of 3 plumose setae.

Male. — Body as in fig. 131e. Total length 11.39 mm (11.16-11.62 mm) (n = 2) specimens. Cephalothorax as in female except smaller, 4.93 mm (4.91-4.95 mm) × 4.81 mm (4.68-4.95 mm).

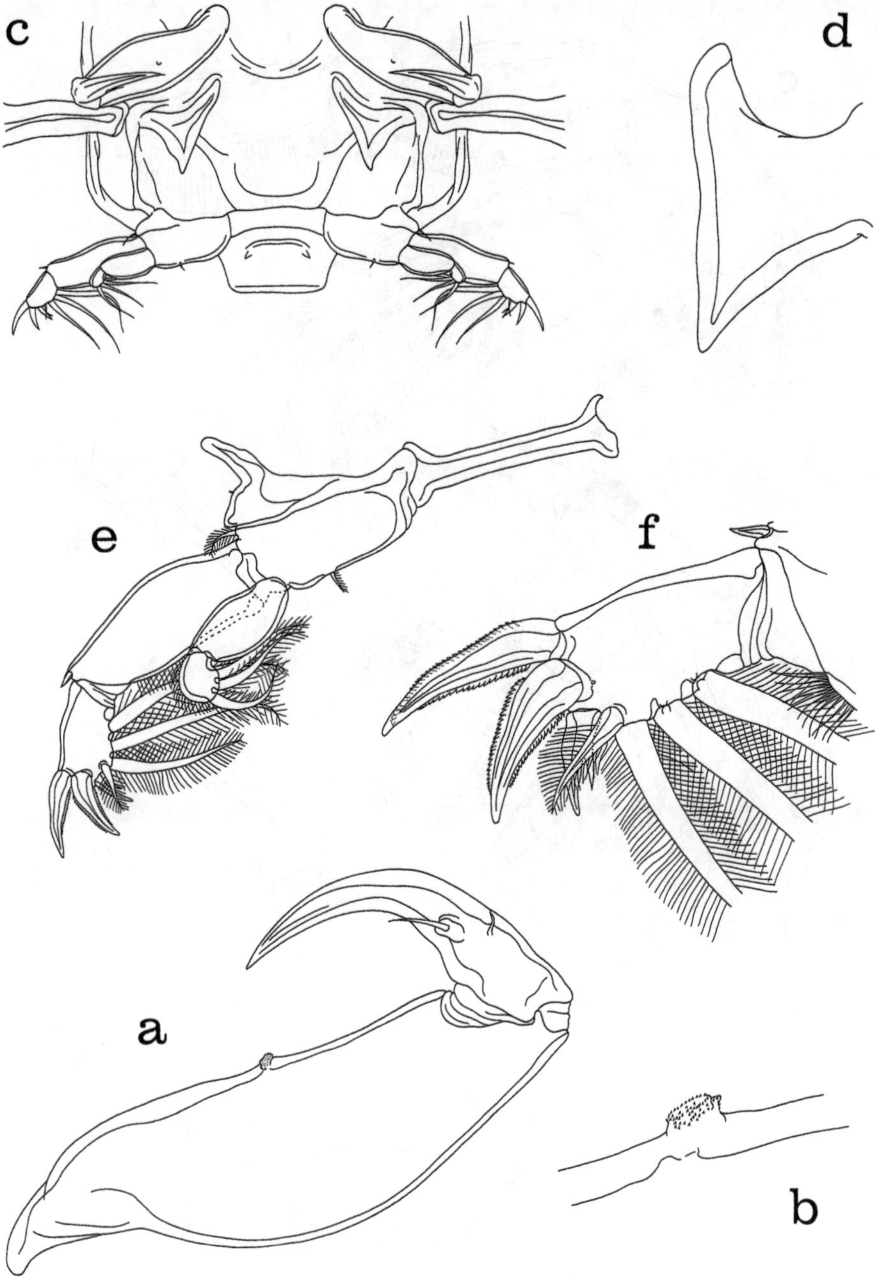

Fig. 128. *Pupulina flores* Beneden, 1892, female. a, maxilliped, ventral (scale C); b, tubercle on inner margin of corpus maxillipedis, ventral (D); c, maxilliped, sternal area, and leg 1, ventral (AA); d, spinelike sternal process (posteromedial to bases of maxillipeds), ventral (F); e, leg 1 and intercoxal plate, ventral (H); f, distal portion of exopod of leg 1, ventral (B).

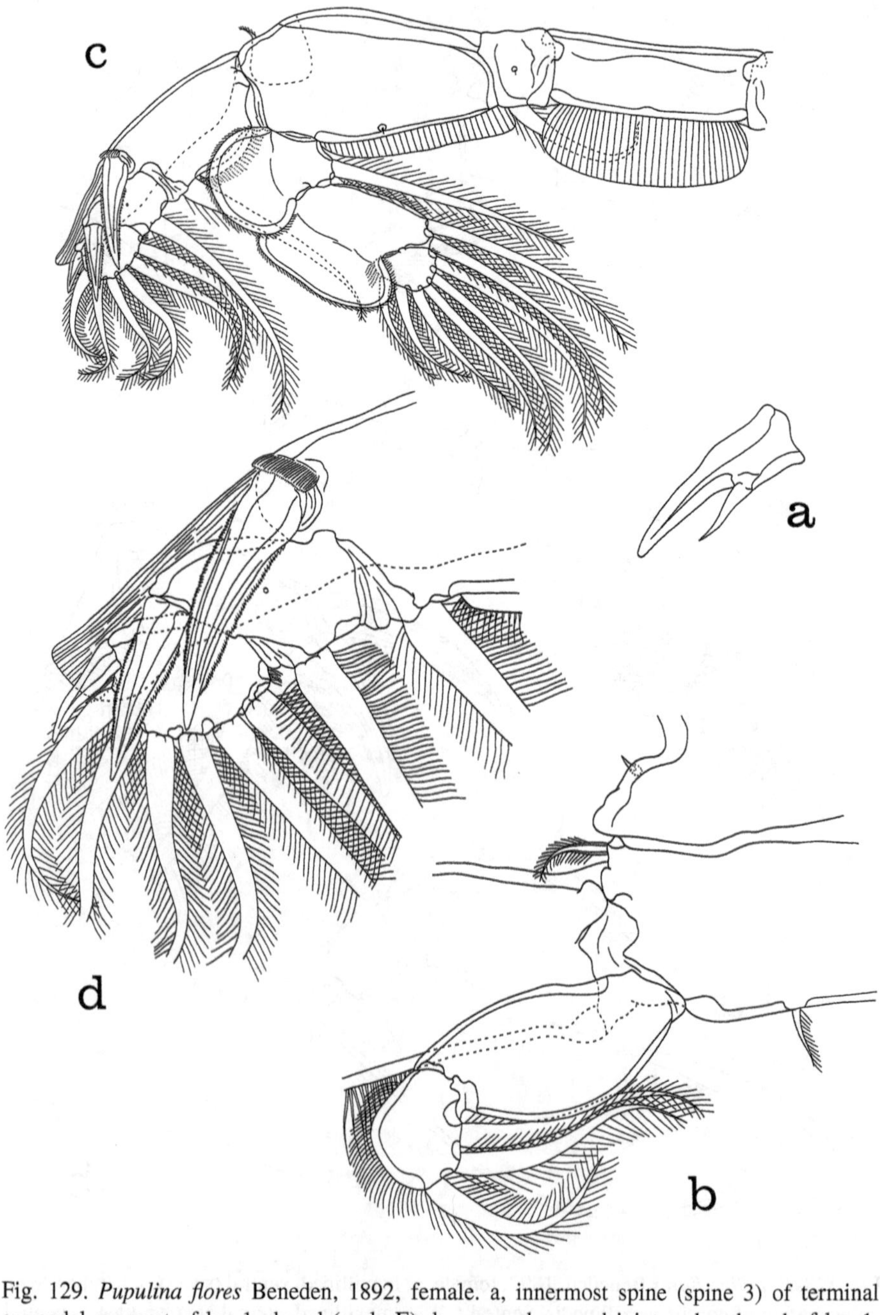

Fig. 129. *Pupulina flores* Beneden, 1892, female. a, innermost spine (spine 3) of terminal exopodal segment of leg 1, dorsal (scale E); b, sympod-exopod joint and endopod of leg 1, ventral (B); c, leg 2 and intercoxal plate, ventral (H); d, exopod of leg 2, ventral (B).

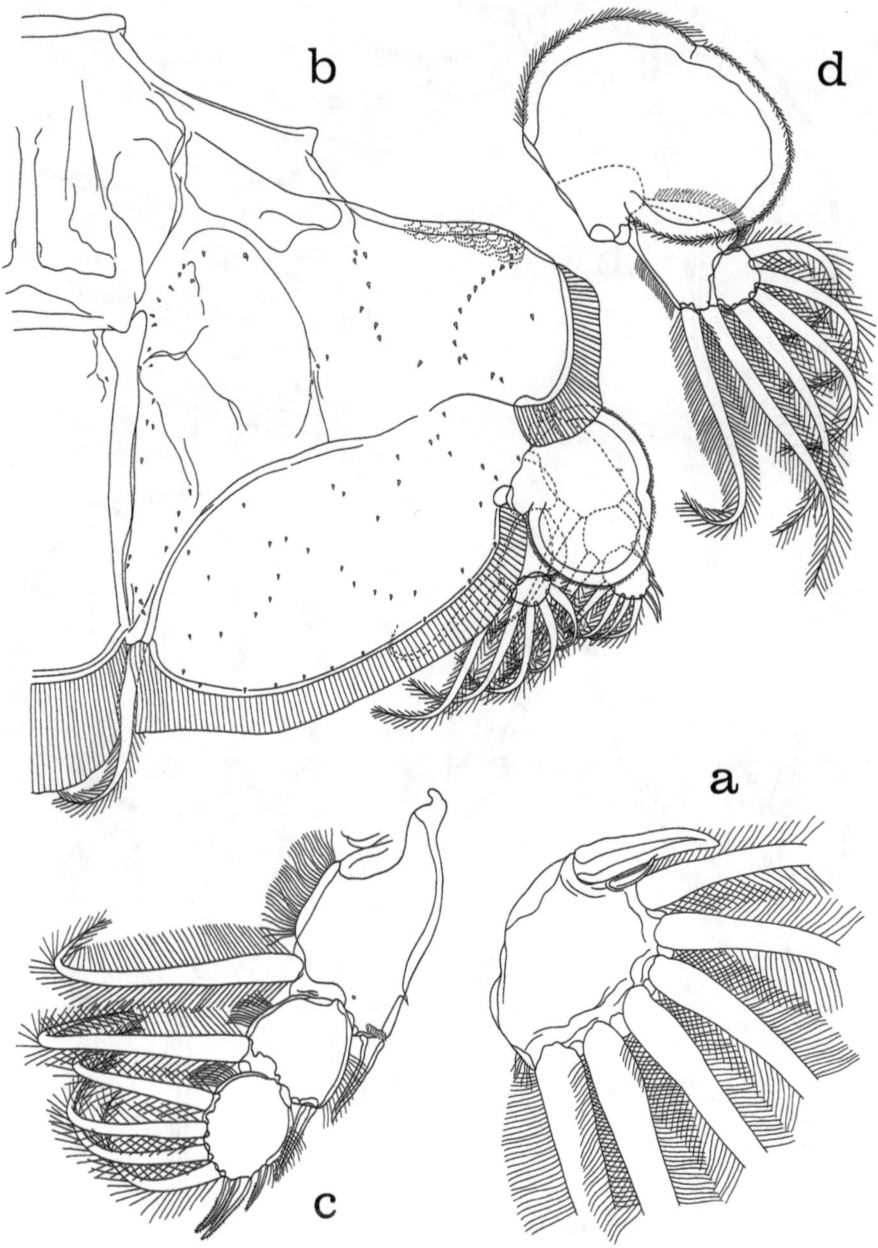

Fig. 130. *Pupulina flores* Beneden, 1892, female. a, terminal exopodal segment of leg 2, dorsal (scale G); b, leg 3, ventral (H); c, exopod of leg 3, ventral (I); d, endopod of leg 3, ventral (C).

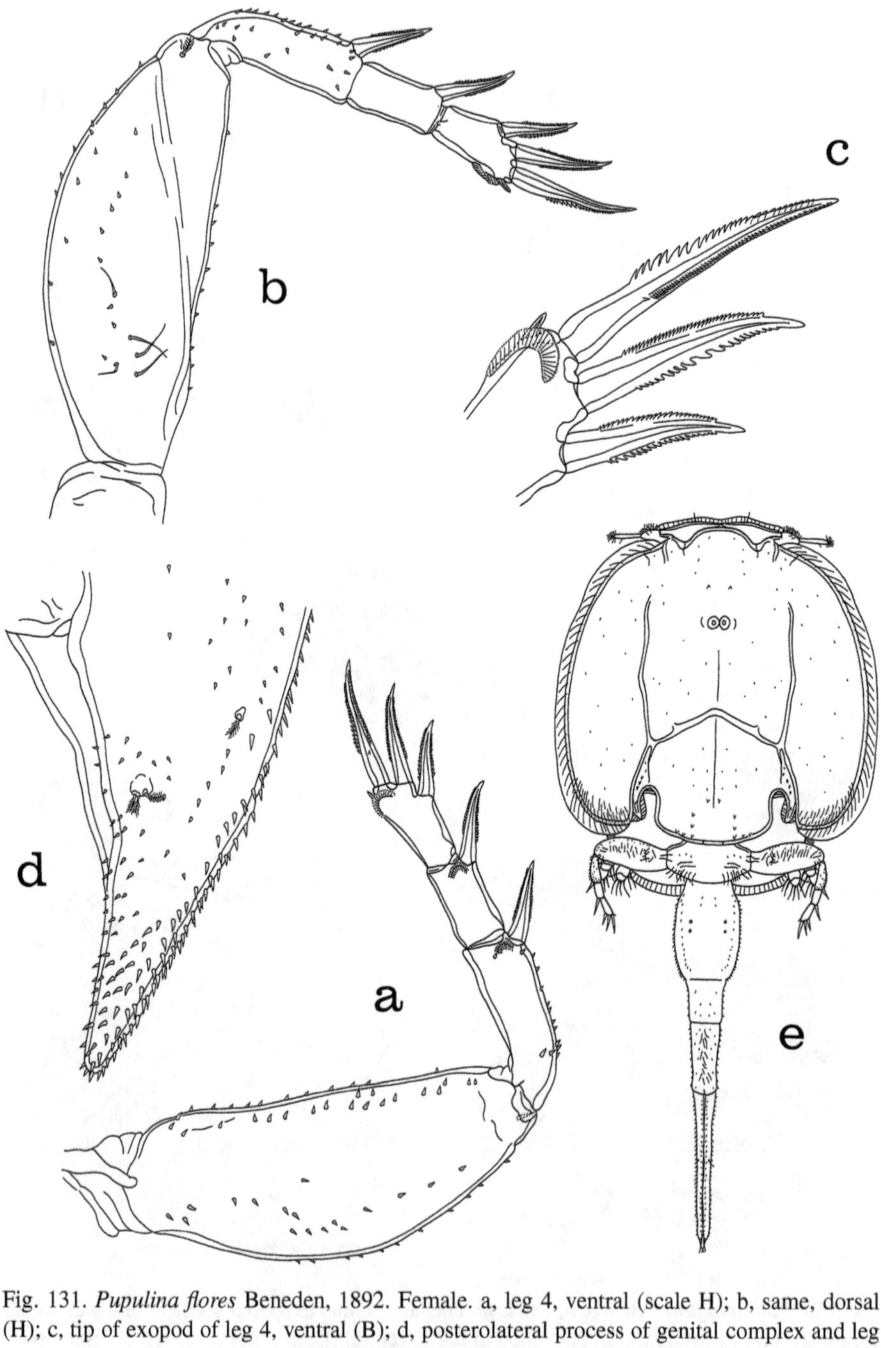

Fig. 131. *Pupulina flores* Beneden, 1892. Female. a, leg 4, ventral (scale H); b, same, dorsal (H); c, tip of exopod of leg 4, ventral (B); d, posterolateral process of genital complex and leg 5, ventral (H). Male. e, body, dorsal (A).

Fourth pedigerous somite wider than long, 0.63 mm (0.60-0.66 mm) × 1.46 mm. Genital complex somewhat oval in outline, 1.48 mm (1.46-1.49 mm) × 1.06 mm (1.03-1.10 mm). Abdomen comprising 2 free somites. First somite 0.65 mm (0.63-0.66 mm) × 0.60 mm. Second somite 1.18 mm (1.16-1.20 mm) × 0.42 mm (0.40-0.43 mm), and carrying several long setules.

Caudal ramus similar to that in female, but longer, 2.76 × 0.16 mm.

Body surface with small spinules similar to that in female.

Antennule as in female. Antenna (fig. 132a-c) different from that of female and apparently 3-segmented. Second segment with large spinelike process and 4 corrugated adhesion pads. Terminal portion a recurved claw with 1 robust proximal seta, 1 rounded lamelliform process, and 1 slender seta at about midlength. Postantennal process as in female.

Mouth tube and mandible as in female. Maxillule (fig. 132d) similar to that in female, but dentiform process bifid in male. Maxilla as in female. Maxilliped (fig. 132e) similar to that in female; however, tubercle (fig. 132f) on inner side of corpus maxillipedis more conspicuous and, although rugose, not spinulated as in female. Legs 1 and 3 as in female. Leg 2 as in female, except second and third exopodal spines as in fig. 133a. Leg 4 (fig. 133b) similar to that in female, except sympod with many more long setules on dorsal surface and inner small spine (fig. 133c) at tip of exopod much more attenuate than in female (compare figs. 131c and 133c). Leg 5 (fig. 133d) consisting of 1 papilla with 3 plumose setae along with 1 plumose seta slightly anterior to this group. Leg 6 (fig. 133d) represented by 2 plumose setae arising from 1 papilla, in addition to 1 plumose seta near base of papilla.

Remarks. — Beneden (1892) was the first to describe *Pupulina flores* as a species representing a previously undescribed genus. According to M. S. Wilson (1952: 246), the host for Beneden's specimens was assumed to be the mobulid ray *Ceratoptera* (= *Manta*). She was probably correct in her assumption, since *P. flores* appears to be specific for species of the ray family Mobulidae.

C. B. Wilson (1935b) described the male of *P. flores* and discovered that Beneden's (1892) description of the male was really that of a young female of another species, *Paralebion elongatus* (refer to C. B. Wilson, 1935a, b).

Presumably due, in part, to Beneden's incomplete original description of *Pupulina flores*, Bassett-Smith (1899) listed this species as *Lepeophtheirus floresi* [sic] and placed *Pupulina flores* as a junior synonym. Bassett-Smith gave no reason for this transfer. C. B. Wilson (1905a), apparently in agreement with Bassett-Smith's synonymy, included *P. flores* in his key to the species of *Lepeophtheirus*. However, in his redescription of this parasitic copepod,

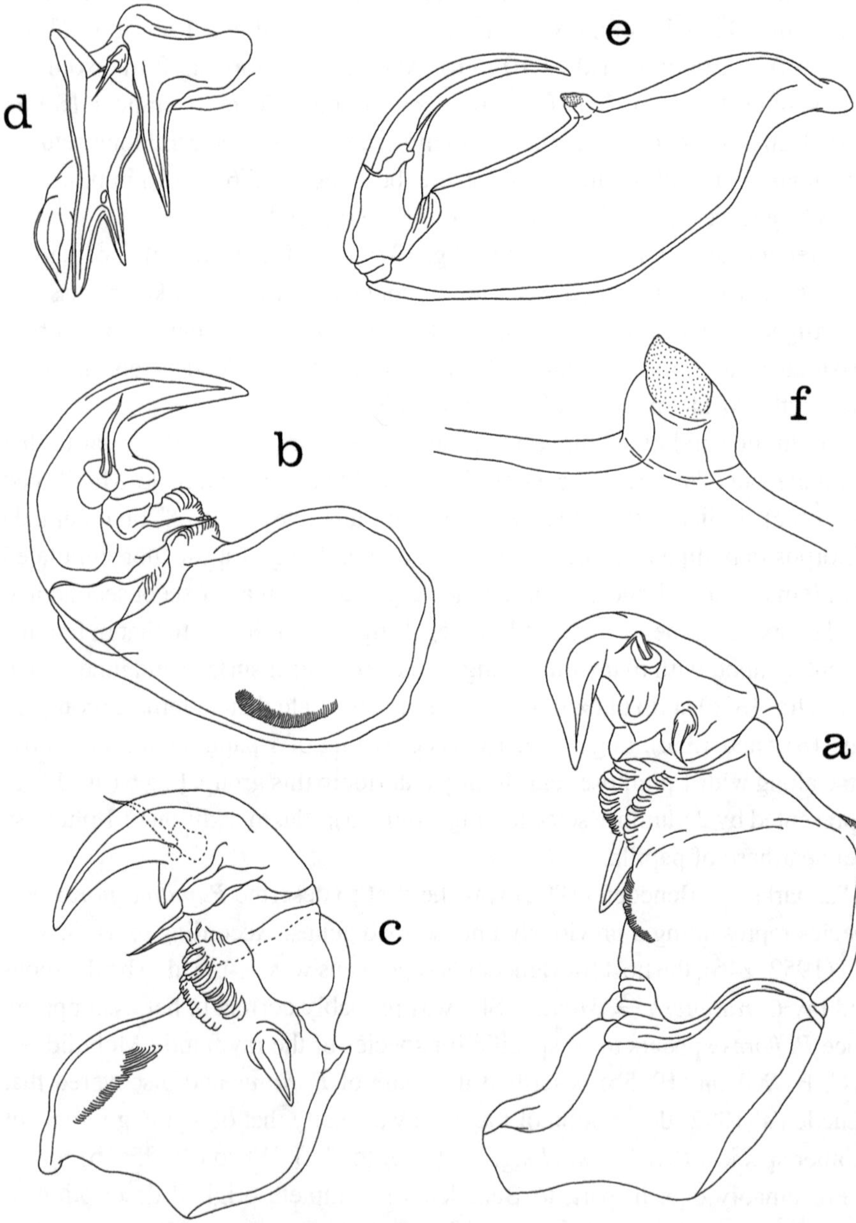

Fig. 132. *Pupulina flores* Beneden, 1892, male. a, antenna, ventro-inner (scale G); b, distal two segments of antenna, ventral (G); c, same, dorsal (G); d, maxillule and cuticular processes, ventral (F); e, maxilliped, ventral (C); f, tubercle on inner margin of corpus maxillipedis, ventral (D).

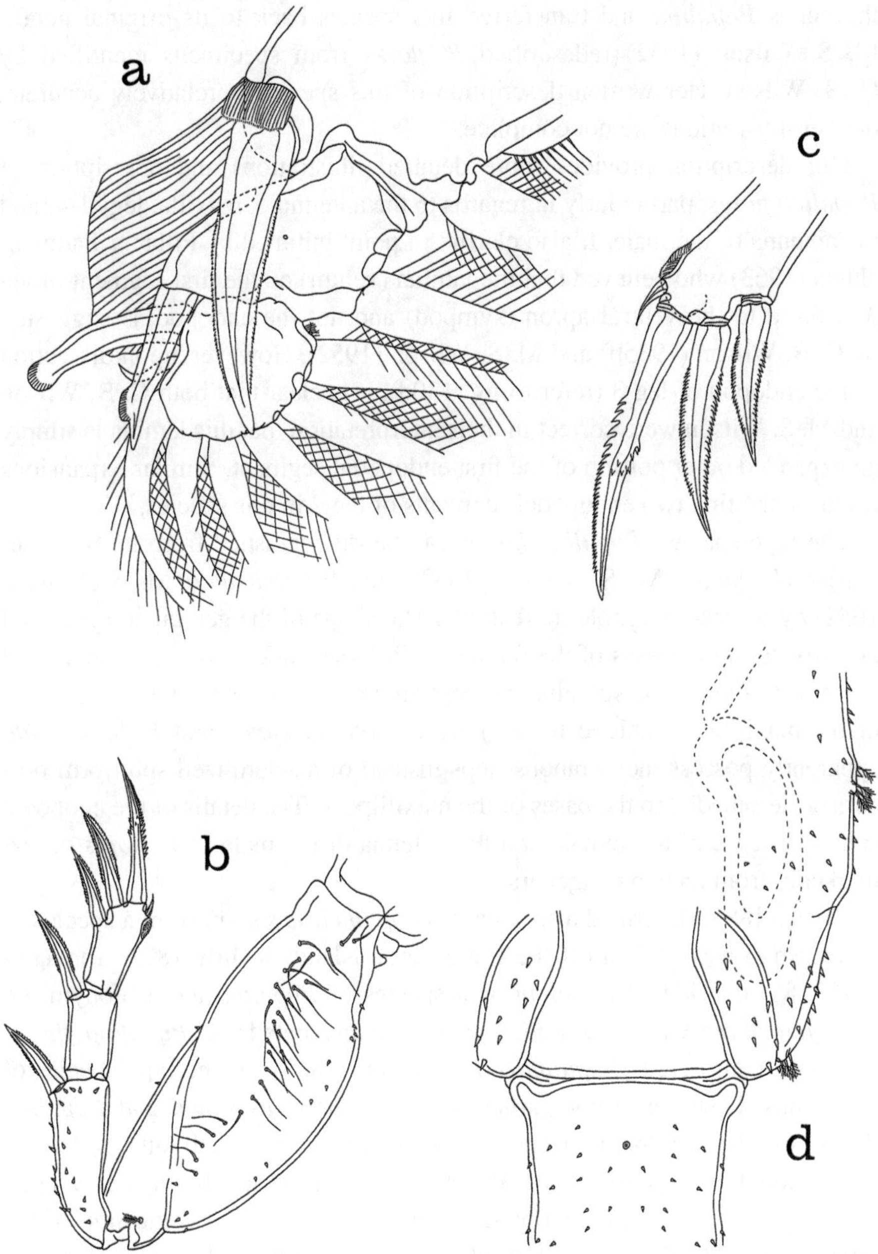

Fig. 133. *Pupulina flores* Beneden, 1892, male. a, exopod of leg 2, ventral (scale G); b, leg 4, dorsal (F); c, tip of leg 4, dorsal (B); d, legs 5 and 6, and genital area, ventral (C).

C. B. Wilson (1935b) reversed his opinion and recognized the validity of the genus *Pupulina* and transferred this species back to its original genus. M. S. Wilson (1952) redescribed *P. flores* from specimens identified by C. B. Wilson. Her written description of this species is relatively accurate, but her illustrations are not complete.

Our description provides more detailed illustrations and description of *Pupulina flores*, particularly in regards to the antennule, maxilla, legs 1-4, and the antenna of the male. It also clarifies a point initiated by Pillai & Padmanabhan (1963) who believed that the lamina (velum) on the first segment of leg 3 is fused to the ventral apron (sympod) and not the endopod as suggested by C. B. Wilson (1935b) and M. S. Wilson (1952). However, upon dissection of the endopod of leg 3 (refer to fig. 130d), it is clear that both C. B. Wilson and M. S. Wilson were correct in their interpretation that this lamina is simply an expanded outer portion of the first endopodal segment. Similar expansions occur on the first two endopodal segments of leg 2 in this species.

The type-species, *Pupulina flores*, can be distinguished from its two congeners (*P. minor* M. S. Wilson, 1952 and *P. brevicauda* M. S. Wilson, 1952) by several morphologic features. The shape of the genital complex and posterolateral processes of the female of *P. brevicauda* is very different from those of *P. flores*. The spinelike process antero-outer to the dentiform process of the maxillule is unique for *P. flores*. *Pupulina minor* and *P. brevicauda* apparently possess membranous flaps instead of a sclerotized spiniform process posteromedial to the bases of the maxillipeds. The details of the exopodal spines of leg 2 of the female and the antenna of the male of *P. flores* differs markedly from its two congeners.

Brady (1883) described a species, *Lepeophtheirus suhmi*, from a species of parrotfish, *Scarus*, taken off the Cape Verde Islands in July 1873. Yamaguti (1963: 85) considered *L. suhmi* as a species of *Pupulina* and synonymized *Pupulina flores* with *L. suhmi*. Since *L. suhmi* antedated *Pupulina flores*, Yamaguti considered "*Pupulina suhmi* (Brady, 1883)" as the type-species of this genus. However, a comparison of *Lepeophtheirus suhmi* and *Pupulina flores* reveals that the two species are distinct from each other. Brady's illustration for *L. suhmi* distinctly shows: (1) a sternal furca; (2) a large clawlike spine of the first exopodal segment of leg 3; (3) a bifid tine of the female maxillule; (4) a very short abdomen, consisting of 1 free somite (anal somite), about 1/3 length of the genital complex; and (5) a caudal ramus as long as the anal somite. Since *L. suhmi* possesses both a sternal furca and a clawlike exopodal spine on leg 3, this species does not even belong to the genus *Pupulina*, but belongs to the genus *Lepeophtheirus* under which Brady had originally described it.

Assuming that Beneden (1892) collected *Pupulina flores* from a species of *Manta*, this species of parasitic copepod has been reported only from species of the Mobulidae from various localities (Azores, Gulf coast of Florida, and Galápagos Islands).

Genus **Sciaenophilus** Beneden, 1852

Sciaenophilus Beneden, 1852: 464; Bassett-Smith, 1899: 453; Scott & Scott, 1913: 62; Yamaguti, 1963: 66; Heegaard, 1966: 191; Kabata, 1979: 188; Dojiri, 1983: 327; Prabha, 1983: 53; Pillai, 1985: 390; Kazachenko, 2001: 27; Kabata, 2003: 66; Boxshall & Halsey, 2004: 726.

Female. — Cephalothorax suborbicular, narrower anteriorly, with relatively shallow posterior sinuses. Frontal plate prominent, with lunules. Fourth pedigerous somite small, without dorsal plates, and incompletely fused to genital complex. Genital complex somewhat triangular in outline or globose, and without processes. Abdomen comprising 1 or 2 free somites, and as long as or longer than other 3 tagmata (i.e., cephalothorax, fourth pedigerous somite, and genital complex) combined. Caudal ramus longer than broad and armed with 6 setae.

Antennule 2-segmented, similar to *Caligus*. Antenna 4-segmented. Second segment without posteriorly directed spinelike process (Rangnekar, 1959 reported a "short blunt spike" on antenna of "*S. inopinus*"). Terminal segment hamate. Postantennal process present. Mouth tube with intrabuccal stylet and strigil. Maxillule bearing dentiform process and adjacent setiferous papilla. Maxilla brachiform, with brachium only slightly longer than lacertus. Maxilliped subchelate, prehensile, and with strongly recurved claw. Sternal furca present or absent.

Leg 1 with 2-segmented exopod and vestigial endopod. Terminal exopodal segment with distal margin angled obliquely; outermost spine (spine 1) shortest, with spines increasing in length from outer to inner margin. Spines 2 and 3 without accessory process. Seta 4 somewhat spiniform, and 3 plumose setae on inner margin. Leg 2 biramous, with 2-segmented rami. Leg 3 with large ventral apron, biramous, with 3-segmented exopod and 2-segmented endopod. Large clawlike spine located distally on basal swelling of first exopodal segment. First segment of endopod with outer margin expanded to form velum. Leg 4 with sympod and 2- or 3-segmented exopod, with total armature of 5 spines. Leg 5 represented by setiferous papilla.

Male. — As described by Kabata (1979).

Type-species. — *Sciaenophilus tenuis* Beneden, 1852.

Remarks. — *Sciaenophilus* was originally described by Beneden (1852) for a single species *S. tenuis* Beneden, 1852. As the generic name implies this species of parasitic copepod was collected from a species of sciaenid, *Sciaena aquila* (Lacépède) [= *Argyrosomus regius* (Asso)]. Since its establishment, the validity of this genus has been questioned many times. Bassett-Smith (1896a, 1898c), for example, demoted *Sciaenophilus* to the status of subgenus within *Caligus*, but later recognized it as a valid genus (Bassett-Smith, 1899). Capart (1941, 1959) considered *Sciaenophilus* a junior synonym of *Caligus*. Finally, Kabata (1979) questioned the validity of *Sciaenophilus*, but retained this genus.

Heegaard (1966) established a new family Sciaenophilidae due to his interpretation that only the first pedigerous somite was fused with the cephalothorax, the second and third somites remaining separate from it. However, the first three pedigerous somites are indeed fused to the cephalothorax as in other genera of the caligids, and therefore the family proposed by Heegaard cannot be considered valid.

Humes (1957) listed three characteristic features of *Sciaenophilus*: (1) presence of lunules, (2) absence of sternal furca, and (3) elongated abdomen. The presence of the sternal furca may be used as a taxonomically important feature when used in conjunction with other characteristics of the organism; however, when used alone, the sternal furca in at least two caligid genera (i.e., *Anuretes* and *Caligus*) may not bear a great deal of taxonomic value.

Also, the taxonomic merit of the elongated abdomen in the female has been questioned (see Kabata, 1979). The basis of the argument stems from the existence of a few species of *Caligus* that possess a relatively long abdomen. Females of species such as *C. kanagurta* Pillai, 1961 and *C. pelamydis* Krøyer, 1863, among others, that possess an abdomen longer than their genital complex, bridge the gap between members of *Caligus* with short abdomens and species of *Sciaenophilus*. There exists a broad range in the lengths of the abdomens of the approximately 250 species of *Caligus*. However, this tagma never reaches the extreme elongation found in *Sciaenophilus*. The abdomen of the female is as long as or much longer than the other three tagmata combined.

Another characteristic that suggests a close relationship within *Sciaenophilus* is the armature of the terminal exopodal segment of leg 1. The reduction in size and displacement of the outermost spine (spine 1) to a subterminal position, the comparatively large size of the innermost spine (spine 3), the

TABLE XVII

List of species of *Caligus* Müller, 1785 with the outermost spine of the terminal exopodal segment of leg 1 smallest of the three spines, and with no accessory processes on spines 2 and 3

C. brevisoris Shen, 1957	*C. multispinosus* Shen, 1957
C. costatus Shen & Li, 1959	*C. pelamydis* Krøyer, 1863
C. cybii Bassett-Smith, 1898	*C. platytarsis* Bassett-Smith, 1898
C. dactylus Ho, Lin & Chang, 2007	*C. polycanthi* Gnanamuthu, 1950
C. diaphanus Nordmann, 1832	*C. priacanthi* Pillai, 1961
C. fistulariae Yamaguti, 1936	*C. reniformis* Prabha & Pillai, 1983
C. kahawai Jones, 1998	*C. robustus* Bassett-Smith, 1898
C. kanagurta Pillai, 1961	*C. rogercresseyi* Boxshall & Bravo, 2000
C. kapuhili Lewis, 1967	*C. scribae* Essafi, Cabal & Raibaut, 1984
C. laticaudus Shiino, 1960	*C. septibensis* Luque & Takemoto, 1996
C. longicaudus Bassett-Smith, 1898	*C. seriolae* Yamaguti, 1936
C. mercatoris Capart, 1941	

absence of the accessory process on spines 2 and 3, and the spiniform appearance of the inner distal seta (seta 4) are all features shared by species of *Sciaenophilus*. Unfortunately, this armature is not unique to *Sciaenophilus*. There are several species of *Caligus* that have a similar armament (table XVII).

Members of *Sciaenophilus* apparently do not possess the posteriorly directed spiniform process on the second segment of the antenna. This process is characteristic for the majority of the species of *Caligus*, but again there are a few exceptions (table III). Rangnekar (1959) reported "...a short blunt spike..." on the proximal segment of this appendage for *S. inopinus* Humes, 1957 [presently considered a junior synonym of *S. pharaonis* (Nordmann, 1832)]. However, according to her illustrations, this process may simply be the sharp inner proximal corner of the robust third segment. In any case, Humes (1957, 1965) does not report it.

There are currently five species of *Sciaenophilus* recognized (Kabata, 1979). The type-species, *S. tenuis* has been reported from numerous species of sciaenids from the eastern Atlantic Ocean, Indian Ocean, and the Gulf of Mexico (table XVIII). *Sciaenophilus pharaonis* (Nordmann, 1832) has been collected from *Sciaena semiluctuosa* Day [= *Paranibea semiluctuosa* (Cuvier)] from India (Rangnekar, 1959); from a serranid, *Lates niloticus* (Linnaeus) from Egypt (Humes, 1957) (both records reported under the binomen *S. inopinus* Humes, 1957); from *Otolithus maculatus* [= *Pterotolighus maculatus* (Cuvier)] from India (Pillai & Natarajan, 1977); and from *Sparus aurata* Linnaeus, *Lates niloticus* (Linnaeus), and *Labeo forskalii* (Rüppell) all from India (Pillai, 1985). The third species, *S. nibeae* (Shen, 1957) (originally described as a species of *Caligus*, but transferred to *Sciaenophilus* by Yamaguti,

TABLE XVIII

Hosts and localities of collections of *Sciaenophilus tenuis* Beneden, 1852, including records of this species reported under its synonyms

Host	Locality	Reference
Sciaena aquila (Lacépède) [= *Argyrosomus regius* (Asso)]	Europe	Beneden, 1861
Sciaena umbra (Linnaeus)	Plymouth, England	Bassett-Smith, 1896a
Sciaena aquila (= *Argyrosomus regius*)	Mauritania	Brian, 1924
Johnius sp.	Vizhingom, India	Pillai, 1961
Pogonias cromis (Linnaeus)	Rockport, Texas	Heegaard, 1966
Otolithoides brunneus [= *Otolithoides biauritus* (Cantor)]	Bombay, India	Rangnekar & Murti, 1967
Umbrina cirrosa (Linnaeus)	Bombay, India	Rangnekar & Murti, 1967
Argyrosomus regius	British waters	Kabata, 1979
Sciaena diacanthus [= *Protonibea diacanthus* (Lacépède)]	Bombay, India	Bassett-Smith, 1898c
Sciaena diacanthus (= *Protonibea diacanthus*)	Palk Strait, Ceylon (Sri Lanka)	Thompson & Scott, 1903
Sciaena angolensis [= *Miracorvina angolensis* (Norman)]	Cabinda (west coast of Africa)	Capart, 1959
Umbrina valida Jordan & Gilbert (= *Umbrina canariensis* Valenciennes)	"baie des Elephants" "baie des Tigres"	Capart, 1959
Sciaena aquila (= *Argyrosomus regius*)	"Pointa do Dande"	Capart, 1959
Johnius hololepidotus (= *Argyrosomus hololepidorua* Lacépède)	Mauritania	Capart, 1941
Larimus fasciatus Holbrook	Lemon Bay, Florida	Bere, 1936
Johnius hololepidotus (= *Argyrosomus hololepidorua*)	Atlantic coast of Africa	Kabata & Gusev, 1966
Pseudosciaena diacanthus (= *Protonibea diacanthus*)	Bombay, India	Rangnekar & Murti, 1967
Sciaena aquila (= *Argyrosomus regius*)	Trivandrum, India	Pillai, 1985
Umbrina cirrosa (Linnaeus)	Trivandrum, India	Pillai, 1985
Johnius maculatus [= *Nibea maculata* (Bloch & Schneider)]	Trivandrum, India	Prabha & Pillai, 1986
Lobotes surinamensis (Bloch)	Grande Isle, Louisiana	Cressey & Nutter, 1987
Pogonias cromis	Grande Isle, Louisiana; Port Aransas, Texas	Cressey & Nutter, 1987
		Cressey & Nutter, 1987
Pogonias cromis	Mississippi Sound	Frasca et al., 2004

1963) was found parasitizing a sciaenid, *Nibea diacanthus* [= *Protonibea diacanthus* (Lacépède)] by Shen (1957), from Kwangtung, China. *Sciaenophilus macrurus* (Heller, 1865) was described from a species of *Lobotes* from Java. This species has not been reported since Heller (1865) described it. Ho & Bashirullah (1977) believed that it may be synonymous with *S. bennetti* (Causey, 1953a), which was also obtained from a tripletail, *Lobotes surinamensis* (Bloch), from the Gulf of Mexico (Grand Isle, Louisiana) by Causey (1953a) and later, on the same species of fish, by Ho & Bashirullah (1977) from Gairia, Venezuela. Ho & Bashirulla (1977) provided a redescription of *S. bennetti* and transferred it back to the originally proposed genus *Caligus*, due to the presence of the sternal furca.

Because of the shared characteristics discussed above, the genus *Sciaenophilus* is retained, particularly since its members can easily be recognized and distinguished from species of *Caligus*.

Sciaenophilus tenuis Beneden, 1852
(figs. 134-137)

Sciaenophilus tenuis Beneden, 1852: 464; Bassett-Smith, 1899: 453; Norman & Scott, 1906: 207; Scott & Scott, 1913: 63; Brian, 1924: 30; Leigh-Sharpe, 1933: 112; Oorde-de Lint & Schuurmans Stekhoven, 1936: 135; Rose & Vaissière, 1953: 85; Capart, 1959: 79; Rangnekar, 1959: 48; Pillai, 1961: 115; Yamaguti, 1963: 66; Heegaard, 1966: 191; Rangnekar & Murti, 1967: 19; Kabata, 1979: 188; Dojiri, 1983: 332; Pillai, 1985; p. 394; Prabha & Pillai, 1986: 22; Cressey & Nutter, 1987: 600; Frasca et al., 2004: 115.

Caligus (Sciaenophilus) tenuis Bassett-Smith, 1896a: 156.

Caligus (Sciaenophilus) Benedeni [sic] Bassett-Smith, 1898c: 9; Gnanamuthu, 1947: 48.

Caligus benedeni Thompson & Scott, 1903: 294; Capart, 1941: 172; Capart, 1959: 79; Kirtisinghe, 1964: 64.

Sciaenophilus benedeni Bassett-Smith, 1899: 453; Bere, 1936: 587; Capart, 1941: 172; Rangnekar, 1959: 48; Yamaguti, 1963: 66; Kabata & Gusev, 1966: 163; Rangnekar & Murti, 1967: 21; Pillai, 1985: 391.

Material examined. — One female (USNM 240765) from inner surface of operculum of *Johnius* sp. (family Sciaenidae) collected at Vizhingom, Kerala, India, by Dr. N. Krishna Pillai. Material on loan from National Museum of Natural History, Smithsonian Institution, Washington, D.C.

Female. — Body as in fig. 134a. Total length (not including setae on caudal ramus) 12.74 mm (n = 1). Cephalothorax (fig. 134b) subcircular in outline, wider than long, 1.13 × 1.20 mm. Dorsal transverse rib situated more posteriorly. Cephalic area larger than thoracic area of dorsal shield. Posterior sinuses shallow and bearing transparent membrane. Free margin of thoracic zone extending well beyond posterior limit of lateral zone. Frontal plate large. Tip of antennule within lateral limit of cephalothorax. Lunules located ventrally on frontal plate.

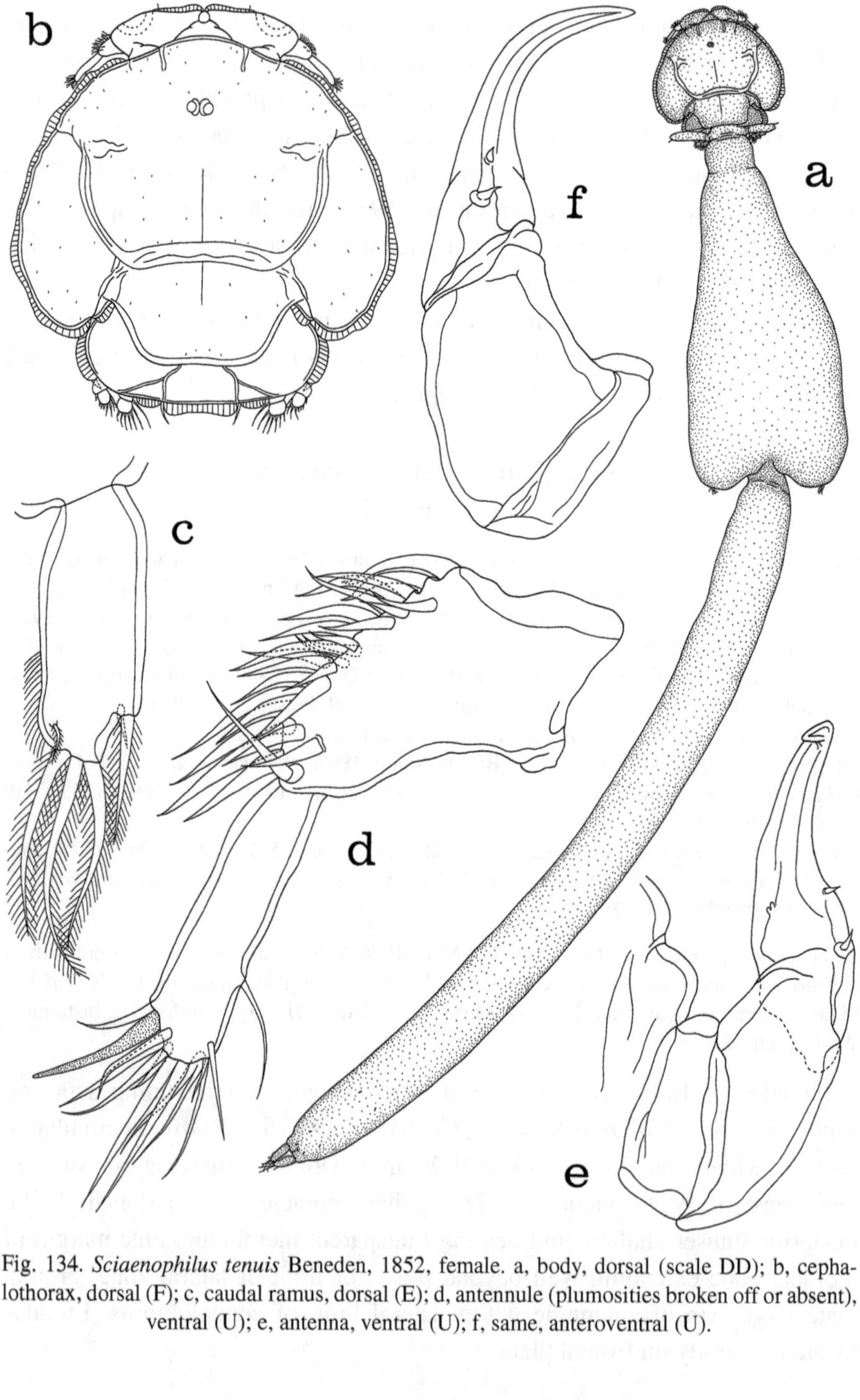

Fig. 134. *Sciaenophilus tenuis* Beneden, 1852, female. a, body, dorsal (scale DD); b, cephalothorax, dorsal (F); c, caudal ramus, dorsal (E); d, antennule (plumosities broken off or absent), ventral (U); e, antenna, ventral (U); f, same, anteroventral (U).

Fourth pedigerous somite wider than long, 0.13 × 0.32 mm, and not clearly delimited from genital complex. Genital complex approximately triangular, widest posteriorly, and longer than wide, 3.20 × 1.53 mm. Abdomen extremely long and cylindrical, 8.07 × 0.60 mm.

Caudal ramus (fig. 134c) longer than wide, 219 × 92 μm, and bearing 6 plumose setae.

Cephalothorax with small setules (sensilla) as in fig. 134b.

Egg sac not observed.

Antennule (fig. 134d) 2-segmented. First segment 78 μm long (measured along nonsetiferous margin), and bearing 27 setae (most of plumosities broken off in specimen). Second segment 116 μm long, cylindrical, with armature of 13 + 1 aesthetasc (2 setae sharing common base). Antenna (fig. 134e, f) apparently 4-segmented. First segment fused to ventral surface of cephalothorax and unarmed. Second segment small and without posteriorly directed spine-like process. Third segment robust and without usual dorsal adhesion pad. Terminal portion a recurved claw with 2 smooth setae. Postantennal process (fig. 135a, b) with relatively straight tine and usual associated, multibranched setule.

Mouth tube (fig. 135c) much longer than wide, 292 × 178 μm, with intrabuccal stylet. Labium with strigil. Mandible (fig. 135d) composed of 4 sections. Terminal section bearing 12 teeth. Maxillule (fig. 135e) consisting of papilla with 3 setae, and curved sharp tine. Maxilla (fig. 135f) brachiform. Flabellum relatively large and near midlength of brachium. Calamus (fig. 135g) longer than canna, and possessing 3 serrated membranes. Canna (fig. 135g) with 2 serrated membranes. Maxilliped (fig. 135h) robust, with inner spine-like process on corpus. Shaft and claw 334 μm long, with seta near junction. Sternal furca absent.

Leg 1 (fig. 135i) biramous. Sympod with 1 outer and 1 inner plumose setae. Exopod 2-segmented. First segment with outer distal spine and inner row of setules. Outer distal spine (fig. 136a) serrated and with small spinules at base. Second segment of exopod with seta 4 smooth (not pinnate as in most caligids). Outermost spine smallest of 3 spines and only unilaterally spinulated. Spines 2 and 3 bilaterally spinulated. Spine 3 longest of 3 spines, and slightly longer than seta 4. All 3 spines with small spinules at bases. Inner margin of second segment of exopod with 3 large plumose setae. Endopod (fig. 136b) greatly reduced, bearing 2 small digitiform processes at tip. Other endopod of same specimen (fig. 136c) with only 1 longer process. Leg 2 (fig. 136d) biramous and with usual armature. Exopod (fig. 136e) with spines as in figure. Sympod

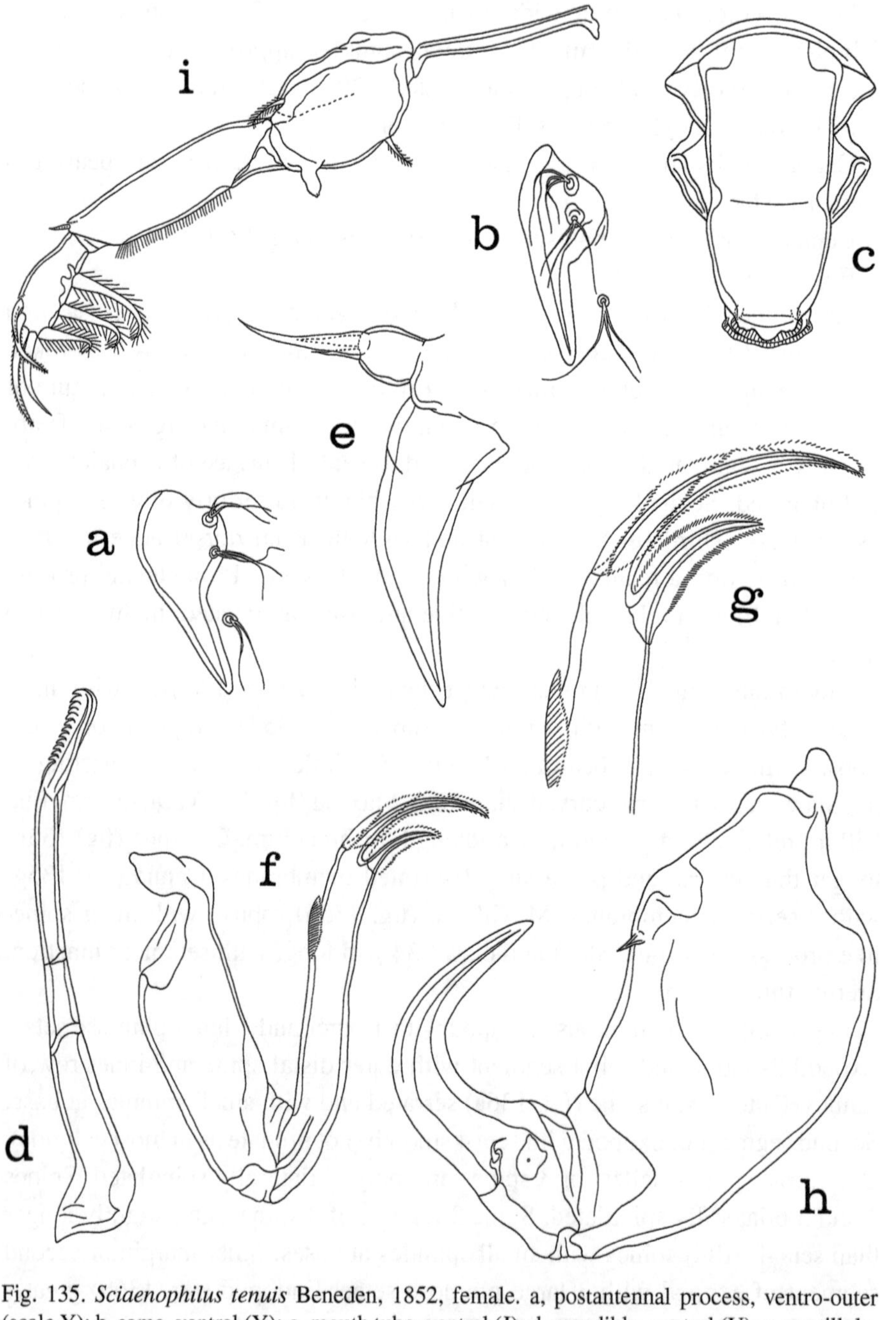

Fig. 135. *Sciaenophilus tenuis* Beneden, 1852, female. a, postantennal process, ventro-outer (scale Y); b, same, ventral (Y); c, mouth tube, ventral (J); d, mandible, ventral (U); e, maxillule, inner (U); f, maxilla, dorsal (D); g, terminal portion of maxilla, dorsal (L); h, maxilliped, ventral (J); i, leg 1 and intercoxal plate (E).

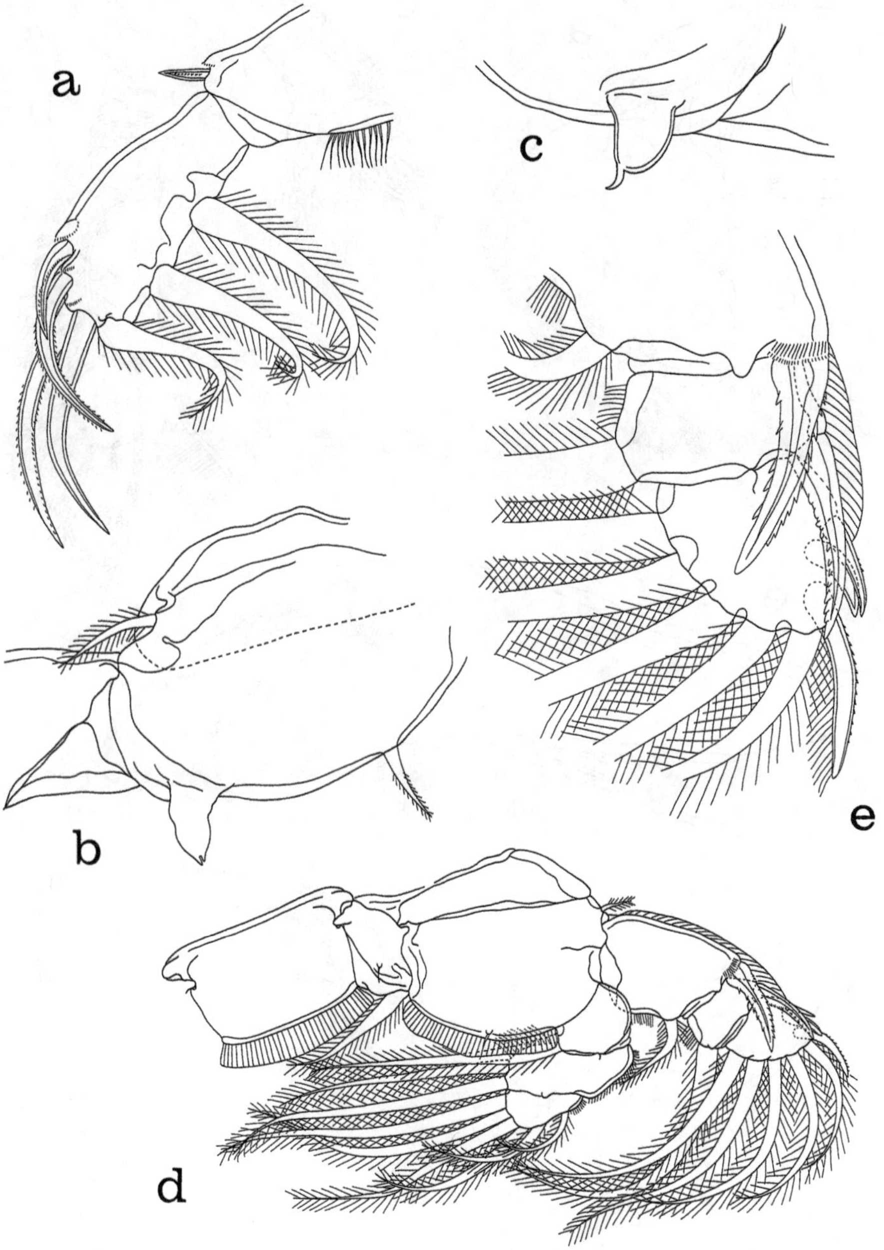

Fig. 136. *Sciaenophilus tenuis* Beneden, 1852, female. a, distal portion of exopod of leg 1, ventral (scale L); b, sympod-exopod joint and endopod of leg 1, ventral (L); c, endopod of leg 1, ventral (Y); d, leg 2 and intercoxal plate, ventral (J); e, exopod of leg 2, ventral (U).

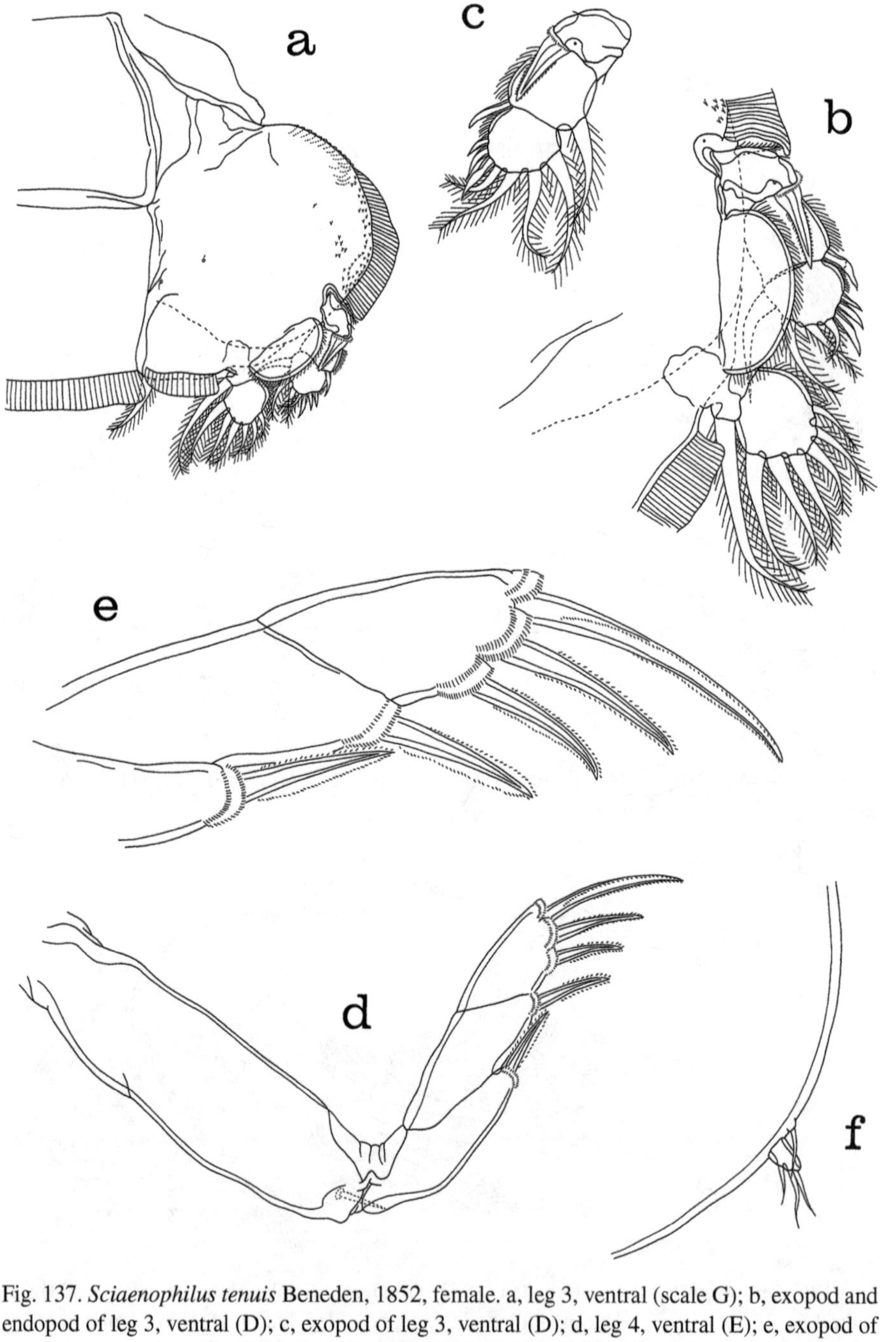

Fig. 137. *Sciaenophilus tenuis* Beneden, 1852, female. a, leg 3, ventral (scale G); b, exopod and endopod of leg 3, ventral (D); c, exopod of leg 3, ventral (D); d, leg 4, ventral (E); e, exopod of leg 4, ventral (U); f, leg 5, ventral (E).

of leg 3 (fig. 137a) with corrugated pad on dorsal surface of anterolateral corner and patch of spinules on lateral (outer) margin. Exopod (fig. 137b, c) 3-segmented. First segment with large clawlike spine; this spine bilaterally serrated and with patch of spinules at base. Second exopodal segment with outer row of setules, 1 smooth outer seta, and 1 plumose inner seta (fig. 137c) (inner plumose seta on second segment absent on other side, see fig. 137b). Third segment with outer row of setules, 3 smooth setae, and 4 plumose setae. Endopod (fig. 137b) 2-segmented. First segment with outer margin expanded into velum and inner margin with plumose seta. Second segment with outer and inner row of setules and 6 plumose setae. Leg 4 (fig. 137d, e) uniramous, brachiform. Sympod with 1 distal smooth seta dorsally. Exopod (fig. 137e) 3-segmented. Spines of terminal segment decreasing in length from innermost to outermost spine. All 5 exopodal spines bilaterally serrated and with pectinate membranes at bases. Leg 5 (fig. 137f) represented by papilla bearing 3 setae and 1 additional seta slightly removed from papilla (plumosities broken off).

Remarks. — *Sciaenophilus tenuis* was first described by Beneden (1852). Subsequent to the discovery of the type-species of this genus, Bassett-Smith (1898c) reported another species of *Sciaenophilus* (cited in the paper as a subgenus of *Caligus*), *Caligus* (*Sciaenophilus*) *benedeni* from *Sciaena diacanthus* [= *Protonibea diacanthus* (Lacépède)] at Bombay. Capart (1959) questioned the validity of this species and raised the possibility that *Sciaenophilus benedeni* (under the binomen *Caligus benedeni*) might be synonymous with *S. tenuis*. In agreement, Kabata & Gusev (1966) also discussed the validity of *S. benedeni* and the difficulty in resolving the problem, since the original descriptions of these nominal species are not very reliable. In addition, the type-material has been lost.

Finally, Kabata (1979) synonymized *S. benedeni* with the type-species, citing that no sufficient taxonomic differences could be found between the nominal species as a basis for the retention of both taxa. Since the descriptions of the species are not very detailed for comparative purposes, it is not possible at this time to make a definitive conclusion regarding this synonymy. However, the description and scanty illustrations of *S. benedeni* are very similar to those of *S. tenuis*. *Sciaenophilus tenuis* is approximately 13-15 mm in total length, greater than the total length of *S. benedeni*. Rangnekar & Murti (1967) redescribed *S. tenuis* and *S. benedeni* in the same paper, but unfortunately did not compare the two. According to their illustrations and measurements, *S. benedeni* possesses a more globular genital complex, shorter abdomen, and shorter total length (7.50 mm) than *S. tenuis* (13.65 mm). As stated by Kabata

(1979) until more is known about *S. tenuis* and *S. benedeni*, the latter should be considered a junior synonym of the former.

Under either binomen, this species has been cited numerous times (Bassett-Smith, 1896a, 1899; Thompson & Scott, 1903; Norman & Scott, 1906; Brian, 1924; Leigh-Sharpe, 1933; Bere, 1936; Rose & Vaissière, 1953; Kirtisinghe, 1964; Heegaard, 1966). Short redescriptions (Bassett-Smith, 1898c; Scott & Scott, 1913; Oorde-de Lint & Schuurmans Stekhoven, 1936; Kabata, 1979) and relatively complete redescriptions (Pillai, 1961; Heegaard, 1966) have been published. Pillai (1961) provided the most complete and detailed description of *S. tenuis*. Our redescription provides a more detailed account of the terminal armature of the exopod of leg 1, leg 3, and leg 4.

The hosts and zoogeographic distribution of *S. tenuis* are provided (table XVIII).

Genus **Sinocaligus** Shen, 1957

Sinocaligus Shen, 1957: 375; Prabha, 1983: 53; Kazachenko, 2001: 27; Boxshall & Halsey, 2004: 726; Ho & Lin, 2004: 307.

Pseudopetalus Pillai, 1962: 296; Cressey, 1967c: 4; Dojiri, 1983: 306; Prabha, 1983: 53; Pillai, 1985: 384; Izawa, 1995: 185.

Female. — Cephalothorax subovate, slender, and narrower anteriorly. Posterior sinuses distinct, but shallow. Frontal plate broad and prominent, with lunules. Fourth pedigerous somite small, and incompletely fused to genital complex. Genital complex triangular in outline and tapering toward anterior end; anteriormost portion forming neck with fourth pedigerous somite. Abdomen consisting of 1 free somite with broad lateral aliform expansions, and approximately as long as genital complex. Caudal ramus short, broad, and setiferous.

Antennule 2-segmented, similar to *Caligus*. Antenna 4-segmented. Second segment with posteriorly directed spinelike process. Terminal segment hamate. Postantennal process present. Mouth tube with intrabuccal stylet and strigil. Mandible comprising 4 sections, with 12 teeth on inner margin of terminal section. Maxillule consisting of dentiform process and adjacent setiferous papilla. Maxilla brachiform. Maxilliped subchelate and prehensile. Sternal furca with relatively sharp, divergent tines.

Leg 1 with 2-segmented exopod and vestigial endopod. Terminal segment of exopod with 3 spines at tip, 1 seta on inner distal corner, and 3 plumose setae on inner margin. Leg 2 biramous, with 3-segmented rami. Outer distal corner of first endopodal segment with patch of spinules. Second endopodal segment with large conical projections on outer margin. Leg 3 biramous, with usual ventral apron. Exopod 3-segmented, with large clawlike spine distally located on basal swelling of first segment. Endopod 2-segmented, with outer margin of first segment expanded to form velum. Leg 4 with sympod and 2-segmented exopod armed with formula I-0; IV. Legs 5 and 6 absent.

Male. — Unknown.

Type-species. — *Sinocaligus formicoides* (Redkar, Rangnekar & Murti, 1949).

Remarks. — Pillai (1962) established the genus *Pseudopetalus* to accommodate *P. formicoides* (Redkar, Rangnekar & Murti, 1949) and *P. caudatus* (Gnanamuthu, 1950), which were transferred from *Parapetalus* (see discussion on *Parapetalus*). Since the establishment of *Pseudopetalus*, Pillai (1968) has transferred another species, *Caligus dussumieri* Rangnekar, 1957, to this genus. However, Shen (1957) described a new species, *S. denticulatus*, and in order to accommodate the type-species, established a new genus *Sinocaligus*, which antedates Pillai's (1962) establishment of *Pseudopetalus*. The generic name *Sinocaligus*, therefore, has priority (Boxshall & Montú, 1997: 49; Ho & Lin, 2004). Unaware of this justified name change, Izawa (1995) reported a new species of this genus, which he named "*Pseudopetalus timorensis* Izawa, 1995".

Sinocaligus is morphologically similar to *Parapetalus*. The females can be distinguished by a combination of several features (table XIV). However, the female of *Parapetalus hirsutus* (Bassett-Smith, 1898) does not possess large lateral aliform expansions on the genital complex that its congeners have. The general habitus of the females of *P. hirsutus* and *Sinocaligus dussumieri* are similar. The generic placement of *P. hirsutus* is questionable; however, the characteristics outlined in table XIV support the retention of this species in *Parapetalus* as suggested by Pillai (1962).

Three of the four species (including two varieties) of *Sinocaligus* appear to have a predilection for fishes belonging to the family Dussumieriidae (genus *Dussumieria*) and the closely related Clupeidae (genus *Sardinella*). However, the specimen described in the following account was taken from a halfbeak, *Hyporhamphus*, and *P. caudatus* var. *rotundus* was collected from a belonid, *Ablennes hians* (Cuvier & Valenciennes), by Pillai (1962). The three species have not been found outside Indian waters, but the fourth species (from an unidentified host) is from Timor Sea, Australia.

Sinocaligus formicoides (Redkar, Rangnekar & Murti, 1949) (figs. 138-140)

Caligus formicoides Redkar, Rangnekar & Murti, 1949: 36.
Parapetalus formicoides Rangnekar & Murti, 1950: 44; Rangnekar, 1956: 48; Yamaguti, 1963: 65.
Sinocaligus denticulatus Shen, 1957: 375.
Parapetalus denticulatus Shen, 1958: 143; Yamaguti, 1963: 65.
Pseudopetalus formicoides Pillai, 1962: 297; Dojiri, 1983: 308; Pillai, 1985: 389.
Sinocaligus formicoides Boxshall & Montú, 1997: 49; Samotylova, 2010: 99.
Synonymy provided above includes synonyms of both *Sinocaligus formicoides* and *P. formicoides* var. *denticulatus* since they are only varieties of the same species.

Material examined. — One female from halfbeak, *Hyporhamphus* sp. (label marked *Hyporhamphus neglectus*), caught at Port Essington, Australia. This specimen was on loan to the first author (M. Dojiri) from the late Dr. Roger F. Cressey, National Museum of Natural History, Smithsonian Institution, Washington, D.C.

Female. — Body modified as in fig. 138a, b. Total length (not including setae of caudal ramus) 8.96 mm (n = 1). Cephalothorax (fig. 138c) longer than wide, 1.26 × 1.03 mm, with shallow posterior sinuses. Free margin of thoracic zone extending well beyond posterior tip of lateral zone. Sensory pit present on posterior end of lateral zone. Lunules located ventrally on large frontal plate. Tip of antennule well within lateral limit of cephalothorax.

Fourth pedigerous somite comparatively small, fused to genital complex. Genital complex (fig. 138a, b) elongate, narrow anteriorly, expanded posteriorly with posterior processes, and 3.45 × 1.56 mm. Abdomen consisting of 1 free somite, expanded laterally by aliform extensions, longer and wider than genital complex, 4.25 × 1.99 mm, and possessing small posterior processes (fig. 138a, b, and d) that flank caudal rami. Aliform extensions slightly curved ventrally at lateral margins (fig. 138b).

Caudal ramus (fig. 138d) longer than wide, 113 × 81 μm, and carrying 7 setae (2 setae broken off in figure).

Egg sacs partially broken off on single specimen examined.

Antennule (fig. 138e) 2-segmented. First segment 78 μm long (measured along nonsetiferous margin), with bifid process on posterodistal corner, and 26 plumose setae. Second segment 59 μm long, with usual 13 + 1 aesthetasc. Antenna (fig. 138f) apparently 4-segmented. First segment fused to ventral surface of cephalothorax, represented by highly sclerotized area. Second segment with large posteriorly directed spinelike process. Third segment with corrugated, dorsal adhesion pad. Claw recurved and with usual 2 setae. Slightly posterior to base of antenna and anteromedial to postantennal process, a posteriorly directed cuticular process present. Postantennal process (fig. 138f)

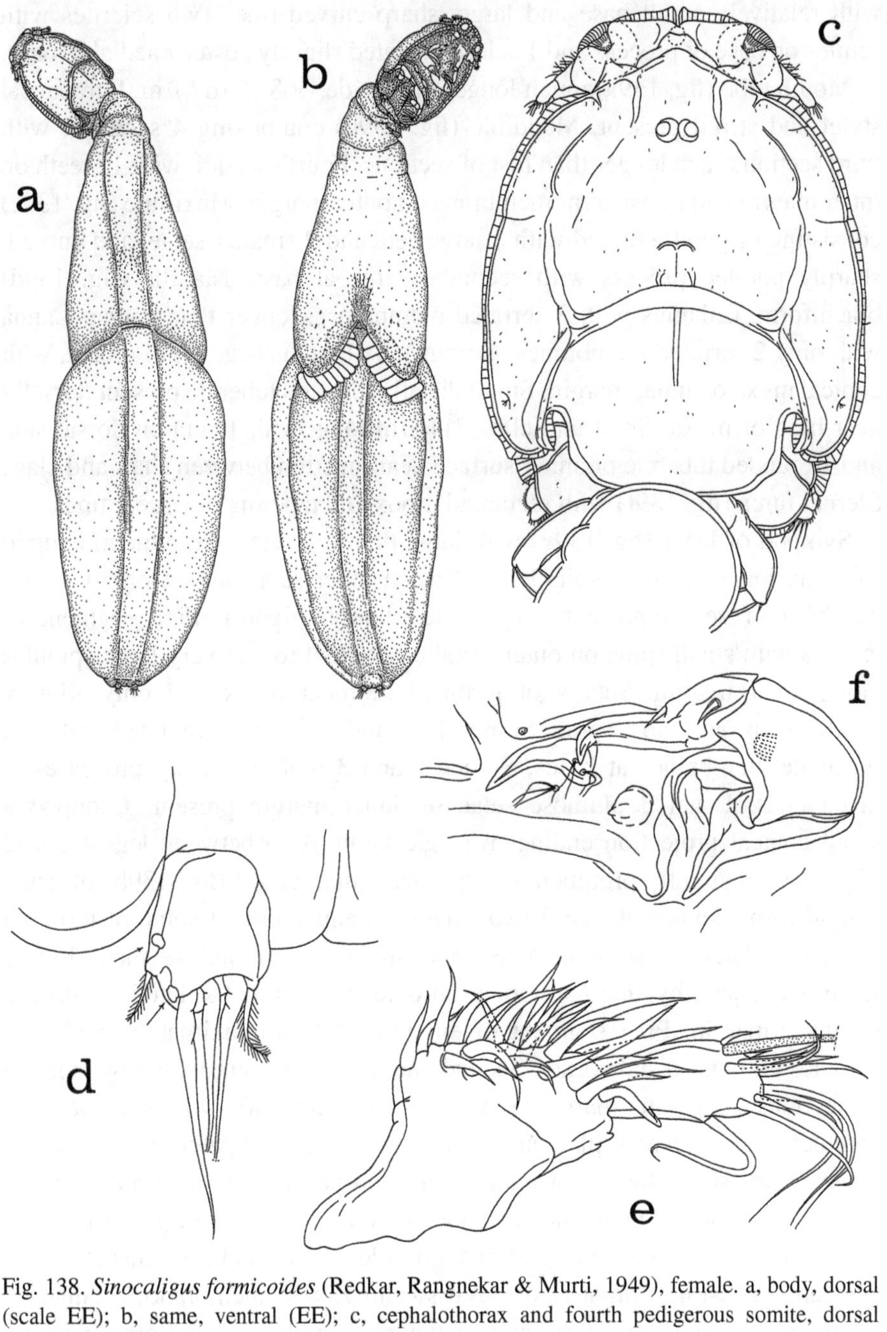

Fig. 138. *Sinocaligus formicoides* (Redkar, Rangnekar & Murti, 1949), female. a, body, dorsal (scale EE); b, same, ventral (EE); c, cephalothorax and fourth pedigerous somite, dorsal (F); d, caudal ramus (arrows indicating positions of broken setae), ventral (E); e, antennule (plumosities not drawn), ventral (Z); f, antenna, postantennal process, and cuticular process, ventral (E).

with relatively small base and large, sharp curved tine. Two sclerites with setules on base of process and 1 sclerite located slightly posteromedial to base.

Mouth tube (fig. 139a) much longer than wide, 368 × 167 μm. Intrabuccal stylet and strigil present. Mandible (fig. 139b) comprising 4 sections, with third section much longer than rest of sections. Fourth section with 12 teeth on inner margin and transparent membrane on outer margin. Maxillule (fig. 139c) consisting of papilla tipped with 1 large seta and 2 smaller setae, and curved, sharply pointed process with secondary tine at base. Maxilla (fig. 139d) brachiform. Calamus with 3 serrated membranes, longer than canna. Canna with only 2 serrated membranes. Corpus maxillipedis (fig. 139e) robust, with conical myxa on inner margin. Small distally directed tubercle present dorsally near base of myxa. Shaft and claw 211 μm long, with 1 seta on dorsal side and 1 rounded tubercle on inner surface near junction between shaft and claw. Sternal furca (fig. 139f) with rectangular box and tapering divergent tines.

Sympod of leg 1 (fig. 139g) with large plumose seta near sympod-exopod joint, an inner patch of spinules and usual inner seta (broken off in figure, but base of seta represented by dotted line in figure). First segment of exopod with small spine on outer distal corner and row of very large spinules along inner margin. Seta 4 of terminal segment of exopod only slightly shorter than outermost spine. Spines 1, 2, and 3 lightly spinulated and with pectinate membranes at bases. Spines 2 and 3 with accessory processes at distal ends. Usual 3 plumose setae on inner margin present. Endopod a small conical projection ending in single knob. Area between legs 1 and 2 (fig. 139h) a slightly protuberant sclerotized area. Leg 2 (fig. 139h) of usual caligid form. Spines of first 2 exopodal segments and first spine of terminal segment bilaterally spinulated. Second spine of terminal segment lightly sclerotized, with hyaline membrane on outer margin and striated membrane on inner margin. First segment of endopod with approximately 5 slender spinules on outer distal corner. Second endopodal segment with large conical projections along outer margin. Sympod of leg 3 (fig. 140a) with large patch of spinules near intercoxal plate and much more dispersed patch of spinules on ventrolateral side. Dorsal corrugated adhesion pad present near anterolateral corner of sympod. Single row of slender spinules located at origin of fringing membrane near basal swelling of exopod of leg 3. Exopod 3-segmented. First segment consisting primarily of basal swelling, with membranous flange on outer distal corner, and terminally situated clawlike spine. Spine carrying membranous flange on outer margin. Second and third segments of exopod as in other caligids. Endopod indistinctly 2-segmented. First segment fused

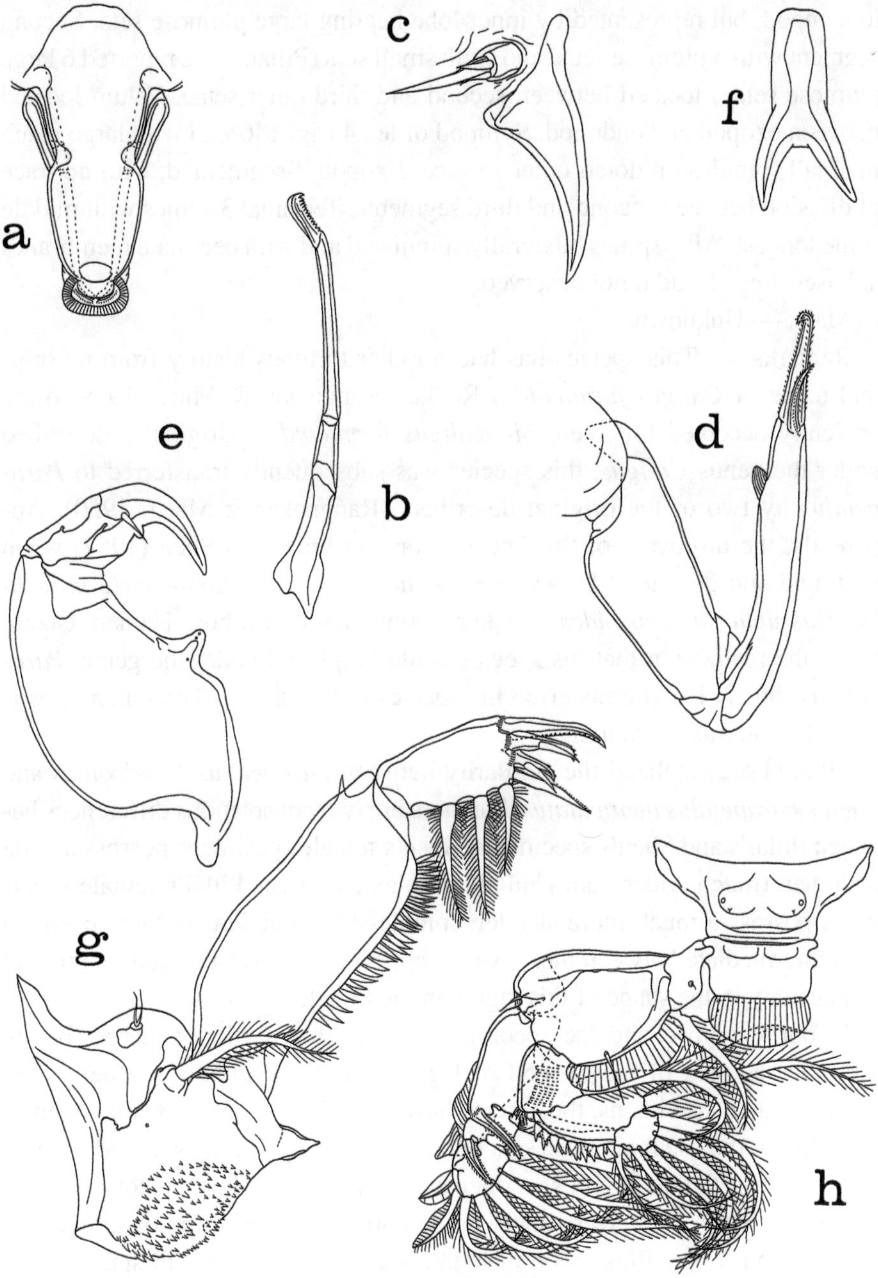

Fig. 139. *Sinocaligus formicoides* (Redkar, Rangnekar & Murti, 1949), female. a, mouth tube, ventral (scale GG); b, mandible, ventral (E); c, maxillule, ventral (E); d, maxilla, dorsal (E); e, maxilliped; ventral (GG); f, sternal furca, ventral (Z); g, leg 1, ventral (Z); h, leg 2, intercoxal plate, and sclerotized area between legs 1 and 2, ventral (GG).

to sympod, but represented by inner lobe bearing large plumose seta. Second segment with 5 plumose setae and sixth small seta (Pillai, 1962 reported 6 large plumose setae) located between second and third outer setae. Velum located between exopod and endopod. Sympod of leg 4 (fig. 140b, c) with large patch of small spinules on dorso-outer surface. Exopod 2-segmented, with no trace of division between second and third segments. Terminal 3 spines with middle spine longest. All 5 spines bilaterally spinulated and with pectinate membranes at bases. Legs 5 and 6 not observed.

Male. — Unknown.

Remarks. — This species has had a rather tortuous history from its original name of *Caligus formicoides* Redkar, Rangnekar & Murti, 1949 to the presently accepted binomen, *Sinocaligus formicoides*. Originally described under the genus *Caligus*, this species was subsequently transferred to *Parapetalus* by two of the original describers (Rangnekar & Murti, 1950). Apparently, the discovery of this species was not known to Shen (1957) when he found and described *Sinocaligus denticulatus* from *Dussumieria hasselti* (= *Dussumieria elopsoides* Bleeker) from San-ah Harbor, Hainan Island. Upon the realization that his species could be placed under the genus *Parapetalus*, Shen (1958) transferred his species to this genus. The binomen was then *Parapetalus denticulatus*.

Pillai (1962) realized the similarity between *Parapetalus formicoides* and Shen's *Parapetalus denticulatus*, but there are two conspicuous differences between Pillai's and Shen's specimens. Shen's female specimens possess a wide abdomen (much wider than genital complex), whereas Pillai's female specimens possess a much more slender, spindle-shaped abdomen. Our specimen falls intermediate between these two. It appears that there is a certain amount of plasticity in the shape of this tagma in the female.

In our specimen and the specimen from China, the inner margin of the terminal segment of the exopod of leg 1 bears three plumose setae, but in Pillai's Indian specimens, these setae have become reduced to 3 small spines. In addition, the usual plumose seta on the inner margin of the sympod of leg 1 is absent in both the present specimen and in Shen's specimens. However, it is presumed that this seta was simply broken off during collection or subsequent examination. In his illustration of leg 2, Shen omitted the small spine on the third segment of the exopod.

Because of the similarity in the remaining structures and the appendages, Pillai (1962) considered *Parapetalus denticulatus* a variety of the species *Pseudopetalus formicoides*, a member of a new genus he established in the same

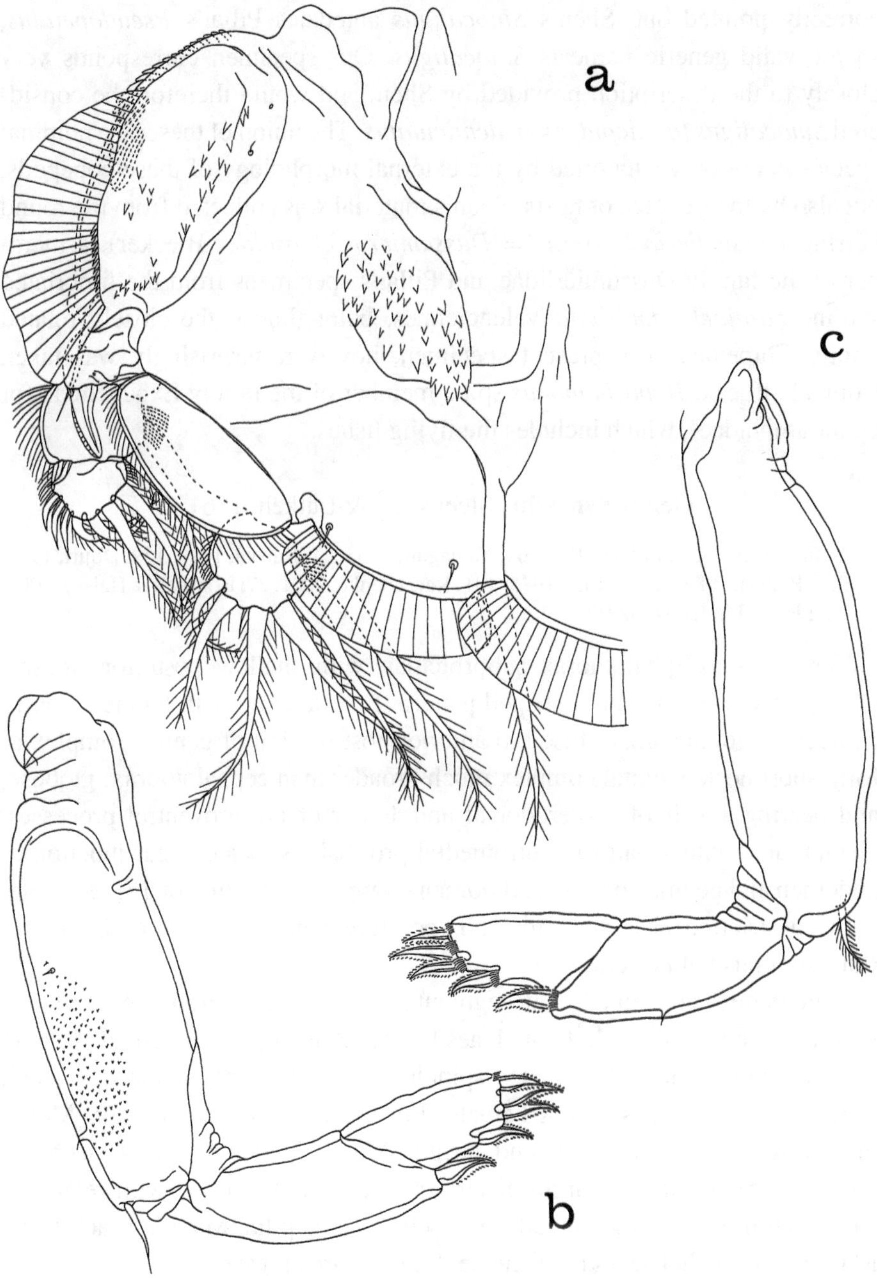

Fig. 140. *Sinocaligus formicoides* (Redkar, Rangnekar & Murti, 1949), female. a, leg 3, ventral (scale E); b, leg 4, dorsal (E); c, same, ventral (E).

paper (see discussion on *Parapetalus*). However, as Boxshall & Montú (1997) correctly pointed out, Shen's *Sinocaligus* antedates Pillai's *Pseudopetalus*, so the valid generic name is *Sinocaligus*. Our specimen corresponds very closely to the description provided by Shen, and would therefore be considered *Sinocaligus formicoides* var. *denticulatus*. The union of these two nominal species is not only supported by the external morphology of these copepods, but also by their choice of hosts. Shen's material was collected from the round herring *Dussumieria hasselti* (= *Dussumieria elopsoides* Bleeker), a member of the family Dussumieriidae and Pillai's specimens from the fimbriated sardine *Sardinella fimbriata* Valenciennes, belonging to the closely related family Clupeidae. The present specimen, however, surprisingly was taken from a halfbeak, *Hyporhamphus* sp., a member of the family Exocoetidae (or Hemiramphidae), which includes the flying fishes.

Genus **Synestius** Steenstrup & Lütken, 1861

Synestius Steenstrup & Lütken, 1861: 64; Yamaguti, 1963: 67; Kabata, 1979: 159; Dojiri, 1983: 337; Prabha, 1983; p. 53; Pillai, 1985: 401; Kazachenko, 2001: 27; Boxshall & Halsey, 2004: 726; Ho & Lin, 2004: 327.

Female. — Cephalothorax suborbicular, with shallow posterior sinuses. Dorsal ribs in prominent H-shaped pattern. Frontal plate with lunules. Fourth pedigerous somite small, fused to anteriormost portion of genital complex to form short neck. Genital complex much broader than cephalothorax, globose, and bearing 1 pair of posterodorsal and 1 pair of posteroventral processes. Genital area with 1 pair of ventromedial protrusions located near junction of abdomen and genital complex. Abdomen elongate, consisting of 1 free somite, with constriction near posterior end (possibly representing second somite). Caudal ramus with 6 setae.

Antennule 2-segmented. First segment robust, with numerous setae. Second segment cylindrical, with 13 + 1 aesthetasc. Antenna 4-segmented. Second segment with posteriorly directed spinelike process greatly reduced to small projection. Terminal segment hamate. Postantennal process present. Mouth tube with intrabuccal stylet and strigil. Mandible comprising 4 sections, with 12 teeth on inner margin of terminal section. Maxillule with relatively small dentiform process and adjacent setiferous papilla. Maxilla brachiform. Maxilliped subchelate and prehensile. Sternal furca present.

Leg 1 with 2-segmented exopod and vestigial endopod. Terminal exopodal segment with 3 terminal spines, 1 seta at inner distal corner, and 3 setae on inner margin. Leg 2 biramous, with 3-segmented rami. Leg 3 with ventral apron, biramous, and possessing 3-segmented exopod and 2-segmented endopod. Large clawlike spine located on basal swelling of first exopodal segment. Velum present. Leg 4 uniramous, brachiform, and with 3-segmented exopod. Leg 5 a setiferous papilla situated on posteroventral process of genital complex. Leg 6 absent.

Male. — Cephalothorax similar to that of female. Genital complex suboval. Abdomen comprising 2 free somites. Antenna with corrugated adhesion pads. Dentiform process of maxillule with digitiform hyaline process. Corpus maxillipedis with rugose myxa and small digitiform process. Legs 5 and 6 represented by setiferous papillae. Other appendages as in female.

Type-species. — *Synestius caliginus* Steenstrup & Lütken, 1861.

Remarks. — *Synestius* Steenstrup & Lütken, 1861 is a monotypic genus containing *S. caliginus*. It can be distinguished from all other caligid genera by the general habitus of the female. The presence of two pairs of long club-shaped processes on the genital complex of the female separates *Synestius* from all caligid genera except *Pseudechetus*. However, the female of *Pseudechetus* exhibits an elongate, slender neck and a pair of long, slender, digitiform abdominal processes not found in *Synestius*.

Synestius caliginus Steenstrup & Lütken, 1861
(figs. 141-145)

Synestius caliginus Steenstrup & Lütken, 1861: 364; Heller, 1865: 179; Bassett-Smith, 1899: 446; Gnanamuthu, 1950a: 281; Gnanamuthu, 1950b: 253; Shen, 1957: 376; Pillai, 1961: 117; Yamaguti, 1963: 67; Kirtisinghe, 1964: 75; Rangnekar & Murti, 1967: 23; Dojiri, 1983: 341; Radhakrishnan & Nair, 1983: 102; Leong, 1985: 47; Pillai, 1985: 401; Piasecki, 1993: 401; Ho & Sey, 1996: 65; Lin & Ho, 2000: 216; Ho & Lin, 2004: 327; Purivirojkul & Areechon, 2008: 42; Samotylova, 2010: 99.

Material examined. — Two females and 1 male (USNM 72638) collected from brown pomfret, *Apolectus niger* [= *Parastromateus niger* (Bloch)], off Java. Specimens on loan from National Museum of Natural History, Smithsonian Institution, Washington, D.C.

Female. — Body as in fig. 141a. Total length (not including setae on caudal armus) 4.22 mm (3.85-4.58 mm) (n = 2). Cephalothorax (fig. 141b) subcircular in outline, almost as wide as long, 1.20 mm (1.13-1.26 mm) × 1.16 mm (1.13-1.20 mm), and incised on lateral margin. Posterior sinuses moderately deep, with transparent membrane along outer edge. Free margin of thoracic zone of cephalothorax at about same level as posterior limit of lateral zone. Lateral margin of cephalothorax rimmed with usual transparent membrane (majority hidden in dorsal view). Frontal plate distinct, with lunules. Tip of antennule almost extending to lateral limit of cephalothorax.

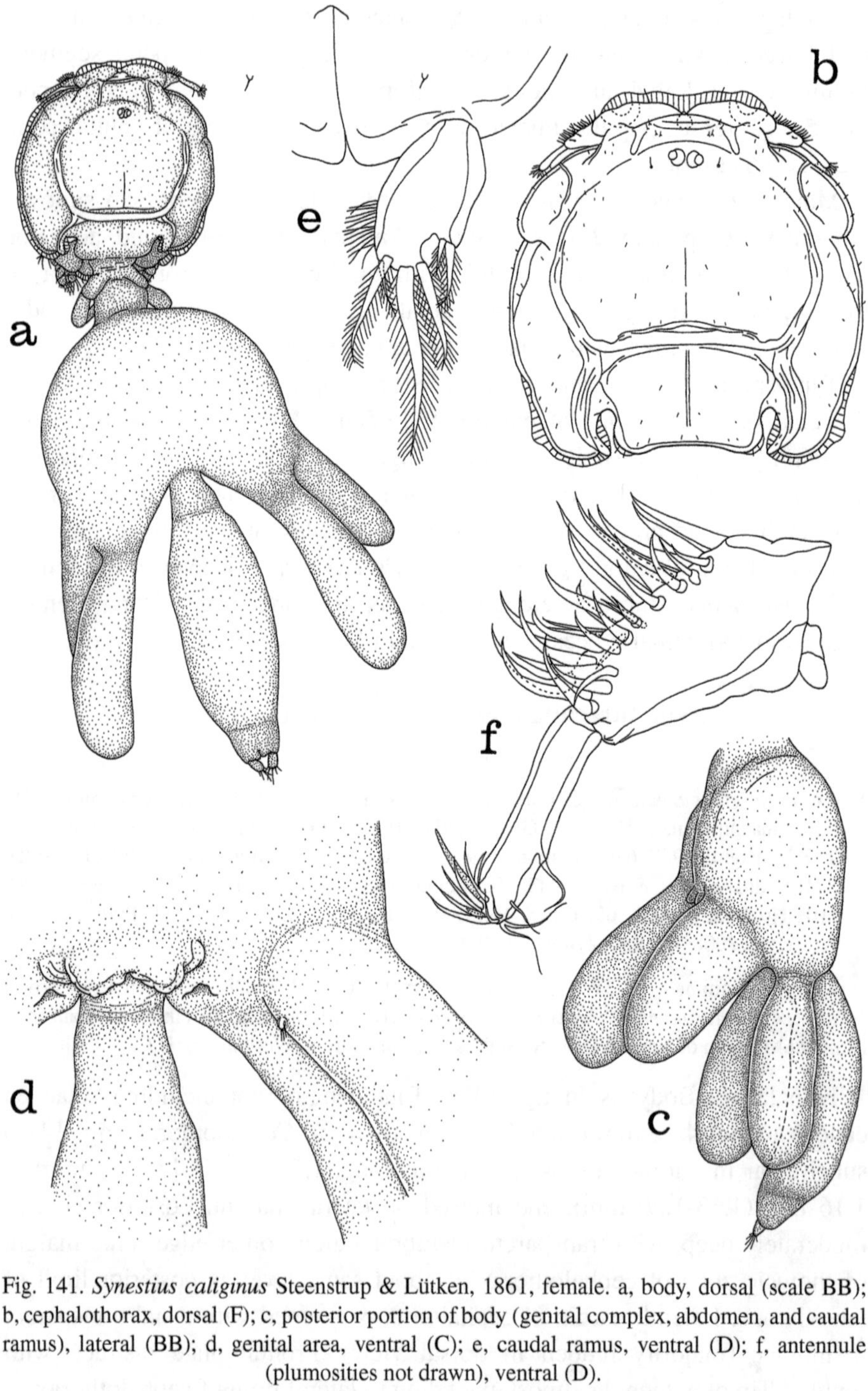

Fig. 141. *Synestius caliginus* Steenstrup & Lütken, 1861, female. a, body, dorsal (scale BB); b, cephalothorax, dorsal (F); c, posterior portion of body (genital complex, abdomen, and caudal ramus), lateral (BB); d, genital area, ventral (C); e, caudal ramus, ventral (D); f, antennule (plumosities not drawn), ventral (D).

Fourth pedigerous somite small, difficult to delimit from genital complex, wider than long, and 0.14 mm (0.10-0.17 mm) × 0.41 mm. Genital complex (fig. 141a, c) possessing 2 pairs of club-shaped processes, globose, and 1.26 mm (1.20-1.33 mm) × 1.39 mm (1.33-1.46 mm) (excluding processes). Posterodorsal process 1.61 mm (1.59-1.63 mm) × 0.33 mm; posteroventral process 0.91 mm (0.86-0.96 mm) × 0.32 mm (0.30-0.33 inm). Genital area (fig. 141d) consisting of 1 pair of ventromedial protrusions located near junction of abdomen and genital complex. Abdomen presumably consisting of 1 free somite, although constricted slightly near posterior end, possessing ventral flexion, and 1.63 mm (1.56-1.69 mm) × 0.58 mm (0.56-0.60 mm).

Caudal ramus (fig. 141e) short, stubby, 100 × 57 μm, and possessing 6 plumose setae and inner row of setules.

Body with few setules (sensilla), except cephalothorax with numerous setules as in fig. 141b.

Antennule (fig. 141f) 2-segmented. First segment 127 μm long (measured along nonsetiferous margin) and possessing 29 plumose setae. Second segment cylindrical, 124 μm long, with 13 + aesthetasc (2 sharing a common base). Antenna (fig. 142a, b) apparently 4-segmented. First segment unarmed; spine-like cuticular process near base most likely not part of first segment. Second segment with very small posteriorly directed spinelike process. Third segment robust, with rounded, corrugated adhesion pad on dorsal side. Fourth segment a claw bearing 1 seta proximally and 1 slender seta at about midlength. Postantennal process (fig. 142c) with small rounded protrusion on outer margin near base and 1 large tine. Usual sclerites bearing setules present.

Mouth tube (fig. 142d) much longer than wide, 265 × 194 μm, with intrabuccal stylet and strigil. Mandible (fig. 142e) comprising 4 sections; tip of mandible with hyaline flange on outer edge and 12 teeth on inner edge. Maxillule (fig. 142f) consisting of papilla tipped with 3 setae (1 much larger than other 2), and 1 curved process. Maxilla (fig. 142g) brachiform. Flabellum (fig. 142h) very large, serrated, and striated. Calamus with 3 serrated membranes. Canna with only 2 serrated membranes. Maxilliped (fig. 142i) with robust corpus. Shaft and claw 380 μm long combined. Seta located near junction of shaft and claw. Sternal furca (fig. 143a) with large rectangular base and short diverging tines. Tines flanked with inner and outer hyaline flanges. Pair of rounded protrusions located medially on base.

Leg 1 (fig. 143b) biramous. Sympod with 1 large outer and 1 small inner plumose setae, and large patch of spinules. Exopod 2-segmented. First segment with outer distal spine and inner row of long setules. Outer distal spine

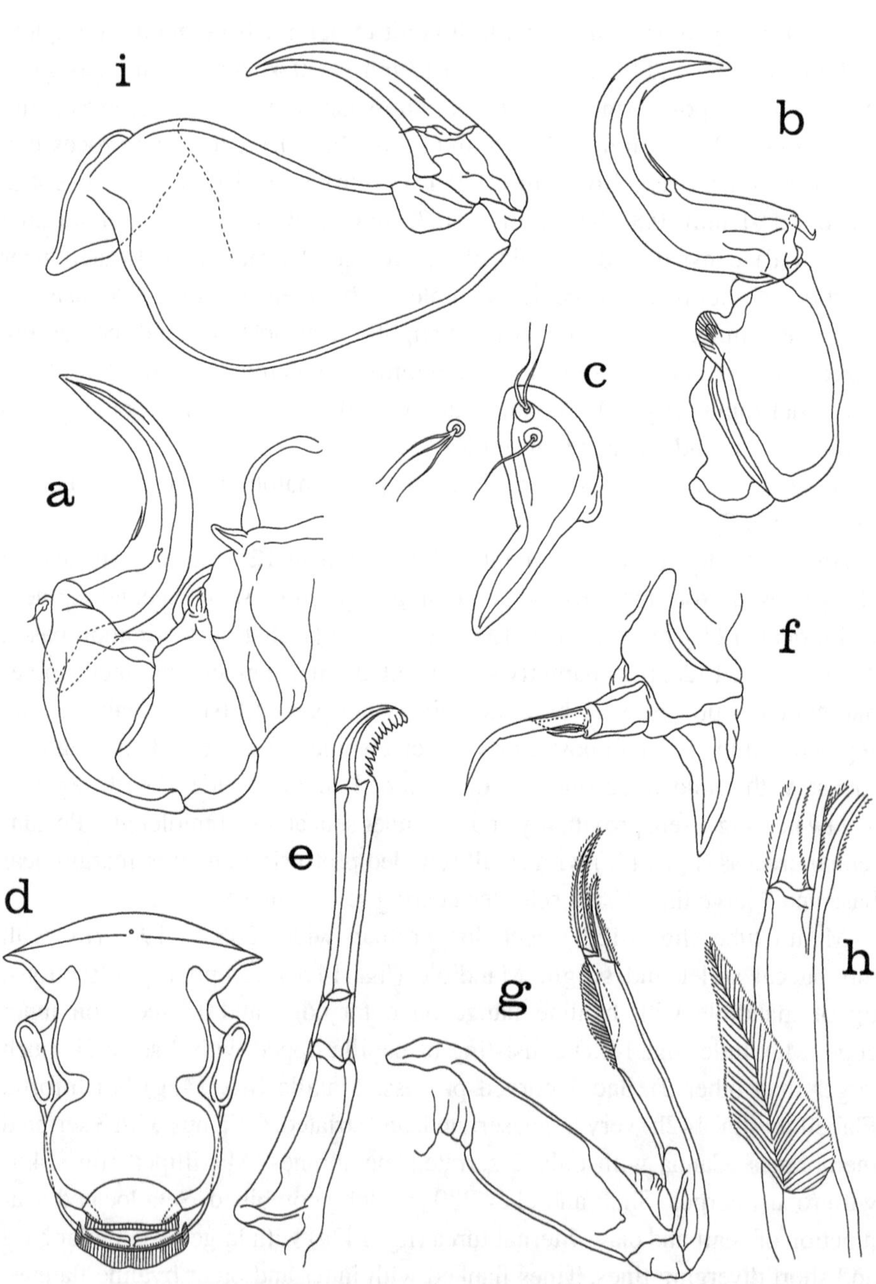

Fig. 142. *Synestius caliginus* Steenstrup & Lütken, 1861, female. a, antenna, ventral (scale D); b, distal three segments of antenna, dorsal (D); c, postantennal process, in situ, ventral (L); d, mouth tube, ventral (E); e, mandible, ventral (U); f, maxillule, ventral (D); g, maxilla, dorsal (E); h, flabellum of maxilla, dorso-outer (L); i, maxilliped, ventral (G).

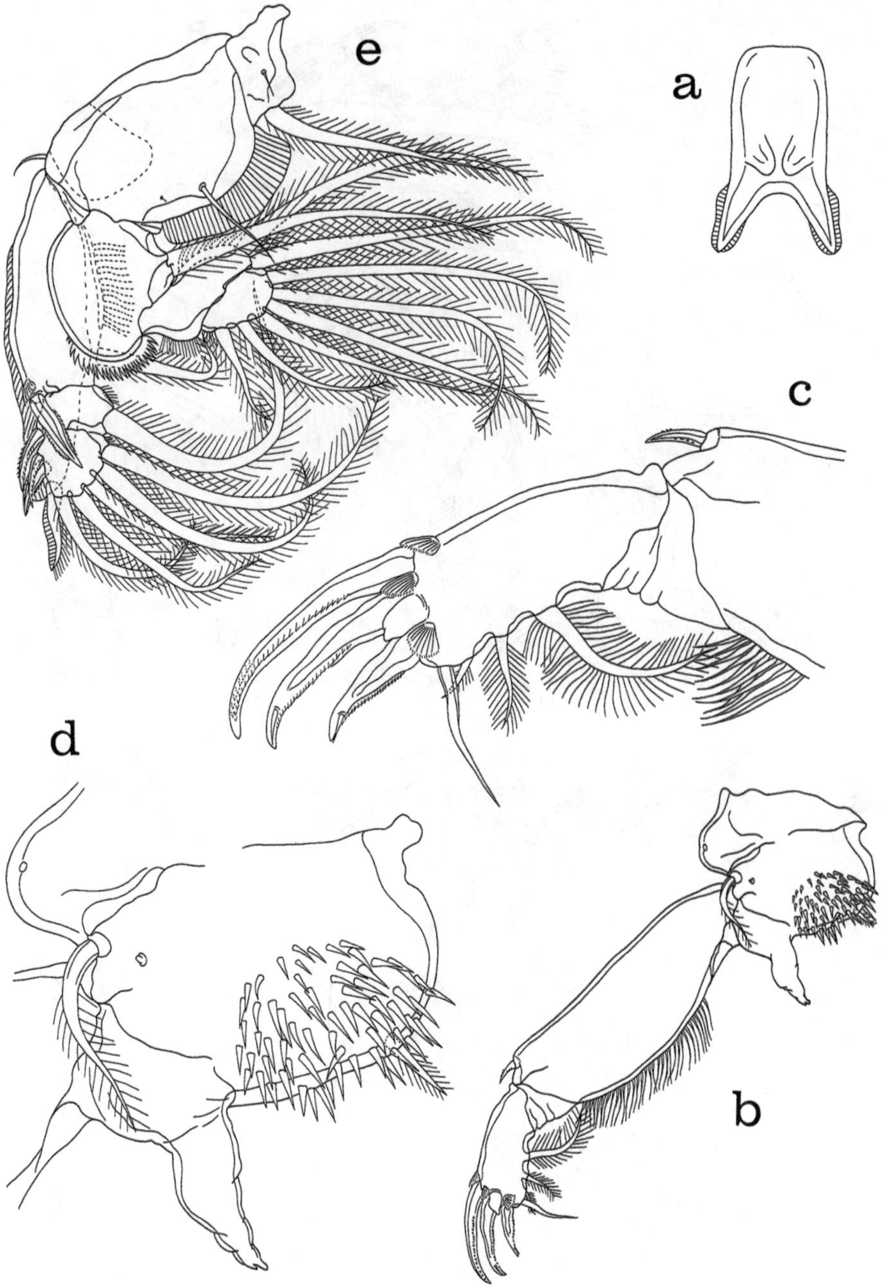

Fig. 143. *Synestius caliginus* Steenstrup & Lütken, 1861, female. a, sternal furca, ventral (scale D); b, leg 1, ventral (E); c, distal portion of leg 1, ventral (L); d, sympod-exopod joint and endopod of leg 1, ventral (L); e, leg 2, ventral (E).

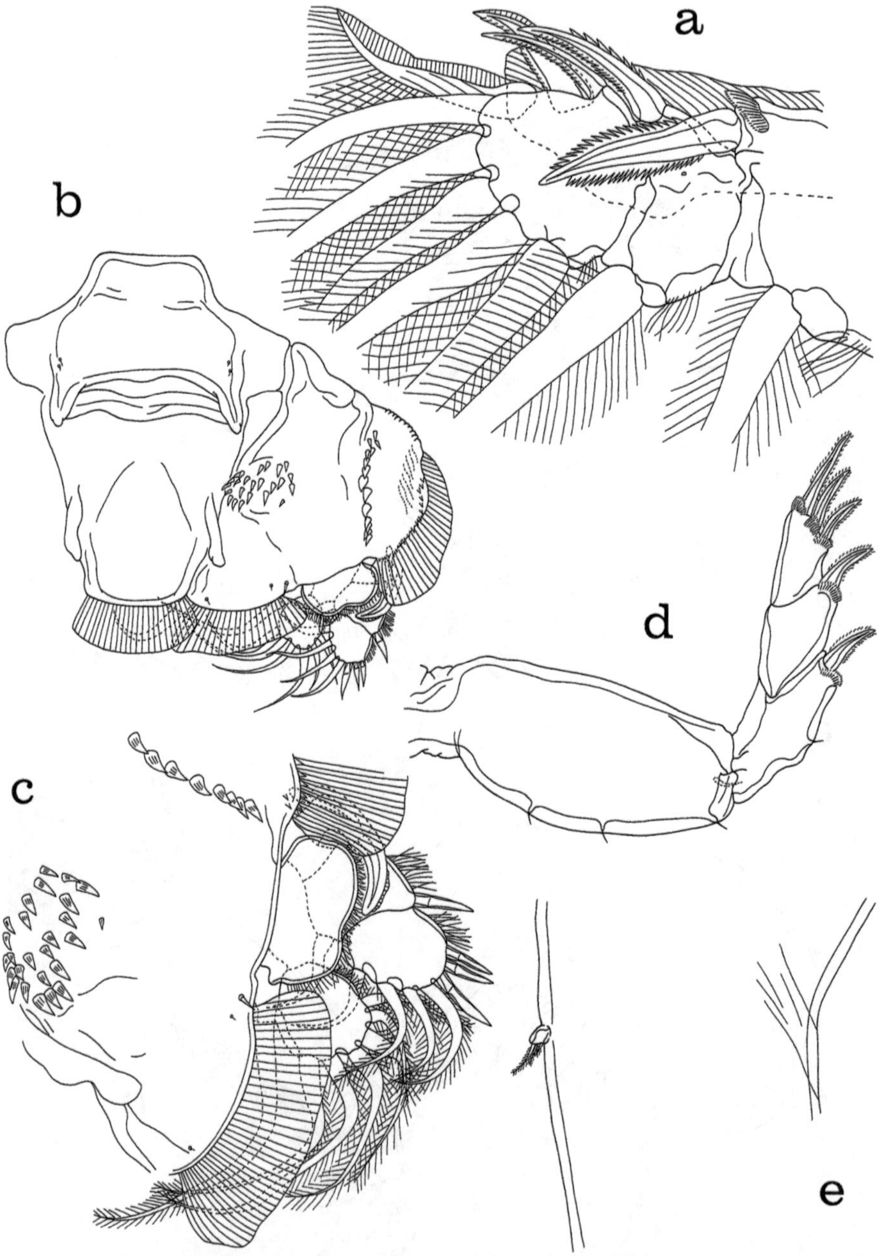

Fig. 144. *Synestius caliginus* Steenstrup & Lütken, 1861, female. a, exopod of leg 2, ventral (scale L); b, leg 3, ventral (G); c, exopod and endopod of leg 3, ventral (D); d, leg 4, ventral (J); e, leg 5, ventral (G).

(fig. 143c) bilaterally barbed and with small spinules at base. Second segment of exopod with seta 4 not pinnate and only slightly shorter than outermost spine. Spine 1 bilaterally spinulated. Spines 2 and 3 spinulated along inner edge and with accessory spine at tip. All 3 spines with pectinate membranes at their bases. Large nipple-tipped, balloonlike structure located between spines 2 and 3. Inner margin of second segment of exopod with 3 plumose setae (increasing in size from distal to proximal setae). Endopod (fig. 143d) reduced and tipped with 2 small digitiform processes. Leg 2 (fig. 143e) biramous, with usual armature. First 2 exopodal spines (fig. 144a) bilaterally serrated; third spine only unilaterally serrated; and fourth spine with inner serrations and outer hyaline flange. First segment of endopod with large spinules along outer margin, otherwise as in other caligids. Sympod of leg 3 (fig. 144b) with corrugated pad on dorsal surface of anterolateral side and 3 groups of spinules (1 row of small spinules, 1 row of large spinules, and 1 patch of large spinules). Intercoxal plate with 2 pairs of cuticular projections (1 pair posterior to first pair). Exopod (fig. 144c) 3-segmented. First segment with basal swelling and large clawlike spine possessing transparent flange. Second segment with 1 outer spine, outer row of setules, and 1 inner plumose seta. Third exopodal segment with outer setules, 3 outer spines, and 4 plumose setae. Endopod (fig. 144c) 2-segmented. First segment with outer expanded margin (velum) and inner plumose seta. Second segment with outer setules and 6 plumose setae. Leg 4 (fig. 144d) uniramous, brachiform. Sympod only slightly expanded, carrying distal plumose seta. Exopod 3-segmented. All 5 spines bilaterally spinulated and with pectinate membrane at base. Leg 5 (figs. 141d, 144e) located on ventromedial margin of posteroventral process of genital complex, and consisting of 1 papilla tipped with 3 plumose setae.

Male. — Body as in fig. 145a. Total length 3.12 mm (n = 1). Cephalothorax as in female, 1.23 × 1.13 mm. Fourth pedigerous somite 0.30 × 0.36 mm. Genital complex ovoid, 0.90 × 0.54 mm. Abdomen comprising 2 free somites. First somite 0.27 × 0.24 mm. Second somite longer than first, 0.37 × 0.23 mm.

Caudal ramus similar to that of female.

Body surface with small setules similar to that in female.

Antennule as in female. Antenna (fig. 145b) different from that of female. First segment small and unarmed. Second segment with ventral corrugated strip; third segment with corrugated strip and ventral corrugated adhesion pad. Claw similar to that of female. Postantennal process as in female.

Mouth tube and mandible as in female. Maxillule (fig. 145c) similar to that in female, except tip of curved tine with indentation and 1 digitiform, hyaline

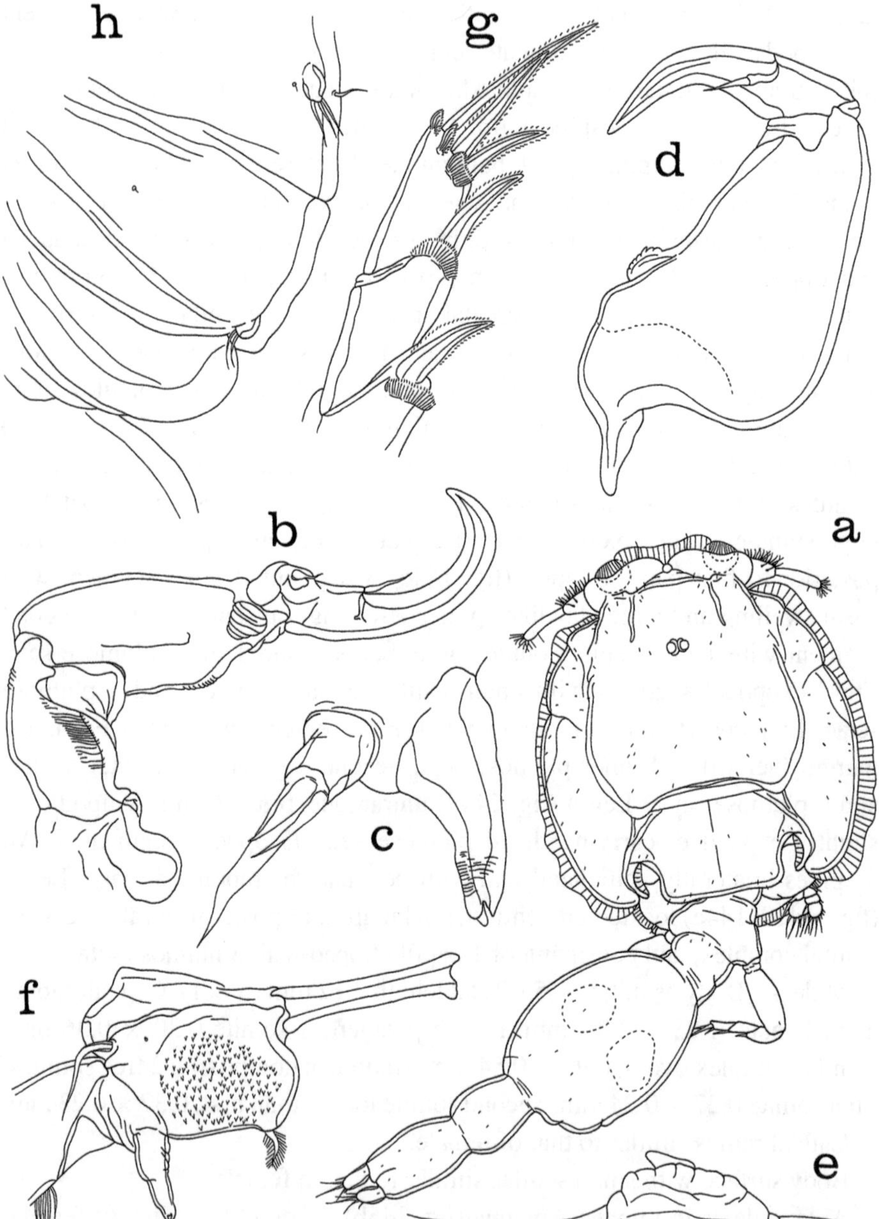

Fig. 145. *Synestius caliginus* Steenstrup & Lütken, 1861, male. a, body, dorsal (scale H); b, antenna, ventral (E); c, maxillule, ventral (U); d, maxilliped, ventral (G); e, myxa of maxilliped, dorsal (L); f, sympod-exopod joint, endopod, and intercoxal plate of leg 1, ventral (D); g, exopod of leg 4, ventral (D); h, legs 5 and 6, and genital area, ventral (E).

process. Maxilla as in female. Maxilliped (fig. 145d) similar to that in female, except myxal area (fig. 145e) rugose and with digitiform process. Sternal furca as in female.

Leg 1 as in female, except patch of small spinules on sympod (fig. 145f). These spinules noticeably larger in female. Outer seta of sympod smooth, but plumosities may be broken off in specimen examined. Legs 2 and 3 as in female. Exopodal spines of leg 4 (fig. 145g) more slender than those of female. Leg 5 (fig. 145h) consisting of 1 lateral seta and 1 papilla tipped with 3 setae. Leg 6 (fig. 145h) consisting of papilla bearing 2 setae.

Remarks. — Since the discovery of *Synestius caliginus* by Steenstrup & Lütken (1861) from *Parastromateus niger* (Bloch) (reported under a different name) collected in the Indian Ocean, all subsequent reports have been from fishes of the family Stromateidae primarily caught in the Indian Ocean (table XIX). The only exceptions are the reports by Shen (1957) from the Kwangtung Province, China (South China Sea) and Lin & Ho (2000) from Taiwan.

Although Gnanamuthu (1950b) provided a redescription of *Synestius caliginus*, it is not as detailed as Shen's (1957) and Pillai's (1961) redescriptions. Pillai (1961) provided accurate illustrations, but the account is not complete, particularly in regards to the male. In addition, the number of exopodal spines of leg 2 of the female reported by Pillai is incorrect. He shows only three exopodal spines, but there are really four exopodal spines present. A detailed redescription of this species, which was collected from the gill filaments of *Parastromateus niger* (Bloch) from Taiwan, was provided by Ho & Lin (2004).

Genus **Tuxophorus** Wilson, 1908

Tuxophorus Wilson, 1908, p. 617; Yamaguti, 1963: 105; Dojiri, 1983: 348; Pillai, 1985: 472; Boxshall & Montú, 1997: 67; Ho, 1998: 253; Boxshall & Halsey, 2004: 726.

Female. — Cephalothorax suborbicular, narrower anteriorly, with distinct, but shallow posterior sinuses. Frontal plate with lunules. Free margin of thoracic zone extending slightly beyond posterior limit of lateral zone. Fourth pedigerous somite small, with 1 pair of dorsal aliform plates. Genital complex quadrangular or globose, occasionally with spikelike projection representing leg 5 on posterolateral corner. Abdomen comprising 1, 2, or 3 free somites. Caudal ramus setiferous, either short and broad, or long and slender.

Antennule as in *Caligus*. Antenna 4-segmented. Second segment with posteriorly directed spinelike process. Terminal segment hamate. Postantennal process present. Mouth tube with intrabuccal stylet and strigil. Mandible

TABLE XIX
Hosts, attachment sites, and localities of collections of *Synestius caliginus* Steenstrup & Lütken, 1861

Host	Attachment site	Locality	Reference
Stromateus argenteus [= *Pampus argenteus* (Euphrasen)]	Gills	Indian Ocean	Heller, 1865
Parastromateus niger (Bloch)	Gills	Indian Ocean	Steenstrup & Lütken, 1861
Stromateus niger [*Parastromateus niger*]	?	Madras, India	Gnanamuthu, 1950b
Stromateoides sinensis [= *Pampus chinensis* (Euphrasen)]	Isthmus	Kwangtung Province, China	Shen, 1957
Parastromateus niger	roof of buccal cavity	Trivandrum, India	Pillai, 1961
Parastromateus niger	gills	Colombo Market, Sri Lanka	Kirtisinghe, 1964
Parastromateus niger	buccal cavity	Trivandrum, India	Radhakrishnan & Nair, 1983
Formio niger [= *Parastromateus niger*]	gill filaments	Malaysia	Leong, 1985
Pampus argenteus	gills	Trivandrum, India	Pillai, 1985
Parastromateus niger	gills	Batavia	Piasecki, 1993
Formio niger [= *Parastromateus niger*]	gills	Kuwait	Ho & Sey, 1996
Parastromateus niger	gill filaments	Taiwan	Lin & Ho, 2000
Parastromateus niger	gills	Gulf of Thailand	Purivirojkul & Areechon, 2008

comprising 4 sections, with 12 teeth on inner margin of terminal section. Maxillule with dentiform projection and adjacent setiferous papilla. Maxilla brachiform. Maxilliped subchelate and prehensile. Sternal furca present; tines single or bifid. Dentiform process occasionally present on either side of sternal furca.

Leg 1 with 2-segmented exopod and vestigial endopod. Terminal exopodal segment with 3 spines, 1 seta on inner distal corner, and 3 plumose setae on inner margin. Leg 2 biramous, with 3-segmented rami. Leg 3 with large ventral apron. Exopod 3-segmented, with large clawlike spine located distally on basal swelling. Leg 4 bearing 3-segmented exopod with formula I-0; I-0; III. Leg 5 represented by setiferous papilla or long spikelike projection bearing setae (leg 5 may be absent in *T. collettei* Cressey & Cressey, 1980). Leg 6 absent.

Male. — Cephalothorax similar to that in female. Fourth pedigerous somite small, with or without dorsal plates. Genital complex suboval. Abdomen and caudal ramus as in female. Antenna with corrugated adhesion pads; terminal claw occasionally with accessory claw. Oral adhesion pad situated posteromedial to bases of maxilla. Legs 5 and 6 represented by setiferous process and setiferous lobe, respectively. All other appendages as in female.

Type-species. — *Tuxophorus caligodes* Wilson, 1908.

Remarks. — *Tuxophorus* was first established by Wilson (1908) for *T. caligodes*, a copepod parasitic predominantly on *Echeneis naucrates* Linnaeus and *Rachycentron canadum* (Linnaeus) (table XX). This genus was first described under the subfamily Caliginae sensu Wilson, 1908, but later transferred to the Euryphoridae by Heegaard (1955). The members of *Tuxophorus* are morphologically very similar to species of *Caligus*. Moreover, the male of *T. caligodes* lacks the dorsal aliform plates on the fourth pedigerous somite; therefore, it is morphologically similar to males of *Caligus*. The main feature that distinguishes females of *Tuxophorus* from *Caligus* is the presence of these aliform plates in *Tuxophorus*. This genus resembles *Caligus*, and must be placed in the Caligidae as noted by Pillai (1961).

There have been seven additional species of *Tuxophorus* described since the discovery of the type-species. Among these, *T. solandri* Kurian, 1961 has been synonymized with *T. cybii* Nunes-Ruivo & Fourmanoir, 1956 by Cressey & Cressey (1980), and *T. tylosuri* Rangnekar, 1956 has been transferred to *Caligus* by Pillai (1961). Besides *T. caligodes*, there remain *T. wilsoni* Kirtisinghe, 1937; *T. cybii* Nunes-Ruivo & Fourmanoir, 1956; *T. cervicornis* Heegaard, 1962; *T. zonichthi* Devi & Shyamasundari, 1978; and *T. collettei* Cressey & Cressey, 1980. These six species can easily be distinguished from each other by a combination of four characteristics (table XXI).

Cressey & Cressey (1980) suggested that *T. cybii* may have been based on immature specimens by Nunes-Ruivo & Fourmanoir (1956). Immaturity of the specimens could possibly account for the absence of a pair of large spikelike posterolateral projections (leg 5) on the genital complex of the female of this species. Since the absence of these projections in the female distinguishes *T. cybii* from *T. cervicornis*, Cressey & Cressey reasoned that the former may be synonymous with the latter. According to the illustration provided by the discoverers (Nunes-Ruivo & Fourmanoir, 1956: 77, fig. 7a), the genital complex is inflated (distended by presence of numerous eggs); the female specimens of *T. cybii*, though nonovigerous, must have been mature. Therefore, *T. cybii* must be considered a valid species, distinct from *T. cervicornis*.

TABLE XX

Hosts and localities of collections of *Tuxophorus caligodes* Wilson, 1908

Host	Locality	Reference
Rachycentron canadum (Linnaeus)	Beaufort, North Carolina	Wilson, 1908
Echeneis naucrates Linnaeus	Beaufort, North Carolina	Wilson, 1908
Trachinotus carolinus (Linnaeus)	Lemon Bay, Florida	Bere, 1936
Trachinotus falcatus (Linnaeus)	Lemon Bay, Florida	Bere, 1936
unidentified species of fish	coast of Sierra Leone (west coast of Africa)	Capart, 1941
Promicrops itiaria [= *Epinephelus itajara* (Lichtenstein)]	Port Aransas, Texas	Pearse, 1952
Rachycentron canadum	Port Aransas, Texas	Causey, 1953b
Rachycentron canadum	Grand Isle, Louisiana	Causey, 1953a
Phycis floridanus [= *Urophycis floridana* (Bean & Dresel)]	Grand Isle, Louisiana	Causey, 1953a
suckerfish	west coast of Africa	Heegaard, 1955
Echeneis naucrates	east coast of southern Africa	Capart, 1959
Rachycentron canadum	Bombay, India	Rangnekar, 1960
Rachycentron canadum	Trivandrum, India	Pillai, 1961
Tylosurus acus (Lacépède)	Trivandrum, India	Pillai, 1985
Mugil platanus (Günther)	Rio de Janeiro, Brazil	Knoff et al., 1994

Pillai (1985) erected a new genus *Tuxophoropsis* to accommodate *T. cybii* and *T. cervicornis* based on two characters that are not found in the remaining congeners: single (versus bifid) tines on the sternal furca and a long, slender (versus short, broad) caudal ramus. As stated by Boxshall & Halsey (2004), neither of these differences appears to be a generic level discriminant. We concur with them and treat *Tuxophoropsis* as a synonym of *Tuxophorus*.

Tuxophorus caligodes Wilson, 1908
(figs. 146-150)

Tuxophorus caligodes Wilson, 1908: 617; Bere, 1936: 587; Capart, 1941: 184; Pearse, 1952: 24; Causey, 1953a: 7; Causey, 1953b: 11; Heegaard, 1955: 45; Rangnekar, 1956: 56; Capart, 1959: 95; Rangnekar, 1960: 20; Pillai, 1961: 98, 120; Yamaguti, 1963: 108; Dojiri, 1983: 352; Pillai, 1985: 473; Cressey & Nutter, 1987: 600; Knoff et al., 1994: 50; Boxshall & Montú, 1997: 57; Luque et al., 1999: 13.

Material examined. — Type-specimens consisting of 2 females and 1 male (USNM 32805) collected from skin of cobia, *Rachycentron canadum* (Linnaeus) from Beaufort, North Carolina.

TABLE XXI

Diagnostic features, hosts, and localities of collections of all six described species of *Tuxophorus* Wilson, 1908. Abbreviations: SF = sternal furca; P_5 = leg 5

Species	Abdomen	Caudal ramus	SF	P_5	Host	Locality
T. caligodes Wilson, 1908	shorter than genital complex	short, broad	bifid tines	short papilla	*Echeneis*; *Rachycentron* (see table XX)	see table XX
T. cervicornis Heegaard, 1962	about as long as genital complex	long, slender	single tines	long spikelike projection	*Scomberomorus commerson* (Lacépède)	Australia; Pakistan
T. collettei Cressey & Cressey, 1980	twice length of genital complex	long, slender	single tines	absent	*Scomberomorus regalis* (Bloch)	Virgin Islands
T. cybii Nunes-Ruivo & Fourmanoir, 1956	about as long as genital complex	long, slender	single tines	2 small spines	*Acanthocybium solandri* (Cuvier)	Sri Lanka; Madagascar
T. wilsoni Kirtisinghe, 1937	shorter than genital complex	short, broad	bifid tines	long spikelike projection	*Chorinemus lysan* [= *Scomberoides lysan* (Forsskål)	India; Sri Lanka]
T. zonichthi Devi & Shyamasundari, 1978	shorter than genital complex	short, broad	bifid tines	short spikelike projections	*Zonichthys nigrofasciata* [= *Seriolina nigrofasciata* (Rüppell)]	Waltair coast

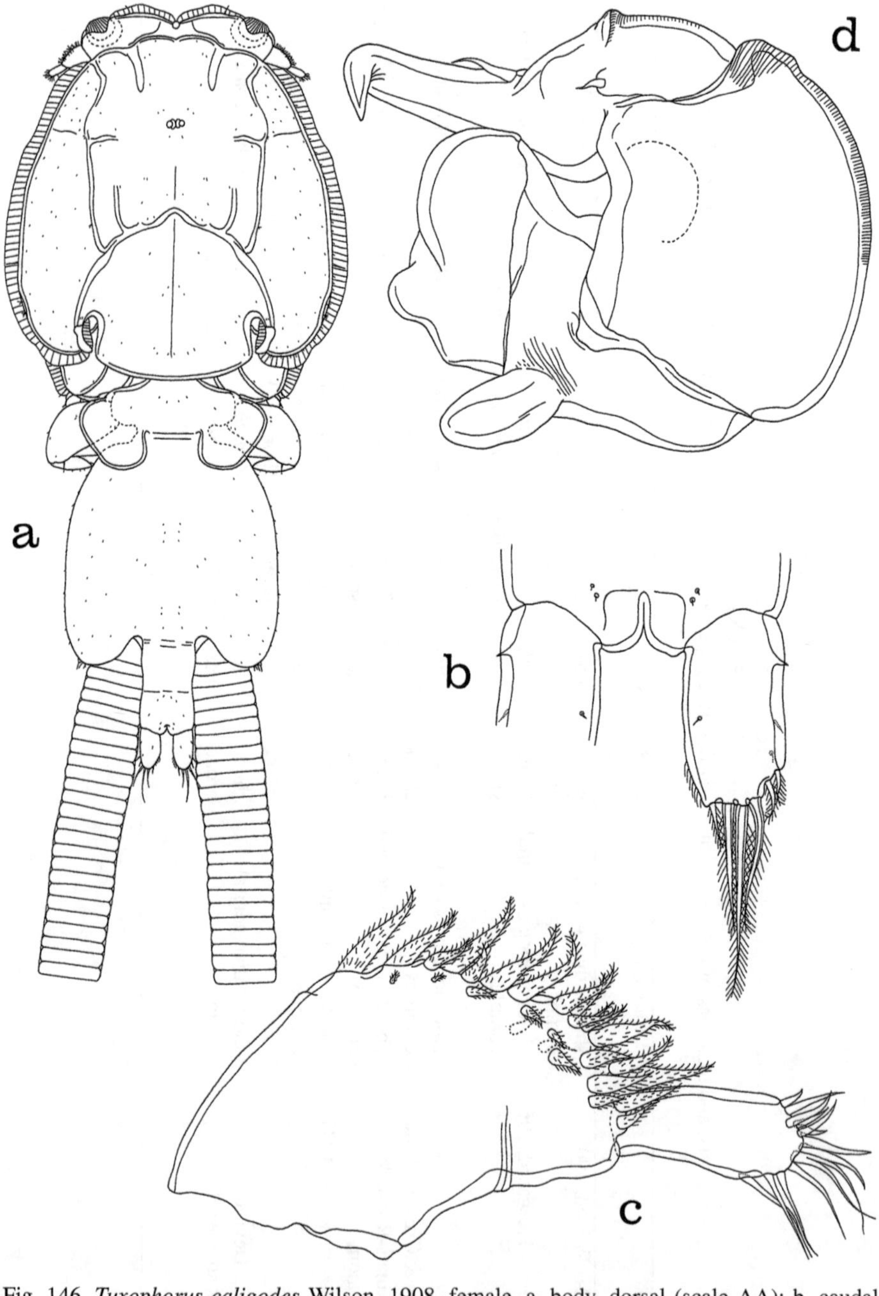

Fig. 146. *Tuxophorus caligodes* Wilson, 1908, female. a, body, dorsal (scale AA); b, caudal ramus, ventral (B); c, antennule, ventral (E); d, antenna, ventral (E).

This material on loan from National Museum of Natural History, Smithsonian Institution, Washington, D.C.

Female. — Body as in fig. 146a. Total length (not including setae on caudal ramus) 4.66 mm (4.55-4.78 mm) (n = 2). Cephalothorax relatively large and suboval, 2.32 mm (2.16-2.49 mm) × 2.09 mm, with shallow posterior sinuses. Free margin of thoracic zone extending slightly beyond posterior tip of lateral zone. Frontal plate wide and carrying pair of lunules ventrally. Tip of antennule not extending beyond lateral limit of cephalothorax. Fourth pedigerous somite more than 2 times as wide as long, 0.40 × 0.86 mm (0.83-0.90 mm), and possessing 1 pair of dorsal plates. Dimensions of fourth pedigerous somite including dorsal plates, 0.48 × 1.28 mm. Genital complex 1.49 mm (1.46-1.53 mm) × 1.43 mm, with posterolateral processes. Abdomen indistinctly comprising 2 free somites; 2 somites combined measuring 0.55 mm (0.53-0.56 mm) × 0.37 mm.

Caudal ramus (fig. 146b) subrectangular, longer than wide, 265 × 127 μm, with 6 plumose setae.

Egg sacs (fig. 146a) 4.15 × 0.43 mm, containing 48 eggs (n = 1).

Antennule (fig. 146c) 2-segmented. First segment robust, 162 μm long (measured along nonsetiferous margin), and carrying 28 plumose setae. Second segment smaller, 119 μm long, and tipped with 14 setae (13 + 1 aesthetasc?). Antenna (fig. 146d) apparently 4-segmented. First segment unarmed. Second segment with large, bluntly rounded, posteriorly directed, spinelike process. Third segment (fig. 147a) large, with dorsal corrugated pad and striations along anteromedial margin. Terminal segment a recurved claw with 2 setae and slightly elevated ridge near base of proximalmost seta. Postantennal process (fig. 147b) strongly curved and tapered, with small knob on lateral surface near base. Groups of setules associated with process present near base (1 sclerite with setules broken off in figure).

Mouth tube (fig. 147c) longer than wide, 297 × 243 μm. Intrabuccal stylet and strigil present. Mandible (fig. 147d) with 4 sections and bearing 12 teeth. Maxillule (fig. 147e) with curved process and sclerotized area carrying 2 small and 1 large setae. Process with small patch of striations on medial surface. Maxilla (fig. 147f) brachiform, with flabellum at midlength of brachium. Calamus much longer than canna. Corpus maxillipedis (fig. 148a) moderately robust. Shaft with depression housing 1 small spinule. Claw with usual seta. Shaft and claw combined measuring 232 μm long. Sternal furca (fig. 148b) with bifid tines.

Sympod of leg 1 (fig. 148c) with sutures indistinctly separating coxa from basis. Base of plumose seta at junction of sympod and exopod not covered by

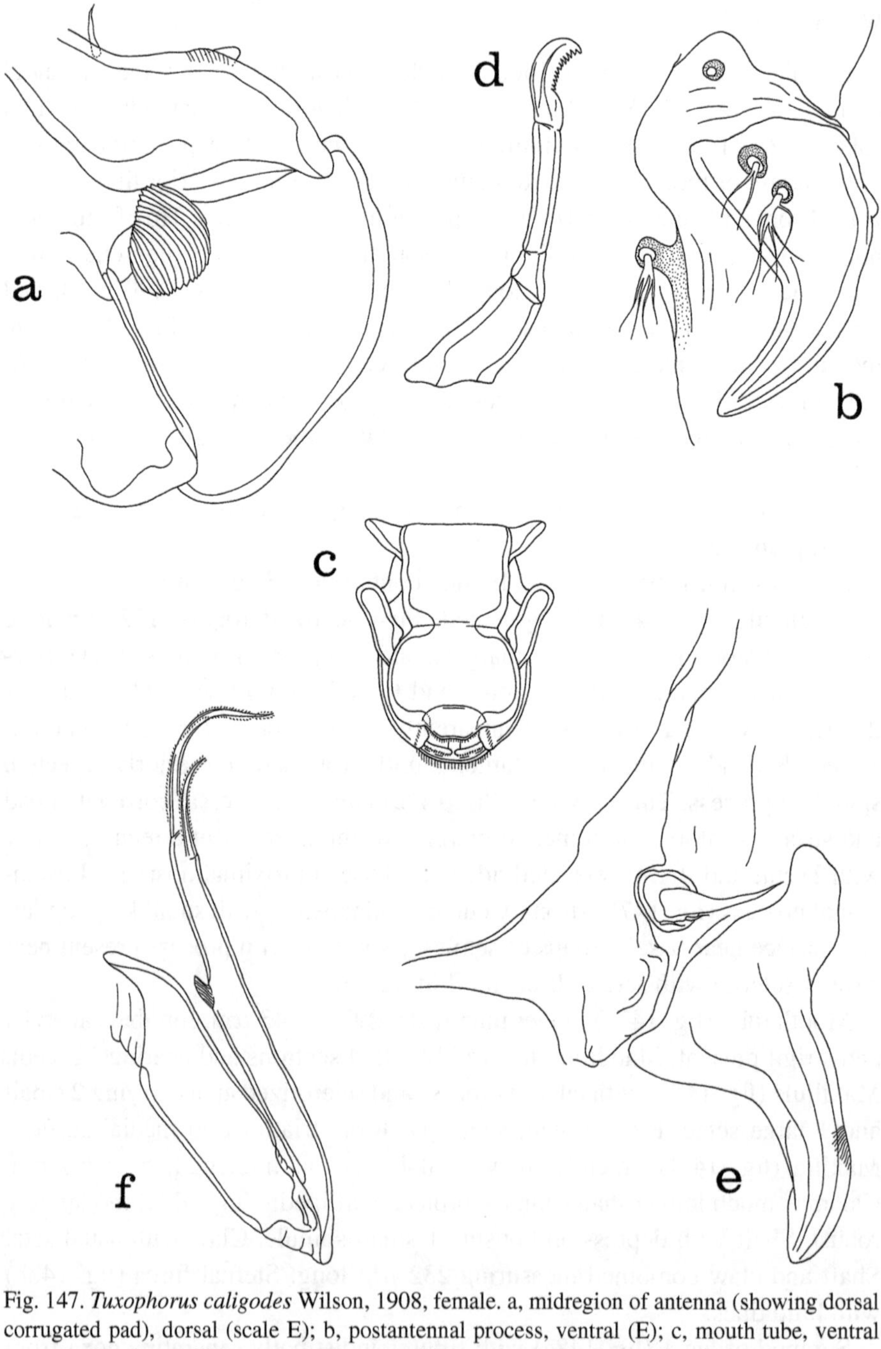

Fig. 147. *Tuxophorus caligodes* Wilson, 1908, female. a, midregion of antenna (showing dorsal corrugated pad), dorsal (scale E); b, postantennal process, ventral (E); c, mouth tube, ventral (B); d, mandible, ventral (E); e, maxillule, ventral (E); f, maxilla, ventral (B).

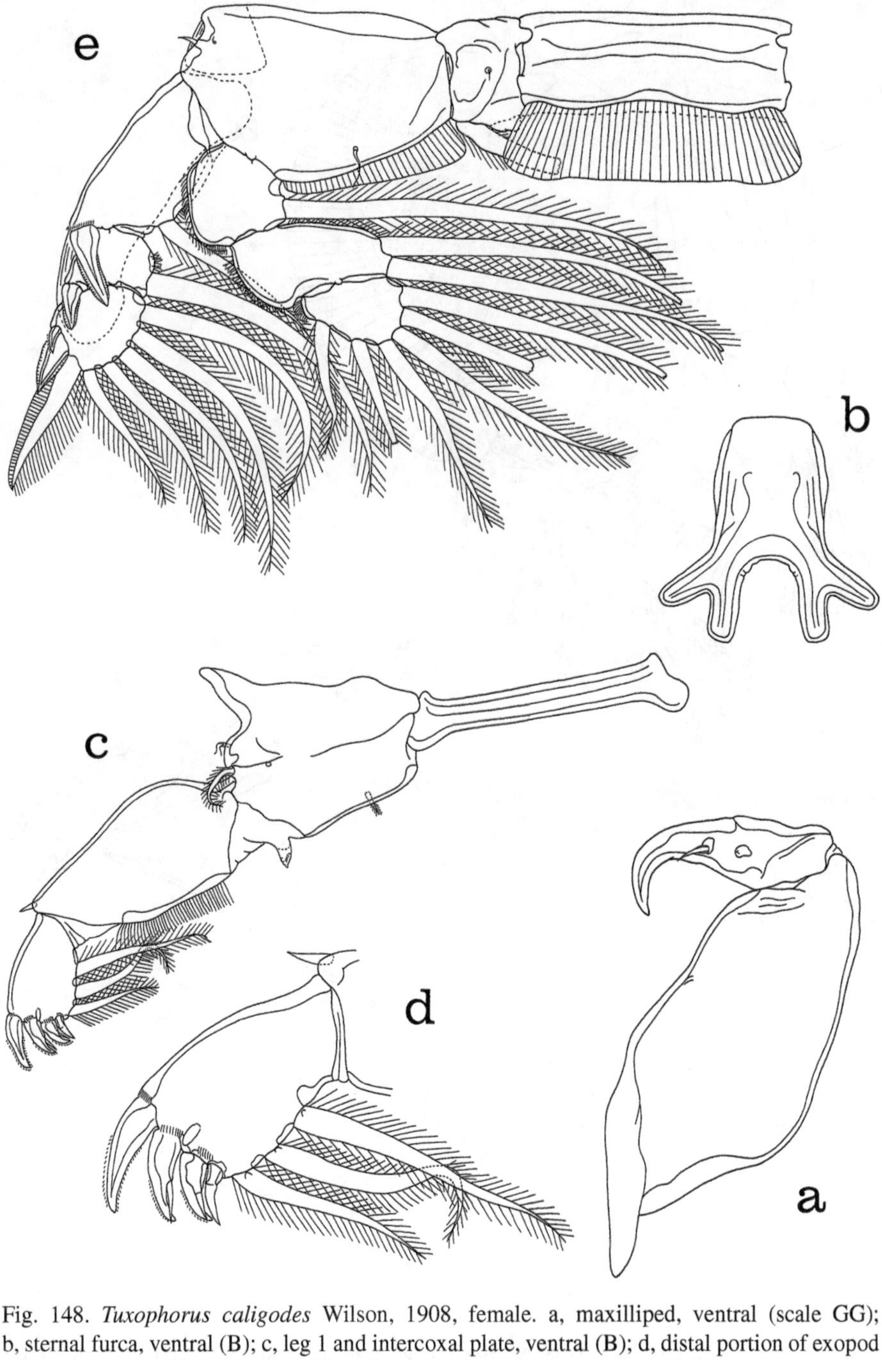

Fig. 148. *Tuxophorus caligodes* Wilson, 1908, female. a, maxilliped, ventral (scale GG); b, sternal furca, ventral (B); c, leg 1 and intercoxal plate, ventral (B); d, distal portion of exopod of leg 1, ventral (E); e, leg 2 and intercoxal plate, ventral (B).

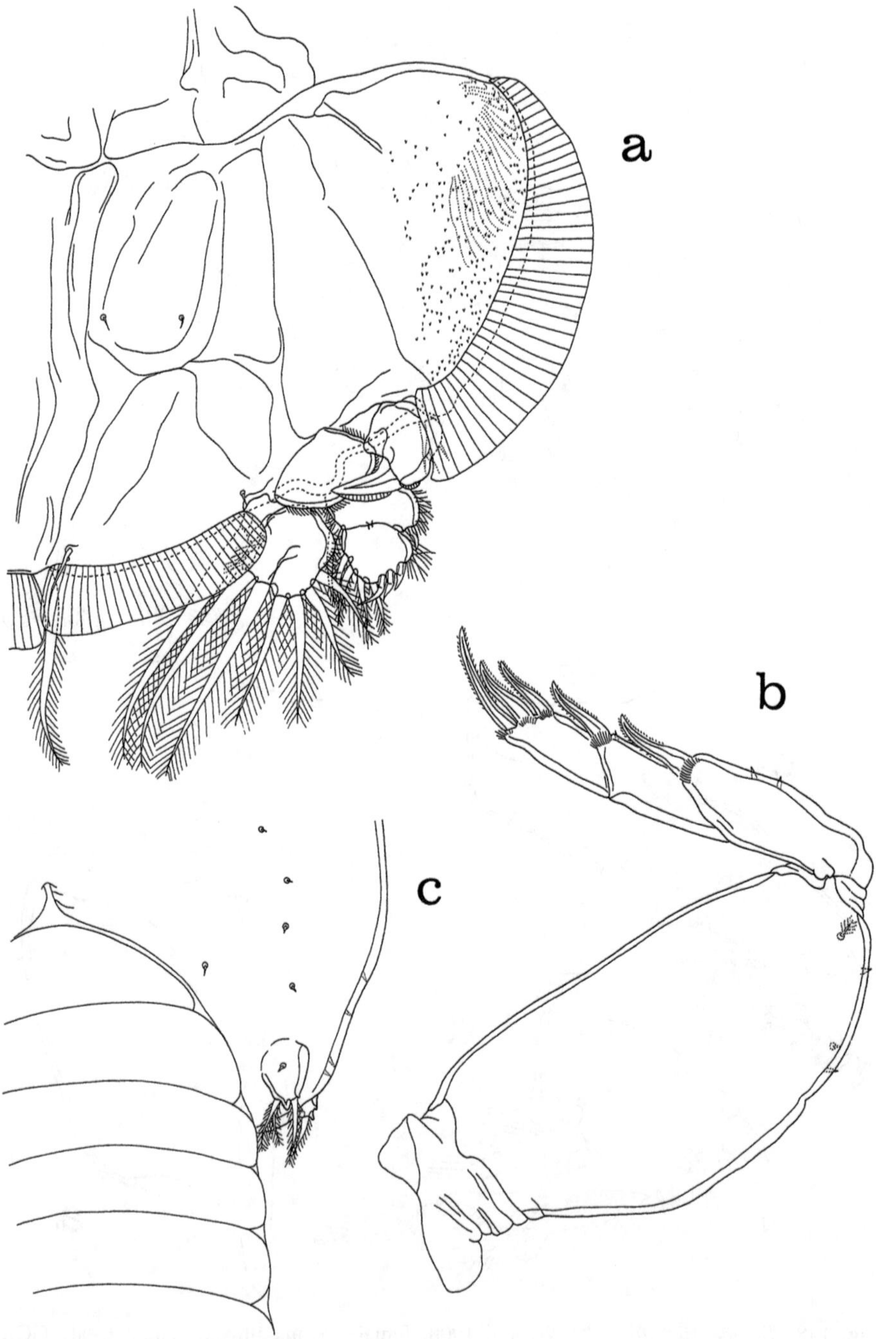

Fig. 149. *Tuxophorus caligodes* Wilson, 1908, female. a, leg 3, ventral (scale B); b, leg 4, ventral (B); c, leg 5 and genital area, ventral (B).

ventral protrusion. First segment of exopod with small spine on outer distal corner and long setules along inner margin. Seta 4 not pinnate, and much shorter than spine 1. Spines 1, 2, and 3 (fig. 148d) stout and spinulated, and with pectinate membranes at bases. Spines 2 and 3 with accessory processes. A small hyaline process located in between bases of spines 2 and 3. Endopod a lobe tipped with small protrusion and row of spinules. Leg 2 (fig. 148e) with short, stout exopodal spines, otherwise typical for caligids. Sympod of leg 3 (fig. 149a) with ventral group of spinules on lateral side. Dorsal adhesion pad present. Exopod 3-segmented. First segment represented by basal swelling and terminally located clawlike spine. Basal swelling and clawlike spine with flange on outer edge. Next 2 exopodal segments characteristic for caligids. First endopodal segment with outer margin expanded to form velum and inner margin with plumose seta. Second endopodal segment with 6 plumose setae, partially divided by furrow. Sympod of leg 4 (fig. 149b) robust. Ventral, plumose seta on outer distal corner small. Exopod 3-segmented, with innermost spine of terminal segment longer than other 2. First segment of exopod with 2 small stout setules on outer margin. All 5 spines with pectinate membranes. Leg 5 (fig. 149c) represented by 2 projections. Smaller one tipped with single plumose seta. Larger one with 3 plumose setae.

Male. — Body as in fig. 150a. Total length 3.05 mm (n = 1). Cephalothorax comparatively large, 1.83 × 1.66 mm, and similar in shape to that of female.

Fourth pedigerous somite wider than long, 0.33 × 0.56 mm. Dorsal plates absent. Genital complex oval, 0.53 × 0.50 mm. Abdomen indistinctly comprising 2 free somites. Both somites 0.37 × 0.29 mm combined.

Caudal ramus longer than in female, 288 × 115 μm.

Antennule and postantennal process as in female. Antenna (fig. 150b) with numerous adhesion pads. Claw with large accessory tine and 2 setae.

Mouth tube and mandible as in female. Maxillule (fig. 150c) different from that of female, with small knob and small setule on ventromedial surface. Pair of corrugated pads located medial to bases of maxilla. Maxilla, maxilliped, and sternal furca as in female. Position of mouth as in fig. 150c.

Legs 1-4 as in female. Leg 5 (fig. 150d) a process terminating in 2 small spiniform tubercles, and bearing 4 plumose setae. Leg 6 (fig. 150d) a rounded lobe possessing 1 small smooth seta, and 2 plumose setae.

Remarks. — Wilson (1908) described *Tuxophorus caligodes* as a representative of a new genus and species of the subfamily Caliginae from the external body surface of the cobia "*Rachycentron canadus*" (= *R. canadum*) and the sharksucker *Echeneis naucrates* Linnaeus. He also collected the chalimus

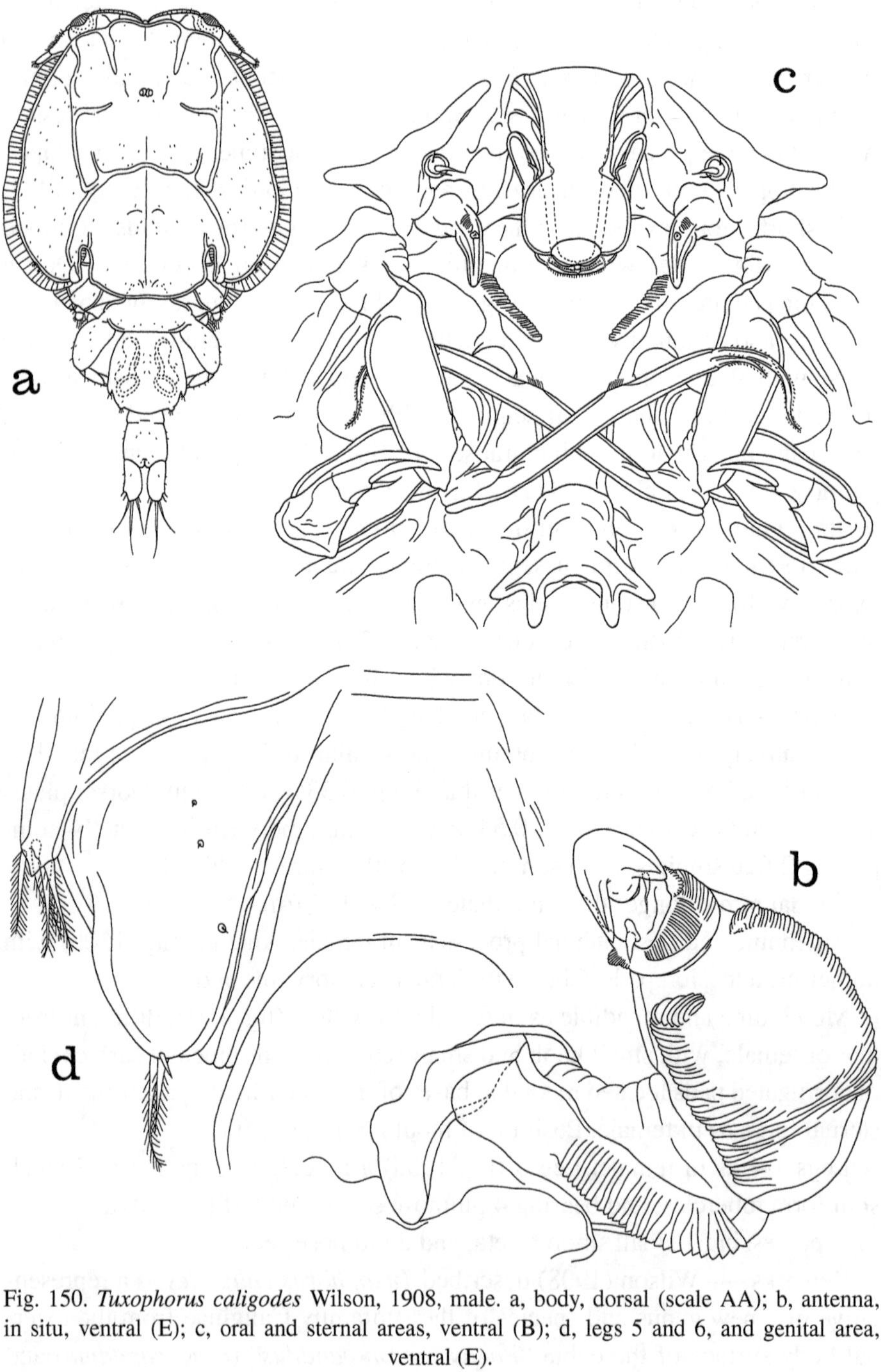

Fig. 150. *Tuxophorus caligodes* Wilson, 1908, male. a, body, dorsal (scale AA); b, antenna, in situ, ventral (E); c, oral and sternal areas, ventral (B); d, legs 5 and 6, and genital area, ventral (E).

stages of this species from the silver gar "*Tylosaurus marinus*" [= *Strongylura marina* (Walbaum)]. Subsequent to Wilson's discovery, a number of new host records (table XX) appeared in the literature including some carangids (Bere, 1936), serranids (Pearse, 1952), gadids (Causey, 1953a) and mugilids (Knoff et al., 1994). Although originally found off the east coast of the United States, it has been reported from the Gulf of Mexico, as far away as the east and west coasts of Africa, and off the coast of India.

Rangnekar (1960) redescribed one female and one male of this species obtained from *R. canadum* caught at Bombay. The details of the appendages, lacking in Rangnekar's description of *T. caligodes*, were described by Pillai (1961).

When Heegaard (1955) reported this species from a suckerfish from West Africa, he stated that Wilson (1908) never assigned *T. caligodes* to any family. So, Heegaard placed this species within the Euryphoridae. The statement was ill-founded since Wilson's entire paper pertains to the "Caliginae". In addition, Wilson specifically stated in his discussion section "...a genus undoubtedly belonging to the Caliginae". Pillai (1961) transferred the species from the Euryphoridae back to the Caligidae.

Capart (1941, 1959) proposed that *Caligus remorae* described by Brian (1905) is probably synonymous with *T. caligodes*. According to Brian's illustration of the genital complex, sternal furca, the armature of the third leg, and the nature of the exopodal spines of the second leg of *C. remorae*, Capart's contention is well founded. Moreover, Brian collected this species from the body of a species of *Remorae*, which is a member of the same family as *Echeneis* (a known host for *T. caligodes*). We are in agreement with Capart that *C. remorae* should most likely be relegated to synonymy with *T. caligodes*, and that the apparent absence of the dorsal plates in the illustrations of *C. remorae* was simply no more than an error in observation by Brian. It is, however, prudent at this time to wait until the type-specimens of *C. remorae* are examined before a definitive decision is made.

MISCELLANEOUS GENERA

Genus **Caligera** Beneden, 1892

Caligera Beneden, 1892: 258; Bassett-Smith, 1899: 462; Wilson, 1907a: 702, 707; Wilson, 1924: 18; Cressey, 1972: l; Dojiri, 1983: 358; Kazachenko, 2001: 34; Boxshall & Halsey, 2004: 839.

Remarks. — *Caligera*, not to be confused with *Caligeria* Dana, 1852, was first established by Beneden (1892) for *Caligera difficilis*. Bassett-Smith (1899) considered this genus to be a synonym of *Alebion*. This synonymy was supported by Wilson (1907a, 1924), Cressey (1972), Kazachenko (2001), and Boxshall & Halsey (2004).

Genus **Caligeria** Dana, 1852

Caligeria Dana, 1852: 57; Dana, 1853: 1360; Yamaguti, 1963: 100; Kabata, 1979: 204; Dojiri, 1983: 359.

Remarks. — *Caligeria* was first erected by Dana (1852) for one species, *Caligeria bella* Dana, 1849, which was collected from the gills of the "Albicore" on 7 May 1842 in the Atlantic Ocean (11°S 14°W). Although mentioned later by Dana (1853) and Yamaguti (1963), this species has never been collected and recorded under the binomen *Caligeria bella* since its discovery as noted by Kabata (1979).

According to the illustrations of Dana (1855), it appears that *C. bella* is synonymous with or closely related to *Euryphorus brachypterus* (Gerstaecker, 1853) as noted by Kabata (1979). The hosts for *E. brachypterus* (listed as *Elytrophora brachyptera* by Cressey & Cressey, 1980) are tunas predominantly belonging to the genus *Thunnus* to which the albacore [*T. alalunga* (Bonnaterre)] belongs. Both Kazachenko (2001) and Boxshall & Halsey (2004) treated *Caligeria* as a synonym of *Euryphorus*.

Unfortunately, Dana (1852) does not mention where he deposited his type-specimens, if he deposited any at all. Our attempts to locate Dana's specimens of *C. bella* proved to be futile.

Genus **Caligopsis** Markewitsch, 1940

Caligopsis Markewitsch, 1940: 14; Markewitsch, 1956: 128; Yamaguti, 1963: 63; Kabata, 1979: 41, 158; Dojiri, 1983: 359; Kazachenko, 2001: 24.

Remarks. — *Caligopsis* was first described by Markewitsch (1940) with its sole species being *C. ponticus* Markewitsch, 1940. The copepods were found in a "crystallizer" (= a crystallizing dish) that contained artificially fertilized eggs of "*Atherina mochon ponticus*". Since subsequent citations of this genus (Markewitsch, 1956; Yamaguti, 1963) are simply listings of the original account, it appears that this species has not been collected since its discovery, and its host remains unknown.

The main taxonomic feature used by Markewitsch to distinguish this genus from *Caligus* is the absence of the fourth leg. Although it has been reported that the members of the genus *Pseudocaligus* (synonymized with *Caligus* herein) all possess a highly reduced leg 4, no other genus lacks leg 4 completely.

It is unfortunate that Markewitsch (1940) did not give a more detailed description of *C. ponticus*. He provided only a cursory description and illustrations of the dorsal views of the female and male (probably an immature female based on the illustration as noted by Kabata, 1979). Apparently the Zoological Museum of the Academy of Sciences, Kiev, Ukraine, no longer houses the type-specimens of *Caligopsis*. According to Titar (personal communication), most of Markewitsch's type-specimens were destroyed during World II.

Although *Caligopsis* was considered to be a monotypic genus since its discovery, Kabata (1979) suggested that *Pseudocaligus apodus* Brian, 1924 be transferred to *Caligopsis*. As noted by Brian (1924), this species lacks leg 4. If leg 4 is absent in *P. apodus*, the transfer would be justified. However, Paperna's (1964) illustration of *Pseudolepeophtheirus mediterraneus* Paperna, 1964, which was later relegated to synonymy with *P. apodus* by Paperna & Lahav (1974), distinctly shows a fourth leg. This description, however, appears to be based on a chalimus stage and not on an adult.

It is not known if the absence of the fourth legs in *C. ponticus* and *P. apodus* is real or simply incorrect observations by the describers. In any case, specimens of both of these species need to be examined in the future to ascertain the validity of *Caligopsis*. Even if leg 4 is absent, can *Caligopsis* be considered valid based solely on the absence of leg 4, a characteristic of questionable generic value? If leg 4 is absent and *Caligopsis* is deemed a valid genus, then a steplike series of the reduction of leg 4 becomes apparent from the brachiform leg 4 of most species of *Caligus* to the reduced lobelike leg 4 of other members of *Caligus* (formerly known as *Pseudocaligus*) to the absence of leg 4 in *Caligopsis*.

The generic name *Caligopsis* was later found to be preoccupied by a butterfly genus, so was renamed by Özdikmen (2008) (see Remarks for *Markevichus*).

Genus **Caligulina** Heegaard, 1972

Caligulina Heegaard, 1972: 303; Kabata, 1979: 156; Ho, 1980: 164; Dojiri, 1983: 361; Kazachenko, 2001: 24.

Remarks. — Heegaard (1972) established the genus *Caligulina* for *C. ocularis* Heegaard, 1972, which was collected in a plankton net. The most distinctive feature of this genus was the presence of large lenses in the eyes of the type-species.

Due to a suspicion of Ho (1980) that *C. ocularis* was synonymous with *Anchicaligus nautili*, the first author (M. Dojiri) borrowed the type-specimens of the former species from the Zoologisk Museum in Copenhagen, Denmark. As suspected by Ho, the holotype male of *C. ocularis* is conspecific with *A. nautili* (see discussion on *Anchicaligus nautili*) and can no longer be considered valid.

Genus **Caligulus** Heegaard, 1962

Caligulus Heegaard, 1962: 171; Ho, 1966: 760; Cressey & Cressey, 1980: 38; Dojiri, 1983: 362; Kazachenko, 2001: 28.

Remarks. — *Caligulus* was first described by Heegaard (1962) who placed this genus within the Euryphoridae. The diagnostic feature that Heegaard used to establish this genus and to include it in the Euryphoridae was the "...presence of dorsal plates on the carapace covering the fourth segment and overlapping the frontal margin of the genital segment...". Heegaard failed to notice that the "dorsa plate" is no more than an elongation of the posterior region of the thoracic zone. He also misconstrued the definition of the Euryphoridae, because at that time euryphorids were considered to possess dorsal plates on the fourth pedigerous somite not the carapace.

To this day, *Caligulus* still only contains one species, *C. longispinosus* Heegaard, 1962, collected from the skin of a scombrid, "*Euthynnus allitteratus*" [= *Euthynnus affinis* (Cantor)], at Howick Islands, North Queensland. Apparently, it has not been collected since its discovery. In their comprehensive study of copepods parasitic on scombrids, Cressey & Cressey (1980) did not collect this species, but examined the allotype male. They suggested that this species was not a euryphorid. It appears that *Caligulus* is probably synonymous with *Caligus*. The elongation of the free margin of the thoracic zone may simply be interpreted as an unusual feature of this species of *Caligus*. Cressey & Cressey (1980) figured leg 2 of the allotype male as possessing two plumose inner setae on the second exopodal segment. This is a somewhat unusual armature for the exopod of leg 2, since only one plumose inner seta is usually found on this segment.

Caligulus was treated as a genus inquirenda by Kazachenko (2001), but considered as a synonym of *Caligus* by Boxshall & Halsey (2004). A definitive decision on the validity of *Caligulus* must be postponed until further collections of *C. longispinosus* are made.

Genus **Calina** Beneden, 1892

Calina Beneden, 1892: 249; Bassett-Smith, 1899: 457; Yamaguti, 1963: 79; Dojiri, 1983: 363; Kazachenko, 2001: 30.

Remarks. — *Calina* was established by Beneden (1892) and presently contains only one species, *Calina brachyura* Beneden, 1892, collected from the skin of "*Ceraptopterus*" (probably meant *Ceratoptera*, a synonym of *Manta*; refer to M. S. Wilson, 1952) in the Azores. This species has not been collected since its discovery. Bassett-Smith (1899) merely cited the original account. Yamaguti (1963) provided illustrations of Beneden's drawings. According to these illustrations, it is apparent that the endopod of leg 1 is 2-segmented and equipped with numerous (about 4?) plumose setae. Leg 3 is difficult to interpret from Beneden's drawings, but appears to be situated on a ventral apron characteristic of caligids. The second and fourth legs are typical for caligids.

The general habitus is very similar to the Caligidae and leg 1 reminiscent of *Pupulina*, *Alebion*, and *Euryphorus*. *Calina brachyura* is placed in the Caligidae. It is, however, difficult to make a positive determination until the type-specimens are examined. Unfortunately, we were not able to locate Beneden's specimens.

Boxshall & Halsey (2004) consider *Calina* a synonym of *Gloiopotes* (Euryphoridae), but Kazachenko (2001) treated it as a valid genus in the Caligidae.

Genus **Calistes** Dana, 1852

Calistes Dana, 1852: 57; Dana, 1853: 1358; Yamaguti, 1963: 79; Dojiri, 1983: 364; Kazachenko, 2001: 30.

Remarks. — *Calistes* was first established by Dana (1852) for one species, *C. trigonis* Dana, 1849, which he collected "in corpus speciei *Trigonis*" from

Rio de Janeiro, Brazil in December 1838. Dana (1853) later repeated his original description, but also provided some illustrations (Dana, 1855) of this species. According to the superficial illustrations, it is difficult to place *Calistes* in any particular family. The postantennal process and sternal furca appear to be present. The endopod of the first leg is reduced and leg 4 is biramous with what appear to be 3-segmented rami. Unfortunately, it is not known with certainty how many pedigerous somites have fused to form the cephalothorax. It seems, according to the ventral view, that the third pedigerous somite of *C. trigonis* is not free and does help to form the posterior margin of the cephalothorax. Therefore, this species appears to belong to the Caligidae. Unfortunately, attempts to locate the type-specimens have failed.

Boxshall & Halsey (2004) considered *Calistes* a genus inquirenda in the Caligidae, but Kazachenko (2001) treated it as a valid genus in the same family.

Genus **Cresseyella** Bezděk & Cressey, 2004

Apogonia Cressey & Cressey, 1990: 45; Boxshall & Halsey, 2004: 725.
Cresseyella Bezděk & Cressey, 2004: 180.

Material examined. — One female holotype (USNM 254430) labeled "*Apogonica stocki*" "sitting on tongue of *Apogon* sp. nov. (AMNH 34481)" and one female paratype (USNM 254429) also labeled "*Apogonica stocki*" attached to tongue of *Apogon mosavi* Dale collected in 1989 from Grand Bahama Island, Bahamas, Caribbean Sea, North Atlantic.

Remarks. — Cressey & Cressey (1990) described a new genus of parasitic copepod from the tongue of the dwarf cardinalfish, *Apogon mosavi* Dale, collected from Grand Bahamas Island in the Caribbean Sea. It was originally named *Apogonia stocki* Cressey & Cressey, 1990 (holotype and paratype labels state "*Apogonica*"), but unfortunately, the generic name of this new species was already preoccupied by a genus belonging to a coleopteran family since 1819 (Bezděk & Cressey, 2004). Therefore, a new generic name *Cresseyella* was proposed by Bezděk & Cressey (2004) to replace the copepod genus *Apogonia*.

This genus was characterized by the absence of a sternal furca and presence of the large posterior lobes of the genital complex extending well beyond the setae on the caudal rami. Unfortunately, neither the absence of the sternal furca nor the presence of the elongate posterior lobes of the genital complex are sufficient for the recognition of this genus. Consequently, *Cresseyella* must be relegated to synonymy and considered to be a junior synonym of *Caligus*.

Genus **Dentigryps** Wilson, 1913

Dentigryps Wilson, 1913: 221; Lewis, 1964a: 203; Lewis, 1964b: 348; Hewitt, 1971b: 333; Dojiri, 1983: 365.

Remarks. — *Dentigryps* was first established by Wilson (1913) to accommodate *D. curtus* Wilson, 1913, which he found infesting the buccal and branchial cavities of the yellowfin grouper *Mycteroperca venenosa apua* (Bloch) in Bermuda. Since then, five species of this genus have been described: *D. bermudensis* (Heegaard, 1943) (later synonymized with the type-species) originally described as *Homoiotes bermudensis*; *D. bifurcatus* Lewis, 1964; *D. litus* Lewis, 1964; *D. ulua* Lewis, 1964; and *D. longicauda* Cressey, 1966. In addition, Ho & Dojiri (1977) transferred two species of *Lepeophtheirus*, *L. lichiae* Barnard, 1948 and *L. spinifer* Kirtisinghe, 1937, to this genus.

Since the establishment of this genus, much discussion has arisen concerning the validity of *Dentigryps*. Pillai (1966) was uncertain about the validity of this genus and suspected that *Dentigryps* may be synonymous with *Lepeophtheirus*. Hewitt (1971b) synonymized the former with the latter. Kabata (1979) agreed with this synonymy. Ho & Dojiri (1977), however, supported the retention of this genus as redefined by Lewis (1964b). The main diagnostic feature distinguishing the members of *Dentigryps* from those of *Lepeophtheirus* is the "strongly projecting fifth legs" of *Dentigryps* (Lewis, 1964b). According to Ho & Dojiri (1977), the females of the species of *Dentigryps* possess fifth legs that are ". . .as long as or longer than the genital complex, and projected posteriorly to or beyond the tip of the caudal ramus".

Although the majority of the females of the species of *Lepeophtheirus* are equipped with only a vestigial fifth leg represented by a small setiferous projection, there are a few species (e.g., *L. goniistii* Yamaguti, 1936; *L. plotosi* Barnard, 1948; and *L. hastatus* Shiino, 1950) that possess longer fifth legs than their congeners. It has been argued (Pillai, 1966; Hewitt, 1971b) that a gradation in the length of the fifth legs exists in the females of *Lepeophtheirus* reaching its maximum length in females of *Dentigryps*. Ho & Dojiri (1977) suggested that the morphological distinction between these two groups was sufficient for the recognition of *Dentigryps* as a genus distinct from *Lepeophtheirus*. Although the differences in length of the fifth legs appears to be present, the taxonomic value of the length of any appendage is highly questionable.

Moreover, there are two genera, *Alebion* and *Tuxophorus*, which have females with long spikelike leg 5 (e.g., *A. carchariae* Krøyer, 1863; *T. cervicornis* Heegaard, 1962; and *T. wilsoni* Kirtisinghe, 1937) and others with a short setiferous papilla representing leg 5 (e.g., *A. glaber* Wilson, 1905 and *T. caligodes* Wilson, 1908) within the same genus. If members of a genus can display a spikelike leg 5 reminiscent of *Dentigryps* and other species of the same genus exhibit a highly reduced setiferous lobelike leg 5, then the retention of *Dentigryps* would not be taxonomically consistent. Since other investigators (Pillai, 1966; Hewitt, 1971b; Kabata, 1979) either have questioned the validity of this genus or abandoned it altogether and since the generic boundary between *Dentigryps* and *Lepeophtheirus* appears to be arbitrary, *Dentigryps* should be considered a junior synonym of *Lepeophtheirus*.

Genus **Diphyllogaster** Brian, 1899

Diphyllogaster Brian, 1899c: 7; Brian, 1906: 148; Rangnekar, 1955: 44; Yamaguti, 1963: 81; Pillai & Padmanabhan, 1963: 94; Laubier, 1968: 159; Dojiri, 1983: 367; Prabha, 1983: 55; Kazachenko, 2001: 31.

Remarks. — *Diphyllogaster thompsoni* Brian, 1899, the type-species of its genus, was first described by Brian (1899c) from the ray *Dicerobatis giornae* Gunt (= *Mobula*). During a complete detailed redescription of *Pupulina brevicauda* M. S. Wilson, 1952, Pillai & Padmanabhan (1963) realized the close morphological similarity between *D. thompsoni* and *P. brevicauda*. These two species are very similar in general habitus and morphology of leg 4. Records suggest that both species prefer host species belonging to the ray genus *Mobula*. The major difference between these two species is the endopod of leg 1. In *D. thompsoni*, the endopod is not figured by Brian (1899c); it is possible that Brian simply overlooked this ramus. In *P. brevicauda* the endopod is 2-segmented with 3 long plumose setae. It is likely, based on available evidence, that these two species are synonymous as suggested by Pillai & Padmanabhan (1963). Dr. A. Carli (personal communication), Instituto di Zoologia, Universita degli Studi, Genova, Italy, informed us that Brian never deposited specimens of *D. thompsoni*.

Rangnekar (1955) described the second species of this genus, *Diphyllogaster aliuncus*, from the body surface of the silver pomfret *Pampus argenteus* (Euphrasen) from Bombay. Although similar in general habitus (namely the posterior processes of the genital complex), *D. aliuncus* differs from the type-species by the morphology of legs 1 and 4. Pillai & Padmanabhan (1963) suggested that *D. aliuncus* is a synonym of *Paralebion*. Upon comparing Rangnekar's illustrations of this species to our description of *Paralebion elongatus* Wilson, 1911, it seems likely that Rangnekar's specimens do indeed belong to *Paralebion*.

Genus **Dysgamus** Steenstrup & Lütken, 1861

Dysgamus Steenstrup & Lütken, 1861: 368; Bassett-Smith, 1899: 460; Wilson, 1907a: 712; T. Scott, 1912: 579; Wilson, 1923: 11; Leigh-Sharpe, 1934: 28; Heegaard, 1943: 24; Wilson, 1944: 531; Wilson, 1950: 201; Pearse, 1952: 23; Rose & Vaissière, 1953: 83; Shiino, 1958b: 161; Yamaguti, 1963: 101; Kabata & Gusev, 1966: 168; Heegaard, 1972: 314; Kabata, 1979: 167, 168, 201-203; Dojiri, 1983: 368.

Remarks. — At one time there were as many as 10 nominal species of *Dysgamus* since the establishment of the genus by Steenstrup & Lütken (1861). Heegaard (1972), in his study of the genus *Euryphorus*, found that all the species of *Dysgamus* were simply juveniles of *Euryphorus* or other genera (see Kabata, 1979). Therefore, *Dysgamus* cannot be considered valid.

Genus **Heniochophilus** Yamaguti & Yamasu, 1959

Heniochophilus Yamaguti & Yamasu, 1959: 124; Kabata, 1965a: 30; Pillai & Mohan, 1965: 273; Pillai, 1967: 362; Dojiri, 1983: 369; Prabha, 1983: 57; Pillai, 1985: 458; Kazachenko, 2001: 31.

Remarks. — *Heniochophilus* was first established by Yamaguti & Yamasu (1959) to contain one species, *H. japonicus* Yamaguti & Yamasu, 1959, that they found on *Heniochus acuminatus* (Linnaeus) in Japan. Kabata (1965a) redescribed some specimens that Mr. P. C. Young obtained from the gills of *Platax teira* (Forsskål) in Moreton Bay, Australia. In the same paper, Kabata pointed out the similarity between *Mappates* Rangnekar, 1958 and *Heniochophilus*. It was thought that these two genera differed mainly in the presence or absence of the sternal furca (see Pillai, 1967). However, the presence or absence of this structure may not be as generically important as once thought (Kabata, 1965a).

After studying a large collection of *Anuretes branchialis* Rangnekar, 1953, Pillai & Mohan (1965) concluded that *H. japonicus* is identical to *A. branchialis*. However, Pillai & Mohan upheld the validity of the genus *Heniochophilus* and so the binomen became *Heniochophilus branchialis* (Rangnekar, 1953).

Pillai (1967) believed that the main distinguishing feature (i.e., the elongated thoracic zone of the cephalothorax in the female) between *Heniochophilus* and *Anuretes* may not be a valid generic discriminant and suggested that the two genera are synonymous. There appears to be no substantial reason for the retention of *Heniochophilus* (see discussion on *Anuretes branchialis*); therefore, it is considered a synonym of *Anuretes*.

Genus **Homoiotes** Wilson, 1905

Homoiotes Wilson, 1905a: 661; Yamaguti, 1963: 83; Kabata, 1973b: 1892; Kabata, 1979: 168; Dojiri, 1983: 378.

Remarks. — *Homoiotes* was first established by Wilson (1905a) with *H. palliata* Wilson, 1905 as the type-species. It was collected from the skipjack tuna "*Thynnus pelamys*" [= *Euthynnus pelamis* (Linnaeus) or *Katsuwonus pelamis*) in Bermuda. Heegaard (1943) subsequently described another species, *H. bermudensis*. However, Heegaard (1945b) synonymized *H. bermudensis* with *Dentigryps curtus* Wilson, 1913. It appeared, therefore, that *Homoiotes* was a monotypic genus. However, Kabata (1973b) examined the type-specimens of *H. palliata* and found that they consisted of three different species of *Lepeophtheirus*: *L. pravipes* Wilson, 1912; *L. breviventris* Fraser, 1920; and *L. bifidus* Fraser, 1920. Therefore, *Homoiotes* cannot be considered valid.

Genus **Indocaligus** Pillai, 1961

Indocaligus Pillai, 1961: 125; Pillai, 1963c: 92; Dojiri, 1983: 371; Prabha, 1983: 54.

Remarks. — Pillai (1961) described the genus *Indocaligus* to accommodate *Indocaligus echinus* Pillai, 1961, which he collected from the branchial cavity of *Ariodes dussumieri* [= *Plicofollis dussumieri* (Valenciennes)] at Trivandrum, India. The diagnostic characters used by Pillai to distinguish this genus from *Caligus* and *Lepeophtheirus* were the large, spiny endopod of leg 1 and the spiny patches on the endopods of legs 2 and 3. In addition, this species does not possess lunules and bears heavily serrated spines on leg 4. In a subsequent paper, Pillai (1963c) synonymized *I. echinus* with *Lepeophtheirus longipalpus* Bassett-Smith. The so-called "peculiarities" or diagnostic features of this species are due to convergent evolution since both *Caligus arii* Bassett-Smith, 1898, and *Hermilius longicornis* Bassett-Smith, 1898, which are also found on the same host, *Pseudarius jatius* [= *Cephalocassis jatia* (Hamilton)], share similar features (Pillai, 1963c). Therefore, the characteristics just mentioned are adaptations to a specific host and are not good generic discriminants in this case. The genus *Indocaligus* must then be relegated to synonymy with *Lepeophtheirus* as Pillai (1963c) suggested.

Genus **Markevichus** Özdikmen, 2008

Markevichus Özdikmen, 2008: 270.

Caligopsis Markewitsch, 1940: 14; Markewitsch, 1956: 128; Yamaguti, 1963: 63; Kabata, 1979: 41, 158; Kazachenko, 2001: 24.

Remarks. — The generic name *Caligopsis* proposed by Markewitsch (1940) to accommodate the type-species *C. ponticus* Markewitsch, 1940 was found by Özdikmen (2008) to be preoccupied by a genus in the butterfly family Nymphalidae. Thus, the preoccupied name was proposed to be replaced with a new name *Markevichus* (Markewitsch's name is often transliterated as Markevich). For a more detailed discussion on this genus, refer to the Remarks section of *Caligopsis*.

Genus **Nogagella** Rose, 1933

Nogagella Rose, 1933: 119; Vaissière, 1961: 80; Dojiri, 1983: 372.

Remarks. — Rose (1933) described a new genus and new species of copepod parasitic on a siphonophore, *Physophora hydrostatica* Forsskål. This new genus was named *Nogagella* and was placed in the family Caligidae.

It is quite apparent that *Nogagella* is definitely not a caligid and should be removed from the Caligidae. The second, third, and fourth pedigerous somites remain unfused with the cephalothorax, a condition completely contrary to all other members of the Caligidae, in which only the fourth pedigeous somite remains free. In addition, the maxillule is not the usual single tine or bifurcate structure characteristic of other caligid genera. Another important reason for removing *Nogagella* from the Caligidae is the morphology of legs 1-4, which are segmented and armed very differently from the remaining caligids. Finally, the third pedigerous somite does not form the ventral apron of the cephalothorax.

Nogagella most likely belongs in the Pandaridae, perhaps a larval stage of this family or a male pandarid (e.g., *Nogagus*). Vaissière (1961) considered this genus with the pandarids in his study of the eyes of copepods.

Genus **Platyporinus** Rao, 1950

Platyporinus Rao, 1950: 306; Cressey, 1972: 5; Dojiri, 1983: 373.

Remarks. — Rao (1950) established a new genus *Platyporinus* containing one species, *P. alata* Rao, 1950 based on three females collected from the body of a hammerhead shark caught in Lawson's Bay, Waltair, South India. The dorsal and ventral views of *P. alata*, the secondary tine on the claw of the maxilliped, and its host (family Sphyrnidae) indicate that *P. alata* is a synonym of *Alebion crassus* Wilson, 1932. Cressey (1972) in his revision of the genus *Alebion* included *P. alata* as a junior synonym of *A. crassus*.

Genus **Pseudocaligus** A. Scott, 1901

Pseudocaligus A. Scott, 1901b: 350; Scott & Scott, 1913: 61; Markewitsch, 1956: 129; Yamaguti, 1963: 65; Kabata, 1965a: 30; Kabata, 1979: 186; Dojiri, 1983: 294; Prabha, 1983; p. 50; Pillai, 1985: 367; Kim, 1998: 723; Kazachenko, 2001: 26; Kabata, 2003: 64; Boxshall & Halsey, 2004: 726; Ho & Lin, 2004: 322.

Remarks. — The genus *Pseudocaligus* was established by A. Scott (1901b) to include *Caligus brevipedis* Bassett-Smith, 1896. The main distinction between *Pseudocaligus* and *Caligus* is the great reduction of leg 4 to a small lobe bearing a few spines or setae. As originally suggested by Kabata (1965a, 1979) the structure of the fourth leg may not be a character of generic importance. He suggested that *Pseudocaligus* be placed into synonymy with *Caligus*, and *Pseudolepeophtheirus* with *Lepeophtheirus*. The latter two genera differ in the same manner as *Pseudocaligus* differs from *Caligus*, namely, in the reduction of the fourth leg.

Since the reduction of the fourth leg is the only feature that distinguishes *Pseudocaligus* from *Caligus*, a comparison of these two genera was conducted. Many species of *Caligus* possess a distinct sympod and a 3-segmented exopod with a spinal formula of I-0; I-0; III (e.g., *C. constrictus* Heller, 1865; *C. brevisoris* Shen, 1957; *C. multispinosus* Shen, 1957; *C. laticaudus* Shiino, 1960; among others). In numerous species, the second and third exopodal segments have undergone fusion with the result that the exopodal spine of the original second segment has become part of the terminal armature; consequently, the formula in these species is I-0; IV with only a 2-segmented exopod (e.g., *C. priacanthi* Pillai, 1961; *C. kala* Lewis, 1964; *C. ligatus* Lewis, 1964; among many others). The second exopodal segment along with its associated spine are completely absent in other species (e.g., *C. curtus* Müller, 1785; *C. flexispina* Lewis, 1964; *C. kalumai* Lewis, 1964; among many others) so that the resultant formula is I-0; III. The fourth leg undergoes more progressive reduction from a sympod and 1-segmented exopod (reduced to a lobe) in *Pseudocaligus fistulariae* Pillai, 1961 (these segments fused in *P. tenuicauda* Shiino, 1964), to a distinct lobe bearing a seta and a spine in *P. fugu* Yamaguti, 1936, and finally to a small setiferous papilla as in *P. indicus* Hameed, 1977 and *P. subparvus* Hameed, 1977. The final step in this evolutionary trend appears to have occurred in *Caligopsis ponticus* Markewitsch, 1940 (= *Markevichus ponticus*), which lacks leg 4 altogether.

It is apparent from the above examples that an evolutionary trend toward a reduction in the segmentation and armature of the fourth leg occurred in the genus *Caligus*. Since there is a progressive reduction in the fourth leg with each step in the evolutionary trend represented by species of the two genera, any generic boundary made between *Pseudocaligus* and *Caligus* would be arbitrary.

The evolutionary trend toward the reduction of the fourth leg is apparently closely tied to the formation of the ventral apron (Parker et al., 1968; Kabata, 1979). The expansion of the intercoxal plate and the sympod of leg 3 resulted in the formation of the ventral apron, which produced the posterior margin of the cephalothorax. With the evolution of the ventral apron, the cephalothorax of the Caligidae functions as an effective suction cup as an aid in attachment to the host. The first two pairs of legs are contained within the cephalothoracic shield, but the fourth legs remain outside of this structure. Moreover, the first and second legs retain their locomotory function, but the fourth pair of legs has become relatively inactive (Kabata & Hewitt, 1971). Apparently the evolution of the fourth leg is closely correlated with the evolution of the third leg (Parker et al., 1968; Kabata, 1979). Furthermore, the trend toward a reduction in the fourth leg in caligid genera appears to be linked to the inactivity of this appendage. There are other caligid genera that also exhibit a trend toward the reduction of the fourth leg, e.g., *Abasia*, *Alebion*, *Hermilius*, and *Pseudanuretes* include members that display this reduction.

There are currently eleven species of *Pseudocaligus* that have been described: *P. brevipedis* (Bassett-Smith, 1896), the type-species, primarily parasitic on gadid fishes; *P. apodus* Brian, 1924 from a scyliorhinid shark of the genus *Galeus*; *P. indicus* from the flying gurnard *Dactyloptena orientalis* (Cuvier); three morphologically similar species, *P. fistulariae*, *P. tenuicauda*, and *P. similis* Lewis, 1968 from cornetfishes of the genus *Fistularia*; four species, *P. parvus* (Bassett-Smith, 1898), *P. fugu*, *P. laminatus* Rangnekar, 1955, and *P. subparvus* from tetraodontid fishes; and *P. uniartus* Ho, Kim, Cruz-Lacierda & Nagasawa, 2004 from a siganid fish. Apart from the species that infest *Fistularia*, the members of *Pseudocaligus* do not appear to form a natural group.

We are in agreement with Kabata (1965a, 1979) that the genus *Pseudocaligus* should be relegated to synonymy with *Caligus*. The morphology of the appendages of the members of *Pseudocaligus* exhibit variation within the known ranges of the current diagnosis for *Caligus*. They do not display any significant morphological features that distinguish them as a group separate from *Caligus*.

Genus **Pseudolepeophtheirus** Markewitsch, 1940

Pseudolepeophtheirus Markewitsch, 1940: 14; Markewitsch, 1956: 142; Yamaguti, 1963: 84; Kabata, 1965a: 30; Dojiri, 1983: 303; Prabha, 1983: 54; Kazachenko, 2001: 33; Boxshall & Halsey, 2004: 726.

Remarks. — *Pseudolepeophtheirus* was first established by Markewitsch (1940) with *P. longicauda* Markewitsch, 1940 as its type-species, which was collected from the starry flounder *Platichthys stellatus* (Pallas). In his key to the species of this genus, Markewitsch (1956) distinguished *Pseudolepeophtheirus longicauda* from *P. schmidti* Gusev, 1951 by a small, but 2-segmented fourth leg in the former species. Unfortunately, he does not mention the armature of this appendage. It is probable, from Markewitsch's description, that *P. longicauda* is a junior synonym of *Lepeophtheirus parvicruris* Fraser, 1920. Unfortunately, Markewitsch (1940) never figured this species and gave only a very superficial account of it. According to Markewitsch, the female of *P. longicauda* possesses a small fourth pedigerous somite resembling a neck, genital complex as long as the cephalothorax, posterior margin of genital complex with an indentation, and an elongate abdomen longer than the genital complex. This description matches that of *L. parvicruris*.

The measurements of the female of *P. longicauda* are also very similar to those of *L. parvicruris*, except the length of the abdomen reported for *P. longicauda* is 4 mm (average length of abdomen in *L. parvicruris* is 2.33 mm). However, there is a discrepancy in the measurements reported for *P. longicauda*. The total body length is reported as 8.5 mm; but, the cephalothorax (3 mm long), genital complex (3 mm long), and the abdomen (4 mm long) add up to a total greater than the reported body length of the organism. In addition, the lengths of the fourth pedigerous somite and caudal ramus have not yet been included. If the abdomen is assumed to be 2.5 mm long (matching that of *L. parvicruris*), the total of the lengths of the tagmata of *P. longicauda* would then equal its total body length.

Both *P. longicauda* and *L. parvicruris* were collected from the branchial cavity of *Platichthys stellatus*, the former from the Motogi River in Vladivostok (near the Sea of Japan) and the latter from Vancouver Island, British Columbia, and Prince William Sound, Alaska.

Titar (personal communication) reported that the type-specimens of *P. longicauda* no longer exist. However, if the above synonymy is valid, then *Pseudolepeophtheirus* is based on a species that clearly belongs in the genus *Lepeophtheirus*, since *L. parvicruris* possesses a normal leg 4. Although this appendage is diminutive in size in *L. parvicruris*, it, nevertheless, is comprised of a sympod and a 2-segmented exopod with four spines.

As mentioned before, Kabata (1965a) suggested that *Pseudolepeophtheirus* may need to be relegated to synonymy with *Lepeophtheirus*. This action would be well founded. There are a number of genera whose members show a progressive reduction in the fourth leg (e.g., *Abasia*, *Alebion*, and *Pseudanuretes*). Other genera exhibit a reduction in the number of exopodal segments and spines (e.g., *Caligus* and *Lepeophtheirus*).

Since members within a genus can show a prominent reduction in the fourth leg, the retention of *Pseudolepeophtheirus* is unwarranted; therefore, we support Kabata's suggestion and relegate *Pseudolepeophtheirus* to synonymy with *Lepeophtheirus*. For a more detailed discussion on this subject see remarks on *Pseudocaligus*.

Genus **Tripartia** Kazachenko, 2001

Tripartia Kazachenko, 2001: 27.

Remarks. — Bassett-Smith (1898c) reported a new species of caligid, *Caligus hirsutus*, from "*Polynemus tetradactylus*" [= *Eleutheronema tetradactylum* (Shaw)] taken from Bombay, India. When Wilson (1912) reported the same species of parasite collected from the same species of host from Batavia (Jakarta), Indonesia, he took the opportunity to transfer *C. hirsutus* to the genus *Parapetalus* Steenstrup & Lütken, 1861. Wilson's (1912) treatment of *C. hirsutus* was subsequently supported by Kirtisinghe (1950), Leong (1985), and Lin & Ho (2000) when they discovered the same species of parasites from Sri Lanka, Malaysia, and Taiwan, respectively.

In the beginning of this century, Kazachenko (2001) created a new genus, *Tripartia*, to accommodate *P. hirsutus*. The proposal was made based on the following characters of *P. hirsutus*: (1) the vestigial leg 1 endopod and distal two segments of leg 2 endopod covered with short "hairs", (2) a patch of long "hairs" at the base of each outer spine on leg 4, (3) a reduced leg 3 ventral apron, (4) one of the 4 setae on the caudal ramus much longer than the other 3, and (5) the claw of maxilliped longer than its corpus maxillipedis. Although the generic placement of this species in *Parapetalus* is questionable due to the absence of large lateral aliform expansions on the genital complex that is characteristic of this genus, we believe this species should be retained in *Parapetalus* based on the characters outline in table XIV, and consider the five characters reported by Kazachenko (2001) to establish a new genus as specific rather than generic discriminants.

PHYLOGENY OF THE CALIGIDAE

Copepods that are parasitic on fishes express a great deal of diversification in specializing toward their parasitic mode of life; this specialization is reflected in their morphology, life history, host specificity, site of attachment, and mode of attachment. Although Heegaard (1945a) suggested that parasitic copepods retain more "primitive" (plesiomorphic) features than their free-living counterparts, it is generally agreed by copepodologists that they do not exhibit as many ancestral features as Heegaard once thought (see Kabata, 1979); consequently, they cannot be considered "primitive", but represent the culmination of the radiation of the Copepoda into the parasitic mode of life (Huys and Boxshall, 1991: 414).

The trends toward a parasitic lifestyle often destroy phylogenetic clues that help identify the affinities among groups of copepods. Moreover, parasitic copepods appear to have evolved along many independent lines (Kabata, 1979); therefore, the evolutionary trends can be expected to be different in each phyletic line. However, within a lineage, certain trends may be followed with helpful clues gleaned from the various taxa beginning with those exhibiting many plesiomorphic characters. In some instances, the tracing of phyletic lineages may be difficult, particularly in highly modified families in which many of the thoracic appendages are absent with the oral appendages remaining as the only external morphological clues to their affinities. The difficulties of establishing a phylogeny are heightened by the ever-present homoplasies that result in a tendency for the establishment of false relationships.

The general morphology, often correlated with the nature of the host-parasite relationships, ranges from the relatively mobile and comparatively less modified species of the Caligidae and some of its relatives (e.g., Dissonidae and Trebiidae) to the permanently attached, nonmobile, and highly modified members of some families (e.g., Chondracanthidae, Lernaeidae, Pennellidae, and Sphyriidae). The evolutionary trend toward a permanent association with the host is seen in the simplification of the body and its appendages involving fusion of body somites (tagmosis) and reduction and/or loss of the locomotory appendages (legs).

As suggested by Wilson (1910), the least conservative character is the body form of the parasitic copepod. The most conservative (least likely to change) is the structure of the mouth parts. The structure of the legs is intermediate between the two extremes and thus correspondingly carries a taxonomic weight that is intermediate. Phylogenetic trends may be identified within the Caligidae based on various characters and their respective states, such as the segmentation and morphology of the appendages and other structures. With the above in mind, we conducted a phylogenetic analysis of the Caligidae.

Materials and methods

The 32 characters used in the phylogenetic analysis of the Caligidae are listed in table XXII, and the character matrix is presented in table XXIII. The use of unordered, multistate characters in the matrix was employed in order not to predetermine the results.

The cladistic analysis was performed with the computer program PAUP* – Phylogenetic Analysis Using Parsimony (*and other methods) version 4.0b10 (Swofford, 2000). Initially, the analysis was conducted using *Dissonus* and *Trebius*, two genera belonging to two different families that are closely related to the Caligidae, as the outgroups. However, since these two genera are members of the caligiform complex along with the Caligidae, the analysis herein utilized the siphonostome genus *Dirivultus* as the outgroup, thereby polarizing the data and rooting the phylogram with this taxon. The character states of *Dissonus* and *Trebius* are included in the character matrix for comparison with the outgroup to show that both genera possess the same plesiomorphic character states for 22 of the 32 characters.

The unordered analysis resulted in 807 equally parsimonious cladograms with one of them depicted as a phylogram (fig. 151), allowing the differential branch lengths to convey the relative amount of anagenetic change between genera. A 50% majority-rule consensus tree (fig. 152) is presented in order to depict the agreement among all 807 trees. Additionally, 32 character trees (figs. 153-160) were generated from Mesquite version 2.74 by employing the "trace character history" module (Madison & Madison, 2010) that mapped every character's distribution onto the phylogram, resulting in an informative visualization of the characters responsible for the groupings of the caligid genera.

Results and discussion

The cladistic analysis of the caligid genera herein using the heuristic search option in PAUP*, with the random addition sequence option set to 1111

TABLE XXII
Characters and character states of the Caligidae

Character number	Characters	Outgroup-*Dirivultus*	*Dissonus* & *Trebius* states	Apomorhphic states
1	2nd pediger fused to cephalothorax	free (0)	free (0)	fused (1)
2	3rd pediger fused to cephalothorax	free (0)	free (0)	fused (1)
3	cephalothorax folded	not folded, flat (0)	not folded, flat (0)	folded (1)
4	lunules	absent (0)	absent (0)	present (1)
5	large lens	absent (0)	absent (0)	present (1)
6	aliform plate on 4th pediger	absent (0)	absent (0)	present (1)
7	4th pediger covered by cephalothorax	not covered (0)	not covered (0)	covered (1)
8	4th pediger forming long neck	not extended to long neck (0)	not extended to long neck (0)	forming long neck (1)
9	aliform expansion on genital complex	absent (0)	absent (0)	present (1)
10	pair of posterior processes on genital complex	absent (0)	absent (0)	1 pair (1); 2 pair (2)
11	aliform expansion on abdomen	absent (0)	absent (0)	present (1)
12	digitiform processes on abdomen	absent (0)	absent (0)	present (1)
13	abdomen diminutive (vestigial)	not diminutive (0)	not diminutive (0)	diminutive (vestigial) (1)
14	caudal setae	6 setae (0)	6 setae (0)	5 setae (1); 4 setae (2); absent (3); 7 setae (4)
15	antennule, first segment armature	numerous segments and setae	27 setae (0)	25-26 (1); 23 (2); 20 (3); 15-19 (4); 15 (5); 28 (6); 29 (7); 31 (8)
16	antennule, second segment armature	12 and 2 + 1 (last 2 segments) (0)	14 (12 + 2a or 13 + 1a) (0)	13 (12 + 1a; 11 + 2a; or 13) (1)
17	antenna segmentation	3 (0)	3 (0)	4 (1); 5 (2)
18	spiniform process of postantennal process	absent (0)	absent in *Dissonus* (0); present in *Trebius* (1)	present (1)
19	spiniform process of maxillule	absent (0)	present (1)	present (1)

TABLE XXII
(Continued)

Character number	Characters	Outgroup-*Dirivultus*	*Dissonus* & *Trebius* states	Apomorhphic states
20	flabellum of maxilla	absent (0)	present in *Dissonus* (1); absent in *Trebius* (0)	present (1)
21	sternal furca	absent (0)	ab/pr in *Dissonus* (0/1); present in *Trebius* (1)	present (1)
22	p1 exopod terminal armature	7 elements (0)	7 elements (0)	6 (1); 5 (2)
23	accessory process on p1 exopodal spines 2 & 3	present on all 3 spines (0)	absent (3)	present on spines 2 & 3 (1); present on spine 3 only (2); absent (3)
24	p1 endopod segmentation	3-segmented (0)	2-segmented (1)	2-segmented (1); lobe (2)
25	p1 endopod armature	6 setae (0)	3 setae (1)	0 setae (2)
26	p2 exopod terminal armature	8 elements (0)	8 elements (0)	7 elements (1); 6 elements (2)
27	p2 endopod terminal armature	6 setae (0)	6 setae (0)	7 setae (2nd & 3rd segments fused) (1)
28	ventral apron	absent (0)	absent (0)	present (1)
29	p3 exopod terminal armature	9 elements (0)	8 elements (1)	8 elements (1); 7 elements (2); 6 elements (3); 4 elements (4); 3 elements (5)
30	p3 endopod segmentation	3-segmented (0)	3-segmented (0)	2-segmented (1); 0-segmented (2)
31	p3 endopod terminal armature	5 setae (0)	4 setae (1)	4 setae (1); 3 setae (2); 2 setae (3); 0 seta (4); 6 setae (5)
32	p4 exopod structure	biramous (0)	biramous (0)	uniramous (1); lobe (2)

Elements = spines and/or setae.

replications with 7 trees held at each step, resulted in 807 equally parsimonious phylograms with a tree length of 119 steps (fig. 151). The phylogram possesses

TABLE XXIII
Character matrix of the Caligidae

Genera	1	2	3	4	5	6	7	8	9	10	11	12	13	14	15	16	17	18	19	20	21	22	23	24	25	26	27	28	29	30	31	32
Dirivultus-outgroup	0	0	0	0	0	0	0	0	0	0	0	0	0	0	0	0	0	0	0	0	0	0	0	0	0	0	0	0	0	0	0	0
Dissonus	0	0	0	0	0	0	0	0	0	0	0	0	0	0	0	0	1	0	1	1	1	0	3	1	1	0	0	0	1	0	1	0
Trebius	1	0	0	0	0	0	0	0	0	0	0	0	0	0	0	0	1	1	1	0	1	0	3	1	1	0	0	0	1	0	1	0
Caligus	1	1	0	1	0	0	0	0	0	0	0	0	0	0	0	0	0	1	1	1	1	0	1	2	2	0	0	1	2	1	5	1
Abasia	1	1	1	1	0	0	0	0	0	0	0	0	0	0	4	0	1	0	1	1	0	2	1	2	2	0	0	1	4	1	3	1
Alanlewisia	1	1	0	0	0	0	0	0	0	0	0	0	0	0	0	1	1	1	1	1	1	0	1	2	1	0	0	1	0	1	5	1
Alebion	1	1	0	0	0	1	0	0	0	0	1	0	0	0	0	0	0	0	1	1	0	0	3	1	1	0	0	1	1	0	1	1
Anchicaligus	1	1	0	1	1	0	0	0	0	0	0	0	0	0	0	0	0	1	1	1	0	0	1	2	2	0	0	1	2	1	5	1
Anuretes	1	1	0	0	0	0	1	0	0	0	0	0	1	1	1	0	0	1	1	1	1	0	1	2	2	0	0	1	2	1	5	1
Arrama	1	1	1	0	0	0	0	0	0	0	0	0	1	2	5	0	1	0	1	1	0	1	3	2	2	1	1	0	5	2	4	2
Avitocaligus	1	1	0	0	0	0	0	0	1	0	1	0	0	0	0	1	1	0	1	1	0	0	1	1	1	0	0	1	2	0	1	0
Belizia	1	1	0	1	0	0	1	0	0	0	0	0	1	2	1	0	0	1	1	1	0	1	1	2	2	0	0	1	2	1	1	1
Caligodes	1	1	0	1	0	0	0	0	0	1	0	0	0	0	2	1	0	0	1	1	1	0	3	2	2	0	1	1	1	1	1	1
Caritus	1	1	0	1	0	0	0	0	0	0	0	0	0	0	0	0	0	0	1	1	0	0	3	2	2	0	0	1	2	1	4	1
Dartevellia	1	1	0	0	0	0	0	0	0	0	1	0	0	3	8	0	0	0	1	1	0	0	1	2	2	0	0	1	4	2	4	1
Echetus	1	1	0	1	0	0	0	1	0	0	0	0	0	0	0	0	0	0	1	1	0	0	3	2	2	0	0	1	1	1	5	1
Euryphorus	1	1	0	0	0	1	0	0	1	0	1	0	0	0	0	0	1	0	1	1	1	0	1	1	1	0	0	1	1	0	1	0
Gloiopotes	1	1	0	0	0	1	0	0	0	1	0	0	0	0	0	1	0	1	1	1	1	0	1	1	1	0	0	1	1	0	1	1
Hermilius	1	1	1	0	0	0	0	0	0	0	0	0	0	0	0	0	2	0	1	1	1	0	1	2	2	0	0	1	2	1	5	1
Kabataella	1	1	1	0	0	0	0	0	0	0	0	0	0	0	3	0	1	0	1	1	0	2	3	2	2	2	0	1	3	2	4	1
Lepeophtheirus	1	1	0	0	0	0	0	0	0	0	0	0	0	0	0	0	0	1	1	1	1	0	1	2	2	0	0	1	2	1	5	1
Mappates	1	1	0	0	0	0	1	0	0	0	0	0	1	1	1	0	0	0	1	1	0	0	1	2	2	0	0	1	2	1	2	1
Metacaligus	1	1	0	1	0	0	0	0	0	0	0	0	0	0	0	1	0	1	1	1	0	0	3	2	2	1	0	1	2	1	5	1
Midias	1	1	0	1	0	0	0	0	0	0	1	1	0	0	0	0	0	1	1	1	1	0	1	2	2	0	0	1	2	1	5	1
Paralebion	1	1	0	0	0	0	0	0	0	1	0	0	0	0	0	0	0	1	1	1	1	0	1	2	2	0	0	1	2	1	5	1
Parapetalus	1	1	0	1	0	0	0	0	1	0	0	1	0	0	6	0	0	1	1	1	1	0	1	2	2	0	0	1	2	1	5	1
Parechetus	1	1	0	1	0	0	0	1	1	0	1	0	0	0	0	0	1	1	1	1	1	0	1	2	2	0	0	1	2	1	5	1
Pseudanuretes	1	1	0	0	0	0	1	0	0	0	0	0	1	1	0	0	0	0	1	1	0	0	1	2	2	1	0	1	2	1	0	1
Pseudechetus	1	1	0	1	0	0	0	1	0	2	0	1	0	0	0	0	1	1	1	1	1	0	1	2	2	0	0	1	2	1	5	1
Pupulina	1	1	0	0	0	0	0	0	0	1	0	0	0	0	7	1	0	1	1	1	0	0	2	1	1	0	0	1	2	0	1	1
Sciaenophilus	1	1	0	1	0	0	0	0	0	0	0	0	0	0	0	0	0	1	1	1	0	0	3	2	2	0	0	1	2	1	5	1
Sinocaligus	1	1	0	1	0	0	0	0	0	0	1	0	0	4	1	0	0	1	1	1	1	0	1	2	2	0	0	1	2	1	5	1
Synestius	1	1	0	1	0	0	0	0	0	2	0	0	0	0	7	0	0	1	1	1	1	0	1	2	2	0	0	1	2	1	5	1
Tuxophorus	1	1	0	1	0	1	0	0	0	0	0	0	0	0	6	0	0	1	1	1	1	0	1	2	2	0	0	1	2	1	5	1

14 + 1a = last two segments combined for *Dirivultus* antennule.

a consistency index of 0.504 and a retention index (a measure of branch support) of 0.638 (maximum value of perfect fit = 1.0). The synapomorphies uniting all the genera of the Caligidae are the fusion of the second and third pedigerous somites with the cephalothorax (fig. 153, character 2) and the

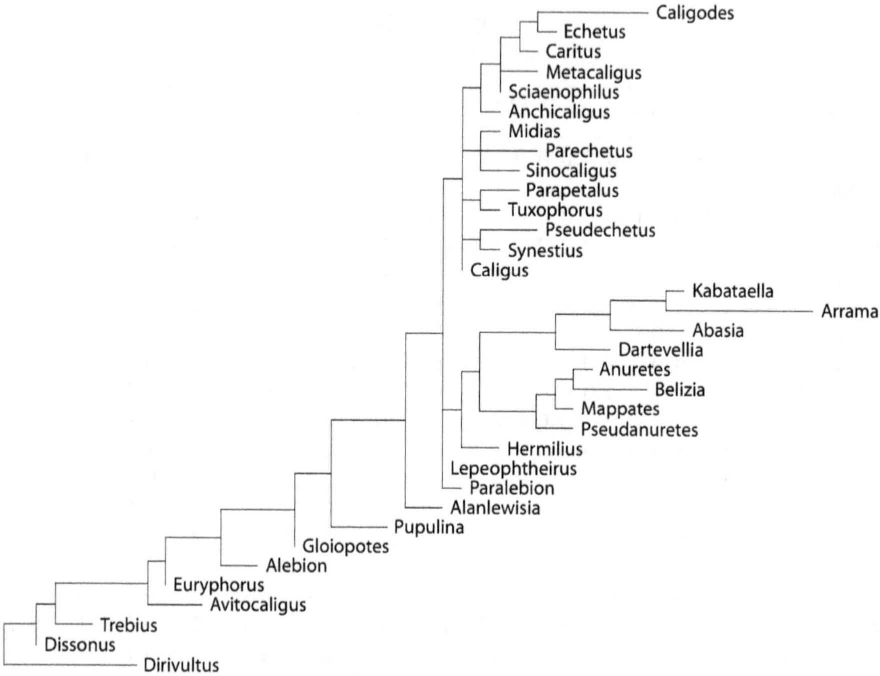

Fig. 151. Phylogram of the Caligidae generated using the matrix in table XXIII and *Dirivultus* as the outgroup.

expansion of the sympod of the third leg to form the ventral apron (fig. 159, character 28).

Although a 2-segmented endopod on leg 1 (fig. 158, character 24) with the armature formula of 0-0, 3, a 3-segmented endopod on leg 3 (fig. 160, character 30), and four (instead of six) plumose setae on the terminal segment of the endopod of leg 3 (fig. 160, character 31) are shared among the five genera *Alebion*, *Avitocaligus*, *Euryphorus*, *Gloiopotes*, and *Pupulina* and are not possessed by any other caligid genus, the phylogram (fig. 151) and consensus tree (fig. 152) resulting from the phylogenetic analysis, using *Dirivultus* as the outgroup, as well as those using *Dissonus* and *Trebius* as the outgroups (not presented here), have revealed that the family "Euryphoridae" is paraphyletic; therefore, the family must be considered synonymous with the Caligidae as originally proposed by Dojiri (1983) in his unpublished Ph.D. dissertation and officially by Boxshall and Halsey (2004). These five genera are basal to the remaining 26 caligid genera (fig. 151), which share a vestigial lobiform endopod of leg 1 (fig. 158, character 24) and a 2-segmented leg 3 endopod (fig. 160, character 30), with the majority of the genera bearing six

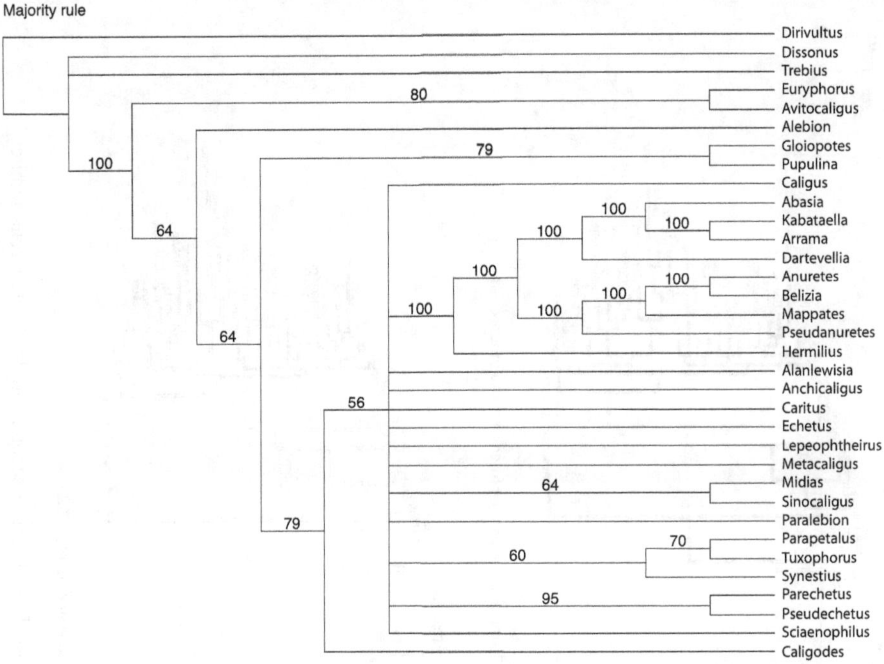

Fig. 152. 50% majority-rule consensus tree of the Caligidae consisting of 31 genera and depicting the former euryphorid genera as paraphyletic and basal to the remaining 26 caligid genera.

setae on the terminal segment of the leg 3 endopod (fig. 160, character 31). The aliform plates on the fourth pedigerous somite (fig. 154, character 6) was previously used as a character uniting the former euryphorid genera; however, this structure has been found only in four former euryphorid genera, i.e., *Euryphorus*, *Alebion*, *Gloiopotes*, and *Tuxophorus* (a genus no longer considered to be included in this basal grouping), while *Pupulina* does not possess this structure. *Avitocaligus*, which is now included in this basal grouping, also does not have these aliform plates (Boxshall and Justine, 2005).

The relationships of the remaining 26 caligid genera are not fully resolved; however, a few groups of genera are evident in this phylogram. *Alanlewisia* is the first offshoot, is the sister group to the *Caligus*-to-*Paralebion* complex, and is characterized by two long setae and a minute spine at the tip of the leg 1 endopod. Within the *Caligus*-to-*Paralebion* complex, which is characterized by an absence of elements (spines and/or setae) on the tip of the leg 1 endopod (fig. 159, character 25), are two major groups of genera that are evident: the *Caligus* group and the *Lepeophtheirus* group.

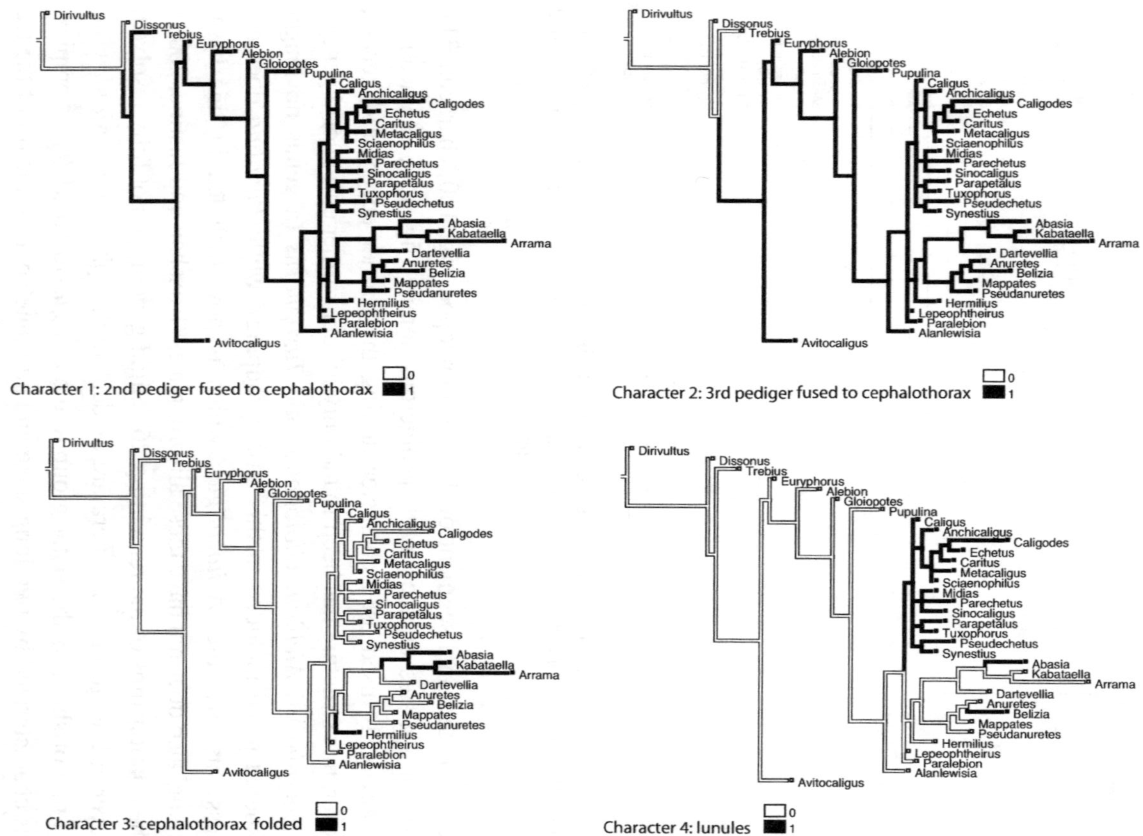

Fig. 153. Character trees for characters 1-4. Abbreviations: 2nd and 3rd pedigers = second and third pedigerous somites.

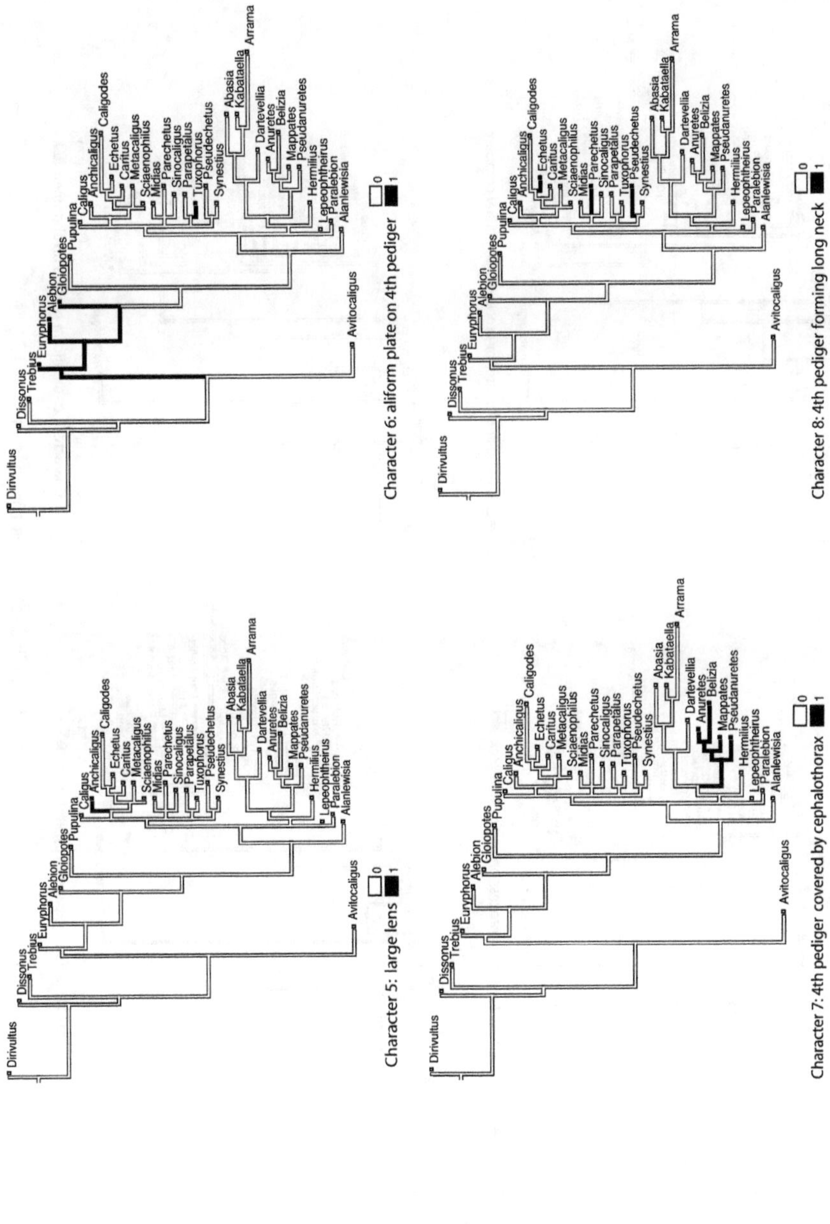

Fig. 154. Character trees for characters 5-8. Abbreviation: 4th pediger = fourth pedigerous somite.

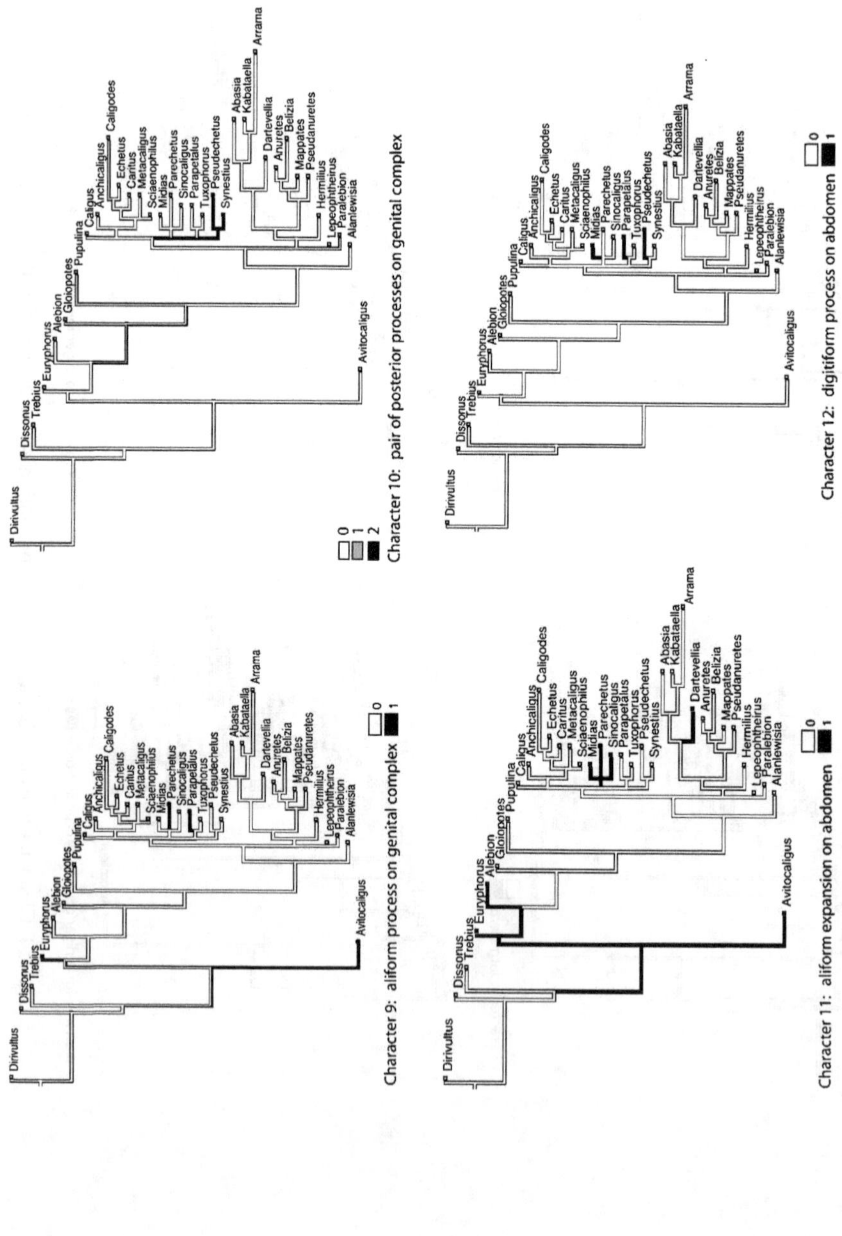

Fig. 155. Character trees for characters 9-12.

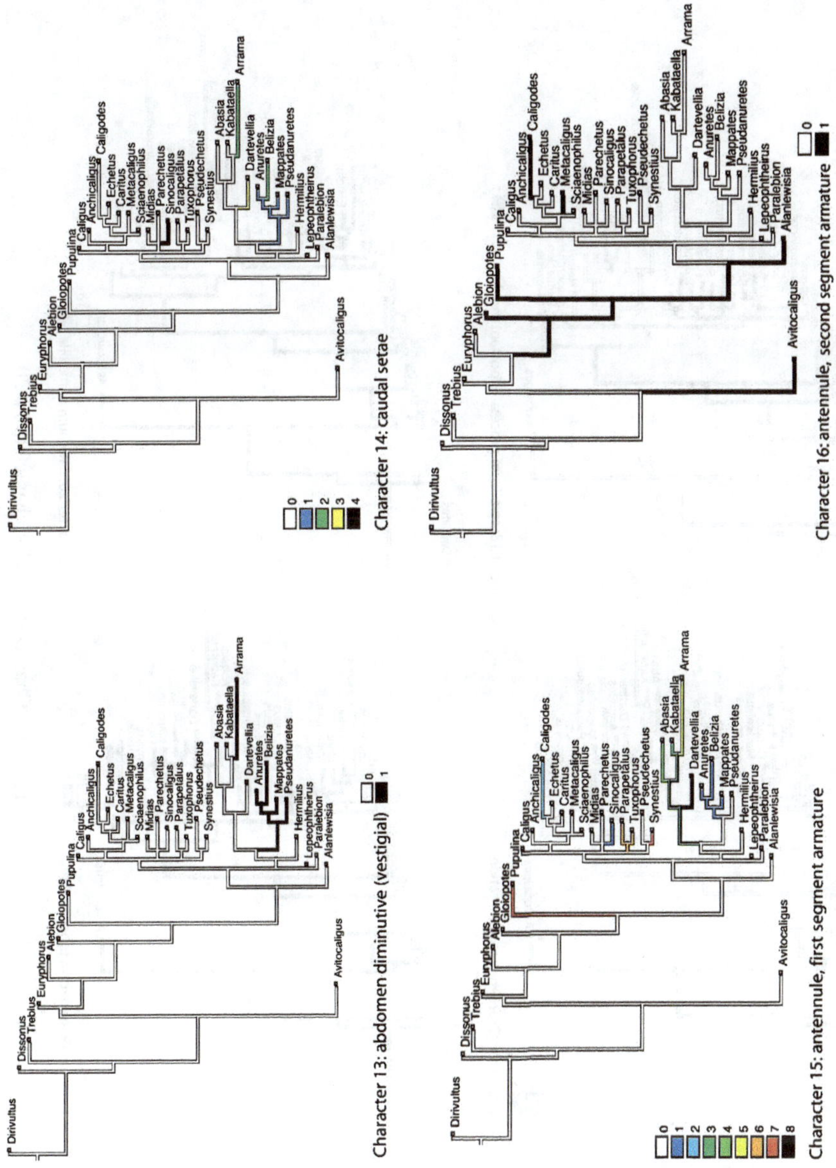

Fig. 156. Character trees for characters 13-16.

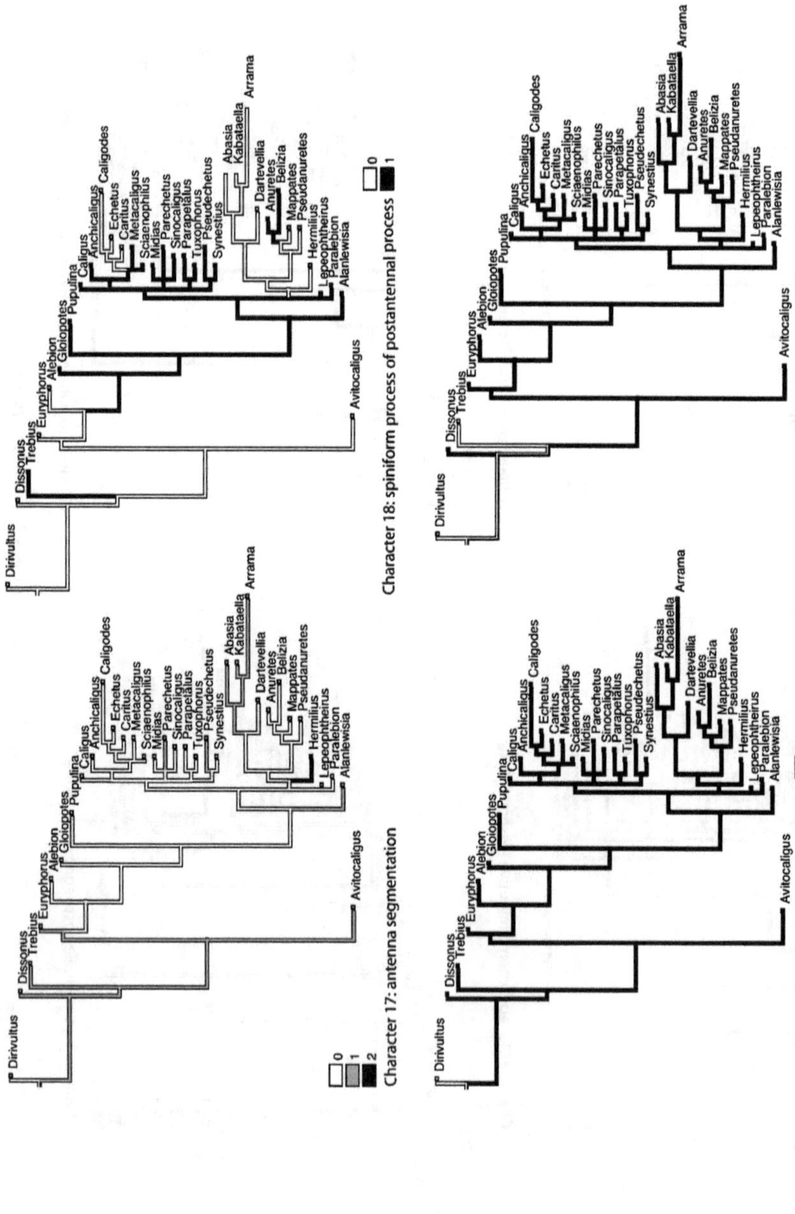

Fig. 157. Character trees for characters 17-20.

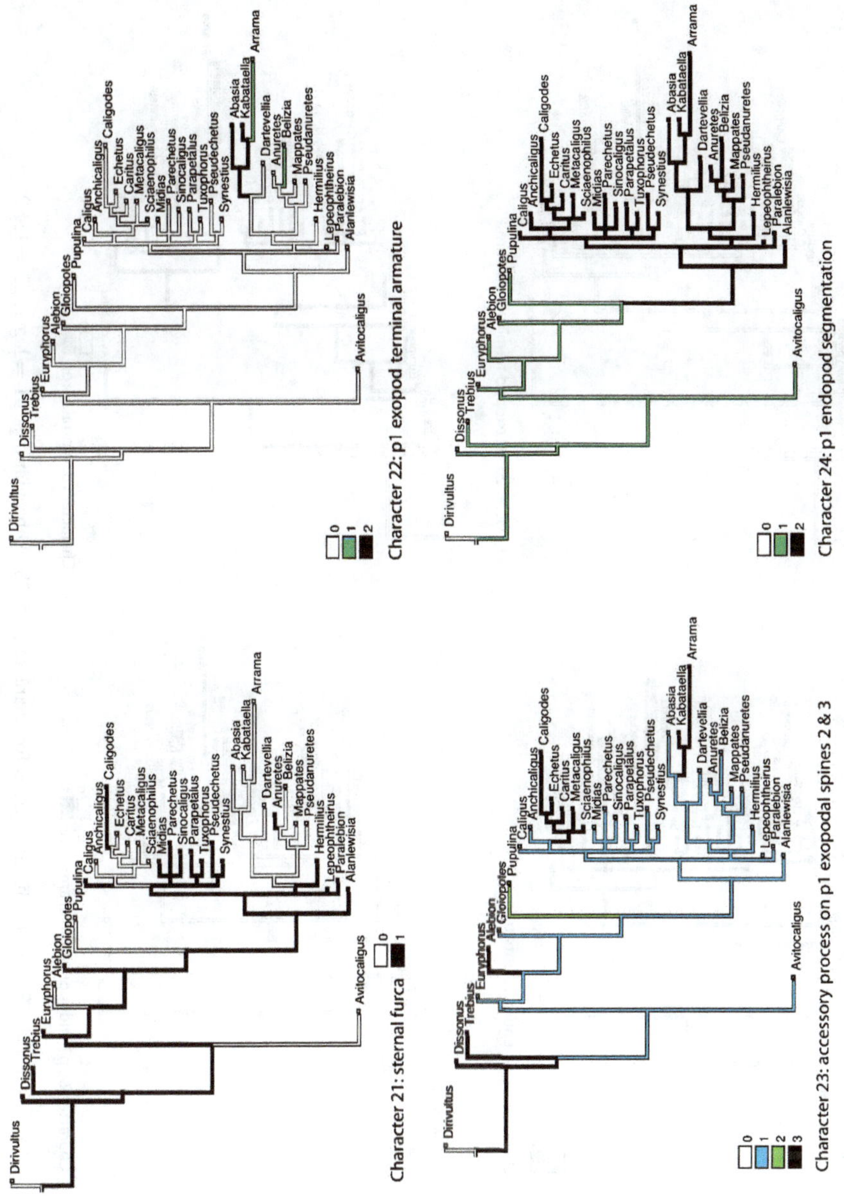

Fig. 158. Character trees for characters 21-24. Abbreviation: p1 = leg 1.

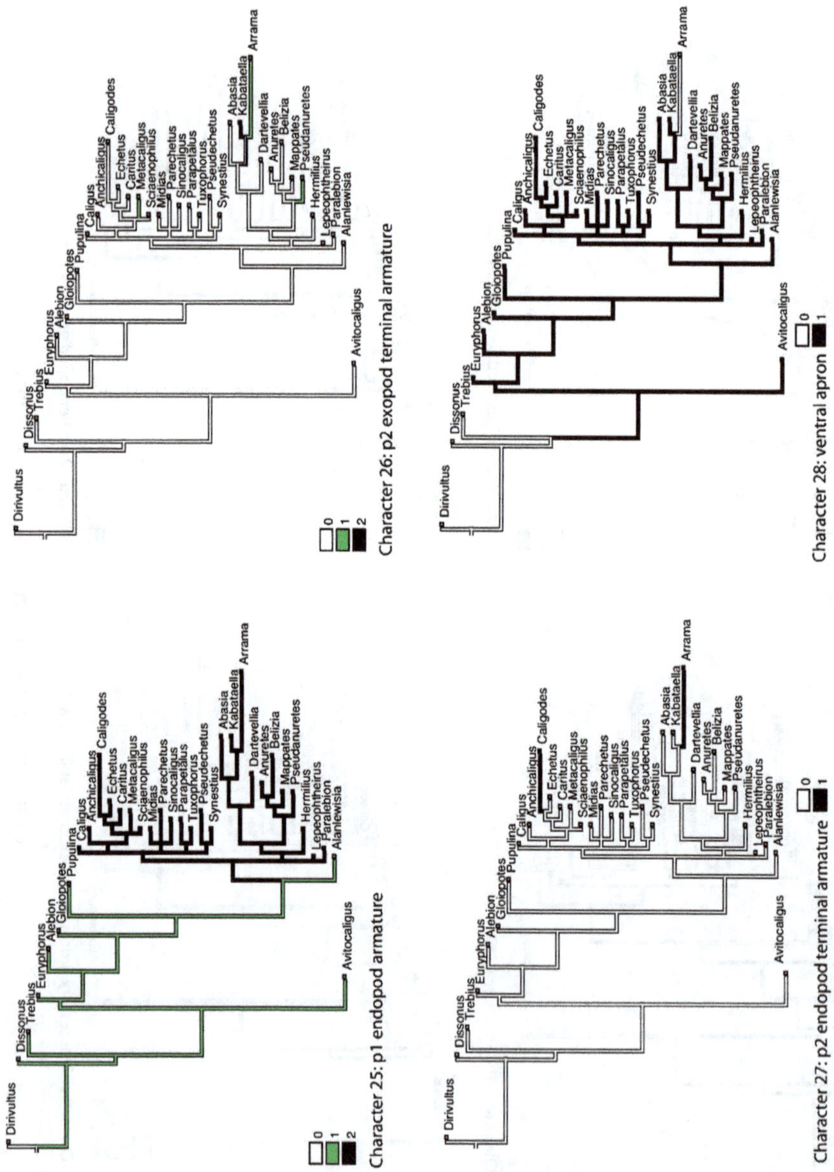

Fig. 159. Character trees for characters 25-28. Abbreviations: p1 = leg 1; p2 = leg 2.

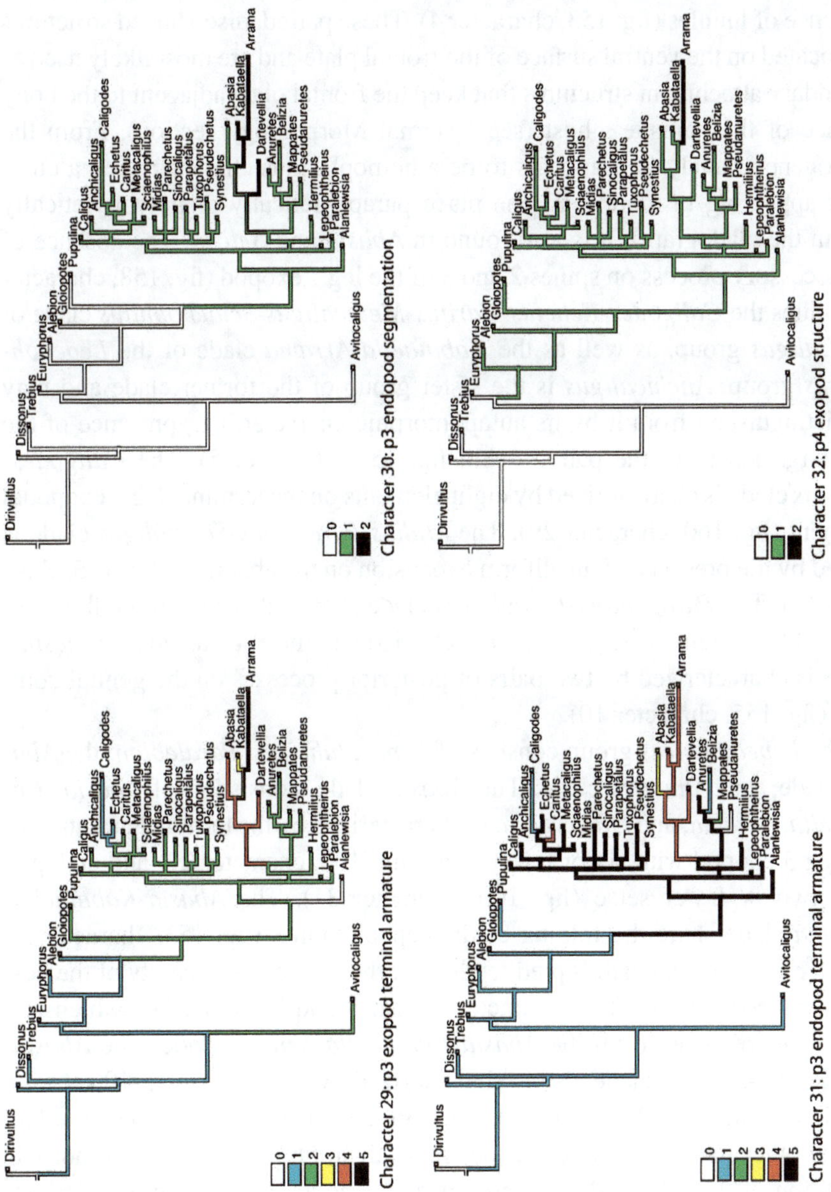

Fig. 160. Character trees for characters 29-32. Abbreviations: p3 = leg 3; p4 = leg 4.

The *Caligus* group, which consists of *Caligus*, *Anchicaligus*, *Caligodes*, *Echetus*, *Caritus*, *Metacaligus*, *Sciaenophilus*, *Midias*, *Parechetus*, *Sinocaligus*, *Parapetalus*, *Tuxophorus*, *Pseudechetus*, and *Synestius*, are united by the presence of lunules (fig. 153, character 4). These paired, disc-shaped structures are located on the ventral surface of the frontal plate and are most likely used as secondary attachment structures that keep the frontal plate adjacent to the body surface of the parasite's host (see External Morphology section). From the phylogenetic analysis it appears to be a homoplasious character (i.e., a character appearing in more than one place paraphyletically or polyphyletically within the cladogram): it is also found in *Abasia* and *Belizia*. The absence of the accessory process on spines 2 and 3 of the leg 1 exopod (fig. 158, character 23) unites the *Caligodes-Echetus-Caritus-Metacaligus-Sciaenophilus* clade of the *Caligus* group, as well as the *Kabataella-Arrama* clade of the *Lepeophtheirus* group. *Anchicaligus* is the sister group of the former clade and may be distinguished from it by its autapomorphic character, i.e., presence of the two large lenses of the paired ocelli fig. 154 (character 5). The *Caligodes-Echetus* clade is characterized by eight elements on the terminal leg 3 exopodal segment (fig. 160, character 29). The *Midias-Parechetus-Sinocaligus* clade is united by the presence of an aliform expansion on the abdomen (fig. 155, character 11). The *Parapetalus-Tuxophorus* clade shares 28 setae on the first segment of the antennule (fig. 156, character 15) and the *Pseudechetus-Synestius* clade is characterized by two pairs of posterior processes on the genital complex (fig. 155, character 10).

The *Lepeoptheirus* group consists of *Lepeophtheirus*, *Paralebion*, the *Abasia* clade, and *Anuretes* clade. The *Abasia* clade, consisting of *Abasia*, *Kabataella*, *Arrama*, and *Dartevellia*, is characterized by the terminal armature of the leg 3 exopod with six or fewer setae (fig. 160, character 29) and endopod with two or fewer setae (fig. 160, character 31). The *Abasia-Kabataella-Arrama* clade share the folding of the cephalothorax (fig. 153, character 3), which reflects the parasite's predilection for the gills or nasal cavity of the host as its attachment site. This character is also found in *Hermilius*, which is a few nodes removed from the *Abasia-Kabataella-Arrama* clade. The *Abasia-Kabataella-Arrama* clade also is characterized by having only 15-19 setae on the first segment of the antennule (fig. 156, character 15), while *Dartevellia*, on the other hand, has 31 setae, four more than the presumed plesiomorphic condition of 27 exhibited by *Dissonus* and *Trebius*. In addition, the *Abasia-Kabataella-Arrama* clade also possesses either six elements (*Abasia* and *Kabataella*) or five elements (*Arrama*) on the terminal segment of the leg 1 exopod (fig. 158, character 22). Also, this clade possesses a 4-partite antenna (fig. 157, character 17) as in *Dissonus* and *Trebius*, but unlike many other caligid genera. However, the segmentation of this appendage is somewhat problematic. The presumed plesiomorphic condition is found in *Dirivultus*, which possesses a 3-segmented antenna, with the basal segment representing the coxa, the second segment representing the basis, and the third segment (a recurved claw) comprised of the endopod fused to the terminal spiniform process (see External Morphology section). However, reports of a 4-partite (e.g., *Abasia*, *Kabataella*, and *Arrama*, among others) and 5-partite (*Hermilius*) antennae (the claw is bipartite) suggest that the basal portion in these genera may consist of the praecoxa and coxa.

The *Anuretes* group, consisting of *Anuretes*, *Belizia*, *Mappates*, and *Pseudanuretes*, is characterized by the fourth pedigerous somite covered by the posterior margin of the cephalothorax (fig. 154, character 7), a diminutive abdomen (fig. 156, character 13), and reduced number of caudal setae (fig. 156, character 14). The *Anuretes*, *Belizia*, and *Mappates* clade shares the character state of having 25-26 setae on the first segment of the antennule (fig. 156, character 15).

Ho & Lin (2004) conducted a cladistic analysis on 28 caligid genera and identified four groupings of genera that are somewhat similar to those depicted in the phylogram (fig. 151) presented here: (1) *Midias-Parapetalus-Parechetus-Sinocaligus* group, (2) *Anchicaligus-Caligodes-Echetus-Caritus-Metacaligus-Sciaenophilus* group, (3) *Dartevellia-Arrama-Kabataella* group, and (4) the *Anuretes-Mappates-Pseudanuretes* group. Although the topology of Ho & Lin's cladogram differs from the phylogram presented here, the similarity of the aforementioned four groups lends some support to the validity of these generic relationships.

The phylogram depicted here reflects numerous homoplasies (see figs. 154-157 & 159; characters 8, 9, 12, 16-18, 26, 27, among others), reversals (e.g., fig. 158, character 21), and autapomorphies (fig. 154, character 5), as well as a few somewhat ubiquitous characters that are found in most, if not all, of the caligid genera (figs. 157 & 160; characters 19, 20, and 32), as evident in the 32 character trees (figs. 153-160). For example, the fourth pedigerous somite forms an elongate neck (fig. 154, character 8) in three genera of three different clades, *Echetus*, *Parechetus*, and *Pseudechetus*, so must be considered a homoplasious character. This is also true for the aliform expansion on the genital complex (fig. 155, character 9), which is possessed by *Euryphorus*, *Avitocaligus*, *Parechetus*, and *Parapetalus*. Additional character data may provide the detailed information required to refine and finally resolve the phylogenetic relationships within the Caligidae. However, even though the phylograms and consensus trees are not fully resolved, one commonality among all the trees is the paraphyletic relationships of the basal five genera (i.e., *Alebion*, *Avitocaligus*, *Euryphorus*, *Gloiopotes*, and *Pupulina*), indicating that they cannot be considered a valid family and they must be included as members of the Caligidae.

The major evolutionary trend within the entire caligiform complex, which currently consists of the families Dissonidae, Trebiidae, Caligidae, Pandaridae, and Cecropidae, is the progressive incorporation of pedigerous somites into the cephalothorax. As in the outgroup *Dirivultus*, *Dissonus* possesses three free pedigerous somites, and all four legs are biramous (see Kabata, 1966a). The segmentation of the rami of the legs, in addition to the bipartite condition of the maxillule (see Kabata, 1979: 31) of *Dissonus* indicates that this genus exhibits many of the features that are found in *Dirivultus*.

Trebius is much like *Dissonus*, except that this genus exhibits only two free pedigerous somites, the second pedigerous somite, in addition to the first, having fused with the cephalothorax (Kabata, 1979). Legs 2-4 are very similar to *Dissonus*, but the trebiid leg 1 is almost identical to that of the genera formerly belonging to the Euryphoridae. The genera of Caligidae display a progressive reduction in the endopod of leg 1, and the structure of leg 4 (refer to discussion on External Morphology). The Trebiidae is closely related to the Caligidae based on the similarity of the appendages among *Trebius* and the caligid genera. This proposed relationship is in agreement with that of Wilson (1907a) who believed in the close affinities among these three families, which are united by the shared possession of several character states.

The culmination of this evolutionary trend is found in the Caligidae in which the second and third pedigerous somites are fused with the cephalothorax. Concomitant with this advancement is a flattened cephalothorax and the formation of the ventral apron, which seals off the posterior margin of the cephalothorax for better prehension (Kabata, 1979). This composite morphology is used to provide a very effective prehensile apparatus that adheres the copepod to the smooth body surface of its fish host (refer to General Habitus and Leg 3 discussions in the External Morphology section herein). Although *Arrama* also displays the first through the third pedigerous somites incorporated into the cephalothorax, this genus has lost its ventral apron and possesses greatly reduced legs 3 and 4, presumably as a result of its predilection for attachment to the gill filaments rather than the body surface of its hosts.

The caligid genera are predominantly found on the smooth, moist (mucous-covered) body or branchial chamber surfaces of their teleost hosts. The spiniform processes, e.g., spiniform process on the base of the antenna, postantennal process, dentiform maxillule, sternal furca, and the suction-cuplike cephalothorax, are well suited for the attachment sites of this copepod family. *Hermilius*, *Kabataella*, and *Arrama* have lost the cephalothoracic suction structure and function, having shifted their attachment site from the body surface to the nasal fossae or gill filaments. These very successful morphological adaptations exhibited by the Caligidae and described above have resulted in an adaptive radiation of the family, which now has over 450 described species.

REFERENCES

ANSTENSRUD, M., 1990a. Effects of mating on female behaviour and allometric growth in two parasitic copepods *Lernaeocera branchialis* (L., 1758) (Pennellidae) and *Lepeophtheirus pectoralis* (Müller, 1776) (Caligidae). Crustaceana, **59**: 245-258.

— —, 1990b. Male reproductive characteristics of two parasitic copepods, *Lernaeocera branchialis* (L.) (Pennellidae) and *Lepeophtheirus pectoralis* (Müller) (Caligidae). Journal of Crustacean Biology, **10**: 627-638.

— —, 1990c. Mating strategies of two parasitic copepods [*Lernaeocera branchialis* (L.) (Pennellidae) and *Lepeophtheirus pectoralis* (Müller) (Caligidae)] in flounder: polygamy, sex-specific age at maturity and sex ratio. Journal of Experimental Marine Biology and Ecology, **136**: 141-158.

— —, 1990d. Moulting and mating in *Lepeophtheirus pectoralis* (Copepoda, Caligidae). Journal of the Marine Biological Association of the United Kingdom, **70**: 269-282.

— —, 1992. Mate guarding and mate choice in two copepods, *Lernaeocera branchialis* (L.) (Pennellidae) and *Lepeophtheirus pectoralis* (Müller) (Caligidae), parasitic on flounder. Journal of Crustacean Biology, **12**: 31-40.

BAIRD, W., 1850. The natural history of the British Entomostraca: 1-364. (Ray Society, London).

BARNARD, K. H., 1948. New records and descriptions of new species of parasitic Copepoda from South Africa. Annals and Magazine of Natural History, (12) **1**: 242-254.

— —, 1955a. South African parasitic Copepoda. Annals of the South African Museum, **41**: 223-312.

— —, 1955b. Additions to the fauna-list of South African Crustacea and Pycnogonida. Annals of the South African Museum, **43**: 1-107.

BARNETT, P. A., 1976. Studies on the host-parasite relationship between the northern anchovy (*Engraulis mordax* Girard) and a caligid copepod (*Caligus klawei* Shiino): 1-106. (Master's Thesis, California State University, Long Beach).

BASSETT-SMITH, P. W., 1896a. A list of the parasitic Copepoda of fish obtained at Plymouth. Journal of the Marine Biological Association of the United Kingdom, **4**: 155-163.

— —, 1896b. Notes on the parasitic Copepoda of fish obtained at Plymouth, with descriptions of new species. Annals and Magazine of Natural History, (6) **18**: 8-16.

— —, 1898a. Some new or rare parasitic copepods found on fish in the Indo-tropic region. Annals and Magazine of Natural History, (7) **2**: 357-372.

— —, 1898b. Further new parasitic copepods found on fish in the Indo-tropic region. Annals and Magazine of Natural History, (7) **2**: 77-98.

— —, 1898c. Some new parasitic copepods found on fish at Bombay. Annals and Magazine of Natural History, (7) **1**: 1-17.

— —, 1899. A systematic description of parasitic Copepoda found on fishes, with an enumeration of the known species. Proceedings of the Zoological Society of London, 1899 (2): 438-507.

BENEDEN, P. J. VAN, 1852. Note sur un nouveau genre de crustace parasite de la famille des Peltocèphales. Bulletin de l'Académie de Belgique, (3) **19**: 462-467.

— —, 1861. Recherches sur la faune littorale de Belgique crustacés. Memoires de l'Académie de Belgique, **33**: 1-180.

— —, 1892. Quelques nouveaux caligides de la côte d'Afrique et de l'archipel des Açores. Bulletin de l'Académie de Belgique, (3) **24**: 241-262.

BEN HASSINE, O. K., 1983. Les copepods parasites de poisons Mugilidae en Méditerranee occidentale (Côtes Francaises et Tunisiennes), morphologie, bioécologie, cycles évolutifs: 1-452. (Doctorate Dissertation, Université de Sciences et Techniques du Languedoc).

BENZ, G. W., 1984. Distribution of siphonostomatoid copepods parasitic upon large pelagic sharks in the western North Atlantic. Syllogeus, **58**: 211-219.

BENZ, G. W., P. J. MOHAN & G. L. CROW, 1992. Developmental stages of *Paralebion elongatus* from aquarium-held reef whitetip sharks (*Triaenodon obesus*) captured in Hawaiian waters. Journal of Parasitology, **78**: 1027-1035.

BERE, R., 1936. Parasitic copepods from Gulf of Mexico fish. American Midland Naturalist, **17**: 577-625.

BEZDĚK, A. & H. B. CRESSEY, 2004. *Cresseyella* nom. nov., a new replacement name for the genus *Apogonia* (Crustacea: Copepoda: Caligidae). Biologia, Bratislava, **59**: 180.

BLAINVILLE, H. M. D. DE, 1822. Mémoire sur les Lernées (*Lernaea*, Lin.). Journal de Physique, de Chimie, d'Histoire Naturelle et des Arts, Paris, **95**: 372-380, 437-447.

BONNET, D., 1948. Some parasitic copepods from Hawaiian fishes. Proceedings of the Hawaiian Academy of Sciences, **23**: 7.

BOSSANYI, J. & H. O. BULL, 1971. The marine fauna of the Cullercoats District. 5. Arthropoda. 3c. Crustacea: Copepoda. Report of the Dove Marine Laboratory, University of Newcastle upon Tyne (3) **17**: 1-59.

BOWMAN, T. E., 1971. The case of the nonubiquitous telson and the fraudulent furca. Crustaceana, **21**: 165-175.

BOXSHALL, G. A., 1974a. *Lepeophtheirus pectoralis* (O. F. Müller, 1776); a description, a review and some comparisons with the genus *Caligus* Müller, 1785. Journal of Natural History, **8**: 445-468.

— —, 1974b. The developmental stages of *Lepeophtheirus pectoralis* (Müller, 1776) (Copepoda: Caligidae). Journal of Natural History, **8**: 681-700.

— —, 1974c. The population dynamics of *Lepeophtheirus pectoralis* (Müller): seasonal variation in abundance and age structure. Parasitology, **69**: 361-371.

— —, 1974d. Infections with parasitic copepods in North Sea marine fishes. Journal of the Marine Biological Association of the United Kingdom, **54**: 355-372.

— —, 1974e. The population dynamics of *Lepeophtheirus pectoralis* (Müller): dispersion pattern. Parasitology, **69**: 373-390.

— —, 1977. The histopathology of infection by *Lepeophtheirus pectoralis* (Müller) (Copepoda: Caligidae). Journal of Fish Biology, **10**: 411-415.

— —, 1986. The comparative anatomy of the feeding apparatus of four orders of copepods. Syllogeus, **58**: 158-169.

— —, 1990. The skeletomusculature of siphonostomatoid copepods, with an analysis of adaptive radiation in structure of the oral cone. Philosophical Transactions of The Royal Society of London, **328**: 167-212.

— —, 2008. A new genus of sea louse (Copepoda: Siphonostomatoida: Caligidae) parasitic on the bluespine unicornfish (*Naso unicornis*). Folia Parasitologica, **55**: 231-240.

BOXSHALL, G. A. & S. H. HALSEY, 2004. An introduction to copepod diversity: 1-966. (Ray Society, London).

BOXSHALL, G. A. & J. L. JUSTINE, 2005. A new genus of parasitic copepod (Siphonostomatoida: Caligidae) from the razorback scabbardfish, *Assurger anzac* (Trichiuridae) off New Caledonia. Folia Parasitologica, **52**: 349-358.

BOXSHALL, G. A., C. L. LIN, J. S. HO, S. OHTSUKA, B. A. VENMATHI MARAN & J. L. JUSTINE, 2008. A revision of the family Dissonidae Kurtz, 1924 (Copepoda: Siphonostomatoida). Systematic Parasitology, **70**: 81-106.

BOXSHALL, G. A. & M. A. MONTÚ, 1997. Copepods parasitic on Brazilian coastal fishes: a handbook. Nauplius, Rio Grande, **5**: 1-225.

BRADY, G. S., 1883. Report on the Copepoda collected by H. M. S. Challenger, during the years 1873-76. Report on the Scientific Results of the Voyage of H. M. S. Challenger, Zoology, **8**: 1-142.

BRANDAL, P. O., E. EGIDIUS, & I. ROMSLO, 1976. Host-blood: a major food component for the parasitic copepod *Lepeophtheirus salmonis* Krøyer, 1838 (Crustacea: Caligidae). Norwegian Journal of Zoology, **24**: 341-343.

BRESCIANI, J. & J. LÜTZEN, 1962. Parasitic copepods from the west coast of Sweden including some new or little known species. Videnskabelige Meddelelser fra Dansk Naturhistorisk Forening i Kjøbenhavn, **124**: 367-408.

BRIAN, A., 1898. Catalogo di copepodi parassiti dei pesci della Liguria. Bollettino dei Musei di Zoologia e Anatomia Comparata della R. Università di Genova, **3**: 1-27.

— —, 1899a. Di alcuni crostacei parassiti dei pesci dell'Isola d'Elba. Bollettino dei Musei di Zoologia e Anatomia Comparata della R. Università di Genova, **4**: 1-8.

— —, 1899b. Crostacei parassiti dei pesci dell'Isola d'Elba. (II. Contribuzione). Bollettino dei Musei di Zoologia e Anatomia Comparata della R. Università di Genova, **4**: 1-11.

— —, 1899c. *Diphyllogaster thompsoni* n. gen. et sp. di Caligidae della *Dicerobatis giornae* Gunt. Atti della Società Ligustica di Scienze Naturali e Geografiche, **10**: 1-11.

— —, 1902. Note su alcuni crostacei parassiti dei pesci del Mediterraneo. Bolletino dei Musei di Zoologia e Anatomia Comparata della R. Università di Genova, **5**: 1-16.

— —, 1905. Nouveau copépode parasite. Archives de Parasitologie, **9**: 564-567.

— —, 1906. Copepodi parassiti dei pesci d'Italia: 1-190. (Genova).

— —, 1908. Note préliminaire sur les copépodes parasites des poissons provenant des campagnes scientifiques de S.A.S. le Prince Albert Ier de Monaco ou déposés dans les collections du Musée Océanographique. Bulletin de l'Institut Oceanographique, **110**: 1-18.

— —, 1912. Copépodes parasites des poissons et des échinides provenant des campagnes scientifiques de S.A.S. le Prince Albert Ier de Monaco (1886-1910). Résultats Campagnes Scientifiques Accomplies sur son Yacht par Albert Ier Prince Souverain de Monaco, **38**: 1-58.

— —, 1924. Parasitologia Mauritanica. Matériaux pour la faune parasitologique en Mauritanie. Arthropoda (1re partie), Copepoda. Copépodes commensaux et parasites des côtes Mauritaniennes. Bulletin du Comité d'Études Historique et Scientifiques de l'Afrique Occidentale Française, **1924**: 1-66.

— —, 1939. Copépodes parasites recueillis par M. E. Dartevelle a l'embouchure du fleuve Congo. Revue de Zoologie et de Botanique Africaines, **32**: 176-198.

BRIGGS, R. P., 1978. Structure of the integument in *Paranthessius anemoniae* Claus, a copepod associate of the snakelocks anemone *Anemonia sulcata* (Pennant). Journal of Morphology, **156**: 293-316.

BRON, J. E., C. SOMMERVILLE, M. JONES & G. H. RAE, 1991. The settlement and attachment of early stages of the salmon louse, *Lepeophtheirus salmonis* (Copepoda: Caligidae) on the salmon host, *Salmo salar*. Journal of Zoology, London, **224**: 202-212.

BRUCE, J. R., J. S. COLMAN & N. S. JONES, 1963. Marine fauna of the Isle of Man and its surrounding seas. Liverpool University Press, Memoir no. **36**: 1-307.

BURGESS, W. E., 1989. An atlas of freshwater and marine catfishes: a preliminary survey of the Siluriformes: 1-784. (T. F. H. Publications, Inc., Neptune, New Jersey).

BURMEISTER, H., 1835. Beschreibung einiger neuen oder weniger bekannten Schmarotzerkrebse, nebst allgemeinen Betrachtungen über die Gruppe, welcher sie angehören. Acta, Verhandlungen der Leopoldinisch-Carolinischen Akademie der Naturforscher, **17**: 269-336.

BURNETT-HERKES, J., 1974. Parasites of the gills and buccal cavity of the dolphin, *Coryphaena hippurus*, from the Straits of Florida. Transaction of the American Fisheries Society, **103**: 101-106.

BYRNES, T., 1986. A new species of *Parapetalus* (Copepoda: Caligidae) parasitic on the Samson fish, *Seriola hippos* from Australia and its similarity with *Caligus spinosus*. Publications of the Seto Marine Biological Laboratory, **31**: 55-62.

CAPART, A., 1941. Résultats scientifiques des croisières du navire-école Belge (Mercator). V. Copepoda Parasitica. Mémoires du Musée Royal d'Histoire Naturelle de Belgique, (2) **3**: 171-197.

— —, 1953. Quelques copépodes parasites de poissons marins de la région de Dakar. Bulletin de l'Institut Français d'Afrique Noire, **15**: 647-671.

— —, 1959. Copépodes parasites. Résultats Scientifiques. Expédition Oceanographique Belge dans les Eaux Cotières Africaines de l'Atlantique Sud (1948-1949), **3**: 57-126.

CARBONELL, E., E. MASSUTI, J. J. CASTRO & R. M. GARCIA, 1999. Parasitism of dolphinfishes, *Coryphaena hippurus* and *Coryphaea equisalis*, in the western Mediterranean (Balearic Island) and central-eastern Atlantic (Canary Islands). Scientia Marina, **63**: 345-354.

CARUS, J. V., 1885. Prodromus faunae mediterraneae. Vol. 1. Coelenterata, Echinodermata, Vermes, Arthropoda: 1-524. (E. Schweizerbart, Stuttgart).

CARVALHO, J. P., 1940. Notas sôbre alguns Caligoida, com a descrição *Pandarus marcusi* sp. nov. Boletim da Faculdade de Filosofia, Ciências e Letras, Zoologia, **19**: 271-289.

— —, 1951. Notas sôbre alguns copépodes parasitos de peixes marítimos da costa do estado de São Paulo. Boletim do Instituto Paulista de Oceanografía, **2**: 135-144.

CAUSEY, D., 1953a. Parasitic Copepoda from Grand Isle, Louisiana. Occasional Papers of the Marine Laboratory. Louisiana State University, **7**: 1-18.

— —, 1953b. Parasitic Copepoda of Texas coastal fishes. Publication of the Institute of Marine Science, University of Texas, **3**: 7-16.

— —, 1955. Parasitic Copepoda from Gulf of Mexico fish. Occasional Papers of the Marine Laboratory, Louisiana State University, **9**: 1-19.

— —, 1960. Parasitic Copepoda from Mexican coastal fishes. Bulletin of Marine Science of the Gulf and Caribbean, **10**: 323-337.

CAVALEIRO, F. I. & M. J. SANTOS, 2007. Survey of the metazoan ectoparasites of the European flounder *Platichthys flesus* (Linnaeus, 1758) along the north-central Portuguese coast. Journal of Parasitology, **93**: 1218-1222.

CLAUS, C., 1864. Beiträge zur Kenntniss der Schmarotzerkrebse. Zeitschrift für Wissenschaftliche Zoologie, **14**: 365-382.

COLLETTE, B. B. & N. V. PARIN, 1970. Needlefishes (Belonidae) of the eastern Atlantic Ocean. Atlantide Report, **11**: 7-60.

COSTELLO, M. J., 2009. The global economic cost of sea lice to the salmonid farming industry. Journal of Fish Diseases, **32**: 115-118.

CRESSEY, R. F., 1966. A new species of *Dentigryps* Wilson (Copepoda, Caligoida) from Madagascar. Proceedings of the Biological Society of Washington, **79**: 91-94.

— —, 1967a. Genus *Gloiopotes* and a new species with notes on host specificity and intraspecific variation (Copepoda: Caligoida). Proceedings of the United States National Museum, **122**: 1-22.

— —, 1967b. Caligoid copepods parasitic on sharks of the Indian Ocean. Proceedings of the United States National Museum, **121**: 1-21.

— —, 1967c. *Caritus*, a new genus of caligoid copepod, with a key to the genera of Caliginae. Proceedings of the United States National Museum, **123**: 1-8.

— —, 1970. Copepods parasitic on sharks from the west coast of Florida. Smithsonian Contributions to Zoology, **38**: 1-30.

— —, 1972. Revision of the genus *Alebion* (Copepoda: Caligoida). Smithsonian Contributions to Zoology, **123**: 1-29.

— —, 1974. A redescription of *Hermilius pyriventris* Heller (Copepoda: Caligoida) with the first description of the male. Proceedings of the Biological Society of Washington, **87**: 235-244.

— —, 1990. *Belizia brevicauda*, a new genus and species of caligid copepod from the western Caribbean Sea. Systematic Parasitology, **15**: 151-154.

CRESSEY, R. F. & B. B. COLLETTE, 1970. Copepods and needlefishes: a study in host-parasite relationships. Fishery Bulletin, **68**: 347-432.

CRESSEY, R. & H. B. CRESSEY, 1979. The parasitic copepods of Indo-West Pacific lizardfishes (Synodontidae). Smithsonian Contributions to Zoology, **296**: 1-71.

— — & — —, 1980. Parasitic copepods of mackerel- and tuna-like fishes (Scombridae) of the world. Smithsonian Contributions to Zoology, **311**: 1-186.

— — & — —, 1990. A new genus of caligoid copepod from the fish genus *Apogon*. Beaufortia, **41**(7): 45-48.

CRESSEY, R. & P. NUTTER, 1987. Reidentification of David Causey's *Caligus* collections (Crustacea: Copepoda). Proceedings of the Biological Society of Washington, **100**: 600-602.

CUVIER, G., 1817. Regne Animal (1st ed.), **4**: 36-37.

DANA, J. D., 1852. Synopsis of characters of parasitic copepods. Proceedings of the American Academy of Arts and Sciences, **2**: 9-61.

— —, 1853. United States exploring expedition during the years 1838-42, under the command of Charles Wilkes, U.S.N., Volume XIII. Crustacea, **2**: 686-1618.

— —, 1855. Atlas. Crustacea. United States exploring expedition during the years 1838-42 under the command of Charles Wilkes, U.S.N. [plates for Dana, 1853].

DAWSON, L. H. J., 1998. The physiological effects of salmon lice (*Lepeophtheirus salmonis*) infections on returning post-smolt sea trout (*Salmo trutta* L.) in western Ireland, 1996. ICES Journal of Marine Science, **55**: 193-200.

DELAMARE DEBOUTTEVILLE, C. & L. NUNES-RUIVO, 1953. Copépodes parasites des poissons Méditerranéens (3e Serie) (1). Vie et Milieu, **4**: 201-217.

DIPPENAAR, S. M., 2005. Reported siphonostomatoid copepods parasitic on marine fishes of South Africa. Crustaceana, **77**: 1281-1328.

DOJIRI, M., 1979. Two new species of *Lepeophtheirus* (Copepoda, Caligidae) parasitic on fishes from southern California waters. Parasitology, **78**: 251-262.

— —, 1983. Revision of the genera of the Caligidae (Siphonostomatoida), copepods predominantly parasitic on marine fishes: 1-721. (Ph.D. Thesis, Boston University, Boston, Massachusetts).

DOJIRI, M. & R. F. CRESSEY, 1991. *Arrama*, new genus (Siphonostomatoida: Caligidae), with two new species, copepods parasitic on Australian fishes. Journal of Crustacean Biology, **11**: 594-606.

DOJIRI, M. & A. G. HUMES, 1982. Copepods (Poecilostomatoida: Taeniacanthidae) from sea urchins (Echinoidea) in the southwest Pacific. Zoological Journal of the Linnean Society, **74**: 381-436.

FERRIS, G. F. & J. M. HENRY, 1949. The nervous system and a problem of homology in certain Crustacea (Crustacea: Copepoda: Caligidae). Microentomology, **14**: 114-120.

FOWLER, H. W., 1912. Crustacea of New Jersey. Annual Report of the New Jersey State Museum for 1911: 29-650.

FRASCA, S. JR., V. L. KIRSIPUU, S. RUSSELL, S. A. BULLARD & G. W. BENZ, 2004. Opercular lesion in wild black drum, *Pogonias cromis* (Linnaeus, 1766), associated with attachment of the sea louse *Sciaenophilus tenuis* (Copepoda: Siphonostomatoida: Caligidae). Acta Ichthyologica et Piscatoria, **34**: 115-127.

FRASER, C. M., 1920. Copepods parasitic on fish from the Vancouver Island region. Transactions of the Royal Society of Canada, (5) (3) **13**: 45-67.

FROESE, R. & D. PAULY, 2011. FishBase. World Wide Web electronic publication. Available from http://www.fishbase.org/.

GERSTAECKER, A., 1853. Über eine neue und eine weniger gekannte Siphonostomen-Gattung. Archiv für Naturgeschichte, **19**: 58-70.

GIESBRECHT, W., 1892. Systematik und Faunistik der pelagischen Copepoden des Golfes von Neapel und der angrenzenden Meeres-Abschnitte. Fauna und Flora des Golfes von Neapel und der angrenzenden Meeresabschnitte, **19**: 1-831.

GNANAMUTHU, C. P., 1947. *Caligus sciaenae* n. sp. parasitic on *Sciaena glauca* from Madras. Proceedings of the Indian Academy of Sciences, **25B**: 43-49.

— —, 1948. Notes on the anatomy and physiology of *Caligus savala*, n. sp., a parasitic copepod from Madras plankton. Proceedings of the Zoological Society of London, **118**: 591-606.

— —, 1950a. Two dichelesthiid copepods from Madras fish. Parasitology, **40**: 276-282.

— —, 1950b. *Synestius caliginus* Steenstrup and Lütken, a copepod parasite of the grey pomfret. Records of the Indian Museum, **47**: 253-258.

— —, 1951. New copepod parasites of sharks. Annals and Magazine of Natural History, (12) **4**: 1236-1256.

GRIMNES, A. & P. J. JAKOBSEN, 1996. The physiological effects of salmon lice infection on post-smolt of Atlantic salmon. Journal of Fish Biology, **48**: 1179-1194.

GURNEY, R., 1933. British fresh-water Copepoda, **III**: 1-384. (Ray Society, London).

— —, 1934. The development of certain Copepoda of the families Caligidae and Clavellidae. Proceedings of the Zoological Society of London, **1934**: 177-216.

— —, 1947. Some remarks on the morphology of the Copepoda. Annals and Magazine of Natural History, **11**: 493-500.

HALLETT, S. L. & F. R. ROUBAL, 1995. Experiments on the infection dynamics of *Caligus epidemicus* (Copepoda: Caligidae) on the small marine fish, *Ambassis marinus* (Günther). Journal of Fish Diseases, **18**: 59-66.

HAMEED, M. S., 1976. Description of a new species of *Anuretes* (Copepoda: Caligidae) from Kerala. Hydrobiologia, **51**: 39-42.

HAMEED, M. S. & N. K. PILLAI, 1973a. Redescription of *Mappates plataxus* Rangnekar (Copepoda: Caligidae). Hydrobiologia, **42**: 403-411.

— — & — —, 1973b. Description of a new species of *Caligus* (Crustacea: Copepoda) from Kerala. Zoologischer Anzeiger, **191**: 114-118.

HANSEN, H. J., 1923. Crustacea Copepoda II. Copepoda parasita and hemiparasita. Danish Ingolf Expedition, **3**: 1-92.

HAVEN, N., 1972. The ecology and behavior of *Nautilus pomilius* in the Philippines. Veliger, **15**: 75-80.

HAYWARD, C. J., N. J. BOTT & B. F. NOWAK, 2009. Seasonal epizootics of sea lice (*Caligus* spp.) on southern Bluefin tuna (*Thunnus maccoyii*) in a long-term trial. Journal of Fish Diseases, **32**: 101-106.

HEEGAARD, P. E., 1943. Parasitic copepods mainly from tropical and Antarctic seas. Arkiv för Zoologi, **34A**: 1-37.

— —, 1945a. Remarks on the phylogeny of the arthropods. Arkiv för Zoologi, **37A**: 1-15.

— —, 1945b. Some parasitic copepods from fishes in the Uppsala University collections. Arkiv för Zoologi, **35A**: 1-27.

— —, 1947a. Contribution to the phylogeny of the arthropods. Copepoda. Spolia Zoologica Musei Hauniensis, **8**: 1-236.

— —, 1947b. Discussion of the mouth appendages of the copepods. Arkiv för Zoologi, **40A**: 1-8.

— —, 1955. Parasitic copepods from tropical West Africa. Scientific Results of the Danish Expedition to the Coasts of Tropical West Africa. Atlantide Report No. 3, **1945-1946**: 41-56.

— —, 1962. Parasitic Copepoda from Australian waters. Records of the Australian Museum, **25**: 149-234.

— —, 1966. Parasitic copepods from Texas. Videnskabelige Meddelelser fra Dansk Naturhistorisk Forening i Kjøbenhavn, **129**: 187-197.

— —, 1972. Caliginae and Euryphorinae of the Dana Expedition (Crustacea, Copepoda, Caligidae). Steenstrupia, **2**: 295-317.

HELLER, C., 1865. Crustaceen. Reise der Osterreichischen Fregatte Novara um die Erde in den Jahren 1857, 1858, 1859 unter den Befehlen des Commodore B. von Wüllerstorf-Urbair. Zoologischen Theil, **2**: 1-280.

— —, 1866. Carcinologische Beiträge zur Fauna des Adriatischen Meeres. Verhandlungen der Kaiserlich-Königlichen Zoologisch-botanischen Gesellschaft in Wien, **16**: 723-760.

HEWITT, G. C., 1963. Some New Zealand parasitic Copepoda of the family Caligidae. Transactions of the Royal Society of New Zealand, Zoology, **4**: 61-115.

— —, 1964a. The postchalimus development of *Lepeophtheirus polyprioni* Hewitt, 1963 (Copepoda: Caligidae). Transactions of the Royal Society of New Zealand, Zoology, **4**: 157-159.

— —, 1964b. A redescription of *Gloiopotes huttoni* (Thomson, 1889), with a key to the species of the genus. Transactions of the Royal Society of New Zealand, Zoology, **5**: 85-96.

— —, 1968. *Elytrophora brachyptera* Gerstaecker (Euryphoridae, Caligoida) from New Zealand waters, with a tentative revision of the genus. Transactions of the Royal Society of New Zealand, Zoology, **10**: 117-126.

— —, 1971a. Two new species of *Caligus* (Copepoda, Caligidae) from Australian waters, with a description of some developmental stages. Pacific Science, **25**: 145-164.

— —, 1971b. Species of *Lepeophtheirus* (Copepoda: Caligidae) recorded from the sunfish (*Mola mola*) and their implications for the caligid genus *Dentigryps*. Journal of the Fisheries Research Board of Canada, **28**: 323-334.

HO, J.-S., 1963. On five species of Formosan parasitic copepods belonging to the suborder Caligoida. Crustaceana, **5**: 81-98.

— —, 1966. Redescription of *Echetus typicus* Krøyer, a caligid copepod parasitic on the red drum, *Sciaenops ocellatus* (Linnaeus). Journal of Parasitology, **52**: 752-761.

— —, 1970. Systematic status of *Eirgos anurus* Bere, 1936, a caligoid copepod parasitic on the spadefish. Crustaceana, **18**: 107-109.

— —, 1980. *Anchicaligus nautili* (Willey), a caligid copepod parasitic on *Nautilus* in Palau, with discussion of *Caligulina* Heegaard, 1972. Publications of the Seto Marine Biological Laboratory, **25**: 157-165.

— —, 1990. Phylogenetic analysis of copepod orders. Journal of Crustacean Biology, **10**: 528-536.

— —, 1998. Maxillopoda – Copepoda, Siphonostomatoida. In: P. S. YOUNG (ed.), Catalogue of Crustacea of Brazil: 251-260. (Série Livros n. 6, Rio de Janeiro: Museu Nacional).

— —, 2000. The major problem of cage aquaculture in Asia relating to sea lice. In: I. C. LIAO, & C. K. LIN (eds.), Proceedings of the First International Symposium on Cage Aquaculture in Asia; 2-6 Nov. 1999, Tungkang, Manila: Asian Fisheries Society and Bangkok: World Aquaculture Society-Southeast Asian Chapter: 13-19.

HO, J.-S. & A. K. M. BASHIRULLAH, 1977. Two species of caligid copepods (Crustacea) parasitic on marine fishes of Venezuela, with discussion of *Metacaligus* Thomsen, 1949. Journal of Natural History, **11**: 703-714.

HO, J. S., S. H. CHENG, W. C. CHIANG & C. L. LIN, 2008. Two species of siphonostomatoid copepods parasitic on marine fishes of Taiwan. Journal of the Fisheries Society of Taiwan, **35**: 87-99.

HO, J.-S. & M. DOJIRI, 1977. Parasitic copepods on the fishes of the Great Barrier Reef, Australia. Part II. Caligoida: *Dissonus*, *Lepeophtheirus*, and *Dentigryps*. Publications of the Seto Marine Biological Laboratory, **24**: 77-97.

HO, J. S. & I. H. KIM, 2000. Copepods of *Hermilius* (Caligidae) parasitic on sea catfish of Kuwait, with a key to the species of *Hermilius*. Parkistan Journal of Marine Science, **9**: 79-90.

HO, J. S., I. H. KIM, E. R. CRUZ-LACIERDA & K. NAGASAWA, 2004. Sea lice (Copepoda, Caligidae) parasitic on marine cultured and wild fishes of the Philippines. Journal of the Fisheries Society of Taiwan, **31**: 235-249.

HO, J. S. & C. L. LIN, 2000. *Anuretes grandis* sp. n., a caligid copepod (Siphonostomatoida) parasitic on *Diagramma pictum* (Pisces) in Taiwan, with discussion of *Anuretes* Heller, 1865. Folia Parasitologica, **47**: 227-234.

— — & — —, 2001. *Parapetalus occidentalis* Wilson (Copepoda, Caligidae) parasitic on both wild and farmed cobia (*Rachycentron canadum*) in Taiwan. Journal of the Fisheries Society of Taiwan, **28**: 305-316.

— — & — —, 2002. New species of *Metacaligus* (Caligidae, Copepoda) parasitic on the cutlassfish (*Trichiurus lepturus*) of Taiwan, with a cladistic analysis of the family Caligidae. Zoological Science, **19**: 1363-1375.

— — & — —, 2004. Sea lice of Taiwan (Copepoda: Siphonostomatoida: Caligidae: 1-388. (The Sueichan Press, Taiwan).

— — & — —, 2010a. The last instar of male *Euryphorus* Milne Edwards, 1840 (Copepoda: Euryphoridae) found on marine fishes of Taiwan. Journal of the Fisheries Society of Taiwan, **37**: 111-120.

— — & — —, 2010b. Two species of *Parapetalus* (Copepoda: Caligidae) parasitic on marine fishes of Taiwan. Journal of the Fisheries Society of Taiwan, **37**: 231-238.

HO, J. S. & K. NAGASAWA, 2001. New records of parasitic Copepoda from the offshore pelagic fishes of Japan. Bulletin of the National Research Institute of Far Seas Fisheries, **38**: 1-5.

HO, J. S. & O. SEY, 1996. Parasitic Copepoda of marine fishes from Kuwait: a preliminary report. Kuwait Journal of Science and Engineering, **23**: 61-69.

HOGAN, W. E. & D. J. TRUDEAU, 1989. Preliminary studies on the biology of sea lice, *Caligus elongatus*, *Caligus curtus* and *Lepeophtheirus salmonis* (Copepoda: Caligoida) parasitic on cage-cultured salmonids in the lower Bay of Fundy. Canadian Technical Report of Fisheries and Aquatic Sciences, **1715**: 1-14.

HUMES, A. G., 1957. Two new caligoid copepods from Egyptian fishes. Journal of Parasitology, **43**: 201-208.

— —, 1965. *Sciaenophilus inopinus* Humes, 1957, a synonym of *Sciaenophilus pharaonis* (Nordmann, 1832) comb. n. (Copepoda: Caligidae). Journal of Parasitology, **51**: 1009-1010.

HUMES, A. G. & M. DOJIRI, 1984. Current concepts of taeniacanthid copepods associated with Echinoidea. In: Studies on Copepoda II, Proceedings from the First International Conference on Copepoda. Crustaceana, (Supplement) **7**: 258-266.

HUMES, A. G. & R. U. GOODING, 1964. A method for studying the external anatomy of copepods. Crustaceana, **6**: 238-240.

HUYS, R. & G. A. BOXSHALL, 1991. Copepod evolution, **159**: 1-468. (Ray Society, London).

HUYS, R., F. FATIH, S. OHTSUKA & J. LLEWELLYN-HUGHES, 2012. Evolution of the bomolochiform superfamily complex (Copepoda: Cyclopoida): New insights from ssrDNA and morphology, and origin of umazuracolids from polychaete-infesting ancestors rejected. International Journal for Parasitology, **42**: 71-92.

HWA, T.-K., 1965. Studies on the life history of a fish-louse (*Caligus orientalis* Gussev) [sic]. Acta Zoologica Sinica, **17**: 48-57.

IZAWA, K., 1969. Life history of *Caligus spinosus* Yamaguti, 1939 obtained from cultured yellowtail, *Seriola quinqueradiata* T. & S. (Crustacea: Caligoida). Report of the Faculty of Fisheries, Prefectural University of Mie, **6**: 127-157.

— —, 1995. A new fish parasite (Copepoda: Siphonostomatoida: Caligidae) from the Timor Sea, Australia. The Beagle, Records of the Museums and Art Galleries of the Northern Territory, **12**: 185-192.

JOHANNESSEN, A., 1978. Early stages of *Lepeophtheirus salmonis* (Copepoda, Caligidae). Sarsia, **63**: 169-176.

JOHANNESSEN, A., P. O. BRANDAL & E. EGIDIUS, 1976. On the biology of the parasitic copepod *Lepeophtheirus salmonis* Krøyer. Norwegian Journal of Zoology, **24**: 467.

JOHN, S. E. & N. B. NAIR, 1974. Functional morphology of the mouth tube of *Hermilius longicornis* Bassett-Smith, a parasitic copepod on *Ariodes dussumieri* Valenciennes. Monitore Zoologico Italiano, **8**: 133-140.

JOHNSON, S. C., R. B. BLAYLOCK, J. ELPHICK & K. HYATT, 1996. Disease induced by the salmon louse *Lepeophtheirus salmonis* (Copepoda: Caligidae) in wild sockeye salmon (*Oncorhynchus nerka*) stocks of Alberni Inlet, British Columbia. Canadian Journal of Fisheries and Aquatic Sciences, **53**: 2888-2897.

JOHNSON, S. C., J. W. TREASURER, S. BRAVO, K. NAGASAWA & Z. KABATA, 2004. A review of the impact of parasitic copepods on marine aquaculture. Zoological Studies, **43**: 229-243.

JONES, J. B., 1991. Parasitic copepods of albacore tuna (*Thunnus alalunga*) in the South Pacific. Bulletin of the Plankton Society of Japan, Special Volume: 419-428.

KABATA, Z., 1958. *Lernaeocera obtusa* n. sp., its biology and its effects on the haddock. Marine Research 1958, **3**: 1-26.

— —, 1964a. Copepoda parasitic on Australian fishes. II. *Mappates alter* sp. nov. (Caligidae). Annals and Magazine of Natural History, (13) **7**: 641-649.

— —, 1964b. Copepoda parasitic on Australian fishes. I. *Hermilius youngi* sp. nov. (Caligidae). Annals and Magazine of Natural History, (13) **7**: 609-618.

— —, 1965a. Copepoda parasitic on Australian fishes. III. Genera *Dentigryps*, *Heniochophilus*, and *Pseudanuretes* (Caligidae). Annals and Magazine of Natural History, (13) **8**: 19-31.

— —, 1965b. Copepoda parasitic on Australian fishes. IV. Genus *Caligus* (Caligidae). Annals and Magazine of Natural History, (13) **8**: 109-126.

— —, 1966a. Copepoda parasitic on Australian fishes. V. Genus *Dissonus* (Dissonidae). Annals and Magazine of Natural History, (13) **9**: 211-226.

— —, 1966b. Copepoda parasitic on Australian fishes. VI. Some caligoid species. Annals and Magazine of Natural History, (13) **9**: 563-570.

— —, 1970. Crustacea as enemies of fishes. In: S. F. SNIESZKO & R. R. AXELROD (series eds.), Diseases of fishes: 1-171. (T. F. R. Publications, Jersey City, New Jersey).

— —, 1972. Developmental stages of *Caligus clemensi* (Copepoda: Caligidae). Journal of the Fisheries Research Board of Canada, **29**: 1571-1593.

— —, 1973a. The species of *Lepeophtheirus* (Copepoda: Caligidae) from fishes of British Columbia. Journal of the Fisheries Research Board of Canada, **30**: 729-759.

— —, 1973b. Taxonomic status of *Homoiotes palliata* Wilson (Copepoda: Caligidae). Journal of the Fisheries Research Board of Canada, **30**: 1892-1893.

— —, 1974. Mouth and mode of feeding of Caligidae (Copepoda), parasites of fishes, as determined by light and scanning electron microscopy. Journal of the Fisheries Research Board of Canada, **31**: 1583-1588.

— —, 1979. Parasitic Copepoda of British fishes, **152**: 1-468. (Ray Society, London).

— —, 1981. Copepoda (Crustacea) parasitic on fishes: problems and perspectives. In: Advances in Parasitology, **19**: 1-71.

— —, 1988. Copepoda and Brachiura. In: L. MARGOLIS & Z. KABATA (eds.), Guide to the parasites of fishes of Canada. Part II. Crustacea. Canadian Special Publication in Fisheries and Aquatic Sciences, **101**: 3-127.

— —, 2003. Copepods parasitic on fishes. In: J. H. CROTHERS & P. J. HAYWARD (eds.), Synopses of the British Fauna, (New Series) **47** (Revised): 1-274.

KABATA, Z. & G. B. DEETS, 1988. *Pseudanuretes papernai* sp. nov. (Copepoda: Caligidae) with comments on the genus *Pseudanuretes* Yamaguti, 1936. Canadian Journal of Zoology, **66**: 680-684.

KABATA, Z. & A. V. GUSEV, 1966. Parasitic Copepoda of fishes from the collection of the Zoological Institute in Leningrad. Journal of the Linnean Society, Zoology, **46**: 155-207.

KABATA, Z. & G. C. HEWITT, 1971. Locomotory mechanisms in Caligidae (Crustacea: Copepoda). Journal of the Fisheries Research Board of Canada, **28**: 1143-1151.

KAZACHENKO, V. N., 1976. Parazititcheskie rakoobraznye (Copepoda) ryb tropicheskoĭ chasti Tikhogo i Indiĭskogo Okeanov. Izvestiya Tikhookeanskogo Nauchno-issledovatel'skogo Instituta Rў̆bnogo Khozyaĭstva i Okeanografii (TINRO), **98**: 211-217.

— —, 2001. Key to families and genera of parasitic copepods (Crustacea: Copepoda) of fishes. Part I: 1-161, Part II: 1-253. (Dalrybtuz, Vladivostk). [In Russian.]

KAZACHENKO, V. N. & G. V. AVDEEV, 1977. Parazititcheskie kopepody (Copepoda, Crustacea) v sborakh 57-go reĭsa nis Vityaz v zapadnoĭ tropicheskoĭ chasti Tikhogo Okeana i moryakh Indo-Malaĭskogo Arkhipelago. Trudy Instituta Okeanologii. Akademiya Nauk SSR, **107**: 30-48.

KENSLEY, B. & J. R. GRINDLEY, 1973. South African parasitic Copepoda. Annals of the South African Museum, **62**: 69-130.

KIM, I. H., 1993. Developmental stages of *Caligus punctatus* Shiino, 1955 (Copepoda: Caligidae). In: G. A. BOXSHALL & D. DEFAYE (eds.), Pathogens of Wild and Farmed Fish: Sea Lice: 16-29. (Ellis Horwood, London).

— —, 1998. Cirripedia, symbiotic Copepoda, and Pycnogonida. In: Illustrated Encyclopedia of Fauna & Flora of Korea, **38**. Cirripedia, Symbiotic Copepoda, Pycnogonida. Ministry of Education, Seoul: 271-835. [In Korean.]

KIRTISINGHE, P., 1937. Parasitic copepods of fish from Ceylon. II. Parasitology, **29**: 435-452.

— —, 1950. Parasitic copepods of fish from Ceylon. III. Parasitology, **40**: 76-86.

— —, 1964. A review of the parasitic copepods of fish recorded from Ceylon with descriptions of additional forms. Bulletin of the Fisheries Research Station, Department of Fisheries, Ceylon, **17**: 45-132.

KNOFF, M., J. L. LUQUE & R. M. TAKEMOTO, 1994. Parasitic copepods on *Mugil platanus* Günther (Osteichthyes: Mugilidae) from the coast of the state of Rio de Janeiro, Brazil. Revista Brasileira de Parasitologia Veterinária, **3**: 45-56.

KRØYER, H., 1837. Om snyltekrebsene, isaer med hensyn til den danske fauna. Naturhistorisk Tidsskrift, **1**: 172-208; 605-628.

— —, 1838. Om snyltekrebsene, isaer med hensyn til den danske fauna. III. Formbeskrivelser (Fortsaettelse). Naturhistorisk Tidsskrift, **2**: 8-52.

— —, 1863. Bidrag til kundskab om snylterkrebsene. Naturhistorisk Tidsskrift, **3**: 75-320.

— —, 1864. Bidrag til kundskab om snylterkrebsene. Naturhistorisk Tidsskrift, **3**: 321-426.

KUBOTA, S. S. & M. TAKAKUWA, 1963. Studies on the diseases of marine-culture fishes – I. General description and preliminary discussion of fish diseases at Mie Prefecture. Journal of the Faculty of Fisheries, Prefectural University of Mie, **6**: 107-124.

LAMARCK, J. B. P. DE, 1818. Histoire naturelle des animaux sans vertébrés, **5**: 1-612. (Déterville et Verdiére, Paris).

LANG, K., 1947. A contribution to the question of the mouth parts of Copepoda. Archiv för Zoologi, **38A**: 1-24.

— —, 1951. On the nature of the so-called sternal furca in the Caligidae. Arkiv för Zoologi, (1) **32**: 499-502.

LAUBIER, L., 1968. Sur deux copépodes parasites de la raie *Mobula mobular* en Méditerranée occidentale. Crustaceana, (Supplement) **1**: 159-171.

LAVERACK, M. S. & M. Q. HULL, 1993. Sensory innervation of the antennule of the preadult male *Caligus elongatus*. In: G. A. BOXSHALL & D. DEFAYE (eds.), Pathogens of Wild and Farmed Fish: Sea Lice: 114-124. (Ellis Horwood, London).

LEIGH-SHARPE, W. H., 1933. A list of British fishes with their characteristic parasitic copepods. Parasitology, **25**: 109-112.

— —, 1934. The Copepoda of the Siboga Expedition. Part II. Commensal and parasitic Copepoda. Siboga-Expeditie Monographe, **29b** (123): 1-42.

LEONG, T. S., 1985. Copepoda of the genera *Parapetalus*, *Synestius* and *Hermilius* (Caligidae) parasitic on Malaysian fishes. Malayan Nature Journal, **39**: 41-56.

LEWIS, A. G., 1963. Life history of the caligid copepod *Lepeophtheirus dissimulatus* Wilson, 1905 (Crustacea; Caligoida). Pacific Science, **17**: 195-242.

— —, 1964a. Caligoid copepods (Crustacea) of the Hawaiian Islands: parasitic on fishes of the family Acanthuridae. Proceedings of the United States National Museum, **115**: 137-244.

— —, 1964b. The caligid copepod genus *Dentigryps* (Crustacea: Caligoida). Proceedings of the United States National Museum, **115**: 347-380.

— —, 1966a. The sternal furca of caligoid copepods. Crustaceana, **10**: 7-14.

— —, 1966b. Copepod crustaceans parasitic on elasmobranch fishes of the Hawaiian Islands. Proceedings of the United States National Museum, **118**: 57-154.

— —, 1967. Copepod crustaceans parasitic on teleost fishes of the Hawaiian Islands. Proceedings of United States National Museum, **121**: 1-204.

——, 1968. Copepod crustaceans parasitic on fishes of Eniwetok Atoll. Proceedings of the United States National Museum, **125**: 1-78.

——, 1969. A discussion of the maxillae of the "Caligoidea" (Copepoda). Crustaceana, **16**: 65-77.

LEWIS, A. G., J. DEAN & E. GILFILLAN III, 1969. Taxonomy and host associations of some parasitic copepods (Crustacea) from pelagic teleost fishes. Pacific Science, **23**: 414-437.

LIN, C. L., 1991. The development of Copepoda – *Caligus* parasitic on sea water cultured *Oreochromis mossambica*: 1-66. (Master's Thesis, National Taiwan University, Taipei, Taiwan). [In Chinese.]

LIN, C. L. & J. S. HO, 1993. Life history of *Caligus epidemicus* Hewitt parasitic on tilapia (*Oreochromis mossambicus*) cultured in brackish water. In: G. A. BOXSHALL & D. DEFAYE (eds.), Pathogens of Wild and Farmed Fish: Sea Lice: 5-15. (Ellis Horwood, London).

—— & ——, 2000. Four species of unrecorded caligid copepods (Siphonostomatoida) parasitic on marine fishes of Taiwan. Journal of the Fisheries Society of Taiwan, **27**: 201-224.

—— & ——, 2002a. Another four species of unrecorded sea lice (Copepoda, Caligidae) parasitic on marine fishes of Taiwan. Journal of the Fisheries Society of Taiwan, **29**: 45-46.

—— & ——, 2002b. Two species of siphonostomatoid copepods parasitic on pelagic fishes of Taiwan. Journal of the Fisheries Society of Taiwan, **29**: 313-332.

LIN, C. L., J. S. HO & S. N. CHEN, 1997. Development of *Caligus multispinosus* Shen, a caligid copepod parasitic on the black sea bream (*Acanthopagrus schlegeli*) cultured in Taiwan. Journal of Natural History, **31**: 1483-1500.

LOPEZ, G., 1976. Redescription and ontogeny of *Lepeophtheirus kareii* Yamaguti, 1936 (Copepoda, Caligoida). Crustaceana, **31**: 203-207.

LÜLING, K. H., 1953. Gewebeschäden durch parasitäre Copepoden besonders durch *Elytrophora bachyptera* [sic] (Gerstaecker). Zeitschrift für Parasitenkunde, **16**: 84-92.

LUQUE, J. L., N. D. CHAVES & A. D. CEZAR, 1999. Novos registros de copépodes caligóideos parasitos de peixes marinhos do Brasil. Nauplius, Rio Grande, **6**: 9-16.

MADDISON, W. P. & D. R. MADDISON, 2010. Mesquite: a modular system for evolutionary analysis. Version 2.74 (Build 550) http://mesquiteproject.org.

MADINABEITIA, I. & K. NAGASAWA, 2011. Chalimus stages of *Caligus latigenitalis* (Copepoda: Caligidae) parasitic on blackhead seabream from Japanese waters, with discussion of terminology used for developmental stages of caligids. Journal of Parasitology, **97**: 221-236.

MANN, H., 1970. Copepoda and Isopoda as parasites of marine fishes. In: S. F. SNIESZKO (ed.), A Symposium on Diseases of Fishes and Shellfishes. Transactions of the American Fisheries Society, Special Publication no. **5**: 177-189.

MARGOLIS, L., Z. KABATA & R. R. PARKER, 1975. Catalogue and synopsis of *Caligus*, a genus of Copepoda (Crustacea) parasitic on fishes. Bulletin of the Fisheries Research Board of Canada, **192**: 1-117.

MARKEWITSCH, A. P., 1940. Novi vidi parazitichnykh Copepoda. Dopovidi Akademiyi Nauk, U.R.S.R., **11**: 13-21.

——, 1956. Paraziticheskie veslonogie ryb SSSR. [Parasitic copepods on the fishes of the USSR]: 1-259. (Izdatel'stvo Akademii Nauk Ukrainskoi SSR, Kiev). National Scientific Documentation Centre, Hillside Road, New Delhi. [Translated from Russian.]

MILNE EDWARDS, H., 1840. Histoire naturelle des Crustacés; comprenant l'anatomie, la physiologie et la classification de ces animaux, **3**: 1-638. (Paris).

MONOD, T. & R. DOLLFUS, 1932. Les copépodes parasites de mollusques. Annales de Parasitologie Humaine et Comparée, **10**: 129-204.

MORALES-SERNA, F. N., S. GÓMEZ & G. P.-P. DE LEÓN, 2012. Parasitic copepods reported from Mexico. Zootaxa, **3234**: 43-68.

MÜLLER, O. F., 1776. Zoologiae Danicae prodromus, seu animalium Daniae et Norvegiae indigenarum characteres, nomina, et synonyma imprimis popularium: i-xxxii, 1-274. (Havniae).

— —, 1785. Entomostraca, seu Insecta testacea quae in aquis Daniae et Novegiae resperit. XI. *Caligus*: 128-134. (Leipzig and Copenhagen).

— —, 1788. Zoologia Danica seu animalium Daniae et Norvegiae rariorum ac minus notorum descriptiones et historia, **1**: 1-52. (Havniae).

NORDMANN, A. VON, 1832. Mikrographische Beiträge zur Naturgeschichte der wirbellosen Thiere. Heft 2, I-XVIII: 1-150. (G. Reimer, Berlin).

— —, 1864. Neue Beiträge zur Kenntniss parasitischer Copepoden. Bulletin de la Société des Naturalistes de Moscou, **37**: 461-520.

NORMAN, A. M. & T. SCOTT, 1906. The Crustacea of Devon and Cornwall: 1-232. (William Wesley and Son, London).

NUNES-RUIVO, L., 1962. Copépodes parasites de poissons des côtes d'Angola. Memórias da Junta de Investigações do Ultramar, (2) **33**: 67-86.

NUNES-RUIVO, L. & P. FOURMANOIR, 1956. Copépodes parasites de poissons de Madagascar. Mémoires de l'Institut Scientifique de Madagascar, (A) **10**: 69-80.

OGAWA, K. & H. YOKOYAMA, 1998. Parasitic diseases of cultured marine fish. Fish Pathology, **33**: 303-309.

OHTSUKA, S., I. TAKAMI, B. A. VENMATHI MARAN, K. OGAWA, T. SHIMONO, Y. FUJITA, M. ASAKAWA & G. A. BOXSHALL, 2009. Developmental stages and growth of *Pseudocaligus fugu* Yamaguti, 1936 (Copepoda: Siphonostomatoida: Caligidae) host-specific to puffer. Journal of Natural History, **43**: 1779-1804.

OLSSON, P., 1869. Prodromus faunae copepodorum parasitantium Scandinaviae. Acta Universitatis Lundensis, **8**: 1-49.

OORDE-DE LINT, G. M. VAN & J. H. SCHUURMANS STEKHOVEN, 1936. Copepoda Parasitica. Tierwelt der Nord-und Ostsee, **31**: 73-197.

O'RIORDAN, C. E., 1966. Parasitic copepods in the collections of the National Museum of Ireland. Proceedings of the Royal Irish Academy, (B) **64**: 371-378.

ÖZDIKMEN, H., 2008. Nomenclatural changes for nine crustacean genera (Crustacea: Copepoda). Munis Entomology and Zoology, **3**: 265-274.

PAPERNA, I., 1964. Parasitic Crustacea (Copepoda and Branchiura) from inland water fishes of Israel. Israel Journal of Zoology, **13**: 58-68.

PAPERNA, I. & M. LAHAV, 1974. Mortality among grey mullets in a seawater pond due to caligid parasitic copepod epizootic. Bamidgeh, **26**: 12-15.

PARKER, R. R., Z. KABATA, L. MARGOLIS & M. D. DEAN, 1968. A review and description of *Caligus curtus* Müller, 1785 (Caligidae: Copepoda), type species of its genus. Journal of the Fisheries Research Board of Canada, **25**: 1923-1969.

PARKER, R. R. & L. MARGOLIS, 1964. A new species of parasitic copepod *Caligus clemensi* sp. nov. (Caligoida: Caligidae), from pelagic fishes in the coastal waters of British Columbia. Journal of the Fisheries Research Board of Canada, **21**: 873-889.

PEARSE, A. S., 1947. Parasitic copepods from Beaufort, North Carolina. Journal of the Elisha Mitchell Scientific Society, **63**: 1-16.

— —, 1948. A second report on parasitic copepods collected at Beaufort, North Carolina. Journal of the Elisha Mitchell Scientific Society, **64**: 127-131.

— —, 1951. Parasitic Crustacea from Bimini, Bahamas. Proceedings of the United States National Museum, **101**: 341-372.

— —, 1952. Parasitic Crustacea from the Texas coast. Publications of the Institute of Marine Science, University of Texas, **2**: 5-42.

PEARSON, J., 1906. A list of the marine Copepoda of Ireland. Part 1. – Littoral forms and fish parasites. Report on the Sea and Inland Fisheries of Ireland for 1904, **2**: 143-170.

PESTA, O., 1934. Krebstiere oder Crustacea. I. Ruderfüssler oder Copepoda. 6. Caligoida. Tierwelt Deutschlands und der Angrenzenden Meeresteile, **29**: 13-34.

PIASECKI, W., 1993. The types of Copepoda parasitic on fishes deposited in the Zoologica Museum Berlin. Mitteilungen aus dem Zoologischen Museum in Berlin, **69**: 165-173.

PIASECKI, W. & B. M. MACKINNON, 1993. Changes in structure of the frontal filament in sequential developmental stages of *Caligus elongatus* von Nordmann, 1832 (Crustacea, Copepoda, Siphonostomatoida). Canadian Journal of Zoology, **71**: 889-895.

PIKE, A. W. & S. L. WADSWORTH, 1999. Sea lice on salmonids: their biology and control. Advances in Parasitology, **44**: 233-337.

PIKE, A. W., K. MACKENZIE & A. ROWAND, 1993. Ultrastructure of the frontal filament in chalimus larvae of *Caligus elongatus* and *Lepeophtheirus salmonis* from Atlantic salmon, *Salmo salar*. In: G. A. BOXSHALL & D. DEFAYE (eds.), Pathogens of Wild and Farmed Fish: Sea Lice: 99-113. (Ellis Horwood, London).

PILLAI, N. K., 1961. Copepods parasitic on South Indian fishes. Pt. 1. Caligidae. Bulletin of the Central Research Institute, University of Kerala, (C) **8**: 87-130.

— —, 1962. A revision of the genera *Parapetalus* Steenstrup and Lütken and *Pseudopetalus* nov. Crustaceana, **3**: 285-303.

— —, 1963a. Observations on the genus *Abasia* (Copepoda) with the description of a new species. Crustaceana, **5**: 1-9.

— —, 1963b. A revision of the genus *Hermilius* (Copepoda: Caligidae). Journal of the Zoological Society of India, **14**: 179-187. [Listed as 1962, but published in 1963.]

— —, 1963c. Copepods parasitic on South Indian fishes – family Caligidae. Journal of the Marine Biological Association of India, **5**: 68-96.

— —, 1964a. A miscellaneous collection of copepods parasitic on South Indian fishes. Journal of the Marine Biological Association of India, **6**: 61-83.

— —, 1964b. A redescription of the copepod *Pupulina minor* M. S. Wilson. Zoologischer Anzeiger, **173**: 237-242.

— —, 1966. Notes on copepods parasitic on South Indian marine fishes. Journal of the Marine Biological Association of India, **8**: 123-140.

— —, 1967. Description of a new species of *Anuretes* (Copepoda: Caligidae) and comments on the validity of a few caligid genera. Zoologischer Anzeiger, **178**: 358-367.

— —, 1968. Additions to the copepod parasites of South Indian fishes. Parasitology, **58**: 9-36.

— —, 1969. Notes on some copepod parasites in the collection of the British Museum (N.H.), London. Journal of the Marine Biological Association of India, **11**: 149-174.

— —, 1977. Copepods parasitic on *Platax tiera* in Indian waters. Aquatic Biology, **2**: 53-60.

— —, 1985. Fauna of India, copepod parasites of marine fishes: 1-900. (Zoological Survey of India, Calcutta).

PILLAI, N. K. & R. S. LAL MOHAN, 1965. Observations on the genus *Heniochophilus* (Copepoda) with a redescription of the type species. Journal of the Marine Biological Association of India, **7**: 270-276.

PILLAI, N. K. & P. NATARAJAN, 1977. Copepods parasitic on fishes of the Kerala coast. Aquatic Biology, **2**: 19-43.

PILLAI, N. K. & K. G. PADMANABHAN, 1963. A redescription of *Papulina* [sic] *brevicauda* M. S. Wilson (Copepoda: Caligoida) with observations on the genus *Diphyllogaster* Brian. Journal of the University of Bombay, **31**: 91-99.

POCHE, F., 1902. Bemerkungen zu der Arbeit des Herrn Bassett-Smith: "A Systematic Description of Parasitic Copepoda found on Fishes, with an Enumeration of the known Species". Zoologischer Anzeiger, **26**: 8-20.

POLY, W. J. & C. L. MAH, 2002. New host and distribution records for parasitic copepods in the northeast Pacific Ocean with a discussion of taxonomy of the genus *Acanthochondria*. Bulletin of Marine Science, **69**: 1121-1127.

PRABHA, C., 1983. A discussion on the genera of Caligidae (Copepoda). In: P. A. JOHN (ed.), Selected Papers on Crustacea. Prof. (Dr.) N. Krishna Pillai Felicitation Volume: Prof. N. Krishna Pillai Farewell Committee, Trivandrum: 49-68.

PRABHA, C. & N. K. PILLAI, 1979. *Pseudechetus fimbriatus* gen. et sp. nov., a caligid copepod from Kerala coastal waters. Parasitology, **79**: 425-429.

— — & — —, 1983. Additions to the copepods parasitic on the marine fishes of India. 1. On twelve species of caligids. Records of the Zoological Survey of India, Occasional Paper No. **46**: 1-49.

— — & — —, 1984. Additions to the copepods parasitic on the marine fishes of India. 2. *Kabataella indica* gen. et sp. nov. (Caligoida). Records of the Zoological Survey of India, **81**: 1-5.

— — & — —, 1986. Additions to the copepods parasitic on the marine fishes of India 4. On twenty-six species of caligids. Records of the Zoological Survey of India, Miscellaneous Publication, Occasional Paper, **79**: 1-139.

PRICE, C. E., 1966. A new species of parasitic copepod from the Caribbean. Caribbean Journal of Science, **6**: 119-122.

PURIVIROJKUL, W. & N. AREECHON, 2008. Survey of parasitic copepods in marine fishes from the Gulf of Thailand, Chon Buri province. Proceedings of the 46th Kasetsart University Annual Conference, Bangkok, Thailand, 29 Jan-1 Feb 2008: 71-80. [In Thai; abstract in English.]

QUIDOR, A., 1913. Affinités des Caligidae et des Lernaeidae. *Caligodes lamarcki*. Bulletin de la Société Zoologique de France, **38**: 191-196.

RADHAKRISHNAN, S. & N. B. NAIR, 1983. Nature of crustacean infestation of fishes along the south-west coast of India. I. Distribution, mode of attachment to the host tissue and incidence and intensity of infestation. Acta Ichthyologica et Piscatoria, **13**: 93-115.

RANGNEKAR, M. P., 1953. *Anuretes branchialis* sp. nov. a copepod parasitic on the fish *Thynnus pelamys*. Journal of the Zoological Society of India, **5**: 239-242.

— —, 1955. *Pseudocaligus laminatus* sp. nov. and *Diphyllogaster aliuncus* sp. nov. (Copepoda), parasitic on Bombay fishes. Journal of the University of Bombay, **23**: 44-52.

— —, 1956. Parasitic copepods from marine fishes of Bombay. Journal of the University of Bombay, **24**: 42-65.

— —, 1957. Copepod parasites of the families Argulidae, Caligidae, Dichelesthidae, and Lernaeopodidae. Journal of the University of Bombay, **26**: 8-20.

— —, 1958. *Mappates plataxus* gen. et sp. nov., a copepod parasite of *Platax teira* (Forsskål). Records of the Indian Museum, **53**: 303-308.

— —, 1959. Parasitic copepods from fishes of the western coast of India with description of four known species. Journal of the University of Bombay, **28**: 43-58.

— —, 1960. *Tuxophorus caligodes* Wilson, a copepod parasitic on a local fish. Journal of Biological Sciences, **3**: 20-25.

— —, 1984. Parasitic copepods from marine fishes of Bombay. In: W. VERVOORT (ed.), Studies on Copepoda II. Proceedings of the First International Conference on Copepoda, Amsterdam, The Netherlands, 24-28 August 1981. Crustaceana, (Supplement) **7**: 344-351.

RANGNEKAR, P. G. & N. N. MURTI, 1950. A note on the transfer of *Caligus formicoides* to the genus *Parapetalus*. Journal of the University of Bombay, **19**: 43-53.

— — & — —, 1967. Redescription of three known species of parasitic copepods. Journal of the University of Bombay, **35**: 19-27.

RAO, T. S. S., 1950. On a new caligid parasite from the Indian hammerhead shark. Proceedings of the Indian Academy of Sciences, **31B**: 302-307.

— —, 1951. On a new caligid parasite, *Gloiopotes zeugopteri* sp. nov., from *Xiphias zeugopteri*, Lawson's Bay, Waltair. Proceedings of the Indian Academy of Sciences, **34B**: 248-255.

RATHBUN, M. J., 1905. Fauna of New England. 5. List of the Crustacea. Occasional Papers of the Boston Society of Natural History, **7**: 1-117.

REDKAR, M., P. G. RANGNEKAR & N. N. MURTI, 1949. Four new species of parasitic copepods from the marine fishes of Bombay. Journal of the University of Bombay, **18**: 36-50.

RENAUD, B. & P. CALS, 1975. Organisation métamérique et identification des structures appendiculaires chez les Crustacés parasites *Lepeophtheirus pectoralis* (O. F. Müller) et *Lepeophtheirus nordmanni* (H. Milne Edwards) (Copépodes Caligides). Compte Rendu Hebdomadaire des Séances de l'Académie des Sciences, (D) **280**: 45-48.

ROBINS, C. R., R. M. BAILEY, C. E. BOND, J. R. BROOKER, E. A. LACHNER, R. N. LEA & W. B. SCOTT, 1980. A list of common and scientific names of fishes from the United States and Canada (4th ed.). American Fisheries Society, Special Publication no. **12**: 1-174.

ROSE, M., 1933. *Nogagella* n. g. *siphonophoriae* n. sp. copépode Caligide parasite des Siphonophores. Annales de l'Institut Ocèanographique, **13**: 119-133.

ROSE, M. & R. VAISSIÈRE, 1953. Catalogue préliminaire des copépodes de l'Afrique du Nord. III. Bulletin de la Société d'Histoire Naturelle de l'Afrique du Nord, **44**: 83-99.

RUSSELL, F. S., 1925. A new species of *Caligus* from Egypt, *Caligus pageti*, sp. n. Annals and Magazine of Natural History, (9) **15**: 611-618.

SAMOTYLOVA, N. N., 2010. Parasitic copepods of family Caligidae (Crustacea, Copepoda, Siphonostomatoida) from fishes of Vietnam. International Scientific and Technical Conference, Vladibostok, 2010: 98-99. [In Russian.]

SARS, G. O., 1903. Copepoda Calanoida. In: An account of the Crustacea of Norway, with short descriptions and figures of all the species, **6**: 1-171.

SCHUURMANS STEKHOVEN, J. H., 1934. Zur Sinnesphysiologie der parasitären Copepoden *Lepeophtheirus pectoralis* (O. F. Müller) und *Acanthochondria depressa* (T. Scott). Zeitschrift für Parasitenkunde, **7**: 336-362.

— —, 1935. Copepoda Parasitica from the Belgian coast. Bulletin du Musée Royal d'Histoire Naturelle de Belgique, **11**: 1-13.

— —, 1936. Copepoda Parasitica from the Belgian coast. II. Mémoires du Musée Royal d'Histoire Naturelle du Belgique, **74**: 1-20.

SCOTT, A., 1901a. On the fish parasites, *Lepeophtheirus* and *Lernaea*. Transactions of the Liverpool Biological Society, **15**: 188-241.

— —, 1901b. Some additions to the fauna of Liverpool Bay. Collected May 1st, 1900 to April 30th, 1901. Proceedings and Transactions of the Liverpool Biological Society, **15**: 342-352.

SCOTT, T., 1897. The marine fishes and invertebrates of Loch Fyne. Part III. Scientific Investigations. Fifteenth Annual Report of the Fishery Board for Scotland, being for the year 1896, 1897: 127-174.

— —, 1900. Notes on some crustacean parasites of fishes. Part III. Scientific Investigations. Eighteenth Annual Report to the Fishery Board for Scotland, being for the year 1899, 1900: 144-188.

— —, 1912. The Entomostraca of the Scottish National Antarctic Expedition, 1902-1904. Transactions of the Royal Society of Edinburgh, **48**: 521-600.

SCOTT, T. & A. SCOTT, 1913. The British parasitic Copepoda. Volumes I and II: 1-257. (Ray Society, London).

SHEN, C. J., 1957. Parasitic copepods from fishes of China. II. Caligoida, Caligidae (1). Acta Zoologica Sinica, **9**: 351-377.

— —, 1958. Parasitic copepods from fishes of China. III. Caligoida, Caligidae (2). Acta Zoologica Sinica, **10**: 131-144.

SHIELDS, R. J., 1967. Frontal filament formation in larval parasitic copepods. Transactions of the American Microscopical Society, **86**: 68-69.

SHIINO, S. M., 1954a. Copepods parasitic on Japanese fishes. 3. On two new species of the genus *Anuretes*. Report of the Faculty of Fisheries, Prefectural University of Mie, **1**: 260-272.

— —, 1954b. Copepods parasitic on Japanese fishes. 4. The family Euryphoridae. Report of the Faculty of Fisheries, Prefectural University of Mie, **1**: 273-290.

— —, 1955. *Alicaligus tripartitus* gen. et sp. nov. a caligid copepod found on the gills of *Sarda orientalis* (T. & S.). Pacific Science, **9**: 56-61.

— —, 1957. Copepods parasitic on Japanese fishes. 13. Parasitic copepods collected off Kesennuma, Miyagi Prefecture. Report of the Faculty of Fisheries, Prefectural University of Mie, **2**: 359-376.

— —, 1958a. Parasitic copepods from fishes collected in the Indian Ocean. Annual Report of the Prefectural University of Mie, **2**: 98-113.

— —, 1958b. Über eine neue Art der schmarotzenden Copepodengattung, *Dysgamus*, aus der Bucht von Sagami. Annotations Zoologicae Japonenses, **31**: 161-166.

— —, 1959a. Ostpazifische parasitierende Copepoden. Report of the Faculty of Fisheries, Prefectural University of Mie, **3**:267-333.

— —, 1959b. Revision der auf Goldmakrele, *Coryphaena hippurus* L., schmarotzenden Caligidenarten. Annual Report of the Prefectural University of Mie, **3**: 1-34.

— —, 1959c. Sammlung der parasitischen Copepoden in der Präfekturuniversität von Mie. Report of the Faculty of Fisheries, Prefectural University of Mie, **3**: 334-374.

— —, 1960. Copepods parasitic on remoras from the Bay of Bengal. Report of the Faculty of Fisheries, Prefectural University of Mie, **3**: 542-552.

— —, 1960b. Parasitic copepods of fishes from the eastern Pacific. Report of the Faculty of Fisheries, Prefectural University of Mie, **3**: 527-541.

— —, 1963. Parasitic copepods of the eastern Pacific fishes. 1. Records of known species. Report of the Faculty of Fisheries, Prefectural University of Mie, **4**: 335-347.

— —, 1965. Parasitic copepods of the eastern Pacific fishes. 6. Euryphoridae. Report of the Faculty of Fisheries, Prefectural University of Mie, **5**: 421-433.

— —, 1976. List of common names of fishes of the world, those prevailing among English-speaking nations. Science Report of Shima Marineland, **4**: 1-262.

SPROSTON, N. G. & P. H. T. HARTLEY, 1941. Observations on the bionomics and physiology of *Trebius caudatus* and *Lernaeocera branchialis* (Copepoda). Journal of the Marine Biological Association of the United Kingdom, **25**: 393-417.

STEBBING, T. R. R., 1900. On Crustacea brought by Dr. Willey from the South Seas. In: Zoological Results based on material from New Britain, New Guinea, Loyalty Islands and elsewhere, collected during the years 1875, 1896, and 1897, Part 5: 605-690.

STEENSTRUP, J. J. S. & C. F. LÜTKEN, 1861. Bidrag til Kundskab om det aabne havs snyltekrebs og Lernaeer samt om nogle andre nye eller hidtil kun ufuldstaendigt kjendte parasitiske copepoder. Saeskit aftrykt af det Kongelige Danske Videnskabernes Selskabs Skrifter, 5te Raekke, naturhistorik og mathematisk Afdeling, **5**: 341-432.

STEPHENSEN, K. H., 1929. Marine parasitic, semi-parasitic and non-pelagic Crustacea Copepoda. Zoology of the Faroes, **30**: 1-18.

— —, 1940. Parasitic and semiparasitic Copepoda. The Zoology of Iceland, **3**: 1-24.

STILLGER, G., 1968. Ein Hüpferling schmarotzt an Flundern: *Lepeophtheirus pectoralis*. Mikrokosmos, **57**: 276-278.

STOCK, J. H., 1981. Parasitic Copepoda of British fishes and British marine Amphipoda: Gammaridea. [Book review]. Crustaceana, **40**: 110-112.

STOSSICH, M., 1880. Prospetto della Fauna del Mare Adriatico. Parte III. Bollettino della Società Adriatica di Scienze Naturali in Trieste, **6**: 178-271.

SWOFFORD, D., 2000. PAUP*. Phylogenetic Analysis Using Parsimony (and other methods) version 4.0b10. Sinauer, Sunderland, Massachusetts.

THOMPSON, I. C. & A. SCOTT, 1903. Report on the Copepoda collected by Professor Herdman, at Ceylon in 1902. Report to the Government of Ceylon on the Pearl Oyster Fisheries of the Gulf of Manaar, **7**: 227-307.

THOMPSON, W., 1847. Additions to the fauna of Ireland. 22. Crustacea. Annals and Magazine of Natural History, (1) **20**: 237-250.

THOMSEN, R., 1949. Copépodos parásitos de los peces marinos del Uruguay. Comunicaciones Zoológicas del Museo de Historia Natural de Montevideo, **3**: 1-41.

THORELL, T. T., 1859. Bidrag til kannedomen om krustaceer, som lefva i arter af slägtet *Ascidia* L. Ofversigt af K. Vetenskapsakademiens Förhandlingar, **3**: 1-84.

VAISSIÈRE, R., 1959. Parasites de poissons de mer ouest-africaines récoltés par J. Cadenat. II. Copépodes; Copépodes caligides. Bulletin de l'Institut Française d'Afrique Noire, **21A**: 534-553.

— —, 1961. Morphologie et histologie comparées des yeau des crustacés copépodes: 1-125. (Ph.D. Thesis, University of Paris; Série A, no. 3510).

VALLE, A. D., 1880. Crostacei parassiti dei pesci del mare Adriatico. Bollettino della Società Adriatica di Scienze Naturali in Trieste, **6**: 55-90.

VENMATHI MARAN, B. A., T. S. LEONG, S. OHTSUKA & K. NAGASAWA, 2009. Records of *Caligus* (Crustacea: Copepoda: Caligidae) from marine fish cultured in floating cages in Malaysia with a redescription of the male of *Caligus longipedis* Bassett-Smith, 1898. Zoological Studies, **48**: 797-807.

VENMATHI MARAN, B. A., S. OHTSUKA & G. A. BOXSHALL, 2008. A new species of *Anuretes* Heller, 1865 (Copepoda: Caligidae) from the yellowbanded sweetlips *Plectorhinchus lineatus* (Haemulidae) off New Caledonia. Systematic Parasitology, **70**: 35-40.

VOTH, D. R., 1972. Life history of the caligid copepod *Lepeophtheirus hospitalis* Fraser, 1920 (Crustacea: Caligoida). Dissertation Abstracts International, B, Sciences and Engineering, **32**: 5547-5548.

WALTER, T. C. & G. A. BOXSHALL, 2008. World of copepod database. Available on line at http://www.marinespecies.org/copepoda consulted on 2011-0208.

WHITE, H. C., 1940. "Sea lice" (*Lepeophtheirus*) and death of salmon. Journal of the Fisheries Research Board of Canada, **5**: 172-175.

— —, 1942. Life history of *Lepeophtheirus salmonis*. Journal of the Fisheries Research Board of Canada, **6**: 24-29.

WILLEY, A., 1896. Letters from New Guinea on *Nautilus* and some other organisms. Quarterly Journal of Microscopical Science, **39**: 145-180.

WILSON, C. B., 1905a. North American parasitic copepods belonging to the family Caligidae. Part 1. The Caliginae. Proceedings of the United States National Museum, **28**: 479-672.

— —, 1905b. New species of parasitic copepods from Massachusetts coast. Proceedings of the Biological Society of Washington, **18**: 127-132.

— —, 1907a. North American parasitic copepods belonging to the family Caligidae. Part 2. The Trebiinae and Euryphorinae. Proceedings of the United States National Museum, **31**: 669-720.

— —, 1907b. North American parasitic copepods belonging to the family Caligidae. Parts 3 and 4. A revision of the Pandarinae and the Cecropinae. Proceedings of the United States National Museum, **33**: 323-430.

— —, 1908. North American parasitic copepods. New genera and species of Caliginae. Proceedings of the United States National Museum, **33**: 593-627.

— —, 1910. The classification of the copepods. Zoologischer Anzeiger, **35**: 609-620.

— —, 1911. North American parasitic copepods: descriptions of new genera and species. Proceedings of the United States National Museum, **39**: 625-634.

— —, 1912. Descriptions of new species of parasitic copepods in the collections of the United States National Museum. Proceedings of the United States National Museum, **42**: 233-243.

— —, 1913. Crustacean parasites of West Indian fishes and land crabs, with descriptions of new genera and species. Proceedings of the United States National Museum, **44**: 189-277.

— —, 1921. New species and a new genus of parasitic copepods. Proceedings of the United States National Museum, **59**: 1-17.

— —, 1923. Parasitic copepods in collection of the Riksmuseum at Stockholm. Arkiv för Zoologi, **15**: 1-15.

— —, 1924. New North American parasitic copepods, new hosts, and notes on copepod nomenclature. Proceedings the United States National Museum, **64**: 1-22.

— —, 1932. Copepods of the Woods Hole region, Massachusetts. United States National Museum, Bulletin, **158**: 1-635.

— —, 1935a. Parasitic copepods from Dry Tortugas. Papers from the Tortugas Laboratory, **29**: 329-347.

— —, 1935b. A parasitic copepod, *Pupulina flores*, redescribed after forty years. Parasitology, **27**: 593-597.

— —, 1936. Plankton of the Bermuda Oceanographic Expedition. IV. Notes on Copepoda. Zoologica, **21**: 89-93.

— —, 1944. Parasitic copepods in the United States National Museum. Proceedings of the United States National Museum, **94**: 529-582.

— —, 1950. Contributions to the biology of the Philippine Archipelago and adjacent regions. Copepods gathered by the United States Fisheries steamer "Albatross" from 1887 to 1909, chiefly in the Pacific Ocean. United States National Museum, Bulletin 100, **14**: 141-441.

WILSON, M. S., 1952. An emended diagnosis of the copepod genus *Pupulina* (Caligoida), with descriptions of new species and a redescription of the genotype. Proceedings of the United States National Museum, **102**: 245-263.

YAMAGUTI, S., 1936a. Parasitic copepods from fishes of Japan. Part 2. Caligoida, I: 1-22.

— —, 1936b. Parasitic copepods from fishes of Japan. Part 3. Caligoida, II: 1-21.

— —, 1963. Parasitic Copepoda and Branchiura of fishes: 1-1104. (Interscience Publishers, New York).

YAMAGUTI, S. & T. YAMASU, 1959. Parasitic copepods from fishes of Japan with description of 26 new species and remarks on two known species. Biological Journal of Okayama University, **5**: 89-165.

YEATMAN, H. C., 1957. A redescription of two parasitic copepods from Bermuda. Journal of the Washington Academy of Sciences, **47**: 346-353.

YÜ, S. C. & H. W. WU, 1932. Parasitic copepods on the flat-fishes from China. Bulletin of the Fan Memorial Institute of Biology, **3**: 55-74.

ZEDDAM, J.-L., P. BERREBI, F. RENAUD, A. RAIBAUT & C. GARION, 1988. Characterization of two species of *Lepeophtheirus* (Copepoda, Caligidae) from flatfishes. Description of *Lepeophtheirus europaensis* sp. nov. Parasitology, **96**: 129-144.

SELECTIVE TERMS & ZOOGEOGRAPHIC LOCALITIES INDEX

COMPREHENSIVE PARASITE INDEX

COMPREHENSIVE HOST INDEX

www.ingramcontent.com/pod-product-compliance
Lightning Source LLC
LaVergne TN
LVHW020631300926
849161LV00004B/666
* 9 7 8 9 0 0 4 2 0 4 2 4 9 *